BEST MANAGEMENT PRACTICES FOR

SALINE AND SODIC TURFGRASS SOILS

ASSESSMENT AND RECLAMATION

BEST MANAGEMENT PRACTICES FOR

SALINE AND SODIC TURFGRASS SOILS

ASSESSMENT AND RECLAMATION

Robert N. Carrow • Ronny R. Duncan

CRC Press is an imprint of the
Taylor & Francis Group, an **informa** business

CRC Press
Taylor & Francis Group
6000 Broken Sound Parkway NW, Suite 300
Boca Raton, FL 33487-2742

International Standard Book Number: 978-1-4398-1474-1 (Hardback)

Library of Congress Cataloging-in-Publication Data

Carrow, Robert N.
Best management practices for saline and sodic turfgrass soils : assessment and reclamation / Robert N. Carrow and Ronny R. Duncan.
p. cm.
ISBN 978-1-4398-1474-1 (hardcover : alk. paper)
1. Turf management. 2. Turfgrasses--Effect of salt on. 3. Soils, Salts in. 4. Saline irrigation. I. Duncan, Ronny R. II. Title.

SB433.C323 2011
635.9'642--dc22 2011010402

Visit the Taylor & Francis Web site at
http://www.taylorandfrancis.com

and the CRC Press Web site at
http://www.crcpress.com

Contents

SECTION III Management BMPs for Saline and Sodic Soil Sites

Preface

THE CHALLENGE

Soluble salts, which can move with gravity-induced movement of water, and salts retained in soils (especially sodium) can adversely influence onsite and offsite natural resources and ecosystems—that is, soils, surface waters, subsurface waters, wildlife, and plant ecosystem sustainability. When a turfgrass site is irrigated with saline irrigation water, salinity management (a) is the most complex stress management issue that the turfgrass manager will confront and (b) requires that daily management for turfgrass performance and site use be tightly coupled with long-term environmental, economic, and social sustainability based on proper science.

One very important lesson that we have learned over the years in dealing with salinity challenges globally is that salinity is often quite subtle in revealing its stress symptoms on plants; but usually by the time those symptoms are recognized, the entire ecosystem is affected. Additionally, fixing the salinity-induced problems usually begins with the secondary plant response and not the primary cause, which means that remediation does not occur rapidly. You must start with the initial sources of salinity additions to the site and develop appropriate reoccurring strategies to overcome the potential limitations.

FEATURES

Barrett-Lennard and Setter (2010, iii) summarized their experience in the vast salt-affected soils of Australia as follows: "It is important to recognize that plant improvement is only part of the solution to the use of salt-land. Better germplasm must be linked to the appropriate management of soils and cropping practices." This quote summarizes the essence of managing turfgrasses or any plant on salt-affected sites—the whole ecosystem must be managed.

The *best management practice* (BMP) concept, considered the gold standard management approach for any individual environmental issue, is used in this book, since it is a whole-ecosystem (holistic), science-based salinity management approach that allows all possible management options to be considered and implemented on a site-specific basis. Features of this book include the following:

- *Provides BMPs addressing both proactive site assessment (initial and ongoing monitoring) and specific individual site management programs* that can be implemented for each type of saline and sodic problem affecting turfgrass performance.
- *Identifies all possible BMP strategies*, including turfgrass and landscape plant selection; irrigation system design; irrigation scheduling and salinity leaching; chemical, physical, and biological amendments; cultivation; topdressing; soil modification; sand capping; surface and subsurface drainage options; nutritional practices; additional cultural practices (plant growth regulators [PGRs], biotic and abiotic stresses, and traffic stresses); and ongoing monitoring. There is no "silver bullet" amendment, treatment, or grass for salinity management—only a holistic BMP approach will be successful and sustainable.
- Includes the role and use of field and laboratory analytical methods for *site assessment approaches for both plant performance and whole-ecosystem assessment* in relation to environmental issues such as soil quality and sustainability, salt disposal, and potential to affect surface and ground waters.
- Since plant and soil nutrient and element deficiencies, imbalances, and toxicities are an integral part of salinity stresses, this book *contains the most detailed information available*

to turfgrass managers specific to turfgrass soil testing (routine and salt package tests), water quality, and plant analyses as well as report interpretation of each of these potential management tools.
- *Presents emerging challenges, technology, and concepts*, including the integration of salinity management into comprehensive site environmental or sustainable management systems; use of halophytic turfgrasses for nontraditional purposes (land reclamation, saline forages, and drainage water reuse schemes); integration of geospatial and geostatistical concepts and technology; and integration of new sensor technology into daily management paradigms.

OUR APPROACH

The *foundational principles* used by the authors in development of this book were as follows:

- *To incorporate both scientific and practical management recommendations (practicum).* Basic scientific principles are necessary to understand specific salinity challenges and to comprehend the logic behind each practical BMP strategy for salinity management.
- *To compile in one source* all the information required *to identify (assessment)* and *manage* the diverse types of salinity stresses.
- *To use a "field problem" approach.* There is not "one" salinity issue but a number of interactive salinity problems, with each salt-affected site encompassing a unique mixture of these multiple challenges. Many times, soil chemical and physical academic courses are not taught from a field problem perspective; but real-world salinity issues that occur on a specific site require a "field problem thought process" that is multidimensional in scope. Color pictures inserted in the center portion of this book will aid in visualizing field problems.
- *To encourage formal and continuing educational programs* in turfgrass science and management at all levels to incorporate more specific *science-based and practical, field problem-solving emphasis on soil salinity and saline water quality*—which are and will continue to be emerging and increasing realities in the turfgrass industry.
- *To encourage the use and development of more salt-tolerant grasses.* Chapter 9 illustrates the science and practicum approach where for individuals interested in turfgrass selection for a salt-affected site, Chapter 9, Section 9.4, titled "Landscape Plants and Salinity Tolerance," is appropriate. Much of the remainder of Chapter 9 plus Section 20.3 ("Grass Selection Issues") are focused on presenting an in-depth review of the current status and challenges for turfgrass breeders, geneticists, and biotechnologists in development of more salt-tolerant grasses in the future.

SUGGESTIONS FOR READERS

Because salinity management is complex and constantly dynamic and since this book presents both the science principles and the practicum, a reader may find it daunting to attempt to comprehend all aspects. Understanding salinity problems and management requires time. Understanding why salinity issues occur and why specific maintenance programs must be implemented are essential for managing the series of events that seem to cascade in sequence to challenge the sustainability of turfgrass ecosystems. The book is developed so that a turfgrass manager or consultant can obtain a basic understanding of salinity issues by reading Chapters 1 to 3. Then, readers can go to the chapter relating to a specific BMP strategy of interest (e.g., sand capping) without reading all site assessment chapters (Chapter 4 to 8) or management chapters (Chapters 9 to 17). For those involved in the development of a new facility that will be using saline irrigation water, the site assessment chapters and the specific salinity management chapters, even in outline form, will provide a detailed overview of preconstruction planning considerations and possible infrastructure concerns.

May the Lord bless each student, turfgrass manager, and consultant as they pursue a productive career in turfgrass management and science!

Robert (Bob) N. Carrow
Ronny R. Duncan

Acknowledgments

The authors wish to acknowledge contributions from those individuals who provided advice, encouragement, and guidance during the development of this book.

Special thanks for the patience and encouragement of our spouses and families. Thanks to Tim Hiers, golf course manager at The Old Collier Golf Club, for providing the opportunity of a salt-challenged site as a "living laboratory" for defining and refining BMP protocols over the past seven years, for his intense interest in fostering cutting-edge science in salinity management, and for his advice and friendship. We also acknowledge the numerous turfgrass managers and consultants who have provided real-world and site-specific salinity challenges, asked the challenging questions, demonstrated success of the BMP approach for salinity management by implementation, and provided feedback for us. This interaction has allowed the coupling of science and practical management in this book.

Authors

ROBERT N. CARROW, PHD

Dr. Bob Carrow is a professor in the Department of Crop and Soil Sciences at the University of Georgia, located at the Griffin Campus in Griffin, Georgia. After graduation from Michigan State University's (1972) Crop and Soil Science Department, he was a faculty member in turfgrass research and instruction at the University of Massachusetts (1972–1976) and Kansas State University (1976–1984) before coming to the University of Georgia in 1984 as a research scientist in turfgrass stress physiology and soil stresses.

Dr. Carrow has coauthored four books: (a) *Salt-Affected Turfgrass Sites: Assessment and Management* (1998); (b) *Seashore Paspalum: The Environmental Turfgrass* (2000); (c) *Turfgrass Soil Fertility and Chemical Problems* (2001); and (d) *Turfgrass and Landscape Irrigation Water Quality: Assessment and Management* (2009). He was coeditor of *Turfgrass* (Agronomy Monograph no. 32; American Society of Agronomy, 1992). Dr. Carrow is a fellow of the American Society of Agronomy, served as vice president of the International Turfgrass Research Society for ten years, and has written numerous scientific book chapters and journal articles (124) and professional turfgrass trade journal articles (224). He has made over 565 invited presentations on a worldwide basis to turfgrass professionals and 28 to scientific audiences in 38 states and several countries, including Australia, Canada, England, Singapore, Japan, and Guam. He has been a cooperator on the release of three Bermuda grasses, four seashore paspalums, and three tall fescues.

Dr. Carrow's research focus has been in the following areas: (a) precision turfgrass management via mobile spatial mapping for improvement of site-specific irrigation design and scheduling, new water audit approach, salinity management, fertilization, and cultivation; (b) management of salt-affected sites; (c) water conservation and irrigation water quality issues; (d) stress physiology research on drought, saline and sodic soils, low soil oxygen, the acid-soil complex, low light, and nutritional stresses; (e) soil fertility and plant nutrition; and (f) traffic stresses on recreational sites.

RONNY R. DUNCAN, PHD

Dr. Ronny R. Duncan is recently retired after many years as professor of Crop and Soil Sciences at the University of Georgia. From August 2003 to May 21, 2010, Dr. Duncan worked as vice president of Turf Ecosystems LLC, a private consultancy and grass development company that was a subsidiary of Collier Investments, LLC, of Naples, FL. He consults on all turfgrass species regarding water quality problems, soil salinity challenges, and ecosystem agronomics on a global basis.

A premier researcher in the breeding and genetics of turfgrasses, Dr. Duncan continues research interests of edaphic and abiotic environmental stresses on turfgrasses. His expertise extends into salt and salinity-related soil and water problems across all turfgrasses and turfgrass management involving water conservation strategies, water quality problems, and alternative water use on all turfgrasses.

Dr. Duncan's education includes a BS in crop production, an MS in crop science breeding, and a PhD in plant breeding at Texas A&M University in 1977. Current teaching commitments include three popular seminars for the Golf Course Superintendents Association of America. He has taught and/or continues to teach workshops for national and international turfgrass organizations, including the GCSAA, such as his current workshops Seashore Paspalum Turfgrass Management, Greens Management of Seashore Paspalum for Tournament Quality Conditions, Advanced Water Quality Assessment and Management, Turfgrass Water Conservation, and Advanced Salt-Affected Turfgrass Sites: Assessment and Management.

His honors and awards achieved include fellow of the Crop Science Society of America, fellow of the American Society of Agronomy in 1992, and the Excellence in Research Award from Seed Research of Oregon recognizing his turf research, which was given in 1998.

Patents (P) and turfgrasses developed by Dr. Duncan include
'SeaIsle 1' seashore paspalum (U.S. PP no. 12,665)
'SeaIsle2000' seashore paspalum (U.S. PP no. 12,625)
'Southeast' tall fescue
'Tenacity' tall fescue
'Bulldog' tall fescue
'Seaspray' hybrid seeded seashore paspalum (U.S. P no. 7,262,341)
'SeaIsle Supreme' (U.S. PP no. 18,869)
'Platinum TE' paspalum (U.S. PP no. 19,224)

Dr. Duncan shares his research and knowledge through the writing and editing of more than ten books held in high regard for their technical quality and practicality (e.g., *Seashore Paspalum: The Environmental Turfgrass*), plus more than 200 refereed journal articles or book chapters and publications. The first water quality book for turfgrasses (written by him with two coauthors) was published in 2009.

Section I

Understanding Characteristics of Salt-Affected Sites

1 Basics of Salt-Affected Soils

CONTENTS

1.1 OVERVIEW AND CLASSIFICATION OF SALT-AFFECTED SOILS

1.1.1 Overview of Salt-Affected Soils

While salinity is often thought of as a "soil" issue characterized by the accumulation of high total soluble salt concentrations, it encompasses plant responses as well as effects of topically applied saline irrigation water (USDA-ARS, 2008). Thus, "salinity" is not a single stress or problem, but there are four major salinity issues that are the primary problems that require intensive site-specific attention on an individual basis in response to each stress (USSL, 1954; Abrol et al., 1988; Rhoades and Loveday, 1990; Rengasany and Olsson, 1991; Pessarakli and Szabolcs, 1999; Keren, 2000; Levy, 2000; Qadir and Oster, 2004; Carrow and Duncan, 2010). These four primary "salinity" stresses are as follows:

1. Excessive levels of *total soluble salts*, that is, total salinity causing salt-induced drought stress (Chapter 2)
2. *Na permeability hazard*: excessive levels of Na on the soil cation exchange sites (cation exchange capacity, or CEC) and precipitated as Na carbonates causing soil structure degradation (Chapter 3)
3. *Ions that are toxic to roots or shoots or may cause other problem ion issues* as they accumulate (Chapters 4, 5, and 6)
4. *Actual nutrition concentrations, interactions, and imbalances* caused by nutrients and other ions in the water or soil (Chapters 4, 5, and 6)

1.1.2 Classifying Salt-Affected Soils

Salt-affected soils are very diverse in characteristics and exhibit a combination of the *salinity stresses* previously noted (Rengasamy, 2010a). Classification of a salt-affected soil is based on two major stresses, namely, (a) total soluble salt concentration, which relates to the potential for salt-induced drought stress and subsequent osmotic adjustments internally in the plant; and (b) the quantity of exchangeable Na^+, which relates to the potential for deterioration of soil physical conditions from accumulated Na and for subsequent ion toxicities from excess Na (Table 1.1). Soil pH is also often listed, but is generally not used in the U.S. Salinity Lab (USSL, 1954) classification scheme, which is the most prevalent classification.

Traditionally, the main types of salt-affected soils are broadly classified as saline, sodic, and saline-sodic soils. *Saline soils* have high concentrations of total soluble salts, but relatively low exchangeable Na levels; they are the focus of Chapter 2 ("Saline Soils"). A *sodic soil* is characterized by high exchangeable Na on the soil CEC, but relatively low total soluble salt levels. *Saline-sodic soils* exhibit both high total soluble salt and high exchangeable Na levels. A unique soil problem that is usually found on coastal areas is *acid-sulfate soils*, which are usually saline-sodic in nature. Sodic and saline-sodic soils are the topic of Chapter 3 ("Sodic, Saline-Sodic, and Alkaline Soils").

A brief overview of the terminology and criteria used to classify salt-affected soils is presented in Table 1.1, but these will also be dealt with in more detail in the soil-testing chapter (Chapter 4). The first criteria for classifying a salt-affected soil is to determine the soil salinity status, with the best measure of *total soluble salt* concentration being the electrical conductivity of a *saturated soil-water paste extract* (*ECe*). In this method, the soil is brought to saturation, it is allowed to equilibrate to dissolve total soluble salts into the soil solution during a specified time frame, the soil solution is vacuum extracted, and then the electrical conductivity is determined in the extract and noted by the designation of ECe. This is the method developed by the USSL (1954) that most accurately reflects the total soluble salt impact on plants in the field. There are other, less accurate methods using more dilute soil:water extracts, such as 1:2 or 1:5 (soil:water, volume basis), to determine total soil salinity; but these will be discussed in Chapter 4 ("Salinity Soil Tests and Interpretation"). A common unit for ECe is dS/m, and ECe in dS/m can be converted to ppm or mg/kg of salt by the formulas:

$$TDS\ (ppm\ or\ mg/kg) = 640 \times ECe,\ when\ ECe < 5\ dS/m$$
$$TDS = 750 \times ECe,\ when\ ECe > 5\ dS/m$$

TABLE 1.1
Classification of Salt-Affected Soils by U.S. Salinity Laboratory

		Total Salinity	Na Permeability Hazard		
Salt-Affected Soil Class	Old Classification	Soil Electrical Conductivity (ECe, dS/m)[a]	Exchangeable Na percentage (ESP, %)[b]	Na Adsorption Ratio (SAR)[c]	Soil PH
Saline	White alkali[d]	≥4.0	<15	<12	<8.5
Sodic	Black alkali[e]	<4.0	≥15	≥12	>8.5
Saline-Sodic	—	≥4.0	≥15	≥12	<8.5

Source: U.S. Salinity Laboratory, *Diagnosis and Improvement of Saline and Alkali Soils, Handbook 60*, U.S. Government Printing Office, Washington, DC, 1954.

[a] ECe = electrical conductivity of saturated paste extract.
[b] ESP (%) = percentage NA on soil cation exchange capacity (CEC) sites.
[c] SAR = best determined from Na, Ca, and Mg concentrations in the saturated paste extract.
[d] White alkali = tendency for white-gray salt deposition on the soil surface.
[e] Black alkali = tendency for black color at soil surface from high pH dissolution of organic matter and subsequent coating of the soil.

where *TDS* represents *total dissolved salts*, or sometimes *total soluble salts* (*TSS*) nomenclature is used. Thus, an ECe of 1.0 dS/m represents 640 ppm total soluble salts in the soil. As total salts accumulate in the soil, ECe also increases. A soil is classified as "saline" if total soluble salts are at ECe > 4d Sm^{-1} (i.e., TDS = 2560 ppm) (Table 1.1). The ECe measurement does not distinguish which specific salts are present in the sample, but does provide a reliable measure of total soluble salts.

The second criterion used for classification of salt-affected soils is the *exchangeable sodium percentage* (*ESP*), which relates to the potential for Na-induced deterioration of soil structure (Na permeability hazard) (Table 1.1). Exchangeable Na percentage is the percentage of Na on the CEC in units of cmol/kg or meq/100g. The CEC consists of all exchangeable cations on the soil CEC sites, such as Ca, Mg, K, Na, H, and Al. Determination of ESP depends on an accurate measurement of CEC and is defined as

$$ESP = \frac{(Exchangeable\ Na)(100)}{Cation\ Exchange\ Capacity}$$

While both organic matter and clay colloids have negatively charged CEC sites, the clay CEC sites are especially important for Na-induced soil structure deterioration. As Na percentage increases in the soil, the potential for soil structure deterioration increases. The USSL (1954) classification uses an ESP of >15% to classify a soil as sodic or saline-sodic (Table 1.1).

Sodium status can also be reported as the *sodium adsorption ratio* (*SAR*), where

$$SAR = \frac{Na}{\sqrt{(Ca + Mg)/2}}$$

The Na, Ca, and Mg cation concentrations are determined in a saturated paste extract solution (the same extract as ECe is determined) and are designated in units of $mmol_c/L$ or meq/L. As the Na^+ concentration increases, SAR increases. The quantities of Ca and Mg relative to Na in the soil solution are considered in SAR, while ESP uses only Na, but in terms of Na present on the soil CEC sites.

The ESP procedure for measuring soil Na status depends on an accurate determination of the cation exchange capacity, which is pH dependent. If CEC is determined at a laboratory pH different from field pH (for example, a highly calcareous soil at pH > 8.5), the CEC value could be misleading since some CEC sites are pH dependent. Also, in salt-affected soils, some of the cations not associated with CEC may be dissolved and reported as CEC and, again, cause error. Thus, some scientists have favored SAR over ESP. However, they exhibit similar numerical values over a wide range of sodium levels with SAR somewhat lower in value. The relationship of SAR and ESP in a saturated paste extract is reported as follows (USSL, 1954; Naidu et al., 1995):

$$ESP = \frac{1.475(SAR)}{1 + 0.0147(SAR)}$$

Based on the USSL (1954) classification, a soil would be classified as sodic if the ESP > 15% or the SAR > 12. Soil pH is included in the table, but is only used as an indicator of the possibility of a soil being sodic based on a pH > 8.5. Highly sodic soils can exhibit pH up to 10.2, while saline and saline-sodic soils can have a wide range of soil pH, but are generally at pH < 8.5.

While the USSL (1954) classification scheme for salt-affected soils is the most widely used, Australian and South African scientists often use a lower ESP > 5 or 6 as a critical level for classifying a soil as sodic (Rengasamy, 2010a). One reason for this difference is that in these countries, their sodic soils tend to contain higher clay content, which results in the soils being more responsive to structural breakdown by Na-induced effects. The actual criteria for classifying a soil as saline (i.e., ECe > 4.0 dS/m) or sodic (i.e. ESP of >15; or ESP > 5.0) are less important than recognizing that as total soluble salts or Na accumulate in a soil, there is a continual progression toward salinity or sodicity. For example, when soil samples are collected and submitted for soil testing, the sample depth is often 8–10 cm (3–4 inches) If the ESP for the soil was ESP = 12% of Na on the CEC sites, it would not be classified as a sodic soil; however, if the source of Na ions was the irrigation water, then the actual ESP at the soil surface (3–5 cm) would likely be at an ESP > 15. Thus, when a soil test reveals an ESP > 4 on a clay soil that has clay types responsive to structural degradation by Na and has appreciable clay content (these issues are discussed in Chapter 3, "Sodic, Saline-Sodic, and Alkaline Soils"), it would indicate that aggressive management should be instituted to prevent further adverse effects of Na accumulation because the actual ESP at the surface would be significantly higher due to the continual infusion of Na-laden irrigation water. The Australian and South African criteria for classification of sodic soils are, therefore, more conservative and based on identification of sodic problems before they become a major soil and plant response stress.

1.2 CAUSES

1.2.1 Salt Ions and Compounds

Soluble salt ions most common in salt-affected soils are Ca^{+2}, Mg^{+2}, Na^{+}, K^{+}, Cl^{-}, SO_4^{-2}, HCO_3^{-}, NO_3^{-}, and CO_3^{-2} (this last one at pH > 9.0). It is the magnitude and balance of these ions (especially imbalances and dominance on the CEC sites or in the soil solution) that are the basis for the various salinity stresses in a particular landscape. These ions arise from the following:

- Dissolution of minerals in the weathering process
- Salt additions by saline irrigation water
- Salts in standard application of fertilizers and soil amendments
- Salts transported into the rootzone by a rising water table (e.g., saltwater intrusion)
- Capillary rise from deeper in the soil
- Seepage zones where saline water moves to another site
- Subsoil migration to lower-topography areas due to gravity
- Flooding, such as in coastal areas
- Saltwater spray
- Use of primary or secondary salinized dredged soils on a site

Inorganic salts can be present in the soil not just as soluble ions but also as compounds that vary in solubility from relatively insoluble (lime) to moderately soluble mineral forms (gypsum dihydrate) (see Chapter 2, Table 2.1). Relatively soluble minerals that can dissolve to release soluble salt ions include (a) various sulfate compounds such as Na_2SO_4, K_2SO_4, $CaSO_4{\cdot}2H_2O$ (gypsum dihydrate), and $MgSO_4$; (b) chloride chemicals, such as KCl, NaCl, and $CaCl_2$; and (c) carbonate or bicarbonate compounds with high solubility such as Na_2CO_3 and $NaHCO_3$. Examples of insoluble salts would be $CaCO_3$, $MgCO_3$, dolomite ($CaCO_3{\cdot}MgCO_3$), anhydrite of gypsum ($CaSO_4$—i.e., without the hydrated water), and soil minerals such as apatites, while dihydrate gypsum ($CaSO_4{\cdot}2H_20$) is a moderately soluble salt.

While the insoluble and moderately soluble mineral forms can influence soils and plants over time, it is the soluble salts that have the most direct and rapid impact on soils and plants, due to their mobility and innate ability to accumulate in soils and plants. When comparing various salt ions versus salt compounds in the soil, individual salt ions can react with other ions to produce salt compounds varying in solubility. The degree of mobility of individual salt ions in response to water movement in the soil is based on whether they reside in the soil solution, as a component of a soluble compound, as a component of an insoluble compound, or on the soil CEC sites that are associated with negative charges on clay and organic matter. Highly mobile ions are Cl, SO_4, and K, while Na, Ca, and Mg are less mobile in the soil.

Ion and nutrient concentrations and mobility in plants are also important in salt-challenged sites. For example, an immobile ion when applied to the turfgrass foliage will not move to lower tissues, including the root system. Thus, if Ca is required in the rootzone of a sodic soil to maintain root viability (to limit Na displacing Ca in root cell walls and causing roots to deteriorate), application to the shoot tissues would not have any effect. Carrow et al. (2001) reported ion mobility within plants as (a) mobile (N, P, K, Mg, and Na), (b) somewhat mobile (S, Zn Cu, Mo, and B), and (c) immobile (Ca, Fe, Mn, and Si).

1.2.2 Primary and Secondary Salinization

Remediation practices to correct salinity stresses and preventative practices to limit recurring salinity stresses are based on a sound understanding of the causes and recognition of the symptoms on the specific site. Salinity caused by soluble salts on a site cannot be ignored, especially when the source of salts is the irrigation water, since soils, plants, and water (surface and subsurface) are all adversely affected with each irrigation cycle (Oster, 1994). *Salinization* (or *salination*) of irrigated land occurs when dissolved salts accumulate in the upper soil layers and impose multiple salinity stresses on the landscape plants and perennial turfgrass ecosystems. Salinization of ground (aquifers) or surface waters occurs when excessive salt loads come into contact with these waters. Causes of soil salinization are classified as (a) *primary or natural salinization* that occurs from natural processes; or (b) *secondary salinization*, which is a result of human activities (Carrow and Duncan, 2010; Duncan et al., 2009). Examples of natural or primary salinization are as follows:

- Accumulation of salts in the soil over long periods of time from weathering of salt-laden parent materials, especially in arid regions where natural leaching is limited by low and often sporadic precipitation. The largest acreage of salt-affected soils in the world is due to this cause, and this would include many turfgrass areas established on naturally salinized soils. In these locations, salt accumulation is normally at the surface, but salt accumulation layers are often found deeper within the soil profile. Subsurface salt zones can contribute to groundwater salinization if the groundwater comes in contact with the salinized zone.
- Old ocean or coastal beds that have evaporated and left salt deposits at the surface or in layers in the soil profile.
- Sea salt carried by wind, rain, or flooding (e.g., storm surges or periodic high tides) onto adjacent land areas.
- Salt movement into the rootzone from a naturally high saline water table or high tidal swells such as in coastal swamps or marshes. Sometimes, the coastal soils are sandy in nature and can be easily reclaimed by leaching. However, coastal marine clays or more fine-textured soils, especially if the clay type is an expansive/contraction 2:1 clay (with high CEC and very susceptible to Na-induced structure deterioration), can be very difficult if one wishes to remove accumulated total soluble salts and Na. As clay content increases, salt removal becomes increasingly difficult.

Secondary salinization results from the activity of humans, especially via irrigation and drainage practices. Understanding the causes of secondary salinization is especially important because preventative measures can often minimize adverse effects. Types of secondary salinization are as follows:

- Irrigation with saline irrigation water where leaching or drainage is insufficient to prevent salt accumulation in the plant rootzone. The percentage of irrigated lands affected by salinization is 20% to 25% in the United States, 13% in Israel, 30% to 40% in Egypt, 15% in China, and 15% to 20% in Australia (Gleick, 1993). For turfgrass sites irrigated with saline irrigation water sources, this is the major cause of secondary soil salinization and is an ongoing management challenge.

- Irrigation with saline irrigation water where surface drainage does occur, but results in salinization of groundwater (aquifers).
- Irrigation with groundwater from saline strata, which is often located below the drinking water strata and often is exposed to some degree of progressive ocean water infusion.
- Irrigation with saline irrigation water, especially in arid or semiarid regions, where evaporative demands continually surpass leaching requirements and irrigation system capability, resulting in upper soil profile salt accumulation and layering.
- *Dryland salinity* is a type of secondary salinization and is a major problem in Western Australia (Pannell and Ewing, 2006) and other areas in the world with similar ecosystems. Land clearing (native deep-rooted trees and shrubs) coupled with the introduction of more shallow-rooted agricultural crops (nonirrigated or irrigated) can result in a rising water table that eventually leads to secondary salinization and waterlogging of the surface soil profile. The deep-rooted vegetation changes result in more water draining past the rootzone since the root systems for agricultural crops are more shallow compared to perennial native vegetation. The drainage water can result in a rising water table that, in turn, mobilizes soluble salts that are located below the rootzone but above the normal water table. If the salt-laden water table rises (upconing of excess salts) into the rootzone or the capillary fringe is within reach of the plant rootzone, rapid and serious secondary soil salinization can occur, often resulting in plant death since the salinity tolerances of those plants are not adequate to withstand this upward surge in localized rhizosphere salt accumulation. Moreover, the groundwater also is often salinized.
- Salt-laden leachate waters intercepted by tile drains may deposit salts into surface water or shallow groundwater strata.
- Drainage water reuse and other *water reuse schemes* (i.e., reuse of water for irrigation purposes) where the water may become increasingly salinized. Water reuse (water recapture and subsequent recycling) on a specific site and these reuse schemes will become more common in the future on turfgrass and landscape sites as water conservation becomes a greater mandated issue. Examples of water reuse on a site-specific basis where salinity in the irrigation water may contribute to secondary salinization and, therefore, must become a management concern are as follows. (a) *Onsite drainage water reuse* is an area of greater interest in some locations where water that has percolated through the soil into tile lines and then tile drainage water is recollected from a containment site for direct blending and/or reuse for irrigation on various areas of the property. This is a form of recycled water use but without additional treatment except for what occurs during soil percolation before collection and reapplication. Tanji and Kielen (2002) and Oster and Grattan (2002) provide detailed discussions for the management of drainage water when reused for irrigation. (b) *Onsite reclaimed water reuse* is where harvested water is collected from stormwater or sewage drainage lines coming off a property, such as a golf course complex surrounded by a housing development or business complex; treated at an onsite treatment facility (i.e., onsite reclamation); and then reused for irrigation and other suitable purposes on the site. In this instance, the drainage line water is usually not coming from the soil drain lines where water percolated through the soil, but from the surrounding surface stormwater and sewage or water drain lines and catchment facilities. (c) *Onsite stormwater collection* is from surface runoff and/or stormwater drain lines from the site that go directly into a collection lake without the discharge intermingling with sewage water in the sewage lines. And (d) *Other water reuse schemes* involve water collected on or near the site and then reused for irrigation. This could be as simple as using swimming pool water for irrigation, air conditioner condensate for landscape plant drip irrigation, or more complex sources such as using industrial water from cooling towers for landscape irrigation (Gerhart et al., 2006).

Much of the attention for secondary salinization of lands by irrigation and drainage practices has focused on rural agriculture land (Rhoades et al., 1992; Ayers and Westcot, 1994; Oster, 1994; Grattan and Oster, 2003; Qadir and Oster, 2004); but with more saline waters increasingly used for landscape irrigation in urban areas with increasing population growth, urban secondary salinization is receiving more attention (Wilson, 2003). Another trend has been the increased interest in the potential effects of salinity contamination of freshwater (drinking water) ecosystems such as aquifers (Hart et al., 1990; Nielsen et al., 2006).

Pillsbury (1981, 54) noted the impact of salinization in history and why the historical lessons must not be forgotten:

> Many ancient civilizations rose by diverting rivers and irrigating arid lands to grow crops. For such projects to succeed, human beings had to learn to work cooperatively toward a common objective. The most fruitful of the ancient systems was created at the southeastern end of the Fertile Crescent, the broad valley formed by the Tigris and the Euphrates in what is now Iraq. From there civilization spread eastward through present-day Iran, Afghanistan, Pakistan, and India and thence into China, where ever rivers disgorged through valleys of recently deposited alluvial soil. At its peak of productivity, each irrigated region probably supported well over a million people. All these civilizations ultimately collapsed, and for the same reason: the land became so salty that crops could no longer be grown on it. The salts that were washed out of the soil at higher elevations became concentrated in the irrigated fields as the water evaporated from the surface and transpired through the leaves of the growing crops. Although floods, plagues and wars took their toll, in the end the civilizations based on irrigation faded away because of salination.

Thus, history and current experience illustrate that the use of highly saline irrigation water greatly enhances the potential to degrade soil, plant, and water resources unless definite infrastructure improvements and skilled management practices are implemented. Accumulation of excess soluble salts and sodium in the soil is more rapid as irrigation water quality declines, unless salts are continuously managed. Adequate infrastructure provides the necessary "tools" for the site manager to address salinity stresses; in other words, the infrastructure tools include surface and subsurface drainage systems; adequate water sources in terms of quantity and quality; irrigation system design for distribution uniformity; irrigation scheduling; sand capping, when needed; irrigation water treatment technology; sensor technology to proactively monitor salts and water; adequate salt disposal systems; appropriate cultivation equipment; the blending of variable saline water sources in order to reduce the overall salt load being applied to the turfgrass ecosystem; and others. With these tools, management must target soils, plants, and water deposition on the site for holistic environmental protection (Rhoades et al., 1992; Duncan et al., 2000, 2009).

The influence of saline irrigation water will be greatest on the site to which it is regularly applied, but application can impact the surrounding environment. Since the percentage of saline-irrigated turfgrass land area versus total community area in most instances is small, this localization aids in reducing the potential for large-scale adverse environmental impacts on localized community surface and subsurface waters as well as secondary salinization of site-specific community landscapes and waters. However, in some locations with numerous golf courses or other large irrigated turfgrass and landscape sites, salinity impacts for turfgrass and landscape areas can be potentially significant if the salinity is not properly managed at all community levels. For example, in arid regions, golf courses or other landscape sites may be major customers for reclaimed water from the public water treatment facilities, but these reclaimed waters may contain appreciable salts passed through the treatment facility. In these instances, the public water treatment facility and government entities must realize that salts coming into the treatment facilities (such as from ion exchange salt-based water softeners or home- or business-specific reverse osmosis units) are a public responsibility involving potential secondary salinization and not one that can be passed to the current end user—the turfgrass or landscape ecosystem—without compromising the environmental sustainability of the ecosystem. The effluent and the specific effluent constituents must be disposed of in an environmentally compatible manner by the treatment facilities regardless of the end-use customer; and if there is no end-use customer, the effluent disposal could end up in surface waters or subsurface aquifers (Clean Water News, 2007). If undue excess salt levels are passed through to public or private landscape areas for irrigation, the applied salts soon become a community problem from the deterioration of natural resources: soils, surface waters, subsurface waters, and plant ecosystem sustainability.

1.3 SCOPE OF SALINITY PROBLEMS

1.3.1 Land Area

Australia and Asia have the greatest area of salt-affected soil with 38% and 34%, respectively, of their land area degraded by salinity; but saline and sodic soils are found on all continents (Table 1.2) (Zinck and Metternicht, 2009). The 2303 million acres (932 million ha) of salt-affected area in the world represent about 7% of the earth's continental surface area, with saline soils comprising 43% of the total and sodic soils 57%. For reference, Texas contains 172 million acres.

Of the 568 million acres (230 million ha) of irrigated lands in the world, about 20% or 114 million acres are salt affected (Pitman and Lauchi, 2002). There are many causes contributing to the primary salinization or secondary salinization of soils, but human activity accounts for the secondary salinization of approximately 189 million acres (76.6 million ha), or 8.2% of the total salt-affected land area.

1.3.2 Management and Environmental Challenges

The scope of salinity issues goes beyond the land area of these soils and entails increasing management and environmental challenges. On turfgrass and landscape sites, soil salinity stresses have increased in scope and concern in recent years due to several factors. One factor is political and societal pressures for water conservation, resulting in turfgrass sites increasingly using poorer-quality, variable saline irrigation water to alleviate the demand for potable water sources (Marcum, 2006; Carrow and Duncan, 2008). Saline irrigation water sources include saline groundwater (water that is naturally saline, salt affected by salt leaching, drainage water reuse, salt affected by rising water tables, or seawater intrusion into aquifers), brackish surface water, stormwater runoff, recycled water (i.e., reclaimed or effluent water), and seawater or seawater blends. Increasingly, more governmental units are mandating the use of reclaimed water or saline groundwater for larger turfgrass sites (Marcum, 2006). This is a trend expected to continue on a worldwide basis so that in

TABLE 1.2
Global Distribution of Salt-Affected Soils (Million Ha)

Area	Saline Soil	Sodic Soil	Total	Total (%)	Human-Induced Salinization (%)
Australasia	17.6	340.0	357.6	38.4	1.2
Asia	194.7	121.9	316.5	33.9	68.8
America	77.6	69.3	146.9	15.8	5.7
Africa	53.5	26.9	80.4	8.6	19.3
Europe	7.8	22.9	30.8	3.3	5.0
World	351.2	581.0	932.2	100	100

Source: Adapted from Zinck, J. A. and G. Metternicht, Soil salinity and salinization hazard, in G. Metternicht and J. A. Zinck (Eds.), *Remote Sensing of Soil Salinization*, pp. 3–20, CRC Press, Boca Raton, FL, 2009.

the future, irrigation water applied on turfgrass and landscape sites will often be more saline than in the past (Miyamoto et al., 2005; Miyamoto and Chacon, 2006).

A second force increasing salinity as an abiotic environmental stress on turfgrass sites is the development of salt-tolerant grasses (Chapter 9), which allows use of more variable saline irrigation sources (Duncan and Carrow, 2000; Marcum, 2002; Loch et al., 2003). While salinity tolerance may be genetically improved in various turfgrass species, this scientifically improved trait only allows the turfgrass manager time to make appropriate management decisions, but it does not negate the absolute and continuous requirement for managing any salt deposition from saline irrigation water on the specific site for long-term environmental sustainability. Halophytic turfgrasses (i.e., salt-tolerant grasses) and landscape species generally strictly regulate salt ion uptake, thereby leaving salts to accumulate in the soils with each irrigation cycle (Munns and Tester, 2008). Halophytic plants are generally not phyto-accumulators of excess salts and do not remove excess salts from irrigation water or saline soils (see Chapter 9).

The development of coastal or wetland golf courses and parklands is a third factor that results in salinity problems from saltwater intrusion in aquifers used for irrigation, dredged soils, acid-sulfate sites, periodic flooding, high tidal influences, and wind-driven persistent salt spray (Carrow and Duncan, 1998; Loch et al., 2006). Resorts and recreational sites are now being developed in arid, semiarid, tropical, and subtropical areas, where potable water is very limited or nonexistent and saline sources are the only irrigation option for grasses and landscape plantings.

The fourth aspect concerning turfgrass salt-impacted sites is increasing environmental concerns over the protection of surface and subsurface waters (such as aquifers) and the sustainable protection of soil quality (Duncan et al., 2000, 2009; FAO, 2009a, 2009b). Secondary salinization of the ecosystem is a constant threat and a whole-system, systematic approach to managing the soil, the water, and the grass or landscape plants must be a primary focus.

Since the ecosystem is a dynamic and constantly changing entity, the introduction of salinity into the system results in constant movement of salts vertically upward (capillary action), downward in the soil profile (infiltration, percolation, and drainage), and horizontally both at the surface and through the subsurface, thereby impacting any plants that are being grown in that ecosystem. Salt migration in the soil is a constant process, depending on climatic changes such as precipitation and evapotranspiration (ET) rates plus gravity. Once salts have accumulated in the soil profile, the subsurface gravity-induced migration of excess salts to low-topography areas is a persistent, recurring challenge to perennial turfgrass species management.

Thus, the "scope" of salinity problems goes well beyond the land area affected by salinity. Rather, it transcends into being "the major management challenge" on turfgrass and landscape sites with the continuous application of saline irrigation water containing moderate to high salinity levels. Management of such areas must be ongoing with flexibility to adjust with the dynamic salinity changes in the ecosystem. These sites normally exhibit reduced plant growth and vigor if salts are allowed to accumulate in the soil; thereby, a high degree of management is required to achieve acceptable plant performance, avoid soil degradation, and prevent salts from affecting surface or subsurface groundwater sources. Maintenance challenges and increasing overall management costs that can escalate on salt-affected sites include labor; amendment selection and costs for soil, plant, and water treatments; infrastructure improvements for drainage; sand capping; irrigation systems distribution uniformity; a high degree of technical salinity management expertise and salt-challenged site assessment capabilities; very dynamic soil fertility and plant nutrition situations; and frequent, proactive monitoring of soil, water, and plant status. When the source of most of the salts is the irrigation water, salinity management becomes a constant ongoing management issue and not a short-term problem that is resolved, after which the manager moves on to other issues (Duncan et al., 2009).

Salinity stresses are complex, and a successful management approach must be science based and holistic: integrating soil chemical, physical, and biological considerations; plant shoot and root requirements and their sustainability; surface and subsurface drainage; irrigation design and scheduling; personnel management; comprehensive and regimented management strategies on a site-specific basis; labor costs with appropriate time allocation; and sustainability of the ecosystem, including soil, plant, water, and wildlife. The central purpose of this book is to address salinity management on turfgrass and landscape sites in all its facets by using a comprehensive, holistic, science-based approach: the *best management practices (BMP) approach to salinity.*

1.4 A SUCCESSFUL SALINITY MANAGEMENT APPROACH

1.4.1 A BMP-Based Environmental Plan

Environmental management challenges are an increasing part of "routine management" on many agriculture and landscape sites. For example, Carrow et al. (2008) noted 17 different environmental issues that golf courses must routinely consider in their management programs (Table 1.3). Each of these issues (e.g., water use efficiency and conservation; pesticide, nutrient, and sediment levels related to surface and subsurface water quality; and wildlife habitat management) is complex and requires a successful *environmental management plan*. For a single environmental issue, regardless of its nature, the most successful environmental plan is a *best management practices* (*BMP*) approach (Carrow et al., 2008; Carrow and Duncan, 2008, 2010). Readers are referred to Chapter 19, "Sustainable and Environmental Management Systems," for a more comprehensive treatment of BMPs and other environmental management approaches.

Carrow and Duncan (2008) noted that the BMP approach initially evolved out of the 1977 Clean Water Act for the protection of surface and subsurface groundwater from pesticides, nutrients, and sediments. Each of these environmental issues required a different set of management strategies to deal with the site-specific problem, but the overall approach or philosophy is the same for each—that is, a science-based and practical BMP approach. For the past 30 years, the BMP approach for

TABLE 1.3
Environmental Issues Often Present on a Golf Course with Each Requiring an Environmental Management Plan (Each Issue Can Be Addressed by an Appropriate Best Management Practices Plan)

1. Environmental planning and site design of golf courses, additions, and renovations for environmental sustainability
2. Sustainable maintenance facility design and operation
3. Adapted turfgrass and landscape plant selection—for reduced input of pesticides, nutrients, and water
4. Water use efficiency and conservation
5. Irrigation water quality management for sustainability
6. Pesticides: water quality management
7. Nutrients: water quality management
8. Erosion and sediment control: water quality management
9. Soil quality sustainability
10. Stormwater management
11. Wildlife habitat management
12. Wetland and stream mitigation and management
13. Aquatic biology and management of lakes and ponds
14. Waste management
15. Energy management
16. Clubhouse and building Environmental Management Systems (EMS) concepts
17. Climatic and energy management

Source: Adapted from Carrow, R. N., F. C. Waltz, and K. Fletcher, Environmental stewardship requires a successful plan: Can the turfgrass industry state one? *USGA Green Section Record,* 46 (2), 25–32, 2008.

pesticide, nutrient, and sediment management to protect our surface and subsurface waters has proven that this "environmental management plan" is the gold standard. Just as the BMP approach or philosophy was used for these three diverse environmental issues, it can be successfully applied to all environmental issues, including salinity, because of its basic underlying principles (Carrow and Duncan, 2008). These foundational principles (a) are science based; (b) are holistic or "whole systems" in nature; (c) are environmentally sustainable; (d) involve strategies that are selected and applied on a site-specific basis; (e) incorporate consideration of all dynamic environmental (direct and indirect) impacts; (f) are economically sustainable in that they consider economic effects on the site and on society; (g) have values educated managers who understand sustainable site management concepts; (h) are flexible in that new concepts and technologies can be incorporated as they evolve out of science and practical experience; (i) incorporate ongoing proactive (rather than reactive) monitoring and revisions for the entire ecosystem; and (j) encompass basic turfgrass and landscape management guidelines that can be feasibly implemented.

Key reference materials related to managing salt-affected sites for general agriculture include USSL (1954); Abrol et al. (1988); Rhoades and Loveday (1990); Rengasamy and Olsson (1991); Rhoades et al. (1992); Ayers and Westcot (1994); Jayawardane and Chan (1994); Naidu et al. (1995); Tanji (1996); Hanson, Grattan, and Fulton (1999); Qadir et al. (2000); and Grattan and Oster (2003). Carrow and Duncan (1998, 2010) and Duncan et al. (2009) reported on salinity management for turfgrass sites.

While authors of these references to salinity management on agriculture and turfgrass landscapes present multiple options for salinity management, it is the goal for this book to present a systematic set of all possible salinity management options that, when combined together, result in a *BMP salinity management plan* (Table 1.4). Successful salinity management entails not only controlling surface soil salinity and subsurface accumulation problems but also maintaining the site sustainability of environmental, economic, and turfgrass performance aspects. The remainder of this book considers the many different BMP strategies for the management of salt-affected sites, including initial site assessment to build the knowledge base to make basic informed soil, water, and plant management decisions.

To integrate and assist in the development of the BMP salinity management plan into turfgrass management, a useful approach is to adopt the evolving *precision turfgrass management* (*PTM*) concept, which is based on using advanced site assessment methods to obtain detailed site information to make site-specific management decisions (Krum et al., 2010; Carrow et al., 2009a, 2009b). It is based on a whole-systems science-based approach; intensive spatial mapping of key soil and plant characteristics using mobile sensor platforms; site-specific precision management on inputs; basic management strategies for implementation (salinity is the most complex environmental stress); and the integration of GIS + GPS + new moisture and salinity sensor technology with proactive monitoring. This concept removes a lot of the subjective decisions in management and utilizes science-specific technology to implement maintenance programs with salinity challenges. Applications of PTM are integrated into salinity management in the following chapters:

- Chapter 4, Section 4.1.2, "Soil Sampling," where spatial mapping is used to identify similar areas for routine soil sampling and *soil sampling* for soil laboratory salt analyses.
- Chapter 4, Section 4.3.2, "Approaches to Field Salinity Monitoring," where the PTM concept is targeted to *spatial salinity mapping* of large turfgrass sites using mobile salinity-monitoring devices currently under final testing that were developed specifically for turfgrass situations.
- Chapter 10, Section 10.3.4, "Precision Turfgrass Management (PTM) Water Audit Approach," where PTM is applied to a new whole-site *water audit approach* to determine the irrigation system distribution uniformity of applied water, which has a major effect on the (a) salinity spatial distribution across the landscape and within the soil rootzone, and (b) capability of site-specific salinity leaching.

TABLE 1.4
Summary of BMP Strategies for Salt-Affected Turfgrass Sites for Environmental Protection, Sustainability, and Turfgrass Management

1. **Site Assessment.** To identify factors that will influence salinity management decisions.
   a. Soil Physical Aspects
      - Construction and renovation considerations: impediments to infiltration, percolation, or drainage such as calcic, clay, or rock layers; deep ripping or deep cultivation requirements prior to establishment; future cultivation equipment requirements; surface and subsurface drainage improvements; drainage outlets and salt disposal options; irrigation system requirements; presence of fluctuating or high water tables; sand-capping needs; and pre-plant physical and chemical amendments to improve soil physical condition
      - Identifying all salt additions: irrigation water; water table; capillary rise from salt-rich subsurface horizon; mixing of salt-laden soil during construction or dredging; fertilizers; and drainage onto the site
      - Other: soil texture; clay type; and soil physical analyses of rootzone media, including water-holding capacity

   b. Soil Chemical Aspects
      - Routine soil test information (normal soil fertility test; saturated paste extract salinity test)
      - Additional soil test information: SAR, ESP, ECe, and free calcium carbonate content

   c. Irrigation Water Quality Assessment
      - Complete irrigation water quality analyses
      - Health aspects if needed
      - Multiple irrigation water sources: blending, drainage water reuse, reliability of each source, and stability of each source in terms of constituents over time
2. **Plant Selection.** Salinity tolerance is a primary consideration along with adaptation to climatic, pest, and site use stresses (mowing height, and traffic).
   a. Turfgrass species and cultivars
   b. Landscape plants
   c. Buffer zone plantings
3. **Irrigation System Design.**
   - Uniformity of application
   - Flexibility: for site-specific water applications to minimize drought stress and salinity stresses (i.e., salinity leaching and management)
   - Chemigation and fertigation: flexibility
4. **Irrigation Scheduling and Salt Leaching.** For normal irrigation needs and for efficient salt leaching.
   - Reclamation leaching programs and considerations
   - Maintenance leaching programs and considerations
5. **Identification of Water and Soil Amendment Needs for Site-Specific Problems.**
   - Acidification
   - Gypsum and hydrated lime injection
   - Organic amendments
   - Inorganic amendments
6. **Determination of Proper Amendment Application Protocols for Site-Specific Problems.** This includes equipment needs, rates, timing, and frequency aspects.
7. **Additional Cultural Programs.**
   a. Cultivation needs and equipment. Cultivation programs are very important on many salt-affected sites in order to effectively leach salts and to avoid layers that impede salt movement.
      - Surface cultivation equipment and programs for surface problems
      - Subsurface (deep) cultivation equipment and programs for subsurface problems

   b. Fertilization. Soil fertility and plant nutrition are very dynamic with the use of saline irrigation water due to the combination of constituents added from the water, water treatment materials, and soil amendments, plus leaching programs that differentially leach nutrients and elements. Of particular importance are soil and plant tissue concentrations of K, Ca, Mg, Fe, Mn, S, and Zn; and ratios and balances between and among competing ions.
      - Fertigation flexibility
      - Foliar feeding equipment and programs

TABLE 1.4 (Continued)
Summary of BMP Strategies for Salt-Affected Turfgrass Sites for Environmental Protection, Sustainability, and Turfgrass Management

c. Climatic and traffic stresses. Salinity enhances certain other stresses, such as drought, high and low temperatures, and wear or traffic. Thus, these must be carefully managed.
   - Rounds of golf or foot traffic
   - Cart traffic

d. Cytokinin. Soil salinity suppresses cytokinin synthesis in the roots of plants, and grasses often respond (e.g., by root system redevelopment or hormone stabilization) to the application of this hormone (in seaweed or kelp extract products) on saline-irrigated sites.

e. Pest management.
   - Preventative application program
   - Curative application program

8. Sand Capping, Topdressing, and Soil Modification.
   - Enhancing water infiltration of irrigation water and precipitation

9. Drainage and Wetting Agents.
   - Drainage to intercept leached salts
   - Interception drainage to control surface salt movement
   - Wetting agents for improved unsaturated flow and to alleviate localized dry spots

10. Monitoring.
   a. Turfgrass root and shoot responses
   b. Soil and plant fertility status
   c. Soil salinity: temporal and spatial across landscape and by soil depth
   d. Irrigation water quality over time
   e. Salinity effects on surface and subsurface waters

Source: Adapted from Duncan, R. R., R. N. Carrow, and M. Huck, *Turfgrass and Landscape Irrigation Water Quality: Assessment and Management,* Taylor & Francis, Boca Raton, FL, 2009.

1.4.2 Primary versus Secondary Problems

For any turfgrass field problem, understanding the causes and nature is the starting place for developing a sound management plan; however, identification of the basic problem is not always straightforward. While pest stresses are often relatively easy to identify, many soil physical and chemical problems are not easy to "see." Additionally, what we often view are the secondary effects of a primary soil physical and chemical problem; and this is usually the case for salinity issues. Thus, a brief overview of *primary and secondary problems* is warranted in the context of salinity problems.

Excess salts cause a combination of stresses, namely, (a) four primary salinity problems, and (b) a number of secondary problems that arise out of these primary stresses. A *primary problem* is the basic underlying stress, while *secondary problems* are those that arise out of the primary problem. It is not unusual for a site to have several primary problems present and challenging the ecosystem, and this is the typical situation in salt-affected soils. When primary problems are alleviated, the secondary ones are also alleviated. However, if all the management effort is toward the secondary problem, the underlying cause (primary issue) is never corrected. For example, one of the primary salinity stresses is Na-induced deterioration of soil structure where the soil macropores (pores > 0.075 mm diameter) important for water infiltration and percolation, soil aeration, and plant rooting are greatly reduced. Prevention or alleviation of this salinity stress requires a relatively soluble Ca amendment (usually gypsum) to replace Na from the soil CEC sites, and then leaching to remove the more soluble Na sulfate from the rootzone, as well as practices to minimize Na additions to the site. However, if the Na ions are allowed to dominate, the resulting soil exhibits an array of secondary problems, such as low water infiltration and percolation; poor aeration and anaerobic conditions; limited plant rooting; during rainy periods, soils become waterlogged and saturated; soils are hard when dry; diseases favored by moist, anaerobic surface conditions are favored; and rootborne pathogens increase in population and often will destroy the root system (such as take-all or decline diseases).

When these secondary problems occur, they need to be addressed with appropriate management; but the long-term solution is to deal with the primary problem—that is, excess Na causing loss of soil macropores. Individuals using a turfgrass site often see the secondary problems, but do not realize the underlying primary problem. Thus, an essential characteristic of a good turfgrass manager is to recognize which issues are primary and which are secondary, and then be able to communicate these potential problems to site users or owners. As noted, normally more than one primary salinity problem (as well as other primary problems) are "pancaked" on a site. Each challenge to the ecosystem must be recognized and then an appropriate management plan formulated for each problem. When formulating the BMPs for salt-affected turfgrass sites, our emphasis is on the primary or basic problems in terms of either prevention or alleviation and flexibility in altering management strategies to address the potential salt-related challenges.

1.4.3 Primary Salinity Problems

In the context of the previous section on primary versus secondary problems on salt-affected sites, it bears repeating that "salinity" is not a single stress or problem; rather, there are *four major salinity issues*. Each of the primary problems present on a site requires intensive site-specific management attention on an individual basis to each stress. These four "salinity" stresses are:

1. Excessive levels of soluble salts in the soil (Chapter 2)
2. Excessive levels of Na on the soil cation exchange sites and precipitated as Na carbonates (Chapter 3)
3. Ions in the soil or irrigation water that are toxic to roots or shoots as they accumulate, as well as ions that may cause other problems (Chapters 4, 5, and 6)
4. Nutritional levels and imbalances caused by nutrient interactions and other ions in the water or soil (Chapters 4, 5, and 6)

It is the mix and severity of these four salinity issues on a specific site that require a BMP plan to prevent, alleviate, or manage those stresses that are present. At the same time, any secondary problems that arise from these primary salinity stresses must be managed. The focus of proactive soil, water, and tissue testing on salinity sites is primarily toward determining the presence and magnitude of these four stresses before they become limitations to turfgrass or landscape performance or overall ecosystem sustainability. With multiple interactions affecting the ecosystem (namely, the specific turfgrass or landscape plant cultivar, the irrigation water quality, the soil profile, and the climatic changes), sustainability challenges are dynamic and must be continuously addressed.

2 Saline Soils

CONTENTS

2.1 OVERVIEW OF SALINE SOIL PROBLEMS

Soluble salts in saline soils are salt forms with high solubility that exist (a) in the soil solution, especially under irrigated conditions; and/or (b) as precipitated salts under drier soil conditions, where they can easily dissolve as soil moisture increases (Table 2.1). Soluble salts in the soil can induce direct stresses (a) by action of *the total of all soluble salts*, that is, total soluble salts or total salinity, which is the sum of primarily Ca, Mg, Na, K, Cl, SO_4, HCO_3, NO_3, and CO_3; (b) as *individual ions that may accumulate in soil or plant tissues* to the point of becoming toxic to plant roots or shoot tissue (Na, Cl, and B) or cause other problems (SO_4, HCO_3, and CO_3); and (c) as *individual nutrient ion concentrations or imbalances* of ions affecting nutrition, such as Ca, Mg, K, P, N, SO_4, Mn, Mo, Zn, and Na. These three salinity stresses are all related to soluble salts and are normally present at the same time on a site. As total soluble salts accumulate, the potential for individual specific ion toxicities and nutritional imbalances also increases. While these individual salinity stresses often occur together, each problem must be assessed individually by soil, tissue, and water tests and individual management strategies selected as part of an overall best management practice (BMP) plan and precision turfgrass management (PTM) strategy (Carrow et al., 2009a, 2009b). Additionally, *secondary stresses* (such as increased disease pressure when environmental conditions are favorable) may evolve from one or more of the above direct stresses.

Total soluble salt stresses can occur on any soil, but the most rapid occurrence is exhibited by sandy soils due to both lower cation exchange capacity (CEC) and soil moisture retention relative to more fine-textured soils. Consequently, sandy profiles will often salinize faster than fine-textured soil profiles. As a result, fewer salts are needed to sequester on the CEC sites to cause salinity problems, and the inherently lower water-holding capacity of sands can result in higher soil solution concentrations of localized salts. If a sandy soil has higher organic matter content, which provides greater soil water-holding capacity to the profile, this organic amendment can reduce the onset of salinity stress due to increased CEC as well as better water retention compared to the same sand with little organic matter. However, sandy soils are also much better able to leach soluble salts from the rootzone compared to fine-textured soils since sands contain fewer micropores that can retain salts. As microporosity increases, salt retention increases and the effective leaching of accumulated salts is a much slower and a more challenging management process.

Due to high total salt levels, without Na being the dominant accumulated salt, saline soils exhibit similar or even better soil physical properties than the same soil in a nonsaline state. Cation salts aid

TABLE 2.1
Composition and Approximate Solubility[a] of Some Common Soil Salts

Chemical Name	Composition	Solubility (g/liter)[a]	Relative Solubility	Mineral Name
Calcium carbonate	$CaCO_3$	0.006	Insoluble	Calcite, lime
Magnesium carbonate	$MgCO_3$	0.039	Insoluble	Magnesite
Calcium hydroxide	$Ca(OH)_2$	1.73	Slightly soluble	—
Calcium sulfate	$CaSO_4 \cdot 2H_2O$	2.64	Slightly soluble	Gypsum
Sodium bicarbonate	$NaHCO_3$	100	Soluble	Baking soda
Potassium sulfate	K_2SO_4	111	Soluble	—
Sodium sulfate	Na_2SO_4	133	Soluble	—
Langbeinite	$2MgSO_4 \cdot K_2SO_4$	240	Soluble	Sul-Po-Mag
Sodium carbonate	Na_2CO_3	220	Soluble	—
Magnesium sulfate	$MgSO_4$	255	Soluble	—
Sodium chloride	NaCl	357	Soluble	Halite
Potassium chloride	KCl	357	Soluble	Sylvite
Magnesium chloride	$MgCl_2$	546	Soluble	—
Magnesium sulfate hydrate	$MgSO_4 \cdot 7H_2O$	710	Soluble	Epson salts
Calcium chloride	$CaCl_2$	750	Soluble	—

[a] Solubility in the water at 20°C. Solubility in the soil is affected by soil moisture, the presence of other salts, the temperature, and other factors, and may differ from these values.

in maintaining soil structure by fostering aggregation and preventing dispersion of clay particles. When saline soils are leached by rain or low-salinity irrigation water, clay dispersion may occur, often causing a collapsing of the soil surface, and may result in sealed layers and reduced permeability to water and air. Soils higher in clay content are the most likely to exhibit this response.

The attention of this chapter on saline soils is to provide an understanding of the (a) direct and indirect stresses arising from total soluble salt ions present in the soil or accumulating from repeated applications of irrigation water or other ion sources (flooding, capillary rise from saline water table, etc.), and (b) visual plant and soil symptoms that may provide insight into the stress-response nature of these field problems. However, soil and water guidelines are not presented in this chapter. Since soil and water quality testing is based on assessing the presence and magnitude of these soluble salt-related problems, including sodic issues discussed in Chapter 3, the soil and water guidelines related to total salinity, specific and problem ions, and nutrient issues are reported in the respective soil-testing (Chapters 4 and 5) and water quality–testing (Chapter 6) chapters. In these chapters, it will be evident that the various soil and water tests are directed toward each of the salinity stresses, and the soil and water report formats often have specific science-based data sections related to these stresses.

2.2 TOTAL SOLUBLE SALTS (TOTAL SALINITY) PROBLEMS

2.2.1 Physiological Drought

When soluble salts accumulate in the turfgrass rootzone, one of the effects is to limit plant-available water (osmotic pressure gradients are often involved) and, thereby, induce water deficits or drought stress on the plant even though the soil profile may be at or near field capacity (Figure 2.1). This is the most prevalent and important of all the salinity stresses in terms of frequency and land area. Increasing total soluble salt concentrations decrease the availability of the soil water for plant uptake; thereby, this salt-induced drought stress is called *physiological drought*.

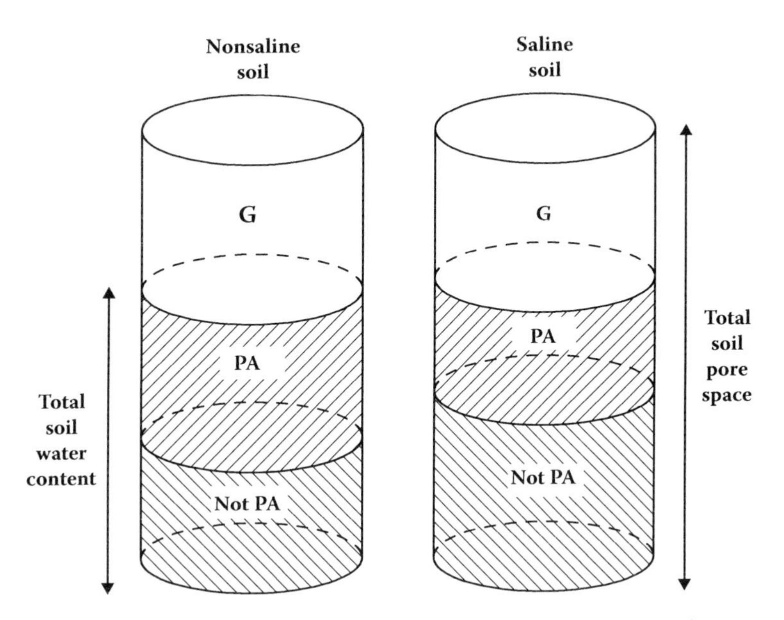

G = **Gravitational** water that drains after rainfall or irrigation. Soil pores in this range are macropores or aeration pores.

PA = **Plant available moisture** retained in the upper (larger) range of micropores.

Not PA = Soil moisture that is **not plant available** since the water is retained in small micropores and as water films around soil colloids and salt ions. The saline soil contains more salts, which reduces the plant-available water.

FIGURE 2.1 Influence of total soluble salts on plant-available water for turfgrass uptake.

Turfgrass plant response to salt-inducted drought stress can range from rather mild symptoms of some green color loss, blue-green coloration resembling regular drought stress plant symptoms, and initial plant wilting; to increasingly intense plant response symptoms under higher salt levels, such as major green color loss, severe wilting, and actual tissue desiccation of leaves, tillers, and whole plants. Soluble salts (primarily Ca, Mg, Na, K, Cl, SO_4, HCO_3, NO_3, and CO_3) may precipitate into soluble compounds as the soil dries, but these compounds then dissolve again into the soluble ions with rainfall or irrigation water applications. Common water-soluble compounds are sulfated salts such as Na_2SO_4, K_2SO_4, $CaSO_4$, and $MgSO_4$; chloride compounds such as KCl, NaCl, and $CaCl_2$; and carbonate or bicarbonate compounds with high solubility such as Na_2CO_3 and $NaHCO_3$. Insoluble compounds like Ca or Mg carbonates or gypsum do not contribute to soluble salts except as they dissolve over longer time periods.

In terms of soil water availability for plant uptake, both the quantity of soil water and its energy status (activity) are important. Water in the soil is influenced by several forces that determine whether it is retained in the soil, taken up by plants, or moved by capillary action or gravity. For example, *total soil water potential* (ψ_t, the total energy status of the soil water) is a function of the sum of various component potentials acting on the water:

$$\psi_t = \psi_m + \psi_o + \psi_p + \psi_z$$

where ψ_m is the *matrix potential* resulting from (a) *adhesion* of water molecules to solid surfaces in the water films close to the solid surfaces, which are called *water of hydration* and are held so tightly that they are retained in the soil unless heated to about 100°C and are generally unavailable for absorption by plant roots; (b) *cohesion* of water molecules with each other in the outer water films around solid particles, which is available for plant root uptake; (c) *surface tension or capillary forces* caused by liquid-gas and liquid-gas-solid interfaces in the irregular-shaped soil pores; and (d) ions on the soil CEC exchange sites that have hydrated water molecules associated with them, which are bound by adhesive and cohesive forces (Figure 2.2). As salt levels increase in a soil, there is more water retained as water of hydration around the salt ions both on the clay colloids as well as on salt ions in the soil solution—this reduces plant-available water, as illustrated in Figure 2.1. The matrix potential is also called the *matrix suction* or *tension*. At saturation, ψ_m is zero, but becomes more negative as the soil dries, resulting in the water becoming increasingly bound and less available for plant uptake or soil movement. The matrix potential is normally the largest component of total potential in partially saturated to unsaturated soils; however, in highly saline soils, the osmotic pressure, discussed below, becomes increasingly important in terms of limiting plant water uptake. The matrix potential can be measured by a tensiometer or in the lab by a pressure plate apparatus.

Osmotic potential (ψ_o), also called *solute potential*, is associated with salts (i.e., solutes) in the soil solution (i.e., water retained in the soil pores, especially the smaller soil micropores of <0.03 mm diameter). Water molecules are attracted to salt ions by adhesive forces in the water films near the salt ion surface and by cohesive forces in the outer water films. As soluble salt concentrations increase in the soil, osmotic potential becomes more negative, thereby reducing the availability of soil moisture for plant uptake. The final result of reduced soil water availability is salt-induced drought stress on the plant even when the soil may appear to be relatively moist or even within hours after the last irrigation cycle. Osmotic pressure is always negative. The relationship between soil osmotic potential and soluble salt concentration in soils is depicted by

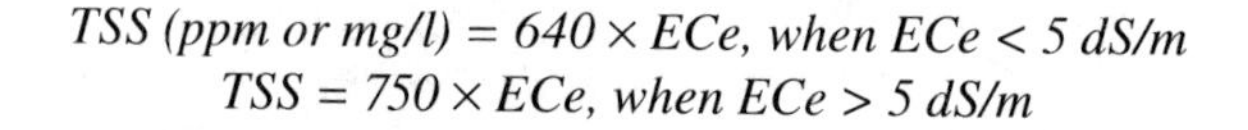

$$\psi_o\,(bar) = -0.36 \times ECe\ (dS/m)$$

where *ECe* is the *saturated paste extract* level of soil salinity expressed in units of dS/m (USSL, 1954). In Chapters 4 and 5, the saturated paste extract method of measuring soil salinity is discussed in detail. For soils, the conversion of ECe in dS/m units to *total soluble salts* (*TSS*) is

$$TSS\ (ppm\ or\ mg/l) = 640 \times ECe,\ when\ ECe < 5\ dS/m$$
$$TSS = 750 \times ECe,\ when\ ECe > 5\ dS/m$$

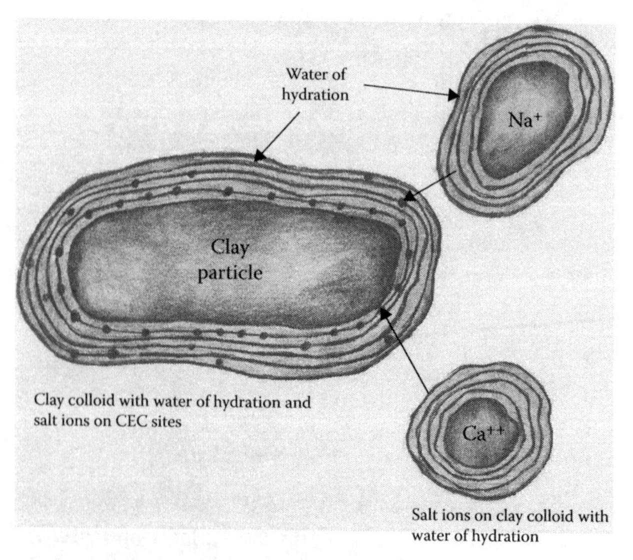

FIGURE 2.2 Clay colloid and salt ions (•) on the clay CEC sites. Water of hydration around clay particles and salt ions (either on CEC sites or in soil solution) is held by adhesion forces (inner water layers nearest the clay or ion surfaces) and cohesion forces (outer layers of water further from the clay or ion surfaces).

In addition to the importance of osmotic potential affecting plant water uptake when soil salts are high, the effects of ψ_o are important in the presence of a selective permeable membrane or a diffusion barrier, which transmits water more readily than salts. Two important diffusion barriers in the soil are (a) the soil–plant root interface, where the root cell membranes are selectively permeable and may selectively take up more water than salts, thus leaving soil salts to accumulate at the soil-root interface or rhizosphere; and (b) the soil water–air interface, where, as water evaporates, the salts are left behind to accumulate at the soil surface.

Pressure potential (ψ_p) is the component of soil-water potential due to hydrostatic pressure exerted by saturating an overlying soil area. When a drainage barrier, such as a soil layer that impedes drainage, allows extra water to be retained in the soil for plant use, the ψ_p is positive below the saturated layer, but is zero at or above the water table. The water table can be a perched water table due to a soil layer, such as in the USGA Green Section construction method for golf greens (USGA, 2010). In an unsaturated soil, $\psi_p = 0$.

Gravitational potential, ψ_z, is due to an elevation difference between two soil points. If the interest is total water potential, ψ_p, at a single point in the turfgrass rootzone, then $\psi_z = 0$, which is the usual situation when concerned with water availability in the rootzone at a particular location. If the concern is about the potential for water to move from a high-topography area of a golf green, for example, to a lower area by gravity (both surface and subsurface movement), then the ψ_z would be considered positive at the low point and the magnitude would be determined by the height difference and a time component (for subsurface migration).

Total soil water potential, ψ_z, is the sum of the above components. It is expressed in various units to describe energy status, but we will use *bar*, where 1 bar = 0.1 MPa = 100 kPa = 100 joules/kg = 100 centibar = 0.987 atm = 1020 cm H_2O. For soils at field capacity or drier, total soil water potential, ψ_z, is considered the sum of the matrix and osmotic potentials:

$$\psi_t = \psi_m + \psi_o$$

At field capacity, $\psi_t = -0.10$ to -0.33 bar (-0.01 to -0.033 MPa) and becomes more negative as the soil dries. At the permanent wilt point, $\psi_t = -15$ bar (-1.5 MPa). Plant-available water is considered between field capacity and the permanent wilt point, but most plant-available water is between $\psi_t = -0.10$ and -4.00 bar.

In nonsaline soils, the matrix potential accounts for the majority of the total soil water potential, while the osmotic potential is low at about $\psi_o = -0.10$ to -0.50 bar. To illustrate the difference between a nonsaline and a saline soil and the effect of ψ_o on ψ_t, first consider a soil with low salinity of ECe = 1.0 dS/m, and at $\psi_m = -1.5$ bar, the total soil water potential would be as follows:

$$\text{Since, } \psi_o\,(\text{bar}) = -0.36 \times \text{ECe (dS/m)} = -0.36 \times (1.0) = -0.36 \text{ bar}$$

$$\psi_t = \psi_m + \psi_o = -1.5 + (-.36) = -1.86 \text{ bar}$$

$$\psi_t \text{ (nonsaline soil)} = -1.86 \text{ bar}$$

Then compare to a saline soil of ECe = 6.00 dS/m and $\psi_m = -1.5$ bar, where

$$\psi_t = \psi_m + \psi_o = -1.5 + (-0.36 \times 6.00) = -1.5 + (-2.16)$$

$$\psi_t \text{ (saline soil)} = -3.16 \text{ bar}$$

These examples illustrate how appreciably soluble salts can greatly reduce total soil water potential and, thereby, the availability of soil water to the plant for uptake. In the above examples, both soils would have the same volumetric soil water content at the matrix potential of $\psi_m = -1.5$ bar, but the availability of this water is substantially less under the saline conditions. Figure 2.1 illustrates the impact in reducing the quantity of water to a plant while the total quantity of water may the same in a nonsaline versus saline soil.

Rengasamy (2010b) found that a soil solution salinity level of 25 dS/m resulted in an osmotic pressure of −9.0 bar, which greatly reduced plant-available water in wheat with 89% to 96% of the field capacity water being unavailable for plant uptake. This was a pot experiment with limited rooting volume, so in field situations where some of the root system may be in less salinized zones the results may be less dramatic. Devitt et al. (1991) noted that in salinized soil, increased irrigation frequency on Bermuda grass with saline irrigation water increased growth. However, on a clay soil, this resulted in less growth because poor soil aeration resulted. This demonstrates the adverse effects of high salinity on water uptake, and corrective measures by maintaining higher soil moisture levels may not work on all soils due to triggering low aeration.

2.2.2 Plant and Soil Symptoms of Total Soluble Salt Stress

Soluble salts are very dynamic spatially across the landscape and vertically within the soil profile, while also being temporally dynamic (horizontally both on the soil surface and in the subsurface) in response to irrigation additions, rain, and microclimate variations in evapotranspiration (ET), soil properties, and plant conditions. When coupled with the spatial and temporal nature of normal soil drought stress and the fact that soluble salts cause the same plant responses as actual water deficit or "normal" drought stress, but in a more rapid manner and to a greater magnitude, it is easy to see why this physiological drought stress is considered as the most important of the salinity stresses. Similar to normal drought responses, physiological drought causes a rapid and direct injury to the plant and becomes increasingly visible unless corrected with appropriate management strategies. Reduced availability of soil moisture for plant uptake under high soil salinity exposure causes a number of plant responses that are the same as with normal drought stress, but in a more aggressive pattern because concentrated salts can easily cause varying degrees of tissue desiccation beyond that of salt-induced drought stress alone. This additional plant response is pointed out later in the section on ion toxicities (Section 2.3.1, "Specific Ion Impact [Root Injury and Shoot Accumulation Injury]").

Important plant responses induced by total salinity that result in *visible plant and landscape symptoms* are:

- Reduced growth rate because cell enlargement requires adequate water volume and osmotic or turgor pressure adjustments.
- Reduced leaf size and leaf area.
- Increased wilting, leaf rolling, and the blue-green coloration evident with drought-stressed grasses.
- Initially, a slight to moderate loss of green color, which may progress into browning or tan color as shoot tissues desiccate.
- Greater potential for desiccation under drought stress since at the same soil volumetric water content, the presence of accumulated salts enhances the degree of the localized water deficit relative to the same soil that is not saline. Tissue desiccation occurs more rapidly under high salinity. The combined discoloration effect of green color loss plus browning or tan discoloration caused by tissue desiccation is often called *leaf firing*.
- If substantial soil drying has occurred to allow salts to concentrate in the soil solution, roots (and especially root hairs) can also be desiccated similar to normal soil drought effects on roots. Roots would appear brown.
- For salt-sensitive plants, sodium levels may be sufficient to cause Na toxicity in the root system, which is exhibited as brown to black, weak, and spindly roots with considerable tissue breakdown.

- Trees and shrubs may show leaf firing and leaf drop, especially in the top branches (a so-called skeleton effect) due to salt accumulation in the leaves.
- The presence of salt-tolerant halophytes in the landscape is another indicator of saline soils. Examples of common halophytes are salt grass (*Distichlis spicata*), cordgrass (*Spatina gracilis*), alkali grass (*Puccinellia nuttalliana*), saltwort (*Salicornia ruba*), marine couch (*Sporobolus virginicus*), kochia (*Kochia scoparia*), or other salt-loving plants (see Chapter 9 for listings of salt-tolerant plants).

In addition to these visible salinity-induced plant shoot and root responses, there are *physiological effects on the plant* that are not visible but are nonetheless important. These physiological effects are similar to normal drought-stressed turf, such as:

- Reduced turgor pressure of shoot cells and cell wall extensibility (pliability or flexibility), where cells become less turgid due to reduced water uptake. Wear injury is greater on plants with reduced turgor pressure. Scalping problems often increase when these symptoms occur on turfgrasses.
- Partial stomatal closure occurs more quickly under saline conditions, which limits CO_2 exchange and transpiration that are important in photosynthesis.
- Photosynthesis may decrease due to reduced leaf area and stomatal closure.
- Transpirational cooling will decline, which is especially important on cool-season grasses in the summer. Plants are generally more sensitive to extreme temperature exposure.
- Reduced cytokinin synthesis in turfgrass roots and subsequent translocation of this hormone to shoots. Cytokinins are involved in many plant processes, including cell division, shoot and root development, chloroplast maturation, cell enlargement, and leaf senescence.
- Respiration increases and the defensive response of the plant is to utilize more stored carbohydrates, which can deplete reserves over time.

Many of these plant responses to high total salinity make the turfgrass more susceptible to other important stresses such as (a) drought stress, (b) indirect and direct high-temperature injury due to reduced water availability for transpirational cooling, (c) wear stress due to reduced turgor pressure and plant vigor for recovery of injured shoot tissues, and (d) predisposition to insect and/or disease attack.

Seedlings or newly established vegetative plant tissues are especially sensitive to these various salt-induced drought stresses due to their roots being confined in the surface zone, where accumulated salts are often concentrated by salt deposition following water loss by ET. Also, the new, juvenile tissues do not have the degree of salt tolerance mechanisms that a more mature plant has. Juvenile tissues are more prone to desiccation stresses. As root volume increases, the plant may be able to obtain water from a less salt-affected soil zone such as deeper in the soil profile and below any salt accumulation zone in the upper soil zones.

The most *visible soil symptom* of high total soluble salts is a white to grayish-white layer on the soil surface. An older term for saline soils was *white alkali* due to the frequent appearance of this white crust of salt arising by soil surface salt deposition as water evaporates or is transpired from the soil surface, leaving behind concentrated total soluble salts. This whitish layer is most apparent on bare soil or on thin turfgrass areas. A soil may have high salinity, but with a good turfgrass stand and canopy density, such as with halophytic (salt-tolerant) grasses, it does not exhibit the surface layer deposition of visible salt accumulation. Obviously, if salts are accumulating at the surface to a point of being visible, even halophytic grasses will exhibit severe salinity stress symptoms, including substantial death of most or all of the turfgrass plants. These saline areas exhibit the same soil structure and permeability properties as a similar nonsaline soil, since Na does not dominate the CEC sites on these saline soils.

Besides the surface crust of salts, saline sites often have subsurface zones of salt accumulation, which may occur from natural origin or be formed at the depth (from downward wetting front migration) of salt leaching by irrigation water or rainfall, especially in arid or semiarid climates. Sometimes these layers or lenses are concentrated enough to be visible in the profile as whitish deposition layers. However, there are other soil types that have white to gray-white layers that are not due to soluble salts, but more often to an insoluble deposition of lime—that is, calcite layers (see Chapter 3, Section 3.4.1). For soil profile layers very high in total soluble salts, it is not desirable for these salt layers to be in contact with the turfgrass root system or to allow salts to rise by capillary action into the root systems and resalinize the rootzone soil. When these total soluble salts move to the turfgrass rootzone or the rootzone is directly in contact with the salt layer, plants exhibit physiological drought symptoms to a pronounced degree and in a rapid manner.

In summary, the soil and plant symptoms caused by high total salt accumulation include the following:

- White or off-white crust on the surface
- Desiccated and dead grass canopy in a random surface zone pattern
- Definite layering in the upper soil profile of cup-cutter plugs
- Wet upper soil profile and somewhat drier soil beneath this salt-accumulated zone
- Rootborne pathogen symptoms on plant roots and at the surface
- Desiccated roots
- Increased incidence of localized dry spots
- Discoloration of the turfgrass or landscape plants
- Decreased wear tolerance in the turfgrass

2.3 ION TOXICITIES AND PROBLEM IONS

2.3.1 Specific Ion Impact (Root Injury and Shoot Accumulation Injury)

As total soil soluble salt level increases, the potential for specific ion toxicity also increases. Soils may contain excessive levels of certain ions that can (a) directly affect root tissues due to multiple soil salt accumulation, and (b) cause injury to plant shoot tissues due to uptake and accumulation in leaves (Duncan et al., 2009). While these ions may be present in the soil at establishment, often the ongoing salt addition source is the irrigation water used on the site. Germinating seeds, young seedlings, and sprigs are especially vulnerable because of their juvenile developing root systems. The ions that most often cause toxicity problems are *Na, Cl, and B*. However, trace elements can also result in toxicity or ion competition (availability for uptake) problems over time in some situations. Guidelines relative to potential for root and shoot injuries from accumulation of these ions are reported in the respective chapters on soil testing (Chapters 4 and 5) and water testing (Chapter 6).

In terms of *foliar uptake, accumulation*, and *injury*, any of the total soluble salts in soil solution can be taken up and potentially accumulate in leaf tissues to cause leaf injury. However, as irrigation water salinity increases, two of the most common salt ions likely to be present are Na and Cl. Chloride is a very common anion in irrigation water and can easily be taken up by plants in considerable excess compared to the very small quantities required as an essential micronutrient. As salts accumulate in leaf tips of grasses or outer margins of landscape plants in the topmost actively growing leaves, the salts can (a) cause tissue osmotic stress by reducing water for cell uptake and inducing dehydration of cells, which can eventually lead to tissue desiccation and leaf firing; and (b) also induce direct toxic effects, depending on the salt ion, since some ions can cause disruption of plant metabolic activities as internal concentrations increase.

Turfgrasses are generally less sensitive to Na and Cl uptake into leaves and foliar injury compared to other landscape plants, primarily because routine mowing removes accumulated ions in the shoot tissues. When trees and shrubs accumulate excessive salts, the initial symptom is leaf wilting, followed by firing on leaf margins (yellow to tan color as tissues die and desiccate), and finally leaves may fall from the plants. For tall trees, the upper leaves and branches often exhibit leaf desiccation stress first from internal salt ion accumulation and localization.

Boron toxicity to shoot tissues can occur if B accumulates to a lethal level within leaves or stems of plants. Boron phytotoxicity on plants may be exhibited by yellowing followed by a dark necrosis on the margins of older leaves. Boron is often associated with saline hydrogeological conditions and is another element that can be a toxicity problem if too high in irrigation water. Due to routine mowing of turfgrass leaves, B accumulation is less likely a problem in turfgrasses compared to other landscape plants.

While Na is one of the contributors to foliar salt accumulation and injury, it is especially a major concern for *direct root toxicity*. Root cells preferentially bind Ca to exchange sites on cell walls (similar to soil CEC sites on clays or organic matter), which is beneficial since cell walls and cell plasma membranes require high Ca content to maintain root viability or function. High Na ion concentration in the soil results in Na competing with Ca ions on cell wall exchange sites, causing root cell wall deterioration, especially near the root tips. The Na-induced Ca deficiency in root tissues is sometimes called *Na toxicity*. Highly salt-tolerant turfgrass and landscape plants possess tolerance to Na-induced root toxicity, but less-salt-tolerant plants exhibit severe root injury from high soil Na accumulation. Highly salt-tolerant plants very strictly regulate the uptake of Na through restriction mechanisms, but salt-sensitive plants do not possess this genetic regulatory capability (see Chapter 9). Thus, Na toxicity to roots in salt-sensitive turfgrass plants and sensitive landscape plants can occur even when the soil is not sodic.

When Na ions displace Ca in the root tissues, the roots start to appear black, especially near the tips. As Na toxicity stress continues, root cells break down, and the roots deteriorate and become increasingly black, thin, and spindly, without structure (i.e., resembling a root rot appearance). With increasing injury and death of root tissues, the root viability for their ability to acquire water and nutrients significantly declines. When Na is sufficiently high in the rootzone to cause salt-induced root rot, salt management corrections must go beyond leaching since Na leaches more slowly than most salt ions due to its tendency to combine into carbonate and sulfate compounds as well as to be retained on CEC sites, thereby making it more difficult to remove from the rootzone.

Duncan et al. (2009) reported on the apparent stripping of Ca ions from shoot cell walls by Na when the irrigation water was high in Na and was consistently applied over the leaf tissues. Shoot cell walls also have negative charge sites that result in leaf CEC sites that preferentially retain Ca similar to roots. Calcium is an essential cation in the plasma membrane to work synergistically with potassium in sustaining turgor pressure in cells and for cell wall integrity; if Na displaces Ca in the plasma membrane, K often leaks out of the cell, and osmotic adjustments to increasing salinity exposure are decreased. Instead of visible shoot tissue deterioration, as happens in roots, the visible symptom that appears is Ca deficiency symptoms (usually exhibited by yellowing of the older leaves down in the turfgrass canopy and eventually progressing to newer leaves), which can be verified with tissue testing. Sodium stripping of Ca from shoots and leaves seems to be most prevalent on halophytic grasses or Bermuda grasses (*Cynodon sp.*), probably because it is only on these grasses that highly saline irrigation water is likely to be applied on a frequent basis.

2.3.2 Direct Foliage Injury and Miscellaneous Problems

Plant leaves of trees and shrubs impacted by overhead sprinkler irrigation water may exhibit direct foliage injury from excess Na or Cl concentrations in the water, and guidelines for these ions are given in Chapter 6 ("Irrigation Water Quality Tests, Interpretation, and Selection"). Leaves that are in the pathway of the irrigation water stream may start to lose color, leaves may start to desiccate, and eventually leaves may die within the irrigation water impact area on the plant.

Bicarbonate (HCO_3) concentrations in irrigation water are not phytotoxic to leaves, but at >8.0 meq L^{-1} or 500 ppm, HCO_3 can cause unsightly whitish deposits on leaves of ornamentals, trees, cart paths, and equipment. Bicarbonate complexes (with Ca, Mg, and Na) can, however, accumulate in shallow layers of the upper soil profile to cause water infiltration or percolation problems.

Residual chlorine (Cl_2) that is used to disinfect wastewater becomes toxic at >5 ppm for many plants. Damage is expressed as leaf tip injury. However, chlorine is normally not a problem on turfgrasses since most irrigation water is applied through overhead sprinkler systems and this gas dissipates rapidly when aerated. Storage of high-chlorine water in lakes, ponds, or lagoons will accomplish the same dissipation phenomenon over time, especially if an aeration system is present in the storage area. However, storage of chlorinated reclaimed water in enclosed tanks without aeration or direct deposition of highly chlorinated water from treatment plants into wet wells or enclosed pump houses can cause corrosion problems as well as potential chlorine-laden water droplets that result in leaf injury (white bleaching symptoms) on turfgrass or landscape plants.

Some soils may accumulate SO_4 ions if the irrigation water has appreciable SO_4, such as some reclaimed water. The primary problem of high SO_4 additions onto turfgrass sites occurs when anaerobic conditions (upper surface layering from compaction or salt deposition layering in the soil profile that seals in a particular zone) develop, which transforms SO_4 into reduced S (Carrow et al., 2001; Duncan et al., 2009). Reduced S can react with reduced forms of Fe and Mn to create FeS and MnS compounds in the soil that are contributors to black layer, and this condition results in additional anaerobic conditions, leading to the sealing of soil pores. Thus, a high S level is normally not the initial cause of an anaerobic condition, but it will greatly amplify the condition and require a more aggressive cultivation program when leaching programs do not move the S compound below the turfgrass root system. The SO_4 ion is one of the easiest salt ions to leach, along with Cl, but on fine-textured soils with low percolation rates leaching can be difficult.

Visible symptoms of excess sulfur or sulfates normally do not appear unless anaerobic conditions occur, after which black layer starts to form. *Black layer* can be a relatively thin zone at or near the soil surface or may be 5–20 cm deep into the soil depending on the specific situation. Since the black, gel-like compounds further seal the soil pores, it may increase in size and severity until aeration is restored. Once formed, the black layer impedes the leaching of soluble salts. Remediation involves cultivation for better aeration, limiting S additions, and leaching SO_4 as a preventative measure. When SO_4 is coming from the irrigation water, application of lime to the soil surface results in chemical transformation in the soil to gypsum, which stabilizes the excess sulfur as long as oxygen is moving through the soil profile. It should be noted that SO_4 in the gypsum compound (molecule) is not water soluble, but it is slowly soluble and does not contribute to black layer.

2.4 NUTRIENTS AND ION IMBALANCES

Chapter 16 ("Nutritional Practices on Saline and Sodic Sites") is devoted to an in-depth discussion of the various nutritional challenges that may arise on salt-affected sites and their associated corrective practices. In the current section, an overview is presented with respect to saline soils. Discussion of plant nutrition as affected by salinity across all plant types is reported by Feigin (1985), Rhoades et al. (1992), Naidu and Rengasamy (1993), Ayers and Westcot (1994), Shalhevet (1994), Oster (1994), Marschner (1995), Grattan and Grieve (1999), Alam (1999), Beltran (1999), Kelly et al. (2006), and Barker and Pilbeam (2007). Salinity nutritional issues in turfgrass systems are addressed by Carrow and Duncan (1998), Waddington et al. (2001), Duncan et al. (2009), and Carrow (2011).

When considering the impact of saline soils and irrigation waters on turfgrass nutrition, the most important factor to recognize is that compared to sites without salinity stress, soil fertility and plant nutritional aspects will be much more dynamic in nature. On saline soils, high inherent levels of Na, Cl, and other ions plus those added by irrigation water have a major impact on soil fertility and plant nutrition relationships. When saline or reclaimed irrigation water with an appreciable nutrient load is used, spatial variation of ions horizontally across the landscape and vertically within the soil rootzone is often substantial. Short-term temporal variation in response to weather patterns and seasonal changes in irrigation water quality is also significant. Additionally, leaching to control salt accumulation is necessary, and this management practice alters nutritional status. With effective leaching of salts, soluble nutrients in the soil are also leached. The combined effect of these factors results in very dynamic soil fertility and plant nutrition programs, which require a proactive approach of more frequent soil-, water-, and tissue-testing programs to avoid deficiencies, excesses, or imbalances. It is the experience of the authors that on saline sites, inattention to the dynamic changes in soil fertility and plant nutritional status can result in rapid changes in the turfgrass quality and overall performance; and without timely science-based test information, the turfgrass manager may be unable to determine the specific cause of the basic stress in a timely manner and remedy the problem. The key questions here are, Are critical nutrients *actually available*, and can the plant absorb concentrations at sufficiency-level requirements?

Greater occurrence of nutrient imbalances is a second nutritional factor inherent on saline sites. High concentrations of a particular nutrient or element can lead to not only overfertilization but also nutrient imbalances with counterions, especially cations (Duncan et al., 2009). Guidelines for the various nutrients are listed in the chapters on soil (Chapter 5) and irrigation water (Chapter 6). For assessing the potential to create a cation imbalance, the key ratios are Ca:Mg, Ca:K, and Mg:K. Also, increasing accumulated Na can easily cause an imbalance with K, Ca, and Mg, but particularly K in the soil and plant. On recreational turfgrasses, adequate shoot K content is important for wear tolerance as well as enhanced cold, heat, and drought resistances; but K is the most easily leached cation in the soil as well as the one nutrient that is most influenced by Na accumulation. Some irrigation water, especially if influenced by seawater, is high in Mg, and a high Mg concentration may influence K availability as well as Ca.

Deficiencies of Ca, K, NO_3, Mg, Mn, SO_4, Mo, Zn, Cu, and P can be induced by high-salinity environments, especially on more sandy soils. The degree of deficiency symptoms varies among plant species and within turfgrass species as well as cultivars. Calcium deficiency on turfgrasses is very rare; however, turfgrasses grown in salt-affected soils may be affected by low *available* Ca even if a true deficiency is not observed (even in calcitic soils if Ca is not available for uptake). Sodium can displace Ca from root cell membranes and reduce their integrity, as well as sometimes displace shoot cell wall Ca (as noted in the previous section). Also, saline-induced Ca deficiency may reduce certain salt tolerance mechanisms such as ion exclusion and selective transport (see Chapter 9). Thus, added Ca often improves the salt tolerance of plants, especially when high Na is present. Acidic salt-affected soils may especially benefit from added Ca. Deficiency of K induced by Na and leaching programs on saline sites is quite common, and sufficient K must be added on a timely schedule to maximize plant osmotic adjustment to salinity stress—an example of fertilization not just to meet basic plant growth requirements but also to maximize the expression of the plant's salt tolerance mechanisms. It has been the observation and experience of the authors that on saline sites, extra Mn and Zn (in both granular soil-applied and liquid foliar-absorbed products) are often required to achieve maximum salt tolerance expression of turfgrasses.

2.5 MANAGING SALINE SOILS

Numerous secondary problems or challenges arise from the primary salinity stresses of high total soluble salts, ion toxicities and problem ions, and nutritional levels and imbalances. A number of articles have been written on the management of salinity in agricultural soils (Abrol et al., 1988; Rhoades and Loveday, 1990; Chhabra, 1996; Keren, 2000; Qadir et al., 2000; Rengasamy, 2002; Qadir and Oster, 2004). Carrow and Duncan (1998) and Duncan et al. (2009) have addressed salinity management in turfgrass situations. The BMPs for salt-affected sites presented in this book address both primary salinity stresses as well as the secondary management challenges that occur at the same time.

While osmotic stress is a consistent factor in all salinity sites, nutrient imbalances and possible toxic ions may vary with soil pH. Thus, management would require adjustment for any nutrient imbalances or excesses. Rengasamy (2010a) summarized saline soil problems as:

- *Acidic pH < 6.0*: osmotic effects; root toxicities of Fe, Mn, and SO_4
- *Near-neutral pH of 6.0 to 8.0*: osmotic effects; possible toxicity from any dominant cation or anion, especially under higher salinity
- *Alkaline pH > 8.0*: osmotic effects; excessive HCO_3 and CO_3; and Fe, Mn, Al, and OH^- toxicities at pH > 9.0

Reclamation of saline soils is not by chemical amendments (e.g., gypsum), but by removal of the excess total soluble salts from the plant rootzone. The three most important salinity management aspects are (a) the leaching of total soluble salts by the application of sufficient water volume to allow net downward movement of salts (Chapter 11); (b) the adjustment of soil fertility programs to correct nutritional deficiencies, imbalances, or toxicities; and (c) to select turfgrasses that can tolerate the salinity levels expected when saline irrigation water is used on a regular schedule (Chapter 9). During dry periods, the background soil salinity will not be any lower than what the salinity level in the irrigation water is, and usually accumulated soil salinity is higher. Some common management considerations on saline soils are as follows:

- Irrigation system distribution uniformity and efficiency become very important since the nonuniformity of irrigation water will affect spatial distribution on not just the water but also soluble salts. Design issues (spacing between sprinklers and wind issues) will become more pronounced, and salinity leaching more difficult (Chapter 10).
- Irrigation scheduling, to avoid normal and physiological drought stress and for salt leaching, is critical (Chapter 11).
- Factors to enhance water infiltration (into soil surface), percolation (through the rootzone), and drainage (past the rootzone), such as in Chapter 14 ("Cultivation, Topdressing, and Soil Modification") and Chapter 15 ("Drainage and Sand Capping"), are important for efficient salt-leaching programs.
- Traffic control programs to avoid total salt-induced wear injury must be carefully designed (Chapter 17).
- Fertilization must be adjusted not just to meet plant growth requirements but also to maximize salt tolerance mechanisms and to correct imbalances (Chapter 16). As irrigation water salinity increases, so do the nutritional interaction issues, and these can become dominant management challenges under high salinity.
- Proactive soil and water quality testing should be performed, as well as tissue testing if needed, to monitor salinity impacts (Chapters 4, 5, and 6).
- Drought and high-temperature stresses are more common on saline sites since high-soluble salts cause physiological drought and may injure roots, and plant tolerance to these stresses is reduced by high salinity (Chapter 17).
- High soil salinity depresses cytokinin synthesis in roots, and this "biostimulant" may be necessary to promote adequate growth (Chapter 17).

3 Sodic, Saline-Sodic, and Alkaline Soils

CONTENTS

3.1 SODIC SOIL PROBLEMS

3.1.1 Sodium Permeability Hazard

In the brief introduction of sodic soils in Chapter 1, *sodium permeability hazard* was noted as the primary problem under sodic conditions, which is assessed in soils by *exchangeable sodium percent (ESP) and sodium adsorption ratio (SAR) criteria* (for discussion of ESP and SAR, see Chapter 1, Section 1.1.2, "Classifying Salt-Affected Soils"; and Chapter 4, "Salinity Soil Tests and Interpretation"). In sodic soils, excess Na on the CEC sites and in Na carbonates precipitated in the soil causes soil degradation, resulting in reduced water permeability (infiltration, percolation, and drainage), decreased gas exchange or permeability (low oxygen, reduced oxygen flux into the soil profile, or aeration), and a less favorable rooting media due to soil structural breakdown (Rengasamy and Olsson, 1991; Naidu et al., 1995; Levy, 2000). Thus, this Na-induced salinity stress is often called an "Na permeability hazard" or "Na hazard."

Accumulation of Na on the soil CEC sites to the point of creating a sodic condition is generally a longer term developmental process than accumulation of total soluble salts for the creation of a saline soil. However, a sodic condition is also a much slower process to remediate relative to total soluble salt removal, especially on fine-textured soils. Protection of soils from Na degradation is a primary component of salinity management, especially when the irrigation water contains significant concentrations of Na (>4.35 meq/L or 100 ppm).

Sodium degradation of soil physical conditions starts with the displacement of Ca and Mg by Na ions on the negatively charged cation exchange capacity (CEC) sites of clay colloids. The soil test for

TABLE 3.1
General Soil ESP and SAR Guidelines for Na-Induced Permeability Problems, Assuming Intermediate Irrigation Water Quality; and Approximate Conversions[a]

	Degree of Permeability Problem from Na		
Soil Parameter	**None**	**Increasing[a]**	**Severe**
Guidelines			
ESP (%)	<3.0	3.0 to 15.0	>15
SAR	<2.1	2.1 to 12.0	>12
Conversions of SAR and ESP			
SAR	ESP (%)		
5	6.9		
5.6	7.7		
10	12.9		
12	15.0		
15	18.1		
20	22.8		
25	27.0		

[a] Situations where Na permeability hazard occurs at lower SAR or ESP values include (a) ultrapure irrigation water containing primarily Na ions as the salt, (b) 2:1 clays, and (c) higher clay content. Sodic conditions appear first at the surface when the source of Na is the irrigation water. The average soil sample SAR or ESP may be relatively low, but the actual SAR or ESP at the surface 0 to 2 cm is much higher.

ESP quantifies this process where a soil with 15% of the CEC sites (i.e., 15% exchangeable sodium percentage [ESP]) occupied by Na has been traditionally classified as a *sodic soil* (see Chapter 1, Table 1.1). However, no single ESP value is adequate to determine the actual Na permeability hazard where water infiltration or percolation and gas diffusion are reduced. General guidelines are noted in Table 3.1; but as noted in the discussion that follows, these guidelines are very general in nature. When the irrigation water is the source of most of the Na ions, it is not the absolute concentration of Na that is important, but the balance and interactions of Na with Ca, Mg, HCO_3, and CO_3 ions in the water. In addition to the balance of these ions in the irrigation water, Na activity is also strongly influenced by total salinity (electrical conductivity of irrigation water, ECw) of irrigation water, clay type (expanding/contracting or nonexpanding), and clay content (soil texture). These relationships are briefly discussed in the following paragraphs to provide a basic understanding of the factors affecting the Na permeability hazard; but further information is presented in Chapters 4 and 6 on soil and water testing, respectively.

3.1.1.1 Balance of Na with Ca, Mg, HCO_3, and CO_3 in Irrigation Water

Irrigation water containing moderate to high Na content (>100 ppm or 4.35 meq/L) is a primary cause of sodic and saline-sodic soil conditions. As noted, the balance and interactions of Na with Ca, Mg, HCO_3, and CO_3 in irrigation water are extremely important, since they directly affect both the chemical and the physical status of the soil. Sodium ions on soil CEC sites have a single ionic charge and can only bind to one colloid rather than enhance colloid aggregation by attraction to two colloids, such as Ca can do with its double charge. Also, the large hydrated ion radius (the nonhydrated ion radius of Na is smaller than that of Ca, but the hydrated ion radius is larger since this salt ion has a priority affinity for attracting water films) tends to disperse soil colloids by weakening the binding forces between colloids. Calcium is the primary ion that stabilizes soil structure due to its high affinity for CEC sites; double ionic charge, allowing it to attract adjacent soil colloids; and smaller hydrated ion size than Na, which allows greater intercolloidal attraction. Magnesium offers secondary structural stability, but to a lesser extent than Ca due to a hydrated ion radius approximately 50% greater than Ca, which allows the clay surfaces to absorb more water than if Ca was present. This tends to weaken intercolloidal forces, thereby contributing to decreased aggregate stability. The result is a collapsing of random zones (scald areas) in the upper soil profile due to this structural breakdown caused by excess Na. These scald areas start wherever a slight depression may allow Na to accumulate more than adjacent areas.

When Na concentration (meq/L basis) in the irrigation water exceeds the sum of Ca and Mg concentrations in meq/L, there is an opportunity for Na to start to dominate on the CEC sites. If the irrigation water contains more HCO_3 and CO_3 than Ca and Mg (again in terms of meq/L units), the chance for Na accumulation is even greater since the Ca and Mg can react with the HCO_3 and CO_3 to form insoluble lime complexes, thereby reducing solution Ca and Mg availability levels that could displace Na on the CEC sites. There is no single Na concentration value in irrigation water that indicates a problem in all situations, but Na levels of greater than 100 to 200 ppm (4.3 to 8.7 meq/L) indicate that a close evaluation of excess Na and its potential to accumulate in the soil profile relative to Ca and Mg concentrations is warranted. Experience of the authors has revealed that the threshold concentration of 125 ppm (5.4 meq/L) Na is a manageable level in perennial turfgrass ecosystems when applying repeated irrigation cycles with Na-laden water.

Some irrigation water may have appreciable Na and also contain high levels of Ca and Mg, but have relatively little HCO_3 and CO_3 to react with the Ca and Mg ions. Soils irrigated with such water may not exhibit much deterioration of structure since the Ca- and Mg-rich water with high ECw limits the ability for Na to dominate on the CEC sites. A saline-sodic soil is more likely to occur with these parameters, which is discussed later in the chapter.

3.1.1.2 Total Salinity of the Irrigation Water (ECw)

On sites with irrigation water containing unusually low or unusually high total soluble salts, the effects of the Na on the CEC sites on water permeability are altered (Figure 3.1). In Figure 3.1, soils irrigated with highly saline irrigation water may not exhibit Na-induced reduction in soil permeability except at SAR well above the normal value of SAR = 12 used to classify a soil as sodic. In contrast, the same soil irrigated with irrigation water low in total soluble salts (ECw < 0.50 dS/m) may show reduced water infiltration at SAR considerably less than SAR = 12. The combination of high Na and high total soluble salts can result in a saline-sodic soil.

3.1.1.3 Clay Type

Clay types differ in response to Na addition. Clays, such as montmorillonite (smectite), illite, and vermiculite, that are 2:1 shrink-swell clays will demonstrate a swelling nature after wetting, and macropores are compressed into micropores. This process essentially destroys soil aggregates by pressing the soil into a more massive structure with fewer macropores within or between aggregates. Soils with a high content of 2:1 clay types will show the greatest swelling response on wetting and exhibit the most distinct shrink and cracking response when dried. The 2:1 notation refers to the clay crystalline structure where these clays contain two Si-O tetrahedrals combined with one Al/Mn octahedral layer or sheet. When these clays start to dry, cracks form (i.e., macropores)—so they are "self-cultivating." However, many irrigated turfgrass areas cannot be dried sufficiently to result in cracking without imposing unacceptable drought stress on the grass. These clays are very sensitive to Na accumulation, which can start to cause structural breakdown on 2:1 clays at ESP of 3%–4% Na. Considering that soil samples are often taken to a 8- to 10-cm depth to determine whether a soil is sodic using the ESP or SAR criteria, the average ESP test value can be well below the actual surface 1- to 2-cm zone ESP when the irrigation water contains significant Na concentrations (an approximate Na content of >100 ppm or 4.3 meq/L).

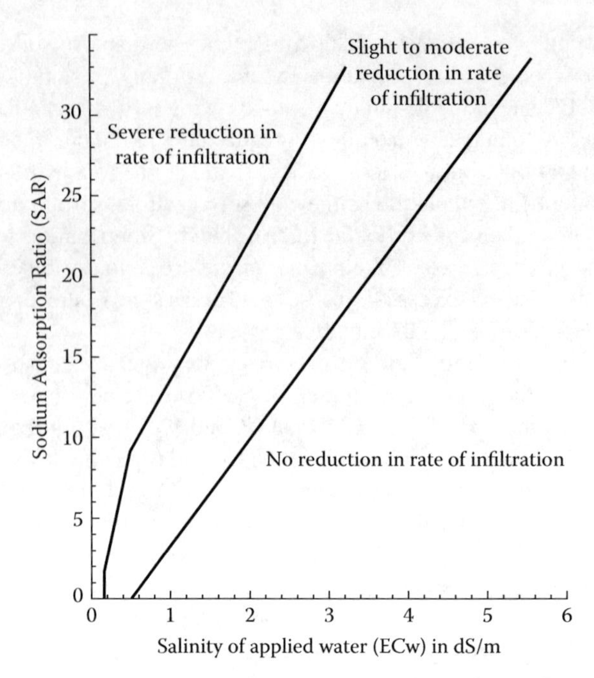

FIGURE 3.1 Effect on water infiltration as influenced by total soluble salts in the irrigation water (ECw) (or soil ECe at the surface) and soil sodium adsorption ratio (SAR). (Adapted from Oster, J. D. and F. W. Schroer, Infiltration as influenced by irrigation water quality, *Soil Sci. Soc. Amer. J.*, 43, 444–447, 1979.)

In contrast to 2:1 clays, 1:1 clay minerals contain one tetrahedral and one octahedral sheet per clay layer. These include kaolinitic clays, allophane, and Fe/Al oxides such as commonly found in the red clays of the U.S. Piedmont region or other humid regions of highly weathered soils. Characteristics of these clays are that they exhibit a nonshrink/nonswell nature, have high soil strength, and do not compact as easily as 2:1 clays. Na does not cause soil structural breakdown as easily, and often 1:1 types are not affected until ESP > 15% to 24% Na, or even at higher ESP if the soil is acidic and Fe and Al complexes are adsorbed and flocculate clay (Kaewmano et al., 2009). Cultivation lasts longer on 1:1 than 2:1 clay types, which is important in terms of the cultivation scheduling of sodic soils.

3.1.1.4 Clay Content

Higher clay content in a soil results in greater expression of Na-induced structural deterioration, regardless of the clay type. A soil with 2%–5% of a 2:1 clay will be much easier to manage compared to one with 30%–40% clay content. Sand and loamy sandy soils contain little clay content and therefore few if any soil aggregates—they have a single-sand grain structure. These sandy soils will not show the effects of Na on structural degradation, which would be observed on more fine-textured soils. However, any colloidal fines (clay or organic colloids) present can be dispersed by Na, and in arid regions these fines may move with irrigation water to the depth of routine irrigation wetting front penetration; or in regions with more rainfall, the colloids may move to the drain lines or deeper into the soil profile.

3.1.2 Process of Physical Degradation of Soil Structure by Na

3.1.2.1 Understanding Good Soil Structure

Water and air movement in soils are critical parameters for management of plants and for sustaining root viability. Soil permeability to water is measured by *soil hydraulic conductivity* (K), which determines the ability of a soil to conduct water (i.e., infiltration, percolation, and drainage), with *saturated hydraulic conductivity* (*Ksat*) as the most common means of assessing soil permeability. Except in sandy soils, Ksat is strongly dependent on soil structure. Additionally, the interchange of gases between the soil and the atmosphere (i.e., gas diffusion) is affected by soil structure. Soil structure is based on the geometry of soil pores and the stability of soil structure.

Geometry of the soil pores is a combination of pore size distribution, pore tortuosity, and connectivity or continuity of pores with each other (see Chapter 8, Section 8.1.2; and Table 8.1 for more detailed discussion of soil porosity). *Macropores* (>0.075 mm diameter pores; also called "aeration porosity" and "noncapillary pores") are important for rapid movement of soil moisture under saturated or semisaturated conditions as well as for gas diffusion to maintain soil oxygen. Roots preferentially grow in macropores. *Micropores* (<0.075 mm diameter; also called "moisture retention pores" and "capillary pores") retain water, but they also retain salts. Tortuosity refers to the nature of macropore channels, with highly contorted (winding or bending) channels having high tortuosity, while a solid cultivation tine would result in a low-tortuosity or straight macropore channel. Pore continuity relates to whether soil pores are connected or isolated. Excess water or oxygen diffusion into the soil would move most rapidly in large-diameter macropore channels with low tortuosity and high connectivity.

Stability of soil structure is based on maintaining flocculation of clay particles and aggregation of soil particles into soil aggregates. Individual clay particles are plate-like in nature. In nonsodic soils, individual clay particles do not stay in an individual (dispersed) state, but flocculate together with other clay particles into *domains* or *tactoids* (Figure 3.2). The single plate-like clay particles essentially flocculate into pancake-like domains with the individual particles held together primarily by ionic bonds from divalent cations. The negative charges within the clay particles attract positively charged cations on the clay surfaces in ion swarms called the *diffuse double layer* of electrical charges, which neutralizes the clay charges. Divalent ions (Ca, Mg) form narrower diffuse double layers than do monovalent (Na, K) ions, thereby allowing closer interparticle distances for more structural integrity. The smaller hydrated ion radius of Ca compared to Mg allows a narrower diffuse double layer of electrical charges for Ca. When clay particles are able to come closer together, a number of bonding mechanisms in addition to ionic bonds occur by interaction of clay particles with inorganic and organic compounds for particle linkage and aggregation (Rengasamy and Olsson, 1991).

For fine-textured soils, the clay domains further combine with silt and sand particles to form structural units or *aggregates* (Figure 3.3). Aggregation is fostered by the action of wetting and drying, freezing and thawing, root penetration pressures, and cultivation operations. Aggregates are held together or stabilized by various mechanisms involving organic matter polysaccharides and gels, divalent (Ca) or trivalent (Al) ions, and Fe/Al/Mn oxides that act as cementing agents. Soil aggregates are important because they contain micropores and macropores within the aggregate, but of most importance, aggregates result in substantial macropores between aggregate units—much like

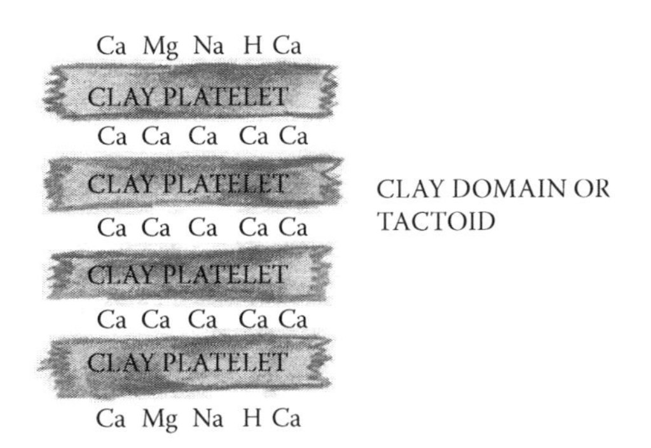

FIGURE 3.2 Clay domain (tactoid) with individual clay platelets flocculated together by Ca ions with small hydrated radius and divalent charge (i.e., Ca^{+2}).

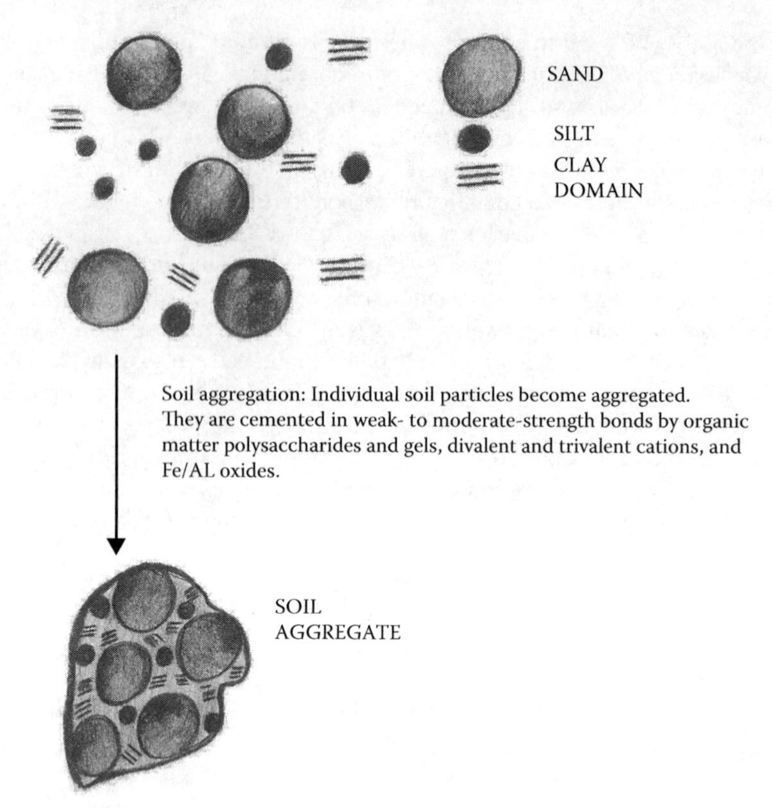

FIGURE 3.3 Illustration of soil aggregation of sand, salt, and clay particles into a soil aggregate.

the macropores between sand particles. In contrast, sandy soils have single-grain structure since they contain few clay and silt particles for aggregation to occur. Thus, Na has much less adverse effects in very sandy soils compared to soils with clay.

Two important factors that cause degradation of soil structure are soil compaction and high Na levels. At the soil surface, compaction from equipment or human traffic destroys structure, while in the subsurface horizons, compaction occurs by the overburdening weight of the surface soil horizon. In either case, structural degradation is caused by high pressures that distort and compress soil aggregates together, thereby destroying the interaggregate macropores. A highly compressed soil exhibits a "massive structure" where total porosity and macroporosity are reduced while microporosity is increased.

3.1.2.2 Sodium-Induced Degradation of Soil Structure

High Na content on the CEC sites in sodic soils is another important factor that degrades soil structure directly and indirectly by making the soil more prone to soil compaction. The most common source of Na is from irrigation water, but Na can also arise by flooding with saline waters such as lands adjacent to oceans, salt water intrusion, or saline water tables rising close enough to the surface to allow capillary rise of Na into the rootzone. When Na is added to the soil by any of these means, the Na can start to displace Ca, Mg, K, and H ions on the CEC sites. The process of deterioration of soil physical conditions then continues through the phases of slaking, dispersion, and particle migration, leading to complete soil structural breakdown.

Slaking is the initial effect of Na on soil structure and can start at well below 15% Na on CEC sites, especially in 2:1 clays (Figure 3.4). Since this process starts at the soil surface when irrigation water is the source of Na, the effects on water and air permeability can soon affect site management.

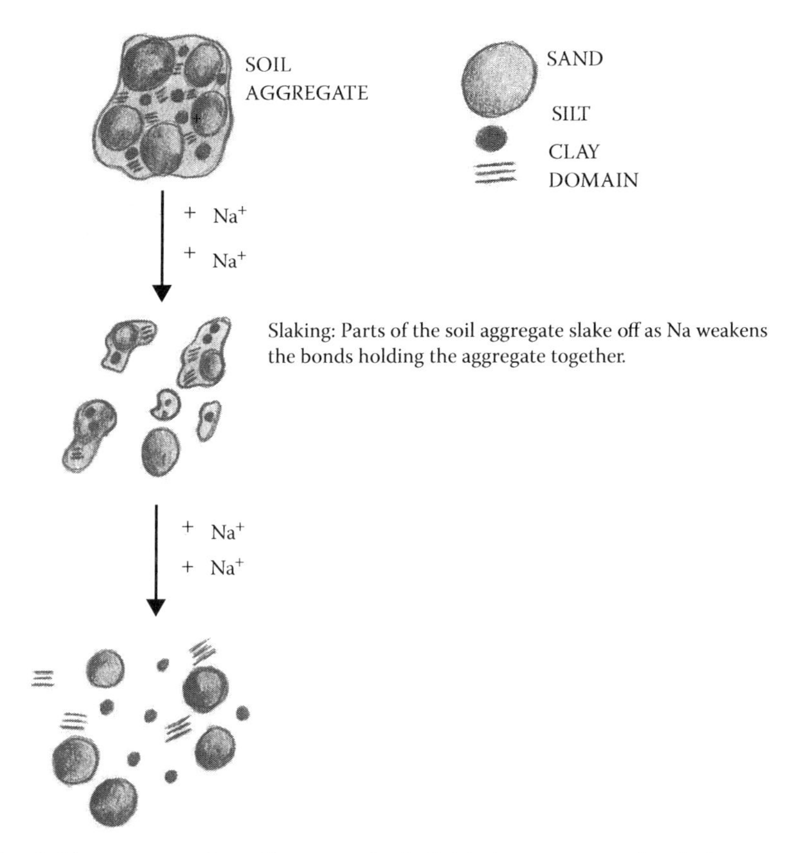

FIGURE 3.4 Slaking process where Na causes the individual aggregate to break down or disperse into smaller aggregates and eventually into individual soil particles.

As Na on the clay surfaces starts to dominate the diffuse double layer, this weakens bonding strength between clay particles and reduces the overall strength of soil aggregates. Soon, parts of the soil aggregate start to slake off. Also, rapid surface wetting of aggregates by rain or irrigation water may cause the weakened aggregates to shatter from entrapped air pressure inside aggregates. The net effect is fewer and smaller aggregates, causing a reduction in macroporosity and changes in water infiltration and percolation through the soil profile.

Dispersion of clay domains into individual clay particles occurs as Na levels increase (Figure 3.5). The single-charge and larger hydrated radius Na ions displace Ca and Mg ions within the domains and result in deflocculation of the domains and eventual destruction (collapse) of the remaining soil structure. With dispersion, total porosity and macroporosity continue to decline and microporosity increases.

A third process is *colloidal particle migration*. In fine-textured soils, particle migration is often thought to be minimal due to lack of macropores for easy movement of individual clay particles. However, Warrington et al. (2007) demonstrated that even in fine-textured soils that are irrigated with Na-rich treated wastewater clay, translocation occurred from the surface zone over time (i.e., 12–15 years in their study). The matrix of sand, silt, organic matter, and clay particles in a more massive structure would inhibit most particle movement, but there is still colloidal particle movement. Near drain lines, it is possible for colloidal particles to move into the lines. In contrast, sandy soils do not have the aggregates to break down, but they do contain some clay and organic colloids that can be dispersed by Na-dispersive forces. These can slowly move in intersand macropore channels and plug pores at the interface between sand particles. Particle migration and layering of fines are most apparent in arid regions where the normal depth of irrigation water penetration is the

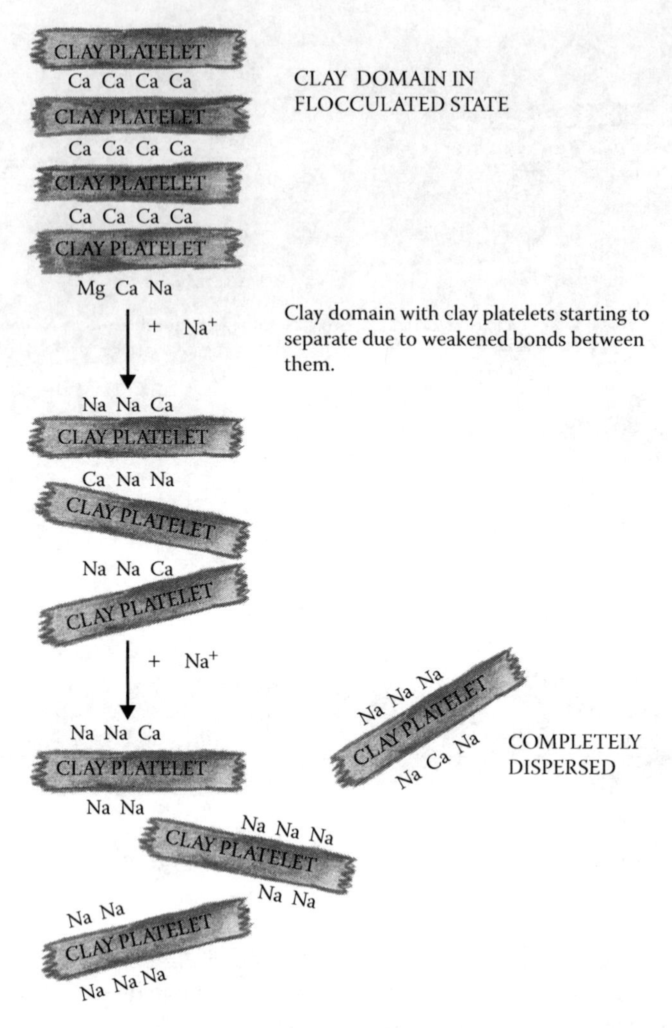

FIGURE 3.5 Dispersion of clay domain into individual clay platelets.

zone of accumulation for precipitated salts and colloids. Over time, this can form a layer that may impede water movement and rooting. Deep cultivation and use of acidic fertilizers or amendments can disrupt the layer.

When sufficient Ca is applied during the reclamation of sodic soils, the reverse process occurs. Available Ca needs to be greater than the Na concentration or applied regularly at levels that will displace the excess Na. First, dispersed clay and organic colloids start to flocculate into domains as the Na is removed from the system by Ca addition plus leaching. Second, the clay domains can then become aggregated again with sand, silt, and organic matter to form individual aggregates, which allows for macropore formation between aggregates.

Reclamation depends on leaching of the displaced Na. Leaching is discussed in Chapter 11 but one point that is important is for leaching of the whole soil mass by insuring that the leaching water flows through all pores—that is, macropores and micropores. For water movement through the micropores, unsaturated flow is required. Applying irrigation water in a manner that results in free or excess water will result in most of the drainage occurring in the macropores due to gravity flow. Using repeated pulse irrigations to foster unsaturated flow conditions will allow for more flow through the micropores and better salt or Na leaching.

3.1.2.3 Na Carbonate and Na Bicarbonate and Soil Physical Conditions

When Na is added via the irrigation water, Na can cause soil physical problems by two major means: (a) by the presence of Na on the CEC site of clay particles that cause structural degradation, as discussed in the previous section; and (b) as precipitated Na carbonate and Na bicarbonate. Both Na carbonate and Na bicarbonate are soluble Na forms and thus can be leached if a sufficient leaching fraction or precipitation occurs. However, in arid regions, Na carbonate forms (i.e., both forms) precipitate along with any Ca or Mg carbonates that may precipitate in the soil from irrigation water constituents, especially at the depth of normal irrigation water penetration. Irrigation schedules are often rather uniform in arid climates, so the precipitation zone can become a zone of accumulation that can start to plug soil pores. In semiarid and humid climates, precipitation is normally sufficient to leach the Na carbonate forms and prevent a distinct zone. An exception would be during prolonged dry periods.

When reclaiming sodic and saline-sodic soils, a source of Ca ions is applied based on the Na content on the CEC. Then, a continuous leaching program that allows downward leaching or the displaced Na (as Na sulfate, a soluble form) will leach not just the Na sulfate but also the Na carbonate forms.

3.1.3 Plant and Soil Symptoms of Sodic Conditions

Plant responses in sodic soils consist of a combination of specific ion effects, nutritional challenges from imbalances or deficiencies, and low-oxygen (deficiency) effects on roots and shoots in response to poor soil physical conditions (often accompanied by increased incidences of localized dry spots, or LDS). Common *visible and physiological (not visible) plant symptoms and stresses* are noted below. Many of the physiological responses are due to Na-induced low soil oxygen stress (Colmer and Voesenek, 2009):

- The most visible plant symptom on sodic areas is poor root growth with weak, spindly roots that are black and exhibit considerable tissue deterioration in response to Na displacement of Ca in root cells and in the plasma membrane, as well as in response to low soil oxygen stress. The lack of a healthy root system can be expressed as frequent and reoccurring drought stress symptoms.
- Plant death or absence of plants in depressed soil surface areas, due to the collapse of soil structure in localized areas that do not dry down rapidly. These become scald areas over time and may appear similar to algae scum.
- Leaf chlorosis and shoot tissue death under low soil oxygen, especially after high rainfall. The turfgrass will initially show yellow-green discoloration common for oxygen-stressed turfgrasses, especially under high temperatures and humidity conditions. Continued low oxygen stress can result in rapid turfgrass death, especially in cool-season grasses.
- Seedlings may not establish well due to soil sealing causing low soil oxygen and also salt accumulation at the surface if the irrigation water is even slightly saline. They may germinate but then die. Soils usually remain very wet and do not dry down rapidly, increasing the incidence of root-rotting organisms.
- Similar to saline soils, specific ions in sodic soils can cause shoot accumulation of Na and other soluble salts as well as trigger soil and plant nutritional challenges from imbalances and deficiencies. These are discussed in Chapter 2, Section 2.3 ("Ion Toxicities and Problem Ions") and Section 2.4 ("Nutrients and Ion Imbalances"), and are often related to irrigation water quality constituents as well as high soil pH associated with sodic soils.
- Reduced root cell permeability to water absorption due to low soil oxygen.
- Wilting tendency from poor rooting or in moist conditions by wet wilt.
- Stomatal closure caused by high abscisic acid hormone (ABA) under low soil oxygen.
- Reduced cytokinin production in roots.

- Root dieback and deterioration of root cells with some aerenchyma (large air spaces within roots) formed.
- Adventitious root (surface roots) formation with high aerenchyma induced by high ethylene concentrations.
- Lack of root regeneration.
- Root tip cells injured by acidosis of cell cytoplasm from H+ leakage in the vacuole.
- Root cell toxicity from CH_4, H_2S, and reduced Mn and Fe.
- Root cell toxicity from anaerobic respiration products within the plant-organic acids, aldehydes, and alcohols.
- Increased incidence of root-borne pathogens and other organisms in the rhizosphere.

Sodic sites often demonstrate certain *soil or landscape field stress symptoms.* These are related to poor soil physical conditions as well as excessively high pH. Sodic soils exhibit soil pH > 8.5 and can show alkaline pH up to 10.4 under highly sodic conditions. As Na_2CO_3 accumulates with Na additions and ESP increases, hydrolysis of the relatively soluble Na_2CO_3 occurs in the following sequence of reactions:

$$\mathbf{Na_2CO_3 + 2H_20 \leftrightarrow 2Na^+ + CO_3^{-2} + 2H^+ + 2OH^- \leftrightarrow}$$
$$\mathbf{2Na^+ + 2OH^- + H_2CO_3 \leftrightarrow 2Na^+ + 2OH^- + 2H_2O + CO_2\ (gas, \uparrow)}$$

At ESP of about 15, the pH may be 8.5; but as ESP Na and Na_2CO_3 increase to higher levels, so does pH up to pH 10.2 to 10.5 (Figure 3.6). It is the OH^- ion concentration that causes pH to increase. Hydrolysis of $CaCO_3$ also occurs but due to its low solubility, the pH does not increase above pH 8.2. Associated with the excessively high pH are as follows:

- *Scald or slick areas* are distinct black-colored areas devoid of plants that appear black and oily when wet. These are often slightly recessed. These areas can look much like algae scald, and, if waterlogged for long periods, surface algae layers may occur along with the Na-induced scald. When dry, the surface may be somewhat powdery due to colloidal dispersion. While the scald may resemble a dried algal layer, during wet periods there may be an actual algal layer present on the soil surface.
- Black coloration of soils, which is due to dissolution of some of the organic matter by OH^- ions under high pH conditions that then precipitates as a black organic coating on the soil surfaces. This symptom is most notable in the scald areas and accounts for these soils being called *black alkali soils.*
- Low infiltration and waterlogging after rain or irrigation due to few surface macropores for water intake and association with low soil oxygen.

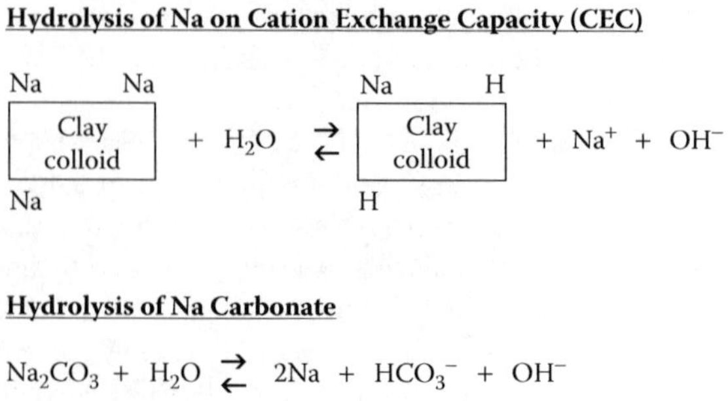

Hydrolysis of Na on Cation Exchange Capacity (CEC)

$$\text{(Na, Na, Na) Clay colloid} + H_2O \rightleftarrows \text{(Na, H, H) Clay colloid} + Na^+ + OH^-$$

Hydrolysis of Na Carbonate

$$Na_2CO_3 + H_2O \rightleftarrows 2Na + HCO_3^- + OH^-$$

FIGURE 3.6 Development of highly alkaline soil pH (pH can be as high as pH = 10.5) from hydrolysis of Na on CEC site and Na carbonate to create OH^- ions.

- When wet, loams and clay loams may demonstrate a putty-like and sticky consistency due to dispersion of clay particles that causes them to move more easily.
- Soils exhibit an absence of soil aggregation when the soil is moist, and when the soil dries, they crack from the shrinkage of the soil (but 2:1 clays exhibit cracking even when not affected by Na).
- Black layer may be more prevalent on sodic soils due to the poor soil aeration and accumulation of excess S.
- On sodic sandy soils in arid regions, subsurface layering may become evident at the depth of routine irrigation water penetration due to dispersed colloids moving into this zone. This is most evident on sandy soils.

3.1.4 Managing Sodic Soils

Similar to saline sites, numerous secondary problems or challenges arise from the primary effects of Na-induced deterioration of soil physical conditions. Except for the book by Carrow and Duncan (1998), most of the extensive publication on sodic and saline-sodic soil management has been focused on agriculture situations (Abrol et al., 1988; Rhoades and Loveday, 1990; Rengasamy and Olsson, 1991; Sumner, 1993; Jayawardane and Chan, 1994; Naidu et al., 1995; Chhabra, 1996; Oster and Jayawardane, 1998; Levy, 2000; Suarez, 2001; Rengasamy, 2002; Qadir and Oster, 2004; Qadir, Noble, et al., 2006; Qadir, Oster, et al., 2006, 2007). While soil permeability problems are present in all sodic soils, except for very sandy soils with single-grain structure, soil pH affects other problems, including nutrient availability. Rengasamy (2010a) noted these as follows:

- *Acidic, pH < 6.0.* Soil structural problems; seasonal waterlogging; and Fe, Mn, and Al toxicities when wet.
- *Near neutral, pH 6.0 to 8.0.* Soil structural problems; seasonal waterlogging; and possible Na toxicity.
- *Alkaline, pH > 8.0.* Soil structural problems; seasonal waterlogging; excess HCO_3 and CO_3; OH− toxicity at pH > 9.5; and Fe, Mn, and Al toxicity at pH > 9.0.

The best management practices (BMPs) for salt-affected sites presented in this book address both primary salinity stresses as well as the secondary management challenges that occur at the same time. Reclamation of sodic soils requires chemical amendments (e.g., gypsum, lime plus S to form gypsum, or some other relatively soluble source of Ca), but removal of the excess total soluble salts from the plant rootzone by leaching is the most important management practice and cannot be omitted. The most important BMPs on sodic soils are as follows:

- Leaching of soluble salts by application of sufficient infiltrated water volume to allow net downward movement of salts (Chapter 11). Even without adding gypsum, leaching can remove Na carbonates and bicarbonates that may be precipitated in the soil since these are soluble salts, but these may cause sodic conditions deeper in the profile. Leaching is also essential to remove the Na displaced from the CEC sites; otherwise, the Na simply goes back on the CEC sites without changing the negative soil physical conditions.
- Application of sufficient quantities of granular Ca is necessary for displacement of Na on the CEC sites so that flocculation of clay and organic colloids can occur (Chapters 12 and 13).
- Routine and vigorous surface and subsurface (deep) cultivation programs are essential to improve water and air permeability. Soils with high clay content of 2:1 clays require the most systematic cultivation to create temporary macropores for water and air movement (Chapter 14). The depth during cultivation must penetrate below any salt accumulation or compaction zones in the soil profile in order to promote effective salt management programs.

It should be stressed that while sodic soils do not have a high concentration of total soluble salts compared to saline soils, irrigation water with only moderate salinity can result in a buildup of high total soluble salts in the surface zone of sodic soils due to their inherent low infiltration and percolation rates where water can accumulate and then evaporate, leaving the salts to layer or migrate upward and deposit at the surface. Thus, salts may not be high in concentration throughout the whole rootzone and therefore not result in high soil ECe; but within the top few centimeters, salts may accumulate at much higher concentrations. This can be an issue especially with new seedlings or newly established sites with vegetatively planted materials and when irrigating with short-duration, frequent irrigation cycles. Cultivation and leaching programs must be developed with this aspect in mind. In addition to these critical BMPs, other management considerations on sodic soils are listed below with several components similar to what would be necessary on saline sites:

- Irrigation system distribution uniformity and efficiency become very important ecosystem infrastructure components since nonuniformity of irrigation water will affect spatial distribution on not just the water applications but also the potential accumulation of total soluble salts. Design issues will become more pronounced and salinity leaching more difficult if not properly designed for salt management (Chapter 10).
- Irrigation scheduling to avoid normal drought and physiological drought stress problems and for effective salt leaching is critical (Chapter 11). The most effective leaching program for removal of Na from the CEC sites is by application of sufficient water volume with each irrigation event to cause slow downward movement of any total soluble salts and activation of Na displacement from the CEC sites by available Ca. Since sodic soils have less than desirable soil physical properties, pulse irrigation cycling is the most effective leaching strategy, and site-specific irrigation is critical—that is, adjusting each zone (single-head zones are best) to achieve adequate irrigation volume to facilitate salt movement.
- Factors to enhance water infiltration (into soil surface), percolation (through the rootzone), and drainage (past the rootzone) such as are discussed in Chapter 14 ("Cultivation, Topdressing, and Soil Modification") and Chapter 15 ("Drainage and Sand Capping"). On high clay content soils with 2:1 clays that are already sodic (e.g., coastal marine clays), sand capping should be strongly considered and can be the most cost-effective BMP in the long term.
- Traffic control programs must be carefully designed and implemented (cart traffic, cart paths throughout the golf course, and approaches to tees and greens) since sodic soils are very prone to soil compaction, and excess Na soil conditions can cause other stresses (drought, excess moisture, nutritional, heat and cold, etc.) that make the plants more susceptible to wear injury—for example, a grass exposed to soil compaction cannot grow and recover as rapidly and cannot tolerate as much traffic before wear damage occurs (Chapter 17).
- Fertilization must be adjusted not just to meet plant growth requirements but also to maximize plant salt-tolerance mechanisms and correct imbalances just as would be true for saline soils (Chapter 16). As irrigation water salinity increases, so do the nutritional issues; and these can become dominant management challenges under high-salinity situations. If you are leaching soluble salts, you are also moving nutrients through the soil profile and often away from the plant root system.
- Proactive soil and water quality testing as well as tissue testing, if needed, to monitor salinity impacts on plant (Chapters 4, 5, 6, and 7).
- Drought, high temperature, and wear stresses are more common on sodic sites since grasses often cannot develop very extensive or viable root systems in sodic conditions (Chapter 17).

- Inherent low oxygen stress on sodic soils depresses cytokinin synthesis in roots just as inherent highly salinity does on saline soils. Thus, this "biostimulant" may be necessary to promote adequate growth and to sustain root regeneration (Chapter 17).
- Secondary biotic problems usually increase on the turfgrass or landscape plants (Chapter 17).

3.2 SALINE-SODIC SOILS

Saline-sodic soils contain high total soluble salts (ECe > 4.0 dS/m) and high Na on the CEC sites (ESP ≥ 15%). All the problems present in saline soils can be exhibited in saline-sodic soils since both contain high total soluble salts. Salt-induced drought stress is the most important and prevalent problem on plants. However, as was discussed earlier in this chapter (Section 3.1.1, "Sodium Permeability Hazard"), the high total soluble salts in irrigation water can inhibit aggregate slaking and clay dispersion; therefore, these soils do not show the poor physical properties of sodic soils.

Total soluble salts are easier and more rapidly leached compared to Na on CEC sites. Thus, if the total soluble salts are sufficiently leached by rain or by irrigation leaching on a saline-sodic soil, the soil may revert to a sodic soil and then exhibit reduced water and air permeability as well as progressive deterioration of soil physical conditions. During reclamation, it is important for gypsum to be applied to the soil before initiating leaching programs to avoid this situation.

Ideally, saline-sodic sites demonstrate physiological plant symptoms and soil symptoms similar to a saline soil; but it is not unusual for a site to show both saline and sodic plant and soil stress symptoms in adjacent areas. Many times, this is exhibited as scald areas (sodic) with saline-sodic conditions in the remainder of the site. This is especially true on irrigated sites due to the spatial and temporal variability of salts and their movement plus tendency to accumulate randomly in the landscape, including during reclamation since adjacent areas may respond differently. Soil tests will aid in identifying the specific salt-affected condition within an area at the time of sampling; however, soil sampling should be conducted in areas that appear to have similar plant and soil responses due to the high degree of spatial variability in saline-sodic situations.

In the sodic soil section, we noted that leaching prior to adding a Ca source to displace Na on the CEC sites could be done and would help leach the soluble Na carbonates and bicarbonates, but not the Na on the CEC. However, on a saline-sodic site, gypsum should be applied prior to any leaching to help prevent the soil changing into predominately a sodic soil with less favorable soil physical properties. BMPs for saline-sodic sites are the same as for saline sites with the exception of requiring a slow-release granular Ca source to displace CEC-bound Na. But, if the soil does change into a sodic condition, more aggressive cultivation will be needed.

Saline-sodic soils can exhibit pH-dependent stresses that influence management. Rengasamy (2010a) summarized these as follows:

- *Acidic, pH < 6.0.* Osmotic effects from soluble salts; Fe, Mn, Al toxicity; and Na toxicity.
- *Near neutral, pH 6.0 to 8.0.* Osmotic effects; toxicity to dominant cations or anions such as Cl or SO_4; and Na toxicity.
- *Alkaline, pH > 8.0.* Osmotic effects; excessive HCO_3 and CO_3; Na toxicity; and Fe, Mn, Al, and OH^- toxicities at pH > 9.0.

3.3 ACID-SULFATE SOILS

3.3.1 Occurrence and Problems

Duncan et al. (2009) noted that highly saline or sodic soils are normally considered to be the most complex soil problems that confront turfgrass managers, but occasionally turfgrass sites are established on pre- or actual *acid-sulfate soils* (*ASS*) that have been aptly described by Dent and Pons

(1995) as "the nastiest soils in the world." High salinity and/or sodic conditions are often present in most ASS soils; thus, a discussion of ASS is presented in this chapter.

Reduced sulfide layers are formed when seawater or sulfate-rich water mixes with land sediments containing iron oxides and organic matter in an anaerobic, waterlogged situation. Iron sulfide layers are most often found under tidal conditions, in low-lying areas near the coast such as in mangrove forests and salt marshes, estuaries, and tidal lakes. Most ASS soils are along coastal marine areas, but they also occur on inland locations (Fitzpatrick and Shand, 2008). Inland ASS environments are diverse, but the most common situation is from saline and sodic scald discharge areas with a rising water table and where the underlying rocks or subsoil contain sulfides. Mine spoil areas may also result in ASS and acidic water drainage. Key references for understanding the complexities of ASS formation, extent (approximately 40 million acres or 16 million hectares globally), problems, and management are presented by Dent (1992), Dent and Pons (1995), Fanning and Burch (2000), Ritsema et al. (2000), Jesmond (2000), Dear et al. (2002), Fitzpatrick and Shand (2008), and QASSIT (2009). The web-based references have color pictures representing various phases of ASS development. ASS coastal areas are most common in Australia and Southeast Asia, but are also present to a lesser degree in some U.S. coastlands.

The most important factors favoring accumulation of sulfidic layers are a supply of organic matter such as plant debris in shallow coastal areas; consistent anaerobic, waterlogged, oxygen-reducing conditions; a supply of sulfate ions, often from the tidewater, that can be reduced by bacteria that decompose the organic matter; and a supply of Fe, usually from sediments, since iron sulfides make up most of the sulfidic material, with *iron pyrite* (FeS_2) the most prevalent (Ritsema et al., 2000). As long as these sulfidic layers remain waterlogged and anaerobic, there are no ASS-related problems; but once these sites are drained for development so that oxidation conditions (from moving or altering the soil profile) can occur, the transformation into more challenging acid-sulfate soil phases starts. Sometimes, this transformation is activated by draining the previously waterlogged sites, while at other times the soils may have been dredged from the ocean tidal area and placed in a better drained site for coastal parks, golf courses, or other uses. The extent and type of problems that may occur on an ASS site depend on the phase of the ASS at the time the ecosystem is altered. There are three general phases of ASS soils: unripe sulfidic soils, raw ASS, and ripe ASS.

Unripe sulfidic soils are still waterlogged due to being underwater or from a high water table. If waterlogged from a high water table, these soils are sometimes called "potential ASS" (PASS), while the next two phases are actual ASS soil conditions. In this PASS phase, no visible problems are apparent. But, once drained or dredged material is used for land application, the next, more serious phase starts to occur. Potential acid-sulfate soils exhibit pH close to neutral; contain iron sulfides; usually are soft, sticky, and saturated; and have muds that may be gel-like but can contain sands and gravel.

Raw acid-sulfate soils contain pyrite at shallow depths. This is the phase that occurs immediately after drainage and can continue for several years. It is the raw ASS phase where high levels of sulfuric acid occur. Jarosite is a distinctly yellow to orange sulfate mineral that forms under pH < 4.0 on ASS soils once oxidation starts, and it is a key indicator of a raw ASS situation. It is normal for the deeper soil horizon to exhibit a wet gray appearance if this zone is anaerobic (i.e., a PASS condition), while the horizon above can demonstrate a mottled yellow to orange appearance from the jarosite (i.e., a raw ASS condition). When the surface soil is fully transformed into a raw ASS, soil pH will be within a 1.5 to 3.5 range, Al toxicity is the primary chemical problem, but other potential toxic ions (Fe, Mn, and metal ions) can occur due to dissolution under extreme acidity depending on the site. Also, nutrient deficiencies (P, Ca, Mg, and K), high salinity, and sodic conditions can be present on the site due to previously persistent waterlogging by ocean water or saline waters. Soil pH ranges of 1.5 to 3.5 are typical of a fully raw ASS, but prior to becoming a fully raw ASS or after this phase when it is transforming into a ripe ASS soil, the pH can be somewhat higher. If there is insufficient $CaCO_3$ in the sediment or added as lime and mixed after drainage, then the extreme acidity of raw ASS can develop within weeks to months with the jarosite becoming evident within the soil profile.

The authors were involved with a golf course in the coastal Carolinas that started to exhibit rapid deterioration of the turfgrass about 12–18 months after establishment. Soil samples revealed that most of the soil was in the raw acid-sulfate phase. This phase had progressed rapidly since no lime was applied to the site before establishment due to the near neutral pH at that time. With aerobic conditions, the pyrites were quickly oxidizing to generate sulfuric acid, and therefore the soil was becoming extremely acid as sulfuric acid was generated. The soil exhibited jarosite in the surface foot of soil. Vigorous surface and subsurface treatments with lime for ASS condition and gypsum for sodic condition were required to stabilize the surface rootzone to prevent turfgrass loss.

Ripe acid-sulfate soils eventually form where there is continuous drainage, there is irreversible loss of water, and the soil becomes firm. This phase occurs as the pyrite in the drained areas becomes exhausted, and the jarosite is hydrolyzed to stable iron oxides. At this stage, the soils will not exhibit the yellow jarosite symptoms in the surface zone, but may develop deeper in the profile; and even deeper in the soil, there may be PASS soil if that zone is waterlogged. However, these soils are still very acidic (pH 4.0 to 5.0 are common) unless high lime rates have been applied, and plants can show Al/Mn toxicities on the roots. Weathering and leaching of a fully ripe ASS will have occurred to more than 3 feet (1 m) deep in the soil in some situations.

3.3.2 Plant and Soil Symptoms of Acid-Sulfate Sites

Prior to draining a site or using dredged soil for a turfgrass upland site, it is important to determine if the waterlogged site or dredged soil is a PASS (i.e., pre-ASS or unripe sulfidic soil). A hydrogeological assessment of the site should be conducted to determine how serious the problem can be. Only then can the true costs be determined and proactive actions taken to deal with the extreme acidity, saline, and sodic conditions prior to planting turfgrasses or landscape plants.

Key indicators of ASS soils are as follows:

- Elevation. ASS are usually less that 5 m above sea level and waterlogged along coastal areas.
- Vegetation may include mangroves, marine couch (*Sporobolus virginicus var. minor*), acid-tolerant plants such as *Phragmites australis*, and she-oaks (*Casuarinaceae* species).
- Vegetation that is stunted or dead, acid scalds, sodic scalds, and salinity at the surface may all be present on a site once oxidation starts.
- If undisturbed, the soils will appear as unripe ASS conditions, but if drained, then the extreme acidity, jarosite, and sometimes a hydrogen sulfide (rotten egg) smell will occur.
- Soil profiles consistent with layering of the various ASS soil zones as described previously.
- Drainage water quality such as highly acidic ($pH < 4.0$) and often a clear blue-green color where colloidal particles are precipitated by the presence of soluble Al and Fe in the water.
- Iron staining and oily-appearing bacterial scums.
- Fish kill in areas receiving the drainage from ASS.

3.3.3 Management of Acid-Sulfate Sites

Conversion of ASS soils to an upland soil suitable for turfgrass growth can be very expensive. A hydrogeological site assessment should be done. An economic evaluation should be conducted based on a sound understanding and knowledge of the factors influencing reclamation before attempting drainage and earth moving (Dear et al., 2002). It is important to realize that the acidity, sodicity, and salinity problems must be addressed with separate BMP-based management programs for each component of stress. Dent (1992), Fanning and Burch (2000), and Dear et al. (2002) provide good discussions of classifications for acid-sulfate soils and reclamation strategies. Also, Loch et al. (2003, 2006) conducted studies of treatment options on coastal parks with soil dredged from the adjacent ocean tidal lands, which then became an ASS upon oxidation and were saline-sodic in nature. They evaluated an array of treatments such as grass selection, liming, fertilization, cultivation, sand capping, and leaching.

Important factors to consider before reclamation are the quantity of pyrite present; location of pyrite in the soil profile; potential for leaching toxic concentrations of Al from the profile; degree of saline and sodic stresses; and potential for control of the water table. Some ripe acid-sulfate soils that have been exposed to considerable weathering and leaching may be reclaimed using as little as 150 lb. $CaCO_3$ per 1000 sq. ft. (7300 kg/ha), while others may require many years of weathering remediation and much higher lime rates (Dear et al., 2002). When applying lime and gypsum prior to establishment, these amendments should be incorporated into the rootzone or surface foot of soil. If application is delayed until after plant establishment, deep injection can be applied with a device such as the WaterWick Turf Drainage System.

Other management factors to consider are the influence of soil mixing and contouring of the landscape on movement of pyrite during reclamation; location of the water table; leaching ability; need for gypsum in the surface zone for establishment and improvement of soil physical conditions; cultivation needs; fate of leachates; and nonuniformity of pyrite within the soil. Organic matter additions can help reduce soluble Al, but should not be added until later in the reclamation program if it hinders initial leaching of toxic levels of Al. Plant nutritional adjustments will also require calcium amendments when Al, Na, and S are all potentially toxic. Any irrigation water pHs < 5.5 should also be adjusted to higher ranges prior to application on turfgrass or landscape plants as well as to protect pumping equipment and any metal in or exposed to the irrigation system.

3.4 ALKALINE SOIL × SALINITY INTERACTION CHALLENGE

Any soil with pH > 7.0 is considered an *alkaline soil*. Degrees of soil alkalinity are recognized where soils with a pH 7.4–7.8 are mildly alkaline, 7.9–8.4 are moderately alkaline, 8.5–9.0 are strongly alkaline, and more than 9.0 are very strongly alkaline. There are many alkaline soils that are not salt affected, but many saline, sodic, and saline-sodic soils are alkaline in pH (Carrow, Waddington, and Rieke, 2001). In arid and semiarid alkaline soil regions, Ca, Mg, Na, and K base cations accumulate due to lack of leaching and dominate the exchange sites, while very few H ions and no Al or aluminum hydroxyl ions are generally present. It is not unusual to have a salt-affected soil also be an alkaline soil that may have chemical or physical problems that are not caused by the salts but must be managed along with the salt issues. Thus, we discuss these soils in this section.

Alkaline soils can be formed from calcium carbonate parent material, where free calcium carbonate accumulates from the irrigation water or from sands formed from coral (or quartz sands mixed with seashell fragments). Irrigation of any soil with water that is high in Ca, Mg, or HCO_3, such as from a limestone aquifer, may precipitate calcium and magnesium carbonates that can increase soil pH. At pH 8.2, calcium carbonate can precipitate out of soil solution, and a pH of 8.2 to 8.3 is a good indication that free calcium carbonate is present in the soil. Problems associated with alkaline soils (Carrow, Waddington, et al., 2001) include the following:

- P deficiency in conjunction with free $CaCO_3$ or abundant $CaSO_4$, forming calcium phosphate complexes.
- Fe and Mn deficiencies, particularly on sandy soils, since these ions are changed to their oxidized states, forming insoluble oxide and hydroxide compounds. Iron chlorosis is a common nutritional symptom.
- Zn deficiency, especially when high P amendments are applied to the soil.
- B deficiency, since this ion tends to bind with soil colloids as pH increases from 7.0 to 9.0.

- Mo deficiency.
- Reduced efficacy of pesticides.
- Caliche layer issues, which are discussed in the next section.

3.4.1 Caliche Formation

Irrigation water high in bicarbonate and carbonate along with high Ca and Mg can result in lime deposition near the surface, usually at the depth of irrigation water penetration, especially in arid regions. This caliche-like layer can become a zone of salt accumulation and hinder leaching. Lowering pH with acidification materials to dissolve the caliche-like layer can be considered. Disruption of this physical barrier can be accomplished with routine cultivation practices (core aeration, solid tine coring, high-pressure water injection, and dry air injection), with dissolved sulfur (usually water acidification) or other acidic compounds (acid-forming fertilizers), by stimulating root production where natural plant-derived organic acids are produced, or by applying weak organic acids (this latter approach can be expensive and provides variable results due to the limited volume of product that can be applied to the ecosystem; see Chapter 17, Section 17.4.6, "Microbial Amendments and Bionutritional Products").

When an arid soil develops from parent material high in lime or gypsum, calcitic or gypsic soils containing thick layers (>6 inches or 15 cm) of $CaCO_3$ or $CaSO_4$, respectively, form over many years. This layer is often a few inches or more below the surface and can be very hard and impervious to water, or in other cases when not fully formed it may allow water percolation. Calcitic layers that are not fractured can impede water movement, thereby fostering salt accumulation and hindering the leaching of soluble salts. If water ponds above these layers, it is not unusual for black layer to occur. When a calcitic layer impedes water percolation necessary for salt leaching, these areas may require deep cultivation (such as Soil Reliever, Vertidrain, or WaterWick) to penetrate through the layers and enhance downward water movement. All soils should be sampled by depth prior to treatment in order to provide the precise management option for effective remediation. Calcitic conditions within the surface few inches may benefit from sand-slitting such as by a Blec SandMaster or by core aeration and filling holes with sand, especially on sites with saline irrigation water.

One additional caution is that extreme care should be exercised when acidifying calcareous sands or calcitic soils or pulverized caliche profiles since excess acid can alter the soil particle configurations and develop a compacted soil physical and chemical profile that will be difficult for sustaining turfgrass growth. Long-term acidification on calcareous sands on greens may alter their physical properties.

Section II

Site Assessment BMPs for Saline and Sodic Soil Sites

4 Salinity Soil Tests and Interpretation

CONTENTS

4.1 BASICS OF SOIL SAMPLING AND TESTING

4.1.1 Importance of Soil Testing

Soil testing is an essential part of good management on saline and sodic sites. On turfgrass areas irrigated with saline irrigation water, soil testing should be conducted on a more frequent basis compared to nonsaline sites due to the high levels of chemical constituents added in the irrigation water plus chemicals added as soil amendments or for irrigation water treatments. Also, salinity leaching programs influence soils nutrient status by leaching nutrients along with the excess salts. The net effect is a high degree of spatial and temporal variability of salts and nutrients on such salt-affected sites.

Primary reasons for soil testing include (a) benchmarking the initial nature and magnitude of salt-related problems on a site; (b) benchmarking the initial soil nutrient status of a soil; (c) assessing potential nutrient deficiencies and imbalances; and (d) monitoring changes over time of salt problems and soil nutrients in response to irrigation applications, addition of soil amendments, fertilizer applications, and leaching programs. When the irrigation water is moderately saline or higher, the largest quantities of chemical constituents added to the soil are often via the irrigation water. These regularly scheduled soil status assessments are essential for sustainable turfgrass decision making relative to site-specific best management practices (BMPs).

Many soil test labs offer a "routine soil test package" and a "soil test package for salt-affected soils." For saline and sodic turfgrass sites, both sets of information are necessary. In this chapter, the emphasis is on information that may be found in soil test packages for salt-affected soils. Readers may wish to refer to Chapter 16 ("Nutritional Practices on Saline and Sodic Sites") for additional information on the nutritional challenges inherent in saline and sodic sites and suggested BMPs. Carrow, Waddington, et al. (2001) provide a detailed discussion of soil fertility issues as related to turfgrass sites.

The *routine soil test package* is targeted to evaluate current soil fertility status on all soils and not just salt-affected soils, and is the subject of Chapter 5 ("Routine Soil-Testing Methods"). Soil test reports can vary to some degree with laboratories, but most labs report the following parameters in their routine or basic package (Westerman, 1990; Sparks, 1996) (see Tables 5.8 and 5.9 in Chapter 5):

- *Soil pH*
- *Buffer pH*
- *Total cation exchange capacity* (meq/100g)
- *Organic matter content* (%)
- *Sufficiency level of available nutrients* (*SLAN*) is the primary soil test method and is based on extractable plant-available nutrients (P, K, Ca, Mg, S, Na, Fe, Mn, Zn, Cu, B, and Al). The chemical extractants are developed to measure what soil fractions would be plant available over a season. Extractable levels are reported in terms of a quantity of plant-available nutrients in units of ppm, mg/kg (1 ppm = 1 mg/kg), or lb./A (lb./A = 2 ppm). A number of different chemical extractants can be used, and numerical values will vary with the extractant used by the laboratory.
- *Base cation saturation on the CEC sites* (Ca, Mg, K, Na, H, and Al) is another method of making soil test recommendations, but is generally less reliable or accurate than SLAN. This method is sometimes called *basic cation saturation ratio* (*BCSR*), and results are reported in percent units. The percent Na on the CEC is the same as the *exchangeable sodium percent* (*ESP*) used to classify the sodium permeability hazard of a soil.
- *Water-extractable nutrients or ions* (the saturated paste extract [SPE] method is best; see Chapter 5, Section 5.4) make up another method sometimes used to make soil fertility recommendations and is the least accurate or scientifically based of all methods. But water is used as the extractant for total soluble salts, and, for this purpose, this is the best method when assessing salinity challenges—of special interest are Na, CL, sulfates, bicarbonates, and carbonates. Bicarbonates and carbonate concentrations should especially be included in this analysis for alkaline soils.

Saline and sodic soils require additional information beyond the basic package information to determine the types of salt problems and their magnitude. Thus, the soil assessment requires a typical *soil test package for salt-affected soils*, which is the subject of the current chapter. Sometimes the soil salinity package is integrated into the saturated paste extract (or dilute paste) report noted in the previous paragraph. The basic salinity soil test package often includes the following (see also Table 5.9 in Chapter 5):

- *Total soluble salts*. If reported as EC, the units are dS/m or mmhos/cm; but, when reported as total soluble salts (TSS), the units are usually ppm or mg/kg. Total dissolved salts (TDS) are sometimes reported by the laboratory.
- *Extractable chlorides* (*Cl*) in ppm. The extractable Cl is often in the routine soil test report in ppm Cl, as are the extractable soluble Na and SO_4 (ppm). These three ions (Cl, Na, and SO_4) are important soluble salts that often make up the bulk of the total soluble salts.

- *The sodium adsorption ratio* (*SAR*), which is a measure of the Na permeability hazard. Since the ESP reported in the percent base saturation of CEC sites in the routine soil test report is also a measure of the Na permeability hazard, the SAR is often not reported.
- The *free $CaCO_3$* (% by weight). If appreciable free $CaCO_3$ is present in a sodic soil, it is possible for S to be added, where the reaction would result in gypsum formation, as discussed in Chapter 12 ("Remediation Approaches and Amendments"). Also, the presence of free $CaCO_3$ precludes the use of acidifying fertilizers or S from decreasing soil pH since additional sulfur will be required to neutralize the free calcium carbonate (Carrow, Waddington, and Rieke, 2001). To neutralize a soil that contains 2% calcium carbonate, for example, requires 275 lb. tons S per 1000 sq. ft. to neutralize the calcium carbonate before a pH change would occur. Obviously, it is economically and functionally impractical to apply enough elemental S to alter soil pH of calcareous soils on a field scale, but as noted, free $CaCO_3$ can be reacted with S or SO_4 to form gypsum in sodic soil situations. Free $CaCO_3$ may arise from soil parent material, accumulation in caliche layers in arid regions, irrigation water where carbonates react with Ca in the water to form lime, sea shell mixed with silica sands, and calcareous sands.

A review of the various soil analyses from the combined routine and special salt tests reveals why both sets of data are important on salt-affected soils. Together they provide a detailed assessment of the four primary salt stresses:

1. Excessive levels of total soluble salts—total soluble salts by EC or TSS (TDS).
2. Na permeability hazard—excessive levels of Na on the soil cation exchange sites (CEC) and precipitated as Na carbonates causing soil structure degradation. The ESP, SAR, and extractable Na all relate to this aspect, with the ESP being the most commonly used.
3. Ions that are toxic to roots or shoots or that may cause other problem ion issues as they accumulate—extractable Na, Cl, B, S, and HCO_3 (this ion would come from an SPE if conducted).
4. Actual nutrient soil levels, interactions, and imbalances caused by nutrients and other ions in the soil—extractable P, K, Ca, Mg, S, Fe, Mn, Zn, Cu, and B; percent of Ca, Mg, and K on CEC sites; total CEC; and pH and lime requirement,

It is important to recognize that maintaining adequate nutrient levels and balances is critical on sites irrigated with saline irrigation water to maintain long-term turfgrass performance. Since soil nutrients exhibit considerable spatial and temporal variation under these conditions, basic soil fertilization is much more challenging. More frequent soil testing relative to nonsaline sites is necessary. In Chapters 1 to 3, brief descriptions of criteria to measure total soluble salts and Na permeability hazard were presented. In the current chapter, these topics are expanded on as well as proper soil-sampling methods. The final subject is a discussion of the growing field of field mapping with mobile devices and the application to salt-affected sites—this is a component of precision turfgrass management (PTM) (Carrow et al., 2009a, 2009b; Krum et al., 2010).

4.1.2 Soil Sampling

4.1.2.1 Current Soil-Sampling Protocols

Most often, the soil salinity tests are conducted on the same samples obtained for the routine soil test analyses. Accurate soil-testing information starts with good soil-sampling protocols. Soils are often very heterogeneous, and if samples are not collected properly, the test results will be of limited BMP value to the site manager. Guidelines to aid in collecting a representative soil sample include sampling locations, number of subsamples, soil depth, and proper handling. Various soil test laboratories have soil-sampling guidelines for turfgrass sites for further information, but key points are noted below.

Collect subsamples from defined areas that have similar characteristics (soil type, drainage, thin grass density areas, and good density areas) in a random pattern. Within a similar area, at least 10 subsamples (preferably 15 to 20) should be combined and mixed to form the composite sample. Especially on larger areas such as a golf course fairway or sod production field, too few subsamples will result in a nonrepresentative sample. Generally, each golf green and tee are sampled separately. On fairways, sod production fields, general grounds, and sports fields, dissimilar areas (such as low, moist sites versus upland, drained sites; or when distinct soil types are evident within a fairway) should each be sampled separately. However, most golf course superintendents do not adhere to this important practice and composite the sample for a whole fairway together. In the later section of this chapter on field mapping, the use of *site-specific management units* (*SSMUs*) to identify areas similar in soil texture and water-holding capacity is discussed relative to soil-sampling protocols for fertility purposes (Carrow et al., 2009a, 2009b).

For routine soil analyses and for most salinity testing, sampling depths of 0 to 3 (0 to 8 cm) or 0 to 4 inches (0 to 10 cm) (and where appropriate, 3 to 6 inches [8 to 15 cm] or 4 to 6 inches [10 to 15 cm]) are used. Thatch (dead and live organic matter with little or no intermixed soil) should be discarded, but mat (dead and live organic layer with soil or sand intermixed) such as found on greens should be retained as part of the sample. If a salt accumulation layer is suspected deeper than the surface 10 cm, then selected soil samples may be taken at the 10- to 20-cm zone to determine salinity levels. Sometimes a surface 0- to 1-inch (0- to 2.5-cm) sampling can be useful on selected sites to determine (a) the ESP and ECe within this zone when sodic or bicarbonate/calcite/sodium complexing conditions are expected to be forming at the surface, or (b) whether surface soil pH is changing if S is applied for pH reduction on soils without free calcium carbonate in this zone.

On sites without salt problems, soil testing is often once per year for high-quality turfgrass sites to once every 2 to 3 years on general grounds. Sampling should not be within 2 to 3 weeks of soil-applied fertilizer, lime, or gypsum applications. The sampling time of year should be the same. But, on salt-affected sites, timing and frequency of soil sampling are different. Frequency as often as monthly or quarterly is often required to obtain the necessary science-based information to make fertility and chemical amendment application decisions. When irrigation water is highly saline or water quality varies considerably over time, then soil testing must be frequent enough to monitor changes. Soils will eventually equate to the water quality parameters, and monitoring those changes is a key strategy in managing salt accumulation. Usually after 1 to 2 years on a site, the turfgrass manager has a better understanding of when changes may occur and the magnitude of those changes so that soil-testing frequency may be reduced to some degree; or frequent soil testing is conducted on a more limited number of key representative areas. Also, stress-challenged key turfgrass sites, such as greens and tees or high-traffic areas on sports fields, are often monitored more often than other areas.

4.1.2.2 Precision Turfgrass Management (PTM) Soil-Sampling Protocols

The concept of PTM is just starting to be introduced into turfgrass management, but it is expected to become more widely used in the future on salt-affected sites as well as nonsaline areas (Carrow et al., 2009a, 2009b; Krum et al., 2010, 2011). As in precision agriculture (PA), PTM is based on obtaining intensive site information by using mobile devices with multiple sensors that can determine key soil properties and subsequent plant performance; and such devices are being developed and tested for turfgrass applications. Since each sample location has a GPS notation and they are closely spaced (i.e., a 5- to 10-foot grid, or 1.6 to 3.1 m), GIS maps and geostatistical methods to analyze the data can be used. One application is to map soils at field capacity for volumetric water content (VWC), which is directly related to soil texture and organic matter content (Krum et al., 2010). This allows the boundaries of areas with similar VWC (i.e., similar soil texture and organic matter content) to be determined, and these are called *site-specific management units* (*SSMUs*).

Much effort in PA has been directed to improving the efficiency of soil and plant sampling for site-specific application of fertilizers and amendments, especially for alternatives to intensive and costly soil sampling on a grid (Bramley, 2009; McCormick et al., 2009; Oliver, 2010). The SSMUs based on soil VWC at field capacity offer the possibility of more site-specific decision making for inputs because they reflect differences in soil texture and organic matter, which relate to cation exchange capacity.

As stated before, a basic tenet of soil sampling for routine soil test analyses is to combine samples from areas with similar soil profiles into a composite sample. Golf courses (and many other complex turfgrass sites) are often sampled by random sampling, and samples are then combined for the whole fairway rather than for specific areas (SSMUs) within it. Thigpen (2007) applied PA principles to golf courses in the Carolinas (United States) for site-specific application of fertilizers using grid sampling (about 70 soil samples per fairway on a 15- to 20-m grid) and digital mapping of the data, but this approach is costly as each soil sample is analyzed separately in soil test laboratories. Shaner et al. (2008) found that SSMU-based soil sampling in agricultural fields was a viable option to grid sampling provided the soil sample locations were selected using appropriate protocols, such as those suggested by Corwin and Lesch (2005). With SSMUs, there are several ways to increase sampling accuracy while reducing costs for site-specific precision application of fertilizers and amendments in PTM, such as the following:

- A typical fairway may have two to three SSMUs. About 10 georeferenced samples per SSMU (30 total for a fairway) should provide as accurate information about the soil condition as a more intensive grid sample. This approach would allow spatial maps to be developed for any soil chemical or physical factors determined during laboratory analysis that were significantly correlated to the soil VWC (Corwin and Lesch, 2005).
- Or, taking 8–10 samples per SSMU to form a composite sample (i.e., three composited soil samples in total per fairway for the above example) would provide another possibility for sampling. This would result in about 54 soil samples from an 18-hole course. A similar approach was used by Johnson et al. (2001), who determined four SSMUs for eight different agricultural fields and sampled three sites per SSMU. They concluded that this sample scheme was an effective basis for soil sampling of field spatial variability.
- Another option would be to obtain soil samples from all of the same SSMUs across the site, similar to the approach of Ikenage and Inamura (2008) in agricultural fields. If there were eight SSMUs across the various fairways, there would then be only eight composite soil samples from the course. This would enable site-specific fertilizer applications based on the specific soil property values within SSMUs. Thus, the effectiveness in turfgrass growth responses from these adjusted fertilizer applications is directly related to the precision in soil sampling.

The above summaries are all versions of sampling to enable site-specific fertilizer application for PTM, but with different cost implications for sampling, fertilizer needs, and energy for applications and labor. The critical component is the original delineation of the SSMUs. In Section 4.3 ("Field Monitoring of Soil Salinity"), the PTM concept will be applied to soil salinity mapping.

4.2 SALT-AFFECTED SOIL TEST PACKAGES

4.2.1 Water: The "Salt Extractant" of Choice

In this section, the focus is on soil-testing packages for salt-affected sites and appropriate laboratory methods for determining the soil EC, SAR, ESP and extractable Na, Cl, SO_4, B, and HCO_3 information. In Chapter 5, routine soil-testing packages will be emphasized, especially the chemical extractants used in routine soil testing to provide the SLAN and BCSR information as well as in interpreting soil test results—where routine soil test packages are for nonsaline as well as salt-affected sites. We have adapted information from a series of articles on soil testing in golf course management by Carrow et al. (2003, 2004a, 2004b) to address key issues related to both salt-related and routine soil testing.

Salts (Na^+, Ca^{++}, Mg^{++}, K^+, SO_4^-, NO_3^-, Cl^-, HCO_3^-, and CO_3^{-2}) and nutrients in the soil solution and soil can be present in various forms or compounds (soluble salts: KCl, Na_2SO_4, K_2SO_4, $CaCl_2$, Na_2CO_3, $MgSO_4$, and others; and less soluble salts: $CaSO_4$, $CaCO_3$, dolomite, calcium sulfate anhydrite, and others) that range from soluble to very insoluble, and can be associated with other soil constituents of soil clays, organic materials, and inorganic sand substitutes. In terms of soil testing, these can be thought of as different chemical fractions, as summarized in Table 4.1. The three chemical fractions of special importance for saline and sodic soils are fractions 1 to 3, which are as follows:

- *Salts dissolved in the soil solution* as cations (positive charged) and anions (negative charged). These ions are readily available for plant uptake; but if the total soluble salt concentration is too high, a saline condition can occur. This fraction is part of the soil electroconductivity (EC) test for total soluble salts.
- *Readily soluble salts that precipitate out of solution* but can easily dissolve (and sometimes be diluted) with rain, with irrigation application, and in the laboratory samples since they are saturated. For example, as the soil starts to dry between irrigation events, Na^+ and Cl^- ions in solution may start to precipitate as NaCl crystals. If high levels of soluble salts are present as precipitated compounds, the soil will be saline. This fraction is also part of the soil EC analyses for total soluble salts, but water-extracted Na, Cl, B, SO_4, and HCO_3 levels are also important for assessing possible toxic or problem ions to roots or foliage, and for determining the potential for black layer problems from excess S or sodium bicarbonate accumulation. Table 2.1 lists common soil salts and approximate solubility that are found in salt-affected soils.
- *Cations that are on the soil cation exchange sites*: measured as the soil cation exchange capacity (CEC). The CEC nutrients are the most important pool of plant-available cations for plants, but Na can easily displace the critical nutrient cations (Ca, Mg, and K). Since a significant part of the total CEC sites is associated with clay minerals, as Na predominates, the adverse effects on soil structure appear. The Na on the CEC sites is measured by ESP and soil SAR.

It's not a surprise that the soil solution and soluble precipitated nutrients/salts described above—the first two groups—are most easily separated from the soil using (you guessed it!) "water" as the extractant. Accurate determination of these two pools (separate from what is on the CEC or other fractions) is very important for salt-affected soils or sites receiving irrigation water with moderate to high salt levels. The salts in these two fractions are what the plant roots experience in a field soil solution situation. However, the procedure and quantity of water used in the process make a difference in the reliability of the test results.

The *saturated paste extraction* (*SPE*) procedure was developed by the U.S. Salinity Laboratory (USSL) in 1954 to determine (a) ECe as the best measure of total soluble salts in the soil; and (b) and SAR as a measure of Na-induced permeability hazard in the soil (USSL, 1954). SPE is the "accepted standard" extract procedure for both chemical determinations, and the vast majority of scientific literature reports the ECe and SAR values based on the SPE extract (USSL, 1954; Rhoades and Miyamoto, 1990; Rhoades et al., 1999). In Chapter 1 ("Basics of Salt-Affected Soils"), the ECe and SAR terminology and methods were discussed in summary form. We now expand on these concepts in this chapter.

4.2.2 Total Soluble Salts

4.2.2.1 Saturated Paste Extract (SPE) and ECe

The best criteria for determining the presence and magnitude of soluble salts in a soil is the EC of a saturated soil-water paste extract (denoted as *ECe*). In this method, the soil is brought just to the point of saturation, it is allowed to equilibrate for at least 2 hours to dissolve total soluble salts into the soil

TABLE 4.1
Soil Fractions That Contain Plant Available Nutrients (PAN) and Soil Fractions Where Plant Unavailable Nutrients Reside

Soil Fraction	Comments
Plant Available Nutrients (PAN) Fractions (Within a 1–12-Month Period)	
1. Soil solution	Normally <1% of total PAN is present at any one time in the soil solution for plant uptake. As plant roots extract nutrients from the soil solution, they are replaced by nutrients from fractions 2–6, which are discussed below.
2. Precipitated as soluble salts (readily soluble salts)	Under normal irrigation and rainfall conditions, precipitated soluble salts are present in low concentrations since they are easily leached. With saline irrigation water, especially under conditions favoring soluble salt accumulation, precipitated soluble salt levels can increase and cause salinity stresses (salt-induced drought, and specific ion toxicities). Any precipitated soluble salts that are dissolved into the soil solution can be taken up by plants. This fraction is very dynamic depending on the potential for accumulation or leaching.
3. Soil cation exchange capacity (CEC) sites	Cation exchange capacity (CEC) is a measure of the number of cation adsorption sites in soil. The CEC is normally the largest pool of plant-available cation nutrients (Ca, Mg, K, and cation forms of micronutrients) as well as Al, H, and Na that is available for plant uptake. CEC sites are associated with clays, organic matter, or some sand substitutes (especially zeolites). On some soils, especially tropical soils, anion exchange capacity is present to hold plant-available anions (P and S anions).
4. Slowly dissolved salts and minerals, plus K and Mg from interlayers of micaceous minerals	This fraction is the second largest pool of PAN cations. Some relatively soluble forms of lime and gypsum (such as very fine particles of lime and gypsum) are included in this fraction. Also, small quantities of relatively soluble Ca-P, Fe-P, and Al-P compounds provide slow release of P over a growing season. For P, this is in the largest pool for plant-available P.
5. Chelated micronutrients	Organic matter can chelate many micronutrients, where the metal is complexed (chelated) with organic compounds that allow the metal to be plant available, but protected from leaching. Organic chelates can be an important pool for micronutrients.
6. Nutrients released from decomposing organic matter	As OM decomposes nutrients within the OM, compounds become available to plants; however, some OM compounds are highly resistant to decomposition, and the nutrients in these substances are not available.
Soil Nutrient Fractions Not Available for Plant Uptake (Within 1–12 months)	
7. Nutrients in very insoluble mineral compounds	Most nutrients in soils reside in insoluble compounds such as sand, silt, and clay particles; Fe and Al oxides; much of the lime particles; some of the gypsum; and most Ca-P, Al-P, and Fe-P compounds.
8. Organic matter resistant to microbial decomposition	This fraction is the humic substances that are very complex compounds and, therefore, very resistant to microbial decomposition. Nutrients in this fraction are not plant available.

Source: Adapted from Carrow, R. N., Surface organic matter in bermudagrass greens: A primary stress? *Golf Course Management,* 75 (5), 102–106, 2004a, http://archive.lib.msu.edu/tic/gcman/article/2004may102.pdf.

solution, the soil solution is vacuum extracted, and then the electrical conductivity is determined in the extract and noted by the designation of ECe (Rhoades and Miyamoto, 1990). This is the method developed by the USSL (1954) that most accurately reflects the total soluble salt impact on plants in the field. A common unit for ECe is dS/m, and ECe in dS/m can be converted to ppm or mg/kg of salt by the following formulas:

$$TDS\ (ppm\ or\ mg/kg) = 640 \times ECe,\ when\ ECe < 5\ dS/m$$
$$TDS = 750 \times ECe,\ when\ ECe > 5\ dS/m$$

where *TDS* represents *total dissolved salts*; or sometimes, *total soluble salts* (*TSS*) nomenclature is used. Thus, an ECe of 1.0 dS/m represents 640 ppm total soluble salts in the soil. As total salts accumulate in the soil, ECe also increases. A soil is classified as "saline" if total soluble salts are at ECe > 4 d Sm^{-1} (i.e., TDS = 2560 ppm) (Table 1.1). The ECe measurement does not distinguish which specific salts are present in the sample, but it does provide a reliable measure of *total soluble salts.* Laboratories and field instruments used to measure salinity in the SPE do not always report in dS/m or ppm. *Some useful conductivity conversions are 1 dS/m = 1 mmhos/cm = 0.1 S/m = 1 mS/cm = 1000 µS/cm = 100 mS/m.*

Electrical conductivity measurements are extremely useful for a variety of reasons. First, they can indicate whether or not your soil is saline (ECe's above 4.0 dS/m are considered saline) and the degree of soluble salts present, and therefore if salinity is likely to cause problems for plant growth (Table 4.2). Second, since plants and turfgrasses vary in their abilities to tolerate high salts, soil ECe's can help you predict how well a specific landscape plant or turfgrass cultivar is going to perform (see Chapter 9, "Selection of Turfgrass and Landscape Plants") (and see Table 4.2, bottom part). Third, regularly measured soil ECe's can help you time appropriate leaching events more

TABLE 4.2
Classification of Saturated Paste Extract ECe of Soil for (a) Total Salinity to Cause Salt-Induced Drought and (b) Plant Soil Salinity Tolerance[c]

Soil Classification Total Salinity Problems[a]	ECe dS/m	TDS ppm
Low	<1.5	<960
Moderate	1.6–3.9	961–2496
High	4.0–5.0	2497–3200
Very high	>5.0	>3200
Plant Soil Salinity Tolerance Classes	**ECe at 0% Yield Reduction (threshold ECe)[b]**	**ECe at 50% Yield Reduction**
Very sensitive plants can tolerate	<1.5[b,d]	1.5–5.0
Moderately sensitive plants must be used	1.6–3.0	5.1–10.0
Moderately tolerant plants must be used	3.1–6.0	10.1–15.0
Tolerant plants must be used	6.1–10.0	15.1–21.0
Very tolerant plants must be used	>10.0	>21.0

Source: For nongypsiferous soils, based on Maas, E. V., Crop salt tolerance, in G. A. Jung (Ed.), *Crop Tolerance to Suboptimal Land Conditions* (ASA Spec. Publ. 32), American Society of Agronomy, Madison, WI, 1978; and Maas, E. V., and G. J. Hoffman, Crop salt tolerance: Current assessment, *J. Irrig. Drainage, Div. ASCE* 103 (IRZ), 115–132, 1977. Classification of saturated paste extract ECe from U.S. Salinity Laboratory, *Diagnosis and Improvement of Saline and Alkali Soils, Handbook 60,* U.S. Government Printing Office, Washington, DC, 1954.

[a] Total salinity problems: as ECe increases due to increasing soil salts, the potential for salt-induced drought problems are enhanced.

[b] The maximum ECe within each tolerance class is called the "threshold ECe," which is the maximum soil salinity that does not reduce yield below that achieved under nonsaline conditions.

[c] Nongypsiferous soil does not contain gypsum. For gypsiferous soils, plants will tolerate ECe of approximately 2 d Sm^{-1} higher than the above table values. When developing the saturated paste extract, gypsum will dissolve sufficiently to increase ECe by about 2 d Sm^{-1}.

[d] ECe can also be used as an approximate guideline to Na and Cl toxicities since most salts contain high Na and Cl. Thus, a plant moderately sensitive to soil salinity can be considered moderately sensitive to Na and Cl ions if the concentration of these ions gave a similar EC reading when more specific data are not available.

accurately so that they take place before a hazardous buildup of soil salts occurs. Finally, the SPE is used to provide the information needed to determine the SAR and soil boron concentrations.

Because ECe is determined at saturated soil conditions, the *actual* salinity level at field capacity that a plant root would experience will be higher due to less soil water with the salts concentrated in this solution. A couple of formulas to estimate *actual* soil water EC (ECsw) from the ECe are

$$ECsw \text{ at field capacity} \approx 2 \times ECe, \text{ in dS/m}$$
$$ECsw \text{ at permanent wilt point} \approx 4 \times ECe, \text{ in dS/m}$$

These values are not included in soil test reports.

Duncan et al. (2009) noted that some sand substitute materials, such as calcined clays, exhibit porosity primarily in the smaller pore range of micropores; these contain water, but it is too tightly retained to be plant available. However, salts in the bulk soil solution would equilibrate with the pore water by slow water movement and diffusion of salts. The small-diameter micropores may shield salts from leaching, but the salts can still reequilibrate with the soil solution following a leaching event to increase salinity in the bulk soil solution. This would be a potential issue on sites that are irrigated with saline irrigation water and contain relatively high levels of these inorganic sand substitute materials (i.e., >12%–15% by volume). If the turfgrass manager suspects that the above situation is occurring and has a salinity meter, the manager may conduct a 1:5 extract (soil:water by volume) analysis for total soluble salts; then filter the soil and collect it while allowing the leachate with soluble salts to be discarded. A new five-parts volume of water needs to be added, and the sample is then allowed to equilibrate over 24 to 48 hours to allow sufficient time for any accumulated salts to move out of the micropores; and soluble salts are again determined by EC measurement of the leachate. If salt levels are again high, this would support the premise that salts are moving out of the inorganic amendment into the bulk soil solution to salinize it. Management of salts in high-microporosity situations is discussed in Chapter 8 ("Assessment of Salt Movement, Additions, and Retention"; see Section 8.1.2, "Inorganic Amendments") and Chapter 11 ("Irrigation Scheduling and Salinity Leaching").

4.2.2.2 Dilute Soil: Water Extracts and Slurries for EC

There are other less accurate methods using more dilute soil:water extracts, such a 1:2 or 1:5 soil to water (volume basis) to determine total soil salinity, but reliability is the issue. These methods are popular with some soil test laboratories because they are easier to conduct than SPE procedures. Compared to the EC values in a SPE, a 1:2 dilution would have lower salinity values; and a 1:5 would have the lowest numerical salinity values due to a more dilute solution. However, dilute water extracts are not as accurate or reliable for quantifying total soluble salts and actual field impact on perennial turfgrasses for the following reasons (USSL, 1954; Westerman, 1990; Rhoades et al., 1999; Shaw, 1999):

- **Measurements are indirect:** Because these methods dilute the soil extracts, results obtained from dilute water extracts must be mathematically converted, using correlation methods, to estimate a correct measurement of ECe. In other words, the dilute extraction methods provide only indirect measurements, and are not as straightforward or accurate as the directly measured SPE-based determinations.
- **Importance of relating to ECe:** With various dilute extract methods used as a substitute for ECe determined by the SPE method, it is important to relate information back to accepted standards (i.e., ECe). Otherwise, the turfgrass manager cannot determine what the EC value means in terms of potential for salinity stress on plants, or use the information to aid in selecting appropriate turfgrasses for the salt-challenged site.
- **Correlations used by labs may vary:** A reasonably accurate conversion of dilute EC to ECe requires an estimate of soil texture or percent clay content and the inclusion of a factor based on Cl content in order to estimate the ratio of soluble to slowly soluble salts (Shaw,

1999). If the estimate of soil texture is incorrect, the conversion will be incorrect. For example, Hazelton and Murphy (2007) reported a range of conversions factors from 5.8 (heavy clay) to 23 (sand, loamy sand, and clayey sand). The primary reason for such a wide range in conversion factors is because different soil types vary in water-holding capacity and also in the quantity of water held at saturation that the soluble salts dissolve into. The SPE is a direct measure of the saturated water-holding capacity, while the conversion factors used in dilute methods are an estimate. Simpler conversion formulas have been developed for a narrow set of soils typical of a certain location that are reasonably accurate, but these do not universally apply to all soils.

- **EC measurements are not standardized:** Some labs that use dilute extractions will report the EC value that results from analysis of the dilute solution, different labs may use a different dilution, and others will adjust their results to estimate ECe. This can result in significant discrepancies in assessing total salinity impact on a specific site. For example, an EC of 3.0 dS/m obtained using a dilute extract method may actually be an ECe of 6 to 7 dS/m, which are very important data when the turfgrass manager has a salt-sensitive plant to manage. In addition, comparing results from one laboratory to the next is very difficult.
- **Overestimates are possible:** The dilute water extract procedure is less predictable in its relationship to actual field soil water content, and is more likely to remove salts from soil exchange sites, including some of the slowly soluble salts. This would cause an overestimate of the concentration of total soluble salts relative to an SPE.
- **Indirect SAR calculations:** The dilute methods often remove more calcium and magnesium than the SPE method, which can lead to calculation of erroneous SAR values. Although some labs take this into account by performing mathematical conversions, the SAR value is obtained indirectly and is subject to error. By definition, SAR is based on SPE data for Ca, Mg, and Na concentrations in $mmol_c/L$ or meq/L.

We hope that the information above has convinced you that SPE is the way to go when you need accurate determinations of ECe and SAR values. *As an important principle in obtaining the most specific soil test information to assess salt-affected sites, we suggest that turfgrass managers insist on the SPE method to most accurately determine ECe and SAR* (see Table 5.1, Questions 1–4, for other important information when dealing with soil test labs).

4.2.3 Na Permeability Hazard (SAR and ESP)

The potential for Na-induced deterioration of soil structure is assessed by ESP and/or SAR, with the ESP most widely used. Exchangeable Na percentage is the percentage of Na on the CEC in units of cmol/kg or meq/100 g (1 cmol/kg = 1 meq/100 g). The CEC consists of all exchangeable cations on the soil CEC sites, such as Ca, Mg, K, Na, H, and Al. Determination of ESP depends on an accurate measurement of CEC and is defined as

$$ESP = \frac{(Exchangeable\ Na)(100)}{Cation\ Exchange\ Capacity}$$

While both organic matter and clay colloids have negatively charged CEC sites, the clay CEC sites are especially important for Na-induced soil structure deterioration. As Na percentage increases in the soil, the potential for soil structure deterioration increases. The USSL (1954) classification uses an ESP of >15% to classify a soil as sodic or saline-sodic (Table 1.1).

Sodium status as a measure of potential for soil permeability problems can also be reported as *sodium adsorption ratio* (*SAR*), where

$$SAR = \frac{Na}{\sqrt{(Ca + Mg)/2}}$$

The Na, Ca, and Mg cation concentrations are determined in a SPE solution (the same extract as ECe is determined) and are designated in units of $mmol_c/L$ or meq./L. As Na concentration increases, SAR increases. The quantities of Ca and Mg relative to Na in the soil solution are considered in SAR, while ESP uses only Na, but in terms of Na present on the soil CEC sites. In Chapter 6 ("Irrigation Water Quality Tests, Interpretation, and Selection"), irrigation water is also tested using the SAR formula with the Na, Ca, and Mg concentrations determined in the irrigation water; and the irrigation SAR is reported as SARw to illustrate that it is from the water and not the soil.

The ESP procedure for measuring soil Na status depends on an accurate determination of the cation exchange capacity, which is pH dependent. Thus, the proper method to determine CEC is essential not only to obtain an accurate CEC estimate but also to obtain an accurate ESP value. In Chapter 5 (Section 5.3, "BCSR Approach, CEC Measurement, and Interpretation"), appropriate CEC methods are discussed for saline and sodic soils, especially those soils that are also calcareous. If CEC is determined at a laboratory pH different from field pH (for example, a highly calcareous soil at field pH > 8.5, but a lab method using a pH 7.0 extractant), the CEC value could be misleading since some CEC sites are pH dependent. Also, in salt-affected soils, some of the cations not associated with CEC may be dissolved and reported as CEC and, again, cause error. Thus, some scientists have favored SAR over ESP. However, ESP is very good as long as the appropriate CEC method is used, and this is also important for nutritional decisions on salt-affected soils. Turfgrass managers may request the proper CEC method to insure optimum results. The relationship of SAR and ESP in an SPE is reported as follows (USSL, 1954; Naidu et al., 1995):

$$ESP = \frac{1.475(SAR)}{1 + 0.0147(SAR)}$$

Based on the USSL (1954) classification, a soil would be classified as sodic if the ESP > 15% or the SAR > 12 (Table 1.1). Soil pH is included in the table, but is only used as an indicator of the possibility of a soil being sodic based on a pH > 8.5. Highly sodic soils can exhibit pH up to 10.2, while saline and saline-sodic soils can have a wide range of soil pHs but are generally at pHs < 8.5. Considerable information has been provided in Chapter 3 ("Sodic, Saline-Sodic, and Alkaline Soils") concerning the interpretation of SAR and ESP soil test results and the influence of irrigation water salinity, soil type, type of clay, and other factors.

4.2.4 Specific Ion Toxicity and Problem Ions

In Chapter 6 ("Irrigation Water Quality Tests, Interpretation, and Selection"), certain ions of concern (Na, Cl, B, SO_4, and HCO_3) in the irrigation water and guidelines are discussed relative to irrigation water quality levels. Generally, more attention is given to the irrigation water quality data for these ions than the soil test information, since the water quality information is predictive relative to ongoing soil salt accumulation problems when the water is used for irrigation. However, the soil test data for these ions provide some insight into soil status and potential problems at the time of sampling. In Chapter 5, Section 5.4 ("Water-Extractable [SPE and Dilute] Nutrients and Ions"), additional information on Na, SO_4, B, Na, Cl, and HCO_3 is presented along with the other nutrients; but here they are discussed in terms of their toxicity or problem ion issues.

4.2.4.1 Sodium (Na)

Sodium contributes to salt-related problems by several mechanisms: (a) Na in the soil solution or in soluble salts is a major contributor to total soluble salts; (b) when in the soluble form, it can cause root deterioration of salt-sensitive plants by displacement of Ca in root cell walls; (c) if on the CEC sites of clay colloids, Na is a major cause of soil structure deterioration; (d) Na can react with carbonate and bicarbonate to form moderately soluble Na carbonates that act as a sink for reactive Na and may cause soil sealing in some cases; and (e) in the soluble form and on the CEC, Na has a significant interactive impact on K, Ca, and Mg nutrition, with Na depressing plant uptake and enhancing the leaching potential of these ions. There are no consistent standards for extractable soil levels of Na, but the irrigation water quality levels in Table 6.8 provide a guideline for irrigation water. Individual ions that contribute most to total soluble salts are Na, Cl, SO_4, HCO_3, Ca, and Mg. If ECe is high and the Na level is high on a soil test relative to the other ions contributing to total salinity, then Na is likely to contribute to any of the above-mentioned salt-related problems. For example, if the soil ECe = 6 dS/m and the extractable Na level is 530 ppm Na (23 meq/L), then the contribution of Na to the total EC can be estimated by

$$\text{ECe (from Na)} = 530 \text{ ppm Na}/ 640 = 0.83 \text{ dS/m}$$

Plants that are sensitive to high soil salinity will also be sensitive to high Na levels in the soil. Normally when plants are evaluated for salinity tolerance, the salts used are high in Na such as using NaCl, seawater, or seawater blends. Thus, a salt-tolerant plant is considered to be tolerant to high Na causing root deterioration (refer to Chapter 9).

In the soil SAR calculation, the Na, Ca, and Mg levels are based on water-extractable concentrations of these ions using the SPE method. The SAR correlates relatively well with ESP as determined by extractants to measure percent Na on the CEC sites (often ammonium acetate), especially at SAR < 12% or ESP < 15%. It is useful to understand the most important Na pools in the soil:

- Na on the CEC sites, which is the same as ESP. The ESP is well related to Na effects on soil permeability problems. While ESP of 15% is the criterion for classifying a soil as sodic, in soil test reports and ESP or percent Na on CEC of 3% to 4% is a good indicator of potential water permeability problems in the surface of soils with clay contents of above about 5%, especially if the clay type is a 2:1 clay.
- Na can also be precipitated out as Na-carbonate, Na-sulfates, and Na-chloride, all of which are soluble (Table 2.1). When irrigation water is high in Na, bicarbonate, and carbonate, but low in Ca and Mg, the Na-carbonate can accumulate if it is not leached. In arid climates, Na-carbonate accumulation can be substantial, and this Na pool must be leached in order to reduce Na on the CEC; otherwise, the Na on the CEC is displaced by Ca, and Na from dissolved Na-carbonate can again go onto the CEC site. In semiarid and humid regions with irrigation water rich in Na and carbonates but low in Ca and Mg, accumulation is less likely except (a) during prolonged dry periods with insufficient leaching to remove the Na-carbonate; and (b) in soils with appreciable clay, which can inhibit effective leaching due to soil structure and high microporosity reducing infiltration, percolation, and drainage.
- Another Na pool can be Na that accumulates in intermediate to very small micropores of native soils or inorganic amendments (this is discussed in Chapter 8, Section 8.1.2, "Inorganic Amendments"). Accumulation of Na in the intermediate to very small micropores can occur regardless of climate region if the irrigation water is moderately to highly saline. This micropore size range is very difficult to leach effectively in order to remove the Na, especially if the water is relatively high in Na. Soluble salts, including Na, can result in high salt concentrations in the pores, and the soil solution and pore solution are always in the process of attempting to equilibrate with each other, but the process is slow for these small pores.

In terms of soil testing for Na, different methods result in extracting different fractions, such as (a) a SPE water-based method determines soil solution + soluble Na salts, and this Na concentration is used in the SAR calculation; and (b) chemical extractions used to determine CEC, such as ammonium acetate, will determine Na in soil solution + soluble salts + Na on the CEC sites. This latter value is the best measure of the total Na active pool (i.e., does not include Na in insoluble forms that do not influence plant salinity stress or soil physical properties, where this would be the inactive Na pool). If monitoring total Na, this Na concentration would be the most useful and would be higher than the SPE Na concentration. Neither of the above soil test procedures is likely to determine Na contained in the internal intermediate and small micropores of inorganic amendments unless the extraction time is relatively long in order for diffusion and equilibration to occur. Research is needed to better determine appropriate equilibration times to quantify the Na pool in these micropores within inorganic amendments.

4.2.4.2 Chloride (Cl)

Chloride (Cl) is an essential micronutrient but is required at very low quantities. In saline irrigation water, Cl is a very common ion that contributes to total salinity, and the greatest concern is for high levels of Cl in the soil that could (a) contribute to total soil ECe and reduce water uptake by plants, and therefore slow down plant growth; and (b) add to soil accumulation and uptake by plants that cause foliar injury. Many woody landscape plants are sensitive to high Cl as are many crop plants. With grasses, Cl accumulation in shoot tissues is not a problem since this highly mobile ion is rapidly translocated to growing points that are subsequently mowed. Maas (1978) reported that many grasses (crested wheatgrass, tall fescue, perennial ryegrass, Bermuda grass, and fairway crested wheatgrass) can tolerate soil Cl levels from 1100 to 2600 kg/mg based on Cl in the SPE. The SPE water-based method provides a very good measure of total chlorides in the soil. To determine the contribution of Cl to total salinity, this Cl concentration can be used.

One valuable use of soil Cl levels is to monitor the effectiveness of leaching programs. Since Cl is the most easily leached ion (highly mobile in the soil) and almost all Cl added via irrigation water stays in soluble forms, the soil Cl level should be near the initial Cl level in the irrigation water assuming that all leaching is by the irrigation water and not rainfall—as is common in arid regions. If rain occurs, then Cl concentrations can be lower than the irrigation water. Thus, if Cl in the soil is appreciably higher than irrigation-level Cl, leaching is not effective, especially in 0–3-inch (0–8-cm) upper soil profile zones.

4.2.4.3 Sulfate (SO_4)

Certain irrigation water sources can be relatively high in SO_4 (Duncan et al., 2009). The major soil accumulation concern is that SO_4 can become reduced to S forms under anaerobic conditions and contribute to black layer formation. On any salt-affected soil, this would hinder leaching and contribute to low soil aeration. The SPE water-based extractable SO_4 is a good measure of the readily soluble S pool that may influence black layer formation, and it reflects plant-available S, especially in arid soils, but is a less accurate indicator of plant-available S in soils with high organic matter content. Since SO_4 is almost as leachable as the Cl ion, it is another good indicator of leaching effectiveness. If SO_4 accumulates to above the irrigation water level, then leaching is not effective. In calcitic soils with an impervious calcite layer near the surface, the SO_4 may not leach due to this barrier. This is especially serious if the soil is sodic. Fracturing the calcitic layer will aid in improving soluble salt leaching. Also, light lime applications that match the SO_4 application in the irrigation water (1 lb. lime per 1 lb. SO_4 added) can help remove SO_4 by conversion to gypsum (1 meq/L will react with 2.43 meq/L SO_4) when combined with appropriate cultivation events (Duncan et al., 2009).

4.2.4.4 Boron

Because boron deficiencies are rare (levels of 0.1–2.0 ppm are considered sufficient), the SPE or hot water extraction soil test measurement is typically used to ensure that boron levels are not too high.

TABLE 4.3
Boron Soil Levels and Plant Tolerance before Yield Starts to Decline

B Sensitivity Class	Soil B (mg kg^{-1})[a]	Plant Examples
Very sensitive plants	<0.5	Lemon citrus
Sensitive plants	0.5–1.0	Fruit trees
Moderately sensitive	1.0–2.0	Many vegetables
Moderately tolerant	2.0–4.0	Kentucky bluegrass
Tolerant plants	4.0–6.0	Alfalfa
Very tolerant plants	6.0–10.0	Most grasses

Source: From Maas, E. V., Crop salt tolerance, in G. A. Jung (Ed.), *Crop Tolerance to Suboptimal Land Conditions* (ASA Spec. Publ. 32), Amer. Soc. of Agron., Madison, WI, 1978; Maas, E. V., Salt tolerance of plants, in B. R. Christie (Ed.), *Handbook of Plant Science in Agriculture* (Vol. 2), CRC Press, Boca Raton, FL, 1987.

[a] B concentration in a saturated soil paste extract.

In arid regions with irrigation water high in B, accumulation of B in the soil can occur and inhibit plant growth. Maas (1978) provided soil B guidelines for plants with most grasses being tolerant (Table 4.3). An extensive list of plant tolerance to B is provided by Duncan et al. (2009). Mowing aids in reducing B accumulation in turfgrasses, but other landscape plants can be affected. If irrigation water is high in B, a good leaching program will aid in B removal.

4.2.4.5 HCO_3

High HCO_3 is primarily a concern due to precipitation of Ca and Mg into lime forms, which allows Na to increase on the soil CEC without these counterions. The SPE is a good measure of soluble carbonate and bicarbonate. If HCO_3 is high in the soil, it usually reflects the application of irrigation water high in this ion. Sodium can react with carbonate and bicarbonate to form relatively soluble Na carbonate, which is a sink that retains Na in the soil. When Ca is added to displace Na from CEC sites on sodic soils, the Na in the Na carbonate also requires leaching, but Ca is not required for the Na displacement when in Na carbonate forms. Sodium carbonate that accumulates in the soil profile reflects an inadequate leaching program and can contribute to the sealing of soil pores if it accumulates in concentrated soil zones—such as the depth of irrigation water penetration in arid regions. Acidification of irrigation water can reduce HCO_3 in the water and thereby in the soil while providing previously complexed Ca and Mg in soluble form and improving their availability to counter excess Na.

4.3 FIELD MONITORING OF SOIL SALINITY

4.3.1 Importance of Monitoring Soil Salinity

Soil salinity is not uniform in the soil, and that is the primary issue in why soil salinity should be monitored and when selecting approaches to monitoring soil salinity (Figure 4.1). *Spatial variability* across the landscape and within the soil profile by depth coupled with temporal variability present challenges in soil sampling for proactively monitoring soil salinity such as where to sample and when to sample. For example, on a golf course in an arid environment that was a salt-affected site prior to turfgrass establishment, had considerable topography changes, and received saline irrigation water over years, the authors observed *five types of surface or subsurface salt movement contributing to such salinity variability:*

- On areas with steep slopes in the rough, the salt flow onto adjacent fairway areas was primarily by *gravity-induced surface flow movement* (#1); however, *subsurface lateral salt movement* (#2) was also occurring at a slower but continuous rate over time.

- Within fairway areas with topographies involving steep slopes such as leading to a surface drain, the bottom of the drain channel and subsequent slopes leading to the drain would be similar to the situation in #1, discussed above—surface flow being the primary means of salt movement, with slower subsurface lateral movement also gradually but continuously occurring. The slopes of the grassed waterways on the fairway could be thought of as potential "seepage" sites with subsurface lateral salts emerging unpredictably on sloped areas—similar to seeing water coming out of a hillside due to lateral subsurface water movement. One factor that contributes to the unpredictability of subsurface salt emergence can be interception in irrigation trenches with subsequent lateral movement due to gravity with salts then emerging in low areas along the irrigation mains or laterals.
- *Capillary or uprising (upconing)* (#3) salt movement from the underlying water table, underground water flow areas, and possibly old arroyo sites where salts were deposited thousands of years ago. These sites can usually be recognized only after the other areas have been remediated or corrected or shrunk in size, and generally reveal the primary localized "salt sink"—whose locations are generally unpredictable, varying from lower topography areas to even the edges of high-sloped roughs, berms, or mounds.
- The flat fairway areas with little or no slope for surface water movement (little to no gravity-induced movement) were prime presodic to sodic scald sites where a small pocket (generally barely noticeable to the eye) develops over a variable time period into a "localized salt accumulation pocket or scald area" that kills all grass in the depressed area. These scald areas always occur on flat areas with no slope for surface drainage. Thus, when irrigation water collects and accumulates in these small localized pocket areas, they eventually develop into scalds. This is an example of *gradual salt movement into small depressions over time* (#4). While these sites are initially very small depressions where sodic scald conditions occur, they become somewhat more depressed due to soil structure collapse.
- Another potential type of salt movement actually involves *nonmovement of salts (or very high salt retention)* (#5), such as can be enhanced by inclusion of excessive quantities of inorganic amendments that are high in microporosity, or micropores that readily retain excessive salt loads, such as the porous ceramics or other water retention products.

In addition to these surface and subsurface water flow issues, nonuniformity of irrigation water application, soil types differing in percolation rates, and differential ET losses contributed to salinity accumulation variability. This example serves to illustrate the importance and challenges in monitoring this important soil stress. Seelig (2000) and Wentz (2000) have illustrated salinity movement in agriculture situations that can create saline seeps, and some of these are presented in Figures 8.1 and 8.2 in Chapter 8, "Assessment for Salt Movement, Additions, and Retention."

Accurate determination of soil salinity stress offers some key benefits to the turfgrass manager. For the current practice of obtaining field soil samples for laboratory analyses, the first and most obvious benefit is to identify areas that are currently under total soluble salt stress or that are in the early stress symptom stages. Secondly, follow-up samples aid in determining the effectiveness of salt leaching by rainfall or irrigation leaching. However, to accomplish this on small subareas, such as sections of a golf course fairway, soil sampling would need to be by subareas rather than across a whole fairway. Site-specific problems require site-specific soil monitoring.

Advances in sensor technology, discussed later in this section, offer the potential for more spatially intensive soil sampling that would better define soil salinity spatial variability at least in the surface part of the rootzone (i.e., the top 4 inches or 10 cm). More spatially intense information would be useful for the following:

- Knowledge of the spatial distribution of salinity levels is essential for site-specific leaching, which could conserve water when compared to the current practice of applying a leaching fraction to a much larger area due to lack of information on salt accumulation patterns and spatial diversity. Devitt et al. (2007) investigated the spatial and temporal variation of salinity in the upper 15 cm in an arid region with *in situ* sensors on a grid pattern on the greens and fairways of nine golf courses where irrigation water salinity varied from 0.80 to 2.22 dS/m. As salinity of the irrigation water increased, soil salinity also increased with ECe values of <1.0 to 24.0 dS/m. These studies showed that although salt retention is affected by soil type, irrigation water quality and its distribution were the most important factors affecting salt distribution and subsequent accumulation. Therefore, any factors affecting water distribution and the quantity of water applied (i.e., the degree of leaching occurring at a specific location) may be as dominant as soil type on patterns of spatial salinity accumulation (Duncan et al., 2009), and thereby affect site-specific leaching programs.
- Areas receiving the most soluble salts are likely to be highest in soil Na (ESP); such areas could be treated with site-specific adjusted applications of gypsum rather than treating the whole area.
- With mobile salinity monitoring devices and handheld units to determine soil salinity by soil depth, it would be possible to evaluate the effectiveness of a leaching program in a saline site-specific management unit (SMMU) and to what depth salts were moving. Again, irrigation leaching requirements would be site specific and not for the whole area, therefore resulting in considerable water savings for leaching.
- Locations that consistently exhibit the most rapid and highest accumulation of soluble salts would be ideal locations for *in situ* soil salinity sensors to provide real-time, multiple-depth data on salt accumulation and leaching. Identification of these sites as well as areas with similar salinity accumulation rates would allow for a science-based approach for sensor placement with the fewest sensor arrays and provide the turfgrass manager insight on the specific areas the sensor would represent across the landscape.
- Sites receiving the most soluble salts are likely to be those with fertility problems because soil nutrient imbalance is one of the consequences of salinity stress (Duncan et al., 2009). Site-specific prescription fertilization with key nutrients such as K, Ca, Mg, Mn, and Zn could result in fertilizer savings compared to uniform application over the whole area.

4.3.2 Approaches to Field Salinity Monitoring

For practical turfgrass situations, assessing spatial salinity distribution is at the site level or ground level. This is the focus of this section by use of soil sampling, handheld devices, mobile sensor platform units, and *in situ* (in place) salinity sensors. However, for determining the scope and location of salinity stress over wide geographic areas, remote sensing technologies are used. This multifaceted technology is beyond the scope of this book, but Metternicht and Zinck (2009) and Fruby et al. (2010) provide a good overview of the approaches and issues associated with satellite-based remote sensing. A couple of issues to consider are (a) the pixel size or area of measurement is much larger for remote sensing than for ground-level methods; and (b) the remote data often have to be compared with selected sites where ground-level data are obtained by the appropriate technologies to "ground-truth" and verify the data.

4.3.2.1 Collecting Soil Samples

Currently, field sampling is the approach used by turfgrass managers, where soil samples are collected and then (a) sent to a laboratory for analysis; or (b) in some cases, turfgrass managers may have onsite field or laboratory devices. Usually, the sampling is based on the same samples and areas used for routine soil testing. For example, on golf courses, this is normally a whole fairway, a green, or tee areas. When samples are sent to soil test laboratories, the salinity determinations in the salinity soil test packages discussed in Section 4.2 ("Salt-Affected Soil Test Packages") would be made—especially ECe and SAR as well as routine soil test information.

A number of turfgrass managers have used the relatively inexpensive soil conductivity meters, which can be used in dilute soil:water mixtures to determine a dilute paste EC. Stowell and Davis (1993) tested one hand-salinity device and reported on procedures to determine the EC of the dilute soil:water paste (ECp). They placed 50 cm^3 (about 2 oz) of the soil sample in a small cup, added water while stirring just to the point of the soil surface glistening, then inserted the electrodes into the paste to obtain the ECp reading. This device or similar ones can also be used for a rapid determination of irrigation water salinity. Based on a number of high-sand golf greens in southern California, the conversion to ECe was

$$ECe\ (dS/m) = 0.8 + 2.7\ ECp\ (dS/m)$$

To obtain a correlation of ECp to ECe for a specific site, soil samples could be split with the above procedure used on part of the sample. The remainder would be sent to a soil test laboratory to determine ECe by the saturated paste extract method. Then a linear correlation could be conducted to obtain the appropriate calibration curve. It does not matter which hand-salinity device is used as long as it is calibrated against known saltwater standards, which is a usual procedure for all devices.

4.3.2.2 General Comments on Field Mapping by Handheld and Mobile Platforms

Prior to discussing handheld and mobile salinity-monitoring devices, some general comments are useful when investigating the potential for soil salinity monitoring. Excellent resources for those interested in the various means of soil salinity measurement are the FAO publication by Rhoades et al. (1999), a book by Allred et al. (2008), and papers by Corwin (2008), Corwin and Lesch (2003), and Hendrick et al. (2002). Robinson et al. (2008) and Munoz-Carpena (2009) review various soil moisture measurement approaches including *in situ*, handheld, and mobile devices. The paper by Munoz-Carpena (2009) is especially useful in providing a concise discussion of the various sensor technologies used in soil moisture plus soil salinity measurements. While their focus was on water, they provide a good review of sensors for both water and salinity determinations.

First, all handheld, mobile, or *in situ* methods for soil salinity determine *apparent soil electrical conductivity* (*ECa*), which should be calibrated to ECe for each soil type to obtain the most reliable information. The ECa is always lower than the ECe value. However, the ECe is the standard for reporting soil salinity (a) because the ECe is used to relate total soil salinity stress to plant responses; (b) to select plants for saline sites since plants are ranked as to their salinity tolerance in ECe units; (c) because classification of salt-affected soils is based on ECe; and (d) because, to compare salt-monitoring methods, salinity soil test results, or site values to the literature, ECe is used to provide a common reporting system. Thus, conversion of ECa for any salinity-monitoring device into ECe should be a standard practice (see Section 4.3.3).

Second, whether by handheld or mobile monitoring instruments, readings should be conducted at field capacity to allow for comparison over time and for more reliable results. For handheld devices that must be pushed into the soil, this usually requires the soil to be at or above field capacity such as after an irrigation application. Handheld probes must exhibit field durability to withstand repeated insertion into the soil.

Third, because salinity levels can vary in short distances in the soil, the soil volume in which a measurement is taken is important. Sensors that measure in very small volumes (i.e., approximately 1 $inch^3$ [16 cm^3]) will require more measurements than those with a larger sample volume to adequately determine soil salinity at one location and then across locations. In contrast, devices that measure in too large of a volume such as the electromagnetic (EM) approach may mask important salt differences by depth (such as salt accumulation layers) that are within the large sensing volume.

Fourth, due to the spatial variability of salts on many sites, the sample grid should be close enough to accurately reflect this variation. When sample sites are georeferenced, the data can be analyzed and presented in spatial maps using GIS systems. The geostatistical approaches in GIS allow (a) an evaluation of whether the sampling spacing was sufficiently close to reflect spatial variability—by use of the geostatistical semivariogram approach (ESRI, 2004a, 2004b; Krum et al., 2010); (b) when samples are close enough to reveal a mathematical relationship, GIS programs can accurately estimate values between sample sites (i.e., interpolation); (c) the spatial variability to be analyzed and reported in various descriptive statistical means (mean, range, coefficient of variability, etc.) (McGrew and Monroe, 2000); and (d) for creation of detailed and accurate spatial GIS salinity maps, which can also reflect topographical influences if topographical data are available (Krum et al., 2010).

Fifth, the instruments must have an adequate salinity response range to map soil salinity within at least the ECe 0 to 20 dS/m range. The actual field instrument reading of apparent soil EC (i.e., ECa) should be sensitive within 0 to 5 or 6 dS/m to allow conversion of ECa into ECe by appropriate calibration curves that will be applicable to various soil types. When converted, the resulting estimated ECe is normally from three to eight times higher in value than ECa field readings depending on the soil type. While less salt-tolerant grasses may require soil salinities of ECe < 6.0 dS/m, there are halophytic grasses that can tolerate much higher ECe, and the salinity-monitoring devices should have sufficient range to deal with all grasses.

4.3.2.3 Handheld Salinity-Monitoring Devices

In this section, the focus is on handheld devices that can be inserted directly into the soil to obtain an ECa reading. We discuss representative devices and are not attempting to present every salinity sensor probe available.

- *Spectrum FieldScout EC 110*™ (http://www.specmeters.com) is an "inserted" probe device pushed into the soil that has a small salinity sensor at the tip with a small sample volume. If data are georeferenced, Spectrum Technologies offers an option to develop spatial maps. Due to the small sample volume, several readings should be made at a specific sampling location to obtain a reasonable average. Multiple depths could be read by obtaining readings at different depths of insertion intervals.
- *Eijkelkamp* (http://www.eijkelkamp.com) offers a handheld mobile unit that uses a single sensor probe with four electrodes in a *four-wenner array* arrangement to determine soil electrical resistivity (ER) and conductance ECa (ECa is the reciprocal of ER). A four-wenner array is a set of four equally spaced probes inserted into the soil or turfgrass canopy to the 1- to 2.5-cm (i.e., <1-inch) depth. The insertion probe must be pushed into the soil to at least 6 inches (15 cm) to determine the surface 6-inch zone, but can be inserted deeper to make measurements deeper in the profile. The zone of salinity determination is larger than for the Spectrum instrument, but still relatively small since the ER measurement is confined to 2–3 cm away from the probe sides. Thus, several readings at each sample location may be required to obtain a good average.
- *Landviser* (http://www.landviser.com) has a handheld mobile device, LandMapper™ ERM-02, that can determine ECa at various depths depending on the probe configurations using a four-wenner array means of determining ER and ECa, with the ECa as the reciprocal of ER (Rhoades et al., 1999). The LandMapper unit contains the power generator, meter, and probe arrays. Electrical current is applied to the two outer electrodes, and the drop in potential is measured on the inner electrodes. The depth of readings is approximately the spacing between two adjacent electrodes. Currently the authors are using this

device to monitor soil salinity in field situations at different depths across landscapes. If the distance between adjacent probes is A, then the sample depth is A and the sample volume is A^3 (Rhoades et al., 1999). A common depth is 0 to 4 inches (0 to 10 cm) using a distance of 4 inches between adjacent probes, which provides a sampling area of approximately $(4 \text{ inches})^3 = 64 \text{ inch}^3 = 1050 \text{ cm}^3$.

- *Hand-SMD Experimental Unit*, also based on the four-wenner array method of measuring ER and ECa, has been developed by the Toro Company and the author (Carrow) for rapid salinity mapping at two depths. Depths can be determined by probe configuration. The unit was designed to be sturdy under repeated field use, obtain very rapid readings, and include software to GPS reference sample sites and easily create GIS maps and statistical analyses. Similar to the LandMapper, probe insertion is to <1.0 inch (<2.5 cm) with mapping at near field capacity recommended.
- *Stevens Hydra Probe II Soil Sensor* (http://www.stevenswater.com) is an impedance sensor capable of determining soil moisture content and salinity. The three-prong probe can be inserted into the soil to 2.2 inches (5.7 cm). The salinity range is reported as ECa of 0 to 15 dS/m. This sensor can be installed as an *in situ* sensor or as a handheld mobile device.

4.3.2.4 Mobile Salinity-Mapping Platforms

Precision agriculture (PA) started to make rapid advancements when mobile sensor platforms were developed in the mid-1990s that could determine key soil parameters, especially electrical resistivity (ER) measurements based on either four-wenner array electrical resistance or electromagnetic (EM) approaches (Rhoades et al., 1999; Corwin and Lesch, 2003, 2005; Allred et al., 2008; Carrow et al., 2009a, 2009b). The ECa is the reciprocal of ER, and these parameters are used to define spatial differences in soil salinity, bulk density, and moisture, and then relate these to plant yield or performance. As development continues in mobile sensor platforms and protocols for turfgrass situations, it is expected that precision turfgrass management (PTM) will advance in a similar fashion as PA (Carrow et al., 2009a, 2009b; Carson et al., 2010; Krum et al., 2010, 2011). Two experimental devices developed specifically for PTM-based salinity mapping on large turfgrass areas are noted below.

4.3.2.4.1 Mobile Electromagnetic Induction (EM and EMI) Sensors

EM units have been widely used in PA for field mapping of salinity (Rhoades et al., 1999). These devices can either be used as handheld devices or incorporated into mobile sensor platforms such as normally used in PA. EM is nonintrusive, and the device induces circular eddy-current loops, where the magnitude of the loops is directly proportional to the ECa of the soil within the region of the loops (Corwin, 2008). An important disadvantage of this technology in turfgrass situations is the depth of measurement, where 12 inches (30 cm or 0.3 m) is the narrowest zone. Essentially, the zone of measurement is too large since the zone of measurement for most EM units would be an approximately 2-m-long × 1-m-wide surface elliptical area × 0.30 m deep (estimated as about 6 ft. long × 3 ft. wide × 1 ft. deep) (Rhoades et al., 1999). Since salinity is not normally uniformly distributed within the soil profile and surface salinity conditions are important for perennial turfgrasses, (i.e., surface salinity within the 0- to 4-inch zone is especially important due to the location of crown, rhizome, and root tissues), the EM approach is less useful compared to direct salinity-monitoring mobile methods with smaller zones of salinity determination discussed in the next sections. However, EM is the most widely used mobile-mapping approach for soil properties in PA, including salinity, where the ECa of the surface 12 inches (30 cm) is sufficient.

4.3.2.4.2 Mobile Four-Electrode Sensor Platform Devices

As noted earlier, apparent soil electrical conductivity (ECa) can be measured by using a four-electrode array (i.e., a four-wenner array), which consists of four equally spaced electrodes, a current generator, and a conductivity meter. When measurements are taken at field capacity, the ECa data can be converted to an ECe basis with a calibration curve as discussed in Section 4.3.3.

Rhoades et al. (1999) provide instructions on how to make several versions of four-electrode sensor probes, including *in situ*, hand-carried, and mobile sensor platform versions. One advantage of this approach is that salinity determination is averaged over a relatively larger area, but still within the zone chosen, rather than a small point within the soil. The zone of measurement would be the same as noted under handheld four-wenner array instruments in the previous section. By adjusting distances between the four equally spaced electrodes, various depths of ECa can be made. Mobile units (and handheld units based on the same theory) operate within the ECe 0–20 dS/m range and higher. Mobile platform devices using this technology include the following:

- *Versis Technologies* (http://www.veristech.com) has cart-mounted units used in PA applications where coulters function as the four-wenner array probes to monitor ECa at 1.0-, 1.5-, or 3-ft. (0.3-, 0.5-, or 0.9-m) depths. The coulters cut into the soil surface. One unit can also monitor soil pH. Two problems with using these units on turfgrass applications are that (a) the soil depths are not narrow enough for turfgrasses where 0–4-inch (0–10-cm) zones are much better for perennial turfgrass sites receiving poor-quality irrigation water that requires frequent monitoring; and (b) the coulters cut into the turfgrass sod.
- The *SMD Mobile Experimental Unit* was developed by the Toro Company (Bloomington, MN) and the author (Carrow) specifically for turfgrass sites based on the four-wenner array electrical resistance method. Detailed spatial salinity maps, based on mapping grids of 5 × 10 feet (1.6 × 3.2 m), have been developed with this unit, including salinity levels at ECe of 50 dS/m in the surface 4 inches (10 cm) of the rootzone (Figures 4.1 and 4.2) (Carrow et al., 2009a, 2009b; Krum et al., 2011). Spectral reflectance was also obtained

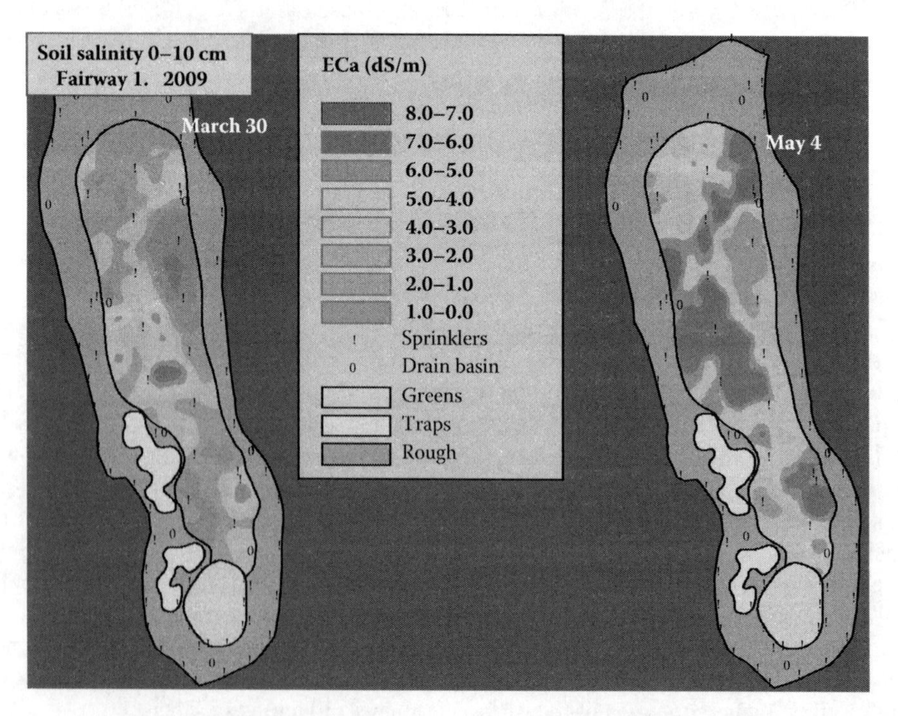

FIGURE 4.1 (See color insert.) Soil salinity mapping with a mobile four-wenner array device on 30 March and 4 May 2009 during a period of very little rainfall and with irrigation water of 5.3 to 9.3 dS m^{-1} applied at Old Colliers Golf Club, Naples, Florida. The approximate conversion of EC_a to EC_e is $EC_e = 1.7\ EC_a + 6.07$. (From Carrow, R. N., J. Krum, and C. Hartwiger, Precision turfgrass management: A new concept for efficient application of inputs, *USGA Turfgrass and Environmental Research Online (TERO)*, 8 (13), 1–12, 2009, http://usgatero.msu.edu/v08/n13.pdf. With permission. Krum, J. M., I. Flitcroft, P. Gerber, and R. N. Carrow, Performance of mobile salinity monitoring device for turfgrass situations, *Agron. J.*, 103, 23–31, 2011.)

FIGURE 4.2 **(See color insert.)** An experimental salinity-monitoring device (SMD) based on a four-wenner array for mapping surface and subsurface salinity, and the plant Normalized Difference Vegetative Index (NDVI). (From Carrow, R. N., J. Krum, and C. Hartwiger, Precision turfgrass management: A new concept for efficient application of inputs, *USGA Turfgrass and Environmental Research Online (TERO)*, 8 (13), 1–12, 2009, http://usgatero.msu.edu/v08/n13.pdf. With permission.)

to determine turfgrass performance based on the Normalized Difference Vegetative Index (NDVI), which is a widely used stress index for many plants (Krum et al., 2010). Good correlation between SMD ECa readings and ECe determined by soil sampling and laboratory analysis was obtained.

- The *Toro Mobile Multisensor Unit (TMM)* (http://www.Toro.com), a mobile platform for use in turfgrass situations, was developed by the Toro Company in 2005. It is compatible with global positioning system (GPS) and geographic information system (GIS) technology, and is capable of rapid measurement of surface zone soil volumetric water content (VWC) at 0 to 4 inches (0 to 10 cm) on a 8- × 10-ft.-grid spacing, turfgrass stress by spectral reflectance to determine NDVI, and soil penetrometer resistance (Carrow, Krum, Flitcroft, and Cline, 2009; Carrow, Krum, and Hartwiger, 2009; Krum et al., 2010). The VWC measurement was based on time domain reflectometry (TDR), which utilizes electromagnetic wave pulses from metal probes transmitted through the soil and is a widely used method for soil moisture (Walker et al., 2004). While TDR has been a possible method to determine ECa, the range has been limited. Recent advances have allowed a much wider range of salinity sensitivity so that the TMM could be used for salinity mapping (Carson et al., 2010). Comparison of the TMM capacitance/frequency domain sensor (Carson et al., 2010) with the electrical resistivity device (Carrow, Krum, Flitcroft, et al., 2009; Carrow, Krum, and Hartwiger, 2009; Krum et al., 2011) with mapping of the same sites on the same day using mobile sensor platforms designed especially for turfgrass situations revealed very similar salinity patterns and levels. With this device, detailed soil salinity distribution patterns can be observed within a single irrigation head zone of influence by using geostatistical approaches for the interpolation of data between sample sites based on their mathematical relationships (Krum et al., 2011; Dr. Van Cline, chief agronomist, Center for

Advanced Turfgrass Technology, Toro Company, Bloomington, MN, personal communication, February 2011). The irrigation head zone of influence can be subdivided into pie-shaped areas of the circle distribution to map both soil moisture and soil salinity patterns. The soil moisture patterns are used to determine head function and also to make decisions of how to adjust for uniformity of application within the zone of influence of each head. After these adjustments, the soil salinity information allows a leaching program to be developed that is specific to that zone (see Chapter 10, Section 10.3.4, "Precision Turfgrass Management [PTM] Water Audit Approach," for an example of VWC spatial mapping where salinity mapping would be similar).

4.3.2.5 *In Situ* Salinity Sensors

In-place sensors to determine soil salinity have been an area of rapid evolution in the turfgrass arena. Charlesworth (2005), Robinson et al. (2008), and Munoz-Carpena (2009) provide reviews and explanations of the soil moisture–sensing technologies and which ones allow salinity determinations to be obtained at the same time as soil moisture. For a number of years, soil solution extractors (called "imbibition-type sensors") and porous-matrix salinity sensors have been available for *in situ* applications (Rhoades et al., 1999; Corwin and Lesch, 2003). However, both devices require a long lag time before salinity changes are observed due to their dependence on the diffusion of ions. The lag time is a problem on more dynamic turfgrass systems compared to agriculture crops. Also, these sensors have small sample volumes that often do not adequately represent spatial variability, especially the imbibition sensors.

The new generation of electromagnetic soil salinity sensors includes TDR, frequency domain (capacitance and frequency domain reflectrometry sensors), and amplitude domain reflectrometry (impedance) sensors (Munoz-Carpena, 2009). Of most interest for complex turfgrass sites, such as golf courses, are "systems," which are multiple-function sensors (soil salinity and soil moisture) installed in the soil normally with at least two sensor depths; readings can be accessed by wireless communication to provide real-time data, and sensors can potentially be connected to the irrigation control system. Currently, systems with all or most of these attributes developed for turfgrass situations that have come to the forefront recently are as follows:

- Turf Guard (Toro Company, http://www.Toro.com). This is a capacitance/frequency domain sensor that has an ECa range of 0 to 6–8 dS/m. This is a complete system with all the attributes previously noted.
- UgMO Sensor (Advanced Sensor Technology, Inc.; http://www.ugmo.com/technology/). This is also a capacitance/frequency domain sensor using high (water) and low (salinity) frequencies. The salinity range is not reported. It is also a complete system.
- Stevens Hydra Probe II (http://www.stevenswater.com). The Hydra Probe II is an impedance-type sensor and has a range of ECa of 0 to 15 dS/m. More than one probe can be installed to monitor salinity by depth.
- Decagon 5TE (Decagon, http://www.decagon.com) uses capacitance/frequency domain technology for soil moisture measurement, and ECa is determined by conductance between two metal screws. Accuracy is at ECa of 0 to 7 dS/m with calibration required at ECa = 7 to 23 dS/m.
- Acclima TDT SDI-12 sensor (Acclima, http://www.acclima.com) is a time domain transmission (TDT) sensor with a sensing volume of 100 cm^3 (6 $inch^3$) and an ECa range of 0 to 5 dS/m.
- Sentek TriSCAN (http://www.sentek.com.au) is based on frequency domain reflectrometry with a high frequency used to determine VWC and a low frequency to estimate soil salinity. Salinity is reported in a propriety volumetric ion content (VIC) value that is related to ECa but not directly interchangeable. The VIC values are more qualitative and may not calibrate well with ECe (Starr et al., 2009).

It is possible to install small four-wenner array sensors for real-time data, but the author does not know of any commercial units. Other salinity sensors are and will come onto the market, but should be carefully evaluated for turfgrass use—just as the ones noted above should. There are "soil moisture" sensors that have the capability to determine soil moisture in saline soils, but they do not measure soil salinity levels; these include the ThetaProbe ML2x and Profile Probe PR2 of Delta-T Devices.

4.3.3 Calibration to Convert ECa to ECe

Conversion from ECa to ECe depends on soil sampling the same location where the ECa reading was obtained and then sending the samples to a laboratory that conducts the SPE method to determine ECe. This should be done with 12–20 soil samples within a soil type (SSMU) area, or, if SSMU boundaries have not been identified, sampling should be within similar soil areas. Samples should be taken at locations representing a range in ECa values from lowest to highest. Linear regression by an Excel program can develop the calibration curve, linear equation, and correlation coefficient.

When mapping has been done with a mobile salinity-mapping unit, locations within a similar soil area (i.e., an SSMU area with similar soil texture and organic matter content) are identified that provide a range of ECa values and soil samples obtained from these sites. Sample locations should be GPS located and be in the exact site where the mobile unit reading was taken. The field sample locations are often marked with a flag during the mobile-mapping process with locations selected to provide a range of ECa. The sampling procedure depends on the mobile salinity device used. For four-wenner array and EM devices, several subsamples are combined to represent the complete sample area due to the larger sample volumes of these units. For the Toro Mobile-Multiple Sensor (TMM) unit, where the sample is a specific location of smaller dimensions, the sample would be obtained directly over the insertion hole. Corwin and Lesch (2003, 2005) discuss soil-sampling issues in PA situations, but these also apply to PTM.

5 Routine Soil Test Methods

CONTENTS

5.1 SOIL TESTING

5.1.1 Confusing Aspects

Knowing the nutritional content of your soil is critical to the development of successful fertilization, irrigation, and soil management programs. But, soil testing can sometimes be made confusing and mysterious, and, as a result, is frequently misused. For example, different testing labs use different reporting forms and units to present their results, the ranking scale for various elemental measurements may not be provided or only partially provided, labs may use different ranking ranges even when using the same chemical extractant, and different labs use different laboratory methods (i.e., extractants or procedures) to achieve their results. Since some soil and water testing is performed by companies who also sell fertilizer, soil, and water amendments, the question arises as to what recommendations can be trusted. In this chapter, we have adapted information from a series of articles on soil testing in golf course management by Carrow (1995) and Carrow et al. (2003, 2004a, 2004b) to address key issues related to routine soil testing.

Regardless of where a soil test is conducted (a university lab, commercial lab, or laboratory associated with a fertilizer company) or who provides the initial interpretation (a laboratory, consultant, or turfgrass manager), it is essential that the turfgrass manager understands the soil test results and makes the final interpretation with consideration for site-specific conditions, or fully understands the recommendations a consultant may make. While this is important on all turfgrass areas, for salt-affected sites, it is critical for implementing the best management programs (BMPs).

5.1.2 Chemical Extractants and Soil Fertility Assessment

We begin by focusing on the different *chemical extraction methods* that laboratories use to measure soil fertility status. Understanding the concept of chemical extractants is important, because use of the right extraction method will give you reliable data that can keep your management programs responsive to the ever-changing microenvironmental conditions at your location. But, use of the incorrect extraction method (and, unfortunately, some labs don't use the right methods) can provide misleading and—to be blunt—just plain wrong information that can result in incorrect management decisions.

In Chapter 4, it was noted that most plant nutrients in the soil are found in insoluble mineral compounds and complex organic matter rather than the soluble or relatively soluble forms, which contribute to plant-available nutrient (PAN) fractions (see fractions in Table 4.1). Thus, determining "total soil content" of a nutrient, such as total P or Ca contents, does not relate to how much of the total nutrient is actually in a plant-available form. A large body of research has been conducted by soil chemists and soil fertility specialists to determine appropriate chemicals (*chemical extractants*) to extract specific soil fractions that are directly correlated to PAN for soils in different locations (Tables 5.1 and 5.2)

TABLE 5.1
Essential Questions for Turf Managers (i.e., the Customer Paying for the Services) to Ask Soil- and Plant-Testing Laboratories in Order to Obtain the Most Specific and Appropriate Test Information on a Turfgrass Site

Soil Test Question 1. Which SLAN extractants are used? Because the extractant plays a critical role in how appropriate the soil test results are on a specific soil, the laboratory should publish the list of extractants used in various analyses so that the turf manager can determine which lab to use. Extractants used in soil testing are not proprietary in nature. This information should be easy to find on lab websites and informational publications. Also, without knowledge of which extractants are used, the turf manager cannot compare the laboratory or consultants' recommended ranking ranges (e.g., the medium range) with those of other laboratories using the same extractant.

Soil Test Question 2. What is the ranking? The extractable level of a specific nutrient, such as Mg, from a soil must be "ranked" within a range from low to high. Soil test labs should report at least their "medium" range for various nutrients. These may vary with soil conditions (acid vs. alkaline soil, low-CEC vs. high-CEC soil, with grass type, and so on) and should be reported for each situation. Without ranking ranges, it is difficult for the turf manager to make the best agronomic and economic decisions from soil testing.

Soil Test Question 3. What is the method used to determine CEC on saline, calcareous soil sites? For salt-affected sites, it is important for soil test laboratories to use the "difference method" proposed by the USSL (1954) and So et al. (2006) to determine CEC and cations on the exchange sites to avoid major errors on saline, calcareous soils. This option may cost more for the customer, but it would insure accurate CEC, percent cations, and cation ratio information. Note: for nonsaline, calcareous soils, the NH_4OAc extractant buffered at pH 8.1 is suitable.

Soil Test Question 4. What is the method used to determine total soluble salts in soils and the soil SAR? (See Chapter 4.2.2, "Total Soluble Salts.") As an important principle in obtaining the most specific soil test information to assess salt-affected sites, we suggest that turf managers insist on the SPE method to most accurately determine ECe and SAR. If labs use one of the dilute pastes or slurry methods, it should be clearly stated as well as the means used to convert data to estimated ECe and estimated SAR.

Plant Analysis Question 5. Do you wash the tissue samples prior to plant analysis, and is the washing process vigorous and with a weak soap and dilute acid wash? See Chapter 7, Section 7.3, "Sample Preparation."

Irrigation Water Quality Question 6. Does the laboratory use the adj SAR (also reported as adj RNA) based on Ca_x adjustment to calculate the Na permeability hazard (Suarez, 1981)? See Chapter 6, Section 6.4, "Sodium Permeability Hazard." The older method to calculate an adj SAR was based on a pHc adjustment. It is acceptable for laboratories to use either the adj RNA or adj SAR terminology as long as the value reported is based on the Ca_x adjustment procedure (see Stowell and Gelernter, 2010a, for online tools to make this calculation).

Note: This soil test information is routinely provided by state and university laboratories, but often is difficult or impossible to obtain from some commercial laboratories.

TABLE 5.2
Common Chemical Extractants Used in Soil Testing for Phosphorous (P)

Chemical Extractant	Comments
Bray P1	Also called Bray-Kurtz P1. 0.03 M NH_4F + 0.025 M HCl. Extracts relatively soluble Ca-P, Fe-P, and Al-P mineral forms and some organic P. Recommended for neutral and acid soils (pH < 7.4), but not for alkaline-calcareous soils (pH > 7.4).
Olsen	0.5 M $NaHCO_3$ at pH 8.5. Extracts various Ca-P fractions and some Fe-P. Very good extractant on alkaline soil (pH > 7.0), but also performs well on acid soils.
Mehlich 3	0.015 N NH_4F + 0.20 M CH_3COOH + 0.25 N NH_4NO_3 + 0.013 N HNO_3 + 0.001M EDTA. This double-acid combination extracts relatively soluble Ca-P, Fe-P, and Al-P mineral forms and some organic P, and is superior on volcanic ash or loess-derived soils. A modification of Mehlich 1 is used for higher CEC Midwestern soils. It is effective across a wide range of pH for P.
Mehlich 1	0.05 N HCl + 0.0125 N H_2SO_4. Relatively soluble Ca-P, Fe-P, and Al-P mineral forms. Removes excessive P in calcareous soils or if apatite mineral is present. Developed for the acid, low-CEC soils, such as those in the Southeast, but is also suitable for neutral soils.
Morgan	0.52 N CH_3COOH + 0.72 N CH_3COONa (pH 4.8). Used in acid soils of Northeast and Northwest. Extracts P fractions dissolved by CO_2 dissolved in water. Developed in the Northeastern states as a more "universal" extractant for acid soil, but is also suitable for P on neutral soils.
Modified Morgan Extractant	0.62 *N* NH4OH + 1.25 *N* CH3COOH (pH 4.8). Similar to Morgan.

(Westerman et al., 1990; Mortverdt et al., 1991; Sparks, 1996; Bartels, 1996; Jones, 1998; Peverill et al., 1999).

Much of this research has been conducted by land grant universities in the United States and similar institutions in other countries and has clarified the influence of regional soil types and environmental conditions. A review of the descriptions of the extractants in Tables 5.1 and 5.2 illustrates that the extractants used must match soil conditions, especially acid, neutral, alkaline, or calcareous soils, since these vary substantially in chemical composition. The crop being grown does not influence which extractant is used, but soil type is very important—the crop consideration involves how the results are interpreted. Some extractants, such as Mehlich III (also written as Mehlich 3), have been formulated to be used on many different soil conditions and for multiple nutrients. These are called "universal extractants," but they are not really universal or always the best extractant in all situations. Other extractants are very specific to certain soil conditions, such as Bray P1 for P on acid and neutral soils but not alkaline-calcareous soils (where the Olsen P test is often used).

The greatest emphasis of soil-testing research has been toward determining the fertilizer needs of agronomic and horticulture crops with an end-of-season yield as the goal. For farmers, fertilizer costs versus yield considerations are paramount. University and state soil test labs use the best extractants for the soils in their state so as to provide the farmers the most specific and reliable information. Commercial labs that process soils from many different locations, and sometimes from different countries, must select chemical extractants appropriate for the majority of the soils they test. Generally they use chemical extractants suitable for multiple soil conditions; or, in the case of P, they may report both Bray P1 and Olsen extractant results where the turfgrass manager can select the data most appropriate to the soil. However, the chemical extractants may not be best for a specific turfgrass site. Thus, *the first principle in obtaining the most specific soil test information is to select a laboratory that uses the best extractant for the soil conditions (see Table 5.1, question 1)*. Laboratories may be able to offer more than one extractant choice to better meet customer needs. The authors strongly suggest that commercial laboratories proactively provide the turfgrass manager the information on what extractants they use or can use by publishing the information on their websites and information packages.

Chemical extractants play a critical role in each of the *three approaches to assess soil fertility status* that are often used for turfgrass situations. The first two, noted below as the sufficiency level of available nutrients (SLAN) and basic cation saturation ratio (BSCR) concepts, are generally included in most laboratory soil test reports, while the third (dilute soil:water-extractable nutrients) is also reported by some labs. A brief view of Tables 5.8 and 5.9 may be useful in understanding the three soil test assessment approaches that are stated below. The three approaches or concepts to soil fertility assessment are as follows:

- *SLAN*, which determines plant-available levels of macronutrients (Ca, Mg, K, P, and S), micronutrients (Fe, Mn, Zn, Cu, and B), and extractable Na and Cl within a soil sample (see Section 5.2 and Table 5.8). The SLAN approach to nutrient recommendations provides important predictive information on the soil's ability to supply a nutrient to the plant over the growing season. Extractants listed in Tables 5.2 and 5.3 are primarily focused on the SLAN concept since this approach is the best means to assess soil fertility.
- The *BCSR* approach is based on determining the percent cation saturation levels on the cation exchange capacity (CEC) sites (% Ca, Mg, K, Na, H, and Al) to provide insight into nutrient balances and some information relative to the nutrient-supplying power of the soil if the CEC is also known (see Section 5.3 and Table 5.8). The BCSR approach also requires an accurate determination of the soil CEC, a measure of the soil's nutrient-holding capacity for cations (positive charged ions). The NH_4OAc extractant at pH 7.0 (noncalcareous soils) or 8.1 (calcareous soils with pH > 7.6) reported in Table 5.3 is often used to determine total CEC and percent cations on the CEC, and is a common SLAN extractant for cations. In Section 5.3, a more detailed discussion of extractants and determination of CEC is presented.
- *Water-extractable nutrients (saturated paste extract [SPE] and dilute pastes)* are a third means of assessing soil fertility status (see Section 5.4 and Table 5.9). In this instance, water is the extractant, but different labs may use variations from the SPE such as 1:1, 1:2, and 1:5 (soil:water) dilution methods, with the SPE being the best and most concentrated and the 1:5 being the most dilute. These dilution methods are not recommended by the authors for plant nutrient recommendations for the reasons presented in Section 5.4; however, the SPE water-extractable levels of certain ions are useful, especially in salt-affected soils, namely, Cl, Na, HCO_3, and sometimes others. Water-extractable ions that impact nutrient availability in saline and sodic stress conditions, such as bicarbonates and carbonates, should also be included in these water extraction procedures.

5.2 SLAN APPROACH, EXTRACTANTS, AND INTERPRETATION

5.2.1 SLAN Approach

By far the best approach for determining the plant-available nutrient status of a soil is by the SLAN concept using appropriate chemical extractants such as those listed in Tables 5.2 and 5.3. This is the traditional method that has a large body of research supporting it. As was noted in Chapter 4 on soil test packages for salt-affected sites, water is the simplest "chemical extractant" and is useful to assess soluble salts in soil solution and precipitated soluble salts as described in Table 4.1. However, more robust chemicals must be used to determine nutrient status on the soil CEC sites, within slowly dissolved salts and minerals, for plant-available K and Mg in clay interlayers, micronutrients on CEC sites and chelated forms, and nutrients within decomposing organic matter or lodged in inorganic amendment micropores—these are described in Table 4.1 in fractions 3 through 6. Commonly used SLAN extractant chemicals are weak acids (acetic, citric, and lactic),

TABLE 5.3
Common Chemical Extractants Used for Soil Testing for Cations, Sulfur (SO_4), and Micronutrients

Nutrient	Chemical Extractant	Comments
		Cations (Macronutrients)
K, Ca, Mg, and Na ion	NH_4OAc, pH 7.0	CH_3COONH_4 (ammonium acetate) (1 M), usually buffered at pH 7.0 (i.e., solution does not change pH when added to soil). Widely used as an SLAN extractant for K, Ca, and Mg in the Midwest and far West. Also, the most common method to determine cation (K, Ca, Mg, Na, H, and Al) saturation on CEC sites and total CEC for BCSR extractant. Extracts primarily cations on CEC sites and in soil solution. On calcareous soils (pH $>$ 7.6) where free $CaCO_3$ is present, the CEC and percent Ca on the CEC sites will be in error—see NH_4OAc at pH 8.1, below. On highly acidic soils (pH $<$ 5.5), the extractant should be unbuffered since the buffered version will greatly inflate the actual field CEC of these soils. Unbuffered extractant allows the extractant pH to adjust to the actual soil pH.
	NH_4OAc pH 8.1	Ammonium acetate buffered at pH 8.1 should be used for calcareous soils, especially those with soil pH approaching pH $>$ 7.6 or expected of having free $CaCO_3$. Many salt-affected soils have free $CaCO_3$, and NH_4OAc at pH 7.0 will extract some of the Ca from the lime, but not at pH 8.1 since CaCO3 is very insoluble at this pH. See comments on NH_4OAc at pH 7.0, above.
	Mehlich 3	0.015 N NH_4F + 0.20 M CH_3COOH + 0.25 N NH_4N0_3 + 0.013 N $HN0_3$ + 0.001M EDTA. This double acid + chelating agent extractant has become widely used as a "universal" extractant for many macro- and micronutrients (P, K, Ca, Mg, Na, B, Cu, Fe, Mn, S, and Zn). It has the chelating agent EDTA.
	Mehlich 1	0.05 N HCI + 0.0125 N H_2S0_4. Double-acid extractant used on acid soils in the Northeast and Southeast for extractable macro- and micronutrient cations and P.
	AB-DTPA	1.0 M NH_4HCO_3 (ammonium bicarbonate). 0.005 M DTPA solution adjusted to pH 7.60. This method estimates the availability of P, K, Ca, Mg, Zn, Fe, Mn, and Cu for soils with neutral to alkaline pH. Used in Western states as a multiple-nutrient extractant.
		Anions
$S(SO_4)$	Water	Hot water is used in many arid or semiarid regions since the sulfate salt form of S often dominates in these regions.
	$Ca(H_2PO_4)_2$	Monocalcium phosphate is used in humid regions since it can extract exchangeable and labile (associated with organic matter) as well as sulfate salt.
	Others	Many other extractants are used such as NH_4OAc (pH 7.0), Mehlich 3, dilute acids. These are often used in humid regions since they can extract exchangeable and labile (associated with organic matter) as well as sulfate salt.
Cl, HCO_3	Water	Saturated paste extract is used.
		Micronutrients
Fe, Mn, Cu, Zn	DTPA	Diethylenetriaminepentaacetic acid (DTPA-TEA) is 0.005 M DTPA + 0.01 M $CaCI_2$ + 0.10 M TEA to buffer at pH 7.3. DTPA is a chelating agent that forms water-soluble complexes with free metal cations and is the most widely used extractant for micronutrients. Adapted to a wide pH range of soils.
	Mehlich 1	See Mehlich 1, above. Used on acidic soils to determine micronutrients but is not used on alkaline or calcareous soils.
	Mehlich 3	See Mehlich 3, above. Double acid with chelating agent; is the most widely used "universal" or multiple-nutrient extractant. Widely used for micronutrients.

diluted stronger acids (hydrochloric, sulfuric, and nitric), double acids (i.e., two different acids used together), bicarbonates (ammonium and sodium), salt solutions, specific ions (fluoride and ammonium), chelating agents, and combinations of these chemicals (Westerman, 1990; Mortverdt et al., 1991; Bartels, 1996; Sparks, 1996; Sumner and Miller, 1996; Jones, 1998; Peverill et al., 1999; Kopittke and Menzies, 2007; St. John and Christians, 2007; Schroder et al., 2009; NRP, 2009; SCSB, 2009). Other countries may use these extractants or develop ones unique to their soils, such as some of the extractants used in Australia (Peverill et al., 1999).

As implied by the term *sufficiency level*, SLAN is based on estimating the *quantity* of *plant-available* nutrient(s) that a soil can supply during the growing season. Assessment of the soil's ability to supply a quantity of a nutrient is important because turfgrass plants require a certain sufficiency quantity of a nutrient for growth and development that must come from the soil or be applied as fertilizer—this distinction will become more apparent when the BCSR approach is discussed, which is based on percentages and ratios rather than quantity or amount. For an example of the SLAN approach, assume that a soil test laboratory uses the Mehlich III extractant and extracts 20 mg/kg (20 ppm; 40 lb./A) of Mg from the soil sample—that would be a *quantity* of plant-available Mg in the soil. In our example of 20 ppm Mg for the Mehlich III extractant, the Mg sufficiency level would be in the low range (Table 5.4), which would suggest that an additional quantity of Mg must be supplied to the plant by application of an Mg fertilizer.

5.2.2 SLAN Extractants for Specific Situations

SLAN extractants developed by soil chemists to determine macronutrient levels (P, K, Ca, Mg, and S) include single-acid (ammonium acetate), double-acid (Bray P1, Morgan, Mehlich I, and Mehlich III), or bicarbonate-based (Olsen) extractants, with specific salts sometimes included (NCR, 1998; Gavlak et al., 2003; NRP, 2009; SCSB, 2009). In recent times, there has been a movement toward developing a *universal extractant* that would extract appropriate levels of the different macronutrients and micronutrients (Schroder et al., 2009). The primary benefit of a universal extractant is cost—it is less expensive for a laboratory to run a sample if only one extractant can be used. What is most important for the turfgrass manager, however, is to obtain accurate estimates of PAN levels given their particular soil and climatic conditions. Mehlich III (a double acid with a chelating agent) has become widely used as the "universal extractant" of choice since it often correlates to other extractants that have been widely used for many years (Schroder et al., 2009). However, on some soils for particular nutrients, the Mehlich III can be less reliable. Some key considerations regarding appropriate SLAN extractants are noted in this chapter for specific nutrients as well as in Tables 5.2 and 5.3.

Sometimes the question arises as to whether decomposed granite (DG) requires any special extractants since DG is often used on golf courses in arid regions. Decomposed granite consists of weathered granite, granodioite, or quartz diorite that can range from relatively unweathered to highly weathered (Rider et al., 2005). While DG is usually sandy to gravel in particle size, it can contain some clays as part of the weathering process, usually vermiculite for intermediately weathered DG, but other clay types can be present depending on the weathering stage. Due to this mineralogy, normal soil test extractants can be used. The vermiculite can result in some K^+ or NH_4^+ fixation.

5.2.2.1 Phosphorus

Of all the nutrients, P is the most complex in terms of soil chemistry, and the extractant used must match the soil conditions to obtain useful results (Table 5.2). The P extractants that have been developed and widely used in specific regions of the United States, and thereby provide a good level of confidence for use in these or similar soil regions, are as follows: (a) for the Northeast and Northwest, Morgan, a weak acid extractant, is used for these acid soils (NRP, 2009); (b) in the Midwest for soils with pH < 7.0 to 7.4, Bray P1 predominates (NCR, 1998); (c) in the Southeast,

TABLE 5.4
Typical Soil Test SLAN "Medium" Sufficiency Ranges for Macronutrients Using Common Extractants and with CEC < 15.0 cmol/kg[a,b]

		Sufficiency Level Range	
Nutrient	**Soil[b]**	**Medium, ppm**	**Extractant**
P	All	15–30	Mehlich I
	All	26–54	Mehlich III
	All	15–30	Bray P1
	All	12–28	Olsen
	All	10–20	Morgan
	All	8–11	AB-DTPA
K	Sands	75–175	NH_4OAc (pH 7.0)
	Others	100–235	NH_4OAc (pH 7.0)
	Sands	50–116	Mehlich III
	Others	75–176	Mehlich III
	Sands	50–100	Mehlich I
	Others	90–200	Mehlich I
	All	155–312	Olsen
	All	120–174	Morgan
	All	61–120	AB-DTPA
Ca	All	200–350[c]	Mehlich I
	All	500–750[c]	Mehlich III
	All	500–750[c]	NH_4OAc (pH 7.0)
	All	500–750[c]	Morgan
Mg	Sands	30–60	Mehlich I
	Others	50–100	Mehlich I
	Sands	60–120	Mehlich III
	Others	70–140	Mehlich III
	Sands	100–200	NH_4OAc (pH 7.0)
	Others	140–250	NH_4OAc (pH 7.0)
	All	>100	Morgan
SO4–S	All	10–20	$Ca(H_2PO_4)_2$
	All	30–60	NH_4OAc (pH 7.0)
	All	15–40	Mehlich III

Source: Carrow, R. N. et al., Clarifying soil testing. III. SLAN sufficiency ranges and recommendation, *Golf Course Management,* 72 (1), 194–198, 2004b, http://www.gcsaa.org/gcm/2004/jan04/PDFs/01Clarify3.pdf.

Note: Conversion for lb./acre: lb./acre = 2 × ppm.

[a] The values in this table are from many sources and are typical. A particular laboratory may vary from these rankings for local soil conditions.

[b] Soils that have high CEC (>15 cmol/kg) may exhibit a higher "medium" sufficiency range. The medium sufficiency range in this table is based on sands or other soils with CEC <15 cmol/kg.

[c] On sites receiving irrigation water high in Na, the upper level of the sufficiency range should be used as a guideline (e.g., for Mehlich III this would be 750 ppm Ca or higher).

the double-acid extractants, Mehlich I and III, are dominant (SCSR, 2009); and (d) in the Western states, the bicarbonate-based Olsen extractant is favored on these primarily calcareous soils (Gavlak et al., 2003). Turfgrass managers can contact their state or university soil test laboratory (most U.S. states have such labs) and ask for information on the best extractants for their soils. Some guidelines for use of other P extractants outside of the "traditional" region of use are as follows:

- *Acid or neutral soils*: The Mehlich III seems to correlate well for P with Morgan, Bray P1, and Mehlich I extractants in acid and neutral soils. The correlation, however, to the Morgan is not straightforward, but must include soil pH and Mehlich III Ca and Al (Ketterings et al., 2001). This illustrates the point that a laboratory testing from a wide geographic range of soils and using a universal extractant must be careful that the correlation of the universal extractant actually provides accurate information for a nutrient.
- *Calcareous soils*: Calcareous soils may occur in many different locations and not just Western states (alkaline soils in general and soils created by pulverizing caliche) and the Caribbean region (where golf courses have been built on pulverized caliche soils). For example, calcareous sands can be found in many states, and the use of irrigation water with high concentrations of Ca and bicarbonate can result in free lime formation in turfgrass soils. For calcareous soils at pH > 7.4, the Olsen is the best to use, especially in situations where P has a history of being limited, where environmental sites are sensitive to P, or when it is important to more accurately determine P status, such as on high-sand-content golf greens. Most acid extractants are not accurate on calcareous soils. An exception is the Mehlich III, which seems to work well on many calcareous soils—but there is still concern that it may not be as good (not as accurate in providing actual P availability) as the Olsen test on some sites. Olsen-extractable P has been correlated to plant growth in many studies, while the Mehlich III has not. It is possible to have good correlation of Olsen versus Mehlich III on some calcareous soils, but not across all calcareous soils in all regions. Thus, a consultant working within a narrower range of soils may use the Mehlich III, while this would be less appropriate when the calcareous soils come from many diverse locations, especially internationally. Our recommendation is that if in doubt, request the Olsen test. If you desire to compare results on your site, request P determined by both extractants.
- *High-sand greens that are not calcareous*: Many golf greens are manmade and are not typical of the general soils of the region. Thus, extractants developed within a region, such as the previously listed ones, for regional soils may not be better on these greens. The Mehlich III can be used on high-sand golf greens that do not contain free calcium carbonate, regardless of location in the United States or world. Mehlich III could be very appropriate for the high-sand greens that are not calcareous, but if fairway or tee soils are calcareous, it may be better to use the Olsen, as discussed in the preceding paragraph on these soils.
- *Volcanic soils and loess-derived soils*: Volcanic soils are composed of ashy and cindery materials (such as pulverized pumice) originally deposited during volcanic activity, while a loess-derived soil has wind-deposited soil as the parent material. Mehlich III has been reported to be the best extractant for these soils.

5.2.2.2 Potassium and Magnesium

Soil test labs generally use the same extractant for Mg as they do for K. Potassium is an important nutrient for turfgrasses for growth and stress tolerance, especially abiotic stresses including salinity stresses. Ammonium acetate, NH_4OAc, at pH 7.0 (buffered) has traditionally been the most widely used extractant for K across all soils and climatic regions, except in the Northeast, where NH_4OAc at pH 4.8 has been used or NH_4OAc at pH 7.0 (not buffered). Mehlich III–extractable K is highly correlated with NH_4OAc at pH 7.0 and has increasingly been used in many labs. Other extractants have been successfully used in some states such as the Mehlich I and Olsen.

5.2.2.3 Calcium

Calcium determination by the SLAN approach for assessing Ca nutritional needs has not been a high priority for most crops since substantially more plant-available Ca is present in the soil than other cations in most situations, except very acid soils. When the soil is acid, lime is generally recommended to increase soil pH, which also provides ample available Ca for plants. However, several turfgrass situations require that Ca be evaluated along with other cations. Such conditions include low CEC sand soils; leaching conditions; when the irrigation water may be adding high levels of other cations that compete with Ca for CEC sites, especially Na that causes sodic conditions; and the presence of high Al/Mn at pH < 4.8 (Carrow, Stowell, et al., 2001). Comments related to Ca extractants are as follows:

- In all soils, Mehlich III and ammonium acetate (NH_4OAc with pH 7.0) are both acceptable extractants and are widely used for SLAN-extractable Ca—which includes Ca on the CEC sites as well as Ca in the solution, in soluble Ca compounds, and in some lime and gypsum particles that dissolve with the acid extractants. However, as will be discussed in Section 5.3, when Ca content only on the CEC sites is desired as a component of determining total CEC and percent Ca on the CEC, these extractants are not appropriate (due to overestimation of Ca) and NH_4OAc with pH 8.1 is recommended.
- Morgan extractant and Mehlich I have also been used, especially in the regions where the same extractants are used for K.

5.2.2.4 Sulfur

Many different extractants have been used for SO_4-S (Table 5.3). The most reliable extractant for the prediction of plant-available SO_4-S across all soil and climate zones is $Ca(H_2PO_4)_2$, but the SPE method is also used in arid regions. In Section 4.2.4, "Specific Ion Toxicity and Problem Ions," the SO_4 soil test was discussed in terms of salt issues (such as leaching and black layer), but the SLAN-extractable level of S (usually presented in the SO_4 form) is also important for assessing soil S availability for turfgrass nutritional needs. In many salt-affected soil sites, especially sodic and saline-sodic soils, ample S for nutritional needs is usually applied due to the use of gypsum or irrigation water acidification with S-based systems. However, on saline sites, particularly high-sand soils, S may be limiting.

5.2.2.5 Micronutrients

The most common extractant for micronutrients in soils is DTPA-TEA (diethylenetriaminepentaacetic acid), a chelating agent; but some labs use other extractants such as Mehlich I and Mehlich III (Table 5.3).

5.2.2.6 Other Ions

In Section 5.4, SPE and dilute soil:water-extractable Na, Cl, HCO_3, and Al are discussed.

5.2.3 Interpretation of SLAN Results

SLAN-extractable results for each nutrient must be ranked because the turfgrass manager is interested in what the SLAN value means—in other words, is the plant-available nutrient level low, medium, or high in the soil? Regardless of the extractant used, the SLAN test result for each nutrient is ranked from low to very high. It is important to recognize that the ranges are based on expected plant species growth responses.

- *Low range* denotes that there is a high probability (80%–100%) of getting a growth response from application of that nutrient.

- *Medium range* implies that there is approximately a 50% chance of getting a plant growth response from application of the nutrient. Thus, if supplemental fertilizer is not applied, there is a good probability that turfgrass growth may become limiting, especially as the season progresses.
- *High range* denotes that there is little or no crop response expected from applying the particular nutrient.
- *Very high range* suggests that further application of the particular nutrient may cause a nutrient imbalance (usually due to competitive inhibition of uptake or domination on the CEC) or reduced growth (from possible toxic interactions) in some cases. From the environmental standpoint, a very high range of nutrients in the soil can be prone to leaching losses, especially with high-rainfall events and on sandy soil profiles.

Tables 5.4 (on macronutrients) and 5.5 (on micronutrients) list common sufficiency ranges (i.e., the "medium range") for different extractants used for the various nutrients based on a review of numerous sources, especially state or university labs, commercial labs with practical agronomic recommendations, and the author's experiences. A specific laboratory may use somewhat different ranges, but these provide a general guideline based on reliable science. Publications by the North-Central (NCR, 1998), Northeastern (NRP, 2009), Southern (SCSB, 2009) and Western (Soltanpour and Follett, 1996) regional soil test laboratories provide the extractants and rankings for states within these regions and serve as excellent guidelines. Ranking ranges may vary with soil conditions (acid vs. alkaline soil, or low-CEC vs. high-CEC soil), and state and university soil test labs often use separate rankings for these situations, especially for K and Mg as illustrated in Table 5.4.

While state and university soil test labs routinely report SLAN extractant rankings, this is not the case in many commercial soil test labs. On the soil test report for the SLAN-extractable level, each nutrient is normally ranked as low, medium, high, or very high; but the ranges that are actually used are not often given. Soil test labs should report at least their "medium" range for various nutrients so that the customer (turfgrass manager) can compare rankings with typical ones used elsewhere, such as in Tables 5.4 and 5.5. The authors have found some laboratories where the recommended medium ranges are consistently higher than the guidelines in Tables 5.4 and 5.5. In fact, in some cases, the medium ranges are two- to fivefold higher, which means that almost all soils tested would have most of the nutrients listed as low using their guidelines, when in fact the soil levels are ample. Oftentimes these labs are associated with companies selling fertilizer. This practice results in greatly increased fertilizer sales, but is not reliable from the turfgrass management, economics, environmental, or ethics viewpoints. With fertilizer a significant input cost, turfgrass managers may wish to evaluate recommendations from the economic and agronomic aspects. Thus, *another principle in obtaining the most specific soil test information is to select a laboratory that provides the ranking ranges for at least the medium level, because this is essential for the turfgrass manager to assess whether the recommendations are reasonable from agronomic and economic viewpoints* (Table 5.1, question 2).

Turfgrass managers of high-maintenance turfgrass sites often target maintaining soil nutrient status within the "medium" range as the desired level and then supplementing with fertilization rather than trying to maintain higher soil test values. However, some consultants or laboratories may use the high range as the "desired range." On high-use recreational sites receiving saline irrigation water, it is best to target toward the high end of the medium range. On more fine-textured soils, it is possible to increase the soil test for various nutrients into the high range since these soils have high CEC and nutrients are less likely to be easily leached. But, for sandy soils, it is difficult to maintain higher nutrient levels due to leaching potential. Fertilizing at lower rates somewhat more often by a spoon-feeding approach to maintain soil nutrients within the medium range is a better practice. Since turfgrasses are perennial and, thereby, have a longer growing season than most production crops, applying high rates of fertilizer for season-long use and to achieve a high soil test range can result in excessive complexation of some nutrients in the soil into unavailable forms or be lost due to elevated leaching problems. Thus, by targeting a medium range and then "spoon-feeding" nutrients as needed, the fertilizer is efficiently used in the turfgrass ecosystem.

One micronutrient of interest on salt-affected sites is Mn. In Table 5.5, the Mn levels are based on nutrient needs for growth and color. However, on salt-affected soils, the grasses, even halophytic grasses, are often under frequent salinity stress, and Mn tends to be low on many alkaline soils. The salinity stress response mechanisms are genetically expressed in the root, crown, and rhizome tissues, where high accumulated soil salts cause injury to these tissues and make the grasses more susceptible to take-all disease (*Gaeumannomyces graminis* var. *avenae*). A high soil level of Mn is important (a) to maximize salinity tolerance mechanisms in the grasses, and high Zn also is synergistically important for the same reason; (b) to suppress take-all and other root-borne pathogens; and (c) because Mn activates over 35 enzymes for growth and development. Thus, on these sites, Mn may be maintained in the high range. Foliar Mn applications do not increase soil levels to any great extent since Mn is not translocated downward in the plant shoots to the tissues in the soil. Either granular or foliar Mn that is washed into the soil is more effective. See Chapter 16 for more discussion on this topic.

How do typical golf course soil test results align with the SLAN ranges in Tables 5.4 and 5.5? In a series of studies, soil samples from golf greens and fairways were analyzed from different regions within the United States using the Mehlich III extractant (Carrow et al., 2001, 2004b) (Table 5.6). Few of the 114 golf courses in this study were on salt-affected sites, so the results reflect sites without the influence of salinity. This can be thought of as a *benchmarking approach* to define current soil test status. The Louisiana, Mississippi, and Florida courses in the United States were all coastal courses from the panhandle of Florida to Louisiana, and therefore the CEC was low due to very little clay or organic matter in these greens and the application of ultrapure irrigation water due to high rainfall and sandy soils; so the aquifers were of high water quality. Observations were as follows:

- For California, Minnesota, and Illinois greens, all macronutrients were well within the sufficiency ranges, except for SO_4 in Minnesota. This is not surprising since the greens all exhibited ample CEC. Higher CEC levels will result in higher extractable levels of most macronutrients. A comparison of the fairway values within a region with the green values illustrates this point. For many of these courses, P fertilization especially could be reduced. In Minnesota, the inherent SO_4 levels appear to be low, and some S may be required if not provided in another source—such as potassium sulfate.
- For the Louisiana and Mississippi greens, the average low SLAN levels for P, K, Ca, Mg, and S reflect the combination of low CEC, very pure irrigation water (this was determined as part of the study), and high rainfall. On these soils, application of high rates of nutrients will not build up the SLAN-extractable levels because of these conditions. Soil incorporation of a good zeolite to raise CEC was suggested to assist in these situations. Based on the average 1.6 CEC of their greens and assuming a zeolite with a CEC of 160 cmol/kg, it would require about 195 lb. per 1000 sq. ft. of zeolite to raise the CEC to 2.6 cmol/kg to a 4-inch (10-cm) depth (Carrow, Waddington, et al., 2001). Other inorganic sand substitute materials have CEC levels of <31 cmol/kg and would require substantially higher rates to achieve acceptable CEC stabilization in the soil profile. Spoon-feeding of all macronutrients on a relatively frequent basis is the most effective means of maintaining adequate nutrients—and the SLAN levels would remain low in most instances unless CEC was increased.
- Data such as in Table 4.6 are valuable for "benchmarking" purposes. This allows golf course superintendents to observe how their soil nutrient levels compare within their regional soil and variable climatic conditions.

TABLE 5.5
Micronutrient Extractants (Fe, Zn, Cu, and Mn) and "Medium" Soil Ranges Used by Many Laboratories for Plant Micronutrient Availability

Micronutrient	Medium[a] ppm
DTPA-TEA[d]	
Fe	10.0–15.0
Mn (pH < 7.0)[c]	2.0–5.0
Mn (pH > 7.0)[c]	5.0–15.0
Zn	>2.0
Cu	>1.0
AB-DTPA	
Fe	>10.0
Mn	>5.0
Zn	>1.5
Cu	>2.0
Mehlich III	
Fe	50–100
Mn (pH < 7.0)[c]	4.0–8.0
Mn (pH > 7.0)[c]	8.0–16.0
Zn	1.0–2.0
Cu	0.4–2.5
Mehlich I	
Fe	50–100
Mn (pH < 7.0)[c]	4.0–10.0
Mn (pH > 7.0)[c]	10.0–18.0
Zn	1.0–3.0
Cu	0.5–1.0
B	>0.3
Hot Water Soluble	
B	0.1–0.2

Source: After Mortverdt, J. J. (Ed.), *Micronutrients in Agriculture* (2nd ed.), Soil Sci. Soc. of America, Madison, WI, 1991; Tisdale, S. L., W. L. Nelson, J. D. Beaton, and J. L. Havlin, *Soil Fertility and Fertilizers*, Macmillan, New York, 1993; Sims, J. T., Soil fertility evaluations, in M. E. Sumner (Ed.), *Handbook of Soil Science*, CRC Press, Boca Raton, FL, 2000; Carrow, R. N., D. V. Waddington, and P. E. Rieke, *Turfgrass Soil Fertility and Chemical Problems: Assessment and Management*, John Wiley, Hoboken, NJ, 2001; Sartain, J. B., *Soil and Tissue Testing and Interpretation for Florida Turfgrasses* (IFAS Extension Publ. SL 181), Institute of Food and Agricultural Sciences, University of Florida, Gainesville, 2008, http://edis.ifas.ufl.edu/pdffiles/SS/SS31700.pdf.

[a] Extractable micronutrient levels are preferred to be within the upper end of the medium range on high-maintenance, recreational turfgrass sites but the lower end of the medium range for nonrecreational grasses.

[b] Rankings for micronutrients are more accurate for plants sensitive to a particular micronutrient, such as vegetable crops, than for turfgrasses, which are not sensitive to micronutrients.

[c] Mn soil tests in this table are for maintaining plant nutrient needs. However, soil Mn levels in the high range are required to suppress take-all disease and to maximize turfgrass salt tolerance—see thee Mn discussion in Chapter 16.

[d] DTPA-TEA is 0.005 M DTPA + 0.01 M CaCI2 + 0.10 M TEA to buffer at pH 7.3; it is the most widely used. DTPA is a chelating agent that forms water-soluble complexes with free metal cations.

5.3 BCSR APPROACH, CEC MEASUREMENT, AND INTERPRETATION

5.3.1 DEPENDENCE OF BCSR APPROACH ON CEC DETERMINATION

The BCSR approach for fertilization decisions is based on determining total CEC and the percent saturation of cations on the CEC sites (Haby et al., 1990; Carrow, 1995; Baird, 2007; Kopittke and Menzies, 2007). Before discussing the BCSR approach, it is useful to understand how total CEC and cation percentages on the CEC sites are determined. Errors in determining cation percentages and total CEC would affect the validity of using the BCSR approach. Errors in CEC determination are not uncommon on salt-affected soils and are associated with incorrect pH of the extractant solution, the presence of soluble salts, or the presence of free $CaCO_3$ or $CaSO_4$. So et al. (2006) noted, "Despite the practical importance of salinity and sodicity, little attention has been directed to the accurate characterization of these soils. This is particularly true for the assessment of exchangeable cations where the most widely used methods have long been known to yield inaccurate results, but few researchers have considered the magnitude of the error in their measurements or its influence on the interpretation of the data" (p. 1820). To be direct, the BCSR method (total CEC plus percent cations on the CEC) is widely used by many commercial labs in making fertilization decisions, but on saline soils with soluble salts at ECe > 4.0 dS/m and/or soils containing free $CaCO_3$ or $CaSO_4$, most of the BCSR soil test data as determined by many laboratories are inaccurate. This is an especially serious issue on salt-affected soils. The BCSR approach to fertilization, whether on salt-affected sites or nonsaline soils, has some real shortcomings relative to the SLAN approach for making fertilization decisions; but when the data are not accurate (which is much more likely on salt-affected soils), then turfgrass managers must be very cautious about using the information as their primary BMP fertility assessment. So et al. (2006) notes how accurate information can be obtained by using

TABLE 5.6
Benchmark of Average Soil Test Values from Different Locations in the United States Based on Mehlich III Extractant

	Mehlich III Extractable							
Location[a]	**P (ppm)**	**K (ppm)**	**Ca (ppm)**	**Mg (ppm)**	**SO_4 (ppm)**	**Na (ppm)**	**Soil CEC**[b]	**Soil pH**
Greens Desired[c]	**26–54**	**50–116**	**500–750**	**60–120**	**15–40**	**<67**	**>3**	**6.0–7.5**
CA (39)	99	156	1346	174	139	174	9.9	7.1
MN (12)	24	146	1660	160	13	10	7.8	7.0
IL (24)	105	170	2726	343	63	41	7.8	7.0
LA/MS (18)[d]	28	37	225	53	4	1	1.6	6.9
FL (21)	85	88	544	91	20	48	2.1	6.5
Fairways Desired	**26–54**	**75–176**	**500–750**	**70–140**	**15–40**	**<67**	**>3**	**6.0–7.5**
MN (12)	52	210	2419	372	34	13.9	16.8	7.1
IL (24)	126	316	3770	910	215	109	29.0	7.1
LA/MS (18)	41	109	774	226	16	1	6.3	6.2

Source: After Carrow, R. N., L. J. Stowell, S. D. Davis, M. A. Fidanza, J. B. Unruh, and W. Wells, Developing regional soil and water baseline information for golf course turf, in *Agronomy Abstracts*, ASA, Madison, WI, 2001; Carrow, R. N., L. Stowell, W. Gelernter, S. Davis, R. R. Duncan, and J. Skorulski, Clarifying soil testing. III. SLAN sufficiency ranges and recommendation, *Golf Course Management*, 72 (1), 194–198, 2004b, http://www.gcsaa.org/gcm/2004/jan04/PDFs/01Clarify3.pdf.

[a] Number in ( )'s is the number of golf courses involved, with two greens and two fairways tested per golf course.

[b] CEC is in cmol/kg.

[c] Based on data in Table 4.4.

[d] The Louisiana and Mississippi soils were all high-sand, low-CEC soils receiving high rainfall and pure irrigation water (ECw < 0.50 dS/m).

laboratory procedures developed by the U.S. Salinity Laboratory (USSL; 1954), as will be discussed in the following subsection.

5.3.2 Determining Total CEC and Exchangeable Cations

A brief overview of how CEC is determined provides a basis for understanding the issues. For estimation of total CEC, a sample is saturated with a cation that replaces into solution all the other cations on the CEC sites (Hendershot et al., 1993). *There are four general procedures for determining CEC and the exchangeable cations on the CEC sites* (USSL, 1954; Hendershot et al., 1993; So et al., 2006; Sumner and Miller, 1996; St. Johns and Christians, 2007).

First, the most commonly used approach is by *summation of exchangeable cations from a single extraction*—an extraction solution of a basic cation is used to displace all the cations from the CEC sites due to the high concentration of the saturation cation where two common saturation chemicals are Ba^{+2} from $BaCl_2$ and NH_4^+ from NH_4OAc (ammonium acetate). The saturation extract may be buffered at pH 7.0, buffered at pH 8.1, or unbuffered at pH 7.0 depending on the soil situation. For most salt-affected soils, the extractant should be buffered at pH 8.1 for reasons given later in this discussion. Ammonium acetate (buffered at pH 7.0) is the extractant most commonly used. Displaced cations would include Ca^{+2}, Mg^{+2}, K^+, Na^+, H^+, Al^{+3}, Fe^{+2}, Fe^{+3}, and Mn^{+2}, but the micronutrients are not included when summing cations due to their low concentrations. The sum of all cations is the total estimated CEC. For soils that are not saline (ECe < 4.0 dS/m) and do not contain free lime or gypsum, this method is accurate—some precautions are noted later to insure accuracy on acidic pH soils. However, the single extraction method is strongly influenced by the presence of soluble salts in soil solution or as precipitates and the dissolution of some lime or gypsum particles, which would appear as CEC cations but are not really extracted from the CEC sites. Thereby, estimated CEC is higher than actual CEC, and the estimated cation percentages will be inaccurate with Ca levels affected the most—and, obviously, any cation ratios would be inaccurate.

Second is the *double-extraction procedure*, which is accurate for the determination of total CEC but does not provide information on percent of each cation on the CEC sites. In this method, the sample is saturated several times with a saturation solution (e.g., NH_4^+) that displaces the CEC cations, any soluble salts, and lime or gypsum that may dissolve. Then the sample is saturated again with another cation solution (e.g., Mg^{++}) to remove the NH_4^+, which is measured as the CEC. Since the double-extraction method does not provide a measure of different cations on the CEC, it would not provide the full BCSR information. St. Johns and Christians (2007) note that this method would be useful in studies to determine the influence of various amendments to increase soil CEC such as inorganic sand substitutes.

Third, on soils with high soluble salts, there are *procedures to remove the soluble salts prior to cation extraction by prewash leaching with water or alcohol* (Tucker, 1985; So et al., 2006). This procedure can give a good estimate of total CEC, but the wash can affect percentage of Ca and Na on the CEC sites—with exchangeable Na considerably underestimated and Ca overestimated. Since the exchangeable Na is used as the ESP value to determine if a soil is becoming sodic, an error in the Na concentration would be serious.

Fourth is the *difference method proposed by the USSL* (1954) and recently evaluated by So et al. (2006). In this procedure, the total CEC is determined on one soil sample using a double-extraction procedure as noted above, but with only one saturation cycle of each extractant. The leachate from the first extraction would contain the soluble salts, any dissolved lime or gypsum, and cations from the CEC sites; the second extractant would accurately determine total CEC. So et al. (2006) recommended a second sample to be equilibrated at field capacity for 48 hours, and then the water is extracted by centrifuge and filtering, followed by determining the cations in the water extract. However, the USSL (1954) used the *SPE method to determine the cations* in the SPE extract, which would be easier for laboratories to do than the method of So et al. (2006).

The soil-exchangeable cations are then calculated as the difference in concentration between the total extractable cations in the first extractant leachate and the soil solution cations from the SPE extract. While this procedure has not been used by commercial soil test labs, *we strongly recommend it on salt-affected turfgrass sites with ECe > 4.0 dS/m* to (a) provide the highest accuracy in determining total CEC and exchangeable cations on the CEC sites for the BCSR approach, (b) give an accurate percent Na (ESP) value, and (c) allow the turfgrass manager to accurately monitor the soil conditions and make wise and valid best management decisions. With the use of halophytic grasses in recent years, it has not been unusual for soil ECe to be much higher than 4.0 dS/m on sites with these perennial grasses.

Because the summation method is the most cost-effective means to provide an estimated total CEC, percent saturation of each cation present on the CEC sites, percent base saturation, and ratios of various cations, this is the one normally used for the BCSR approach for fertilization decisions. However, care should be taken to insure that potential errors are minimized, especially since turfgrass managers may submit soil samples to university or commercial labs that are in different regions or countries. Knowledge of potential errors in CEC determination is useful. Previously, it was noted that errors in the CEC measurement can arise from inappropriate extractant pH, soluble salts in the sample, or dissolution of fine $CaCO_3$ or $CaSO_4$ by the extracting solution. Some additional considerations are as follows:

- Most labs use either $BaCl_2$ or NH_4OAc (ammonium acetate) buffered to pH 7.0 to obtain estimated CEC and percent cations on the CEC sites. This is acceptable for soils without free $CaCO_3$ or $CaSO_4$ that are near neutral in pH, and for soils at ECe < 4.0 dS/m. But, for soils with saline conditions and free $CaCO_3$ or $CaSO_4$, it will dissolve all the soluble salts and some of these lime and gypsum materials and overestimate total CEC and Ca on the CEC sites. Also, on acid soils, the pH 7.0 buffered extract (extractant is buffered to not change in pH to what the pH of the extracted soil is) will not accurately remove Al or Fe on the CEC sites; and it does not accurately measure the pH-dependent CEC prevalent on acid soils with the result of overestimating the CEC by raising the sample pH to 7.0 (Sumner and Miller, 1996; Carrow, Waddington, et al., 2001). Thus, when a lab is testing both near-neutral and acid soils using NH_4OAc at pH 7.0, it would be best to use an unbuffered solution so that the extractant pH can adjust to the soil pH.
- For nonsaline alkaline soils with free lime or gypsum, the use of NH_4OAc (ammonium acetate) buffered to pH 8.1 will minimize errors due to dissolution of $CaCO_3$ or $CaSO_4$, which can be considerable, but not in the presence of soluble salts (St. John and Christians, 2007).
- For acid soils, especially at pH < 6.0, either $BaCl_2$ or NH_4OAc (ammonium acetate) at pH 7.0 but *not* buffered will provide accurate data as long as soluble salts are at ECe < 4.0 dS/m, which is the normal case for acid soils except for acid-sulfate soils. The nonbuffered extractant allows the extractant to adjust to the actual soil pH and give a better estimate of CEC and exchangeable cations on these acid soils. Buffered solutions (i.e., maintained at pH 7.0 or another high pH level) in acid soils are less effective in the removal of Al^{+3} and Fe^{+3}, which are often present on the exchange sites. Also, pH-dependent CEC is more prevalent in acid soils, and a pH 7.0 buffered solution will overestimate actual CEC in the field. Thus, acidic soil CEC is most accurate if measured at the same pH as in the field instead of in a buffered solution. This overestimation of CEC by the use of a higher pH-buffered solution is particularly a problem on soils containing considerable amorphous clay colloids.

Considering the previous discussions on CEC, *a third soil test principle for saline, calcareous soil sites is for soil test laboratories to use the "difference method" proposed by the USSL (1954) and So et al. (2006)* (see Table 5.1). This option may cost more for the customer, but would insure accurate CEC, percent cations, and cation ratio information. If the site is nonsaline but calcareous, the NH_4OAc extractant buffered at pH 8.1 is accurate.

5.3.3 Interpretation of BCSR Data

The following discussion assumes that the BCSR data have been accurately obtained using the appropriate laboratory procedures. On most routine soil test reports from commercial labs, total CEC is reported along with the percent saturation level for each cation, with each cation ranked relative to the percent saturation of an "ideal" soil. Furthermore, two other items are generated from the percent saturation data and used for nutrient recommendations: (a) the total base cation (Ca, Mg, K, and Na) saturation, with the "ideal" base saturation as >80%; and (b) nutrient ratios, such as Ca/Mg, Mg/K, and Ca/K, reported. The cation ratio concept evolved from work in New Jersey in 1945 that suggested an ideal soil should have exchange cation levels of 65% to 85% Ca, 10% to 20% Mg, 2% to 7% K, 0% to 5% Na, and 0% to 5% H (Haby et al., 1990; Kopittke and Menzies, 2007). Using these percent values, the ideal ratios suggested were Ca/Mg of ~6.5:1, Ca/K of ~13:1, and Mg/K of ~2:1. Kopittke and Menzies (2007) provide an excellent review of the history of development of the BCSR concept.

The cation ratios as discussed in the previous paragraph are based on using the percent of Ca, Mg, and K on the CEC sites; and these are the ones that are reported in some soil test reports—or used to make nutrient recommendations. However, if plant-available nutrient levels in ppm (i.e., SLAN) of Ca, Mg, and K are used to determine cation ratios, the SLAN-based ratios will differ considerably from the BCSR ratios. The SLAN-based ratios really better reflect cation ratios than the BCSR ones since the SLAN ratios are based on actual soil levels of these nutrients in plant-available form. The BCSR ratios are very general. If the BCSR ratio suggests a potential problem, then view the SLAN-extractable levels of each nutrient and then the ratios based on SLAN data. Normally, making fertilization adjustments based on SLAN data will bring the SLAN and BCSR within good ranges for the particular soil—whenever within or outside the general guidelines.

The BCSR method has gained increased use in turfgrass ecosystems, especially from commercial labs, primarily because it tends to generate more recommendations for Ca, Mg, and K (Baird, 2007; Kopittke and Menzies, 2007). Kopittke and Menzies (2007, 260) noted a study in Nebraska on agronomic crops by Olson et al. (1982), where "over a period of 8 years, the cost of purchasing fertilizer according to the recommendations of the BCSR concept was generally double that of when the fertilizer was purchased based on SLAN recommendation" (p. 260). The authors of the study also noted, "Not to be overlooked are environmental implications nor the waste of energy and resources from any approach responsible for excessive fertilizer use" (Olson et al., 1982, 492). Recommendations may be made to bring a particular cation up to a certain percentage on the CEC sites, to adjust a particular ratio, or to raise the percent base saturation to 80 percent. For example, some consultants may recommend lime additions to maintain percent Ca on the CEC sites within the 65% to 85% range even when extractable Ca levels are in the upper part of the medium SLAN extractant range, which indicates adequate Ca. While most commercial laboratories report BCSR data and use the information for fertility decisions, university or state soil-testing laboratories (where the vast majority of the research on these methods has been conducted) do not.

Many researchers have questioned the usefulness and validity of the BCSR approach; as stated by Haby et al. (1990), "Numerous experiments over the past 40 years … have demonstrated that the use of the BCSR approach alone for making fertilizer recommendations is both scientifically and economically questionable" (p. 214). This statement is the result of several factors. Scientists have found that wide variations in percent CEC saturation for each cation and BCSR ratios occur and do not correlate well with plant response. In addition, little field evidence can be found for ideal cation ratios or for an ideal percent base saturation level of 80 percent. Furthermore, measurement of the *percent* of a cation of the CEC does not provide a measure of the total *quantity* available to a plant. For example, a golf green with 10 percent K saturation and a CEC of 1.0 meq/100 g will have far less total plant-available K than a soil with 10 meq/100 g of CEC but only 3 percent K saturation. Plants require a quantity of nutrients from the soil to provide their nutritional sufficiency needs, and percentages are very misleading.

In the context of salt-affected sites, the reason for discussing the SLAN versus BCSR methods is that soil fertility and plant nutrition are very critical issues on these sites. Fertilization decisions must be made on the best scientific information. It is more than an issue of cost on salt-affected sites, but an issue of maintaining good turfgrass performance in a system that is a very dynamic perennial ecosystem with high performance expectations.

Some people have even promoted BCSR as more accurate than the SLAN method on golf greens with high sand and low CEC for nutritional recommendations. Actually, the SLAN approach is more accurate, but SLAN readings for K are often very low because it is easily leached in such soils and very little is present, especially when Na is being added via irrigation water. However, trying to adjust a 2 percent K saturation to 6 percent K when the CEC is <3.0 meq/100 g would still result in insufficient K for plant growth and would be frustrating for the turfgrass manager to apply due to high leaching challenges. When nutrients are added based on the SLAN approach, percent saturation and the nutrient ratio automatically adjust to acceptable levels on all but the sandiest sites—where spoon-feeding or CEC enhancement by zeolites would be appropriate.

While not a good means to make fertilization decisions as to quantity of nutrient needed for the plant, *the BCSR information does provide useful information, such as the following:*

- *Tracking or monitoring.* Since the data are in percentages, it is a quick snapshot of changes made by fertilizer, lime, or gypsum applications. It is easy to track changes from one test to another by percent on the CEC. But the most accurate tracking is by comparing SLAN values over time. BCSR might be used as an initial indicator to check other actual nutrient data to see what concentrations are potentially available for turfgrass uptake.
- *ESP.* The percent Na on the CEC site when determined by an accurate method does provide a good measure of Na permeability hazard and whether a sodic soil condition is occurring. Since many soil test labs report BCSR information in their routine soil test report, this gives a good means of following Na influences and fluctuations. Thus, sodic soil development or alleviation can be tracked by the ESP.
- *Nutrient balance.* Even though the ranges of each cation can vary greatly without a nutrient deficiency occurring and the ratios can vary considerably without an imbalance, the ratios calculated by percent of cations on the CEC provide a quick view of potential imbalances. If an imbalance is suggested by a particular ratio being out of the expected range, attention to proper pH (if the soil is acid) and SLAN levels, followed by making nutrient adjustments by the SLAN data, will result in adequate nutrients quantities and their balances. A caution is that published ratios are based on the percent on the CEC sites; but if calculated by using extractable Ca, Mg, and K reported as meq/kg, these ratios can be substantially different.
- On acid sites, the percent Al on the CEC illustrates whether Al toxicity may be likely, especially at pH < 4.8, including acid-sulfate soils. At low concentrations (20–40 μM), Al can be beneficial to plant growth by impeding metal cation movement through channels in root cell plasma membranes, by forming insoluble precipitates with phosphate, or by inhibiting the rapid influx of potentially toxic P, Cu, or Zn concentrations internally in the plant (Hull, 2004). Percent H also increases on CEC as the soil becomes more acid.

5.4 WATER-EXTRACTABLE (SPE AND DILUTE) NUTRIENTS AND IONS

5.4.1 Water-Extractable Approach

Many commercial laboratories, but not state or university labs, have over the past 10 to 15 years reported water-extractable nutrients and selected ions as a third means of assessing soil nutritional status—the other two being the SLAN and the BCSR approaches. Since different labs may use SPE, 1:1, 1:2, or 1:5 water extraction methods, the actual value will depend on the dilution with the SPE giving the highest or most concentrated values. The SPE used to determine total soluble salts is also used to determine water-extractable Na, Ca, and Mg for the soil SAR calculation. Especially on salt-affected soils, the turfgrass manager faces serious soil fertility and plant nutrition challenges; and the very best soil tests should be used to make decisions. How do the water-extractable nutrients relate to the "gold standard" SLAN method for soil testing?

Macro- and micronutrient concentrations can be measured in the SPE or other more dilute pastes—but these will not accurately reflect the soil's nutrient-supplying power and are inferior to more traditional and powerful SLAN extractants developed to assess the soil's ability to supply plant-available nutrients over the growing season, such as acids, bicarbonates, and chelating agents in a variety of methods like Mehlich, Bray, and Olsen procedures. Water-based procedures, whether the SPE or dilute ones, remove little or none of the two most important plant-available nutrient pools, namely, nutrients on the CEC and those in the slowly soluble fractions. It requires stronger chemicals (acids, bicarbonates, and chelating agents) and sufficient extraction times to accurately assess these important fractions. Thus, SPE is *not* the best method for determining soil fertility levels and can, in fact, be very misleading. For most nutrients, <1% of plant-available nutrients are in the solution or soluble form at any one time for extraction by water; but what the turfgrass manager desires is information that predicts the future nutrient-supplying power of the soil (i.e., SLAN data do this quite accurately).

However, SPE-extractable levels of Cl, Na, SO_4, B, HCO_3, and Al can be of value. As discussed in the SLAN section (Section 5.2), water is often used as an extractant for SO_4, B, and sometimes Cl in routine soil tests; thus, SPE or dilute pastes of these nutrients provide similar results. Some comments on SPE or dilute paste results for these ions are as follows:

- *Bicarbonate and carbonate*: Although not a nutrient per se for turfgrass nutrition, bicarbonates and carbonates indirectly affect nutritional balances and actual nutrient availability in the soil. These compounds will complex Ca-P, Ca-Fe, as well as Ca individually, rendering these critical nutrients insoluble and unavailable for uptake by turfgrass roots. Both bicarbonates and carbonates will bind with sodium in soil profiles. High levels of these ions indicate high potential for reacting with Na to form sodium carbonate, which contributes to sodic conditions. Water extraction of the bicarbonates and carbonates will usually need to be requested when sending in calcareous soil samples to the laboratory for analysis.
- *Chloride (Cl)*: If not included in the routine soil test report, the Cl levels in the SPE or dilute paste extract aids in determining potential for root or foliage toxicity and for assessing the effectiveness of leaching programs (chlorides are the most easily leachable ions).
- *Sodium (Na)*: Normally given in the routine soil test report based on the same extractant used for SLAN K, Ca, and Mg; but if not, then the water-based levels give indications of Na soil solution levels. Some true C-4 halophytes may utilize Na at 1–2 μM concentrations as a nutrient.
- *Aluminum (Al)*: If high, then P can be easily tied up as Al-P compounds. In acid soils, Al may be high as well as percent Al on CEC sites. As pH decreases from 5.5, the potential for Al toxicity to plant roots increases.

5.4.2 Comparison of SLAN versus Water-Extractable Nutrients

In 1998, Steve Davis, in cooperation with Larry Stowell (PACE Turf, San Diego, California) and R. N. Carrow, organized a study on 24 golf courses in the Midwest from Chicago to St. Louis, where soil samples were obtained from sites the golf course superintendent considered the best and worse greens, and best and worst fairways (Carrow et al., 2003). With Dr. Stowell organizing the soil testing, the 96 soil samples were split, with analyses using Mehlich III extractant for half of the sample (as representative of a traditional, more vigorous acid-based extractant) and the other half of the sample extracted with the SPE water-based procedure. Table 5.7 provides the extractable levels of each nutrient for both extractants and the linear correlation information for comparing the methods. Some comments relating to the data in Table 5.7:

- Mehlich III extracted much higher quantities of plant-available nutrients than SPE since it is a double acid with a chelating agent versus water. Na, K, B, and SO_4 levels are nutrients or elements where the absolute quantities by both Mehlich III and SPE were most similar. This is because, except for any Na or K on the CEC sites, the other common Na and K compounds in soil are primarily water soluble and therefore more extractable by water. Much of the S that is plant available also resides in solution or in soluble compounds that are water extractable.
- Sometimes a significant relationship (F-test is <0.05) can be observed between the two extractants, but it was not a strong relationship (i.e., there is a statistical relationship, but it is weak and not closely correlated). A strong relationship between Mehlich III and SPE-extractable nutrient would require an $r^2 > 0.80$, but only Na exhibited such a strong relationship.
- The high CV values from the split soil samples are another indication of the poor relationship between plant-available nutrients by Mehlich III extraction compared to the water-extracted nutrients by the SPE. A good CV would be <10%, which indicates a good relationship.

TABLE 5.7
Soil-Extractable Nutrients Using Mehlich III (0.015 M NH_4F + 0.20 M CH_3COOH + 0.25 M NH_4NO_3 + 0.013 M HNO_3 + 0.0005 M EDTA Chelating Agent) Compared to Saturated Paste Extract (SPE) Water Extraction from Split Soil Samples from 96 Golf Courses

	Green		Fairways		Coefficient of Determination	
	Mehlich III	SPE	Mehlich III	SPE	(r^2)[a]	CV[b]
Parameter	ppm		ppm		1.0 = ideal	%
P	106	1.5	71	0.5	<1	59
Ca	2716	38	3771	103	.35**	29
Mg	343	13	907	36	.52**	38
K	170	37	318	19	<1**	50
Na	39	22	108	42	.86**	34
SO_4	62	38	215	118	.61**	39
B	1.0	0.10	1.5	0.13	.51**	34
Fe	248	0.77	247	0.71	<1	42
Mn	33	0.03	67	.02	<1*	94
Cu	4.4	0.03	5.6	.02	<1	56

[a] Coefficient of determination (r^2)—a measure of the strength of the correlation, where a 1:1 relationship (ideal) would have $r^2 = 1.00$. An acceptable r^2 for a good, strong correlation should be $r^2 > 0.80$.

[b] Coefficient of variability—a measure of scatter of the data. Ideal is for CV to be <10%.

*, **Significant probability of F-test at 5% and 1% probability levels, respectively.

TABLE 5.8
Soil Test Report Information

Soil Analysis	Typical Units	Comments
1. Soil pH[a]	pH units, 1–14	Soil pH is a measure of H^+ activity in the soil solution. Usually determined in a 1:1 (soil:water) slurry.
2. Buffer pH[a]	pH units	Buffer pH is used to determine the soil lime requirement recommendation. The lower the buffer pH, the higher the lime requirement. Used only on soil with soil pH < 7.0.
3. Organic Matter	%	Percentage of organic matter in the soil sample.
4. Extractable (SLAN) Nutrients and Ions[c]	ppm; mg/kg; pounds per acre available nutrient[b]	Sufficiency levels of available nutrients (SLAN) = quantity of soil nutrients available to plants.
P	—	Often, extractable P may be reported using two extractants such as the Bray P_1 and Olsen methods.
K	—	—
Ca	—	—
Mg	—	—
Na	—	Although Na is beneficial for halophytes, the primary reason to report SLAN-extractable Na (the same extractants used for K, Ca, and Mg) is to determine the level of Na that can impart total soluble salts, the Na permeability hazard, and Na as a toxic ion to the roots and foliage of some plants, and to assess the effectiveness of Na-leaching programs. The Na reported here is from soil solution + CEC + soluble Na compounds precipitated in soil (Na-carbonates, Na-sulfates, and Na-chloride).
Fe	—	—
Mn	—	—
Zn	—	—
Cu	—	—
B	—	Water-extractable B is also used.
$S(SO_4^-)$	—	Water-extractable SO_4^- is also used.
$N(NO_3^-)$	—	May not be reported. Accurate but not predictive of soil-available N except possibly nonirrigated, arid soils.
5. Percent CEC Saturation[e]	%	Basic cation saturation ratio (BCSR) information on percent of each cation on the CEC sites. Ammonium acetate is a common extractant used. See Section 5.3.2.
% Ca	—	Typically 65–85%
% Mg	—	Typically 10–20%
% K	—	Typically 2–7%
% Na	—	Typically 0–1%; % Na = ESP;[d] desire <3%
% Al	—	Typically 0–5%; present on highly acid soils
% H	—	Typically 0–5%; increases in acid soils
6. Cation Exchange Capacity (or Total Exchange Capacity)[a,e]	meg./100 g soil; cmol/kg	CEC < 2.5 meg/100 g presents challenges in stabilizing nutritional programs, especially on salt-affected sites. Must be accurately determined (see Section 5.3.2).
7. Cation Ratios[e,f]		
Ca : Mg	—	<6.5 suggested, but this is very general. See Section 5.3.3 for means of calculation cation ratios by BCSR or SLAN data.
Ca : K	—	<13:1. See Ca : Mg comments.
Mg : K	—	<2:1. See Ca : Mg comments.

TABLE 5.8 (Continued)
Soil Test Report Information

Note: Example of information commonly included in routine turfgrass soil test reports. Some laboratories include salinity information and saturated paste extract (SPE) or dilute paste soil tests information; both are summarized in Table 5.9.

[a] For a discussion of pH, buffer pH, and CEC concepts, see Carrow, R. N., D. V. Waddington, and P. E. Rieke. 2001. *Turfgrass Soil Fertility and Chemical Problems: Assessment and Management*. Hoboken, NJ: John Wiley.

[b] ppm = 1 mg/kg = (pounds per acres) / 2.

[c] See Table 5.2 for P extractants; Table 5.3 for extractants used for Ca, Mg, K, Na, $S(SO_4)$, Fe, Mn, Cu, and Zn; and Tables 5.4 and 5.5 for typical "medium" levels of extractable nutrients by various extractants.

[d] Exchangeable sodium percentage (ESP) = percentage of Na saturation of CEC. Sodic soil has ESP > 15.0%, but an ESP > 3.0% to 4.0% indicates the potential for surface Na-induced permeability problems.

[e] These parameters are considered BCSR-based soil test information.

[f] These cation ratios are based on percent cations on CEC sites.

This study demonstrates that SPE-extractable levels of nutrients differ substantially from the traditional Mehlich III extractant and that using SPE data for nutrient recommendations is not a science-based approach—except for Na, B, and SO_4, and for providing a very rough approximation of K, primarily on sandy soils. St. John and Christians (2007) reported similar results. If 1:2, 1:5, or other dilute soil:water ratios had been used, similar results would be expected, but with lower water extraction levels relative to the SPE.

Some have suggested that the SPE extraction would provide a better estimate of nutrient status on low-CEC sands than traditional extractants. We would disagree based on two problems with this approach. First, even on low-CEC sands, the nutrients on the CEC are much greater than what is actually found in soil solution or that is readily soluble at any point in time, with SPE-extractable levels often <1% of what would be extracted by a Mehlich III procedure. If salts in the soil solution and soluble fractions that water extractants remove were allowed to build up as soluble salts, the turfgrass would be exposed to a salt stress—that is the reason we measure ECe and leach any excess salts. Second, the soil solution and soluble precipitated salts can be readily leached by rain or irrigation applications, and are, therefore, no longer plant available. Thus, the water-based extractions are not predictive of soil nutritional status for an entire growing season compared to the more powerful extractants.

5.5 UNDERSTANDING THE SOIL TEST REPORT

In Tables 5.8 and 5.9 are listed the various soil test information that may be included in a routine soil test report (Table 5.8) and combined salinity package and SPE or dilute paste soil test report (Table 5.9). Sections of the report in Table 5.8 include (a) soil nutrient status based on the use of SLAN chemical extractants that provide information on the quantity of specific nutrients that are available to plants over the season, and (b) information based on the BCSR approach. Sections in Table 5.9 contain information that may be in a soil salinity test report as well as SPE or dilute paste tests. Which items appear in a particular soil test report depends on the soil tests requested and what is included in the laboratories' soil test packages. University and state soil test laboratories concentrate on the necessary information to make wise soil fertility decisions, while commercial laboratories tend to add more items to their reports—but not all the information may be useful or used in the proper manner. For example, the inclusion of water-extractable nutrients (P, K, Ca, Mg, Fe, Zn, Cu, and Mn) is questionable as to its value for making fertilization decisions.

In summary, routine soil-testing information is important on all turfgrass sites; but for salt-affected sites, it is especially critical due to the opportunities for nutrient imbalances, leaching, and

TABLE 5.9
Soil Test Reports with the Typical Salinity Soil Test Package Information

Soil Analysis	Typical Units	Comments
Saturated Paste Extract Parameters		Some laboratories may not use the SPE but a 1:1, 1:2, or 1:5 dilute soil:water slurry. The SPE is strongly preferred. See Section 5.4 for a discussion of water-extractable nutrients or ions.
Total Soluble Salts		
ECe	dS/m, mmhos/cm	SPE-based soil EC (i.e., ECe)
TDS	ppm, mg/kg	TDS = total dissolved salts
SAR	meq/l	Sodium adsorption ratio based on SPE-extractable (dilute pastes are not accurate) levels of Na, Ca, and Mg. Used to determine soil Na permeability hazard. A soil is determined sodic at SAR > 12 = ESP of 15%.
Chloride	ppm, mg/kg	Chloride is one of the major salts, and SPE water-extractable Cl is a good indicator of soil Cl levels.
Sodium	ppm, mg/kg	SPE water-extractable Na is not as accurate an indicator as Na determined by a SLAN extractable (see extractable Na in Table 5.8) since less Na is removed by water compared to the SLAN extractants, which can determine CEC Na in addition to any soluble Na present.
B, SO_4	ppm, mg/kg	Water (SPE method) is an acceptable extractant for these nutrients to predict the soil-supplying capability of a season.
Bicarbonate, Carbonate	ppm, mg/kg	High SPE water-extractable levels are normally associated with the presence of Na carbonates, which are common in arid regions.
Other Nutrients (P, K, Ca, Mg, Fe, Mn, Cu, Zn)	ppm, mg/kg	Water-extractable levels of these nutrients are of little value in making nutrient recommendations and can lead to inaccurate recommendations. Use SLAN levels, not SPE levels.
Free $CaCO_3$	% weight	Free $CaCO_3$ in the soil can be reacted with S or SO_4 to create gypsum.

Note: Saturated paste extract (SPE) or dilute paste items often include the items listed in this table, but only the SPE method is reliable for those indicated in the table.

rapid changes with time. Only with proper BMP soil test approaches, chemical extractants, and procedures will the soil test results be meaningful, especially when the soil may be saline and/or calcareous. The suggestions made in this chapter may require changes in practices by soil test laboratories, which would likely be more costly and time consuming. However, the price of soil testing is minimal when compared to the benefits on salt-affected sites where chemical constituents, soil chemical amendments, water amendments, and leaching programs' effects on soil fertility must be balanced by correct BMP fertilization decisions. Current turfgrass performance expectations dictate the need for these changes when faced with ecosystem salt challenges.

6 Irrigation Water Quality Tests, Interpretation, and Selection

CONTENTS

6.1 WATER QUALITY TESTING

6.1.1 Importance of Testing

In Chapter 6, the focus is on the chemical constituents that are typically requested by turfgrass managers in a *routine irrigation water analyses* as well as metals that are sometimes analyzed. At times, other biological, physical, and chemical constituents in water that are not included in a routine irrigation water quality analysis may need to be assessed for specific water quality problems—whether in a treatment facility, a well, a lake, irrigation lines, or a canal, or after application on a site. Readers are referred to the book by Duncan et al. (2009) titled *Turfgrass and Landscape Irrigation Water Quality: Assessment and Management*, which provides an extensive view of turfgrass and landscape irrigation water quality problems that may confront the turfgrass manager within the whole spectrum of water irrigation movement—from the initial source location, onsite storage, delivery system, turfgrass plant, soil, and underlying hydrology.

In a routine irrigation water quality test, the basic salt ions that are dissolved in water and their relative concentrations (Table 6.1) have special importance, since they strongly influence potential salinity problems. When discussing various water constituents, problems associated with the particular parameter will be summarized; but preventative and corrective measures to address the

TABLE 6.1
Soluble Salt Ions Common in Irrigation Water

Cations	Anions
Calcium (Ca^{+2})	Bicarbonate (HCO_3^-)
Magnesium (Mg^{+2})	Carbonate (CO_3^{-2})
Sodium (Na^+)	Chloride (Cl^-)
Potassium (K^+)	Sulfate (SO_4^{-2})
Hydrogen (H^+)	Nitrate (NO_3^-)
	Boron (BO_3^{-2})

problem will be dealt with in various chapters related to management of site-specific salinity situations related to the parameter.

Irrigation water quality assessment is one of the more confusing and complex problems facing turfgrass managers, especially with saline or reclaimed irrigation waters. Variable distribution of salts by irrigation systems, seasonal changes within some irrigation sources, differences in soil physical properties, coupled with extreme environmental conditions (high prolonged heat and humidity, severe drought, and traffic) magnify poor water quality problems by contributing to considerable spatial and temporal variability in soil salinity. Water analysis reports often report data in confusing units or with no reference points. Also, some of the terms can be confusing such as adj SAR, adj RNa, RSC, alkalinity, specific toxic ions, and so on. However, knowledge of water quality parameters and the ability to access information in a water quality report are essential, especially for turfgrass managers using reclaimed irrigation water or salt-affected sources.

Understanding the types and quantities of chemicals that are applied to the turfgrass system through irrigation water is critical because these have a dramatic influence on soil health, surface and subsurface waters, irrigation distribution systems, and sustained turfgrass performance. Additionally, many management and economic decisions will be based on water quality aspects. A few examples illustrate the implications of water quality in the decision process:

- Water treatment equipment and amendments depend on the specific levels and balance of Na, Ca, Mg, SO_4, HCO_3, and CO_3.
- Several soil amendment types and quantities are determined by levels and balances of water quality constituents. Included would be decisions on equipment to apply amendments.
- Fertilization becomes very challenging, especially on low cation exchange capacity (CEC) soils, when large quantities of nutrients and competitive ions are added with each irrigation cycle. Soil and tissue testing to proactively monitor soil fertility and plant nutrient status become more frequent. If salts are leached, nutrients will also be leached.
- With saline irrigation water, leaching of excess salts is necessary. Leaching requires a well-designed irrigation system for uniform water distribution, it needs good irrigation scheduling and irrigation-scheduling tools, often a good drainage system is necessary, and cultivation is frequently more common (thereby requiring acquisition of the correct cultivation equipment).

These examples demonstrate that costly infrastructure improvements are often required as irrigation water declines in quality. A lack of understanding of water quality test parameters can translate into costly mistakes in both management and infrastructure decisions.

The challenge is to know what key water quality components to look for when you receive laboratory analysis data, to utilize key data points of concern to make initial best management decisions (BMPs) *based on science*, and to take a *holistic approach to management*, realizing that salinity challenges are dynamic and not static. Turfgrass managers must be flexible in making and implementing decision plans and not hesitate to ask questions. As water quality decreases, short-term management strategies will be reactive, adjusting to environmentally induced changes in turfgrass density, cosmetic appearance, playability, or pest infestation. Long-term management strategies based on water quality information, however, should be proactive, utilizing regularly scheduled activities, such as cultivation (aeration), application of amendments (such as gypsum), irrigation scheduling, and continuous monitoring (water, soil, and tissue) of the entire turfgrass ecosystem.

6.1.2 Units and Conversions

Water quality data are reported in various chemical units (Table 6.2). There is no "standard" requirement for laboratories to report water analyses data in particular units, so it is important to understand common conversions, relationships, and terminology (Table 6.3). The equivalent weight (meq/L) unit is especially useful since 1 meq/L of one ion is chemically equivalent to 1 meq/L of another ion.

6.1.3 Routine Irrigation Water Quality Report Information

Table 6.4 notes the chemical constituents that are typically found in a routine irrigation water quality report. Not all laboratories use the same format or categorization of constituents that are used in Table 6.4. A number of agronomic challenges can occur due to irrigation water chemicals, including salinity-related problems. The most important problems addressed in a water quality report and the constituents used to determine the magnitude of the problems are summarized below—each of these will be discussed in more detail later in the chapter:

- *General water quality characteristics*: Water pH, alkalinity, hardness, HCO_3, and CO_3.
- *Total soluble salts*: The potential for salinity stress on plants and salt accumulation in soils. ECw, TDS.
- *Sodium permeability hazard*: The assessment of whether Na level, in balance or imbalance with Ca, Mg, HCO_3, and CO_3, will cause water infiltration and percolation problems (i.e., soil permeability problems). Sodium can cause deterioration of soil structure by aggregate slaking and colloidal dispersion of clays. No single water quality parameter can be used alone to determine the magnitude of the Na permeability problem. Instead, a combination

TABLE 6.2
Units of Measure Used in Water Quality Testing

Unit	Comments
ppm	Parts per million
mg/L	Milligrams per liter
mmhos/cm	Millimhos per centimeter
Mmhos/cm	Micromhos per centimeter
dS/m	Decisiemens per meter
meq/L	Milliequivalents per liter. An *equivalent weight* of an element or radical is its atomic or formula weight in grams divided by the valence (charge) it assumes in compounds. Thus, for Ca^{+2} with an atomic weight of 40 gms, the equivalent weight is 40/2 = 20 gms. Equivalent weights are important when considering how much of one element or radical will react or displace another; for example how much Ca^{+2} is necessary to displace Na^{+1} (1 milliequivalent of Ca^{+2} = 1 milliequivalent of Na^{+})? Thus, it is a measure of chemical equivalency.

Note: See Table 6.3 for additional information on relations of mg/L or ppm versus meq/L, and for conversion factors.

TABLE 6.3
Conversion Factors Important in Water Quality Analysis

General

1 ppm = 1 mg/L
1 meq/L = 1 $mmol_c$/L
1 dS/m = 0.1 S/m = 1 mS/cm = 1000 μS/cm = 100 mS/m
1 dS/m = 1 mmhos/cm = 1000 Mmhos/cm
1 dS/m = 640 ppm = 640 mg/L
1% concentration = 10,000 ppm = 10,000 mg/L = 15.6 dS/m
TDS (ppm) = ECw × 640; with ECw in dS/m
ECw (dS/m) = TDS (ppm) divided by 640 = TDS/640. Use 640 conversion when ECw is <5.0 dS/m and 750 when ECw is >5.0 dS/m.
Convert mM of a salt to ppm. To convert mM to ppm, multiply the concentration in mM by the molecular weight of the solute. For 1mM $CaCl_2$, it works out to 110.9 ppm. The expanded math is below:
1 mM $CaCl_2$ = 0.001 mol/L
Formula wt = 110.9 g/mol
0.001 mol/L × 110.9 g/mol = 0.1109 g = 110.9 mg
Since mg/L = ppm, 110.9 mg/L = 110.9 ppm = 1 mM $CaCl_2$
For 100 mM NaCl solution in 1 L = 5844 mg/L = 5844 ppm = 7.8 dS/m

Chemical Conversions

Element or Radical	Charge or Valence	Equivalent Weight	Atomic or Formula Weight
Ca^{+2}	2	20	40
Mg^{+2}	2	12.2	24.4
Na^{+1}	1	23	23
K^{+1}	1	39	39
Cl^{+1}	1	35.4	35.4
HCO_3^-	1	61	61
CO_3^{-2}	2	30	60
SO_4^{-2}	2	48	96
$CaCO_3$	—	—	100
$CaSO_4 \cdot 2H_2O$	—	—	172
H_2SO_4	—	—	98
C	—	—	12
O	—	—	16
H^+	1	1	1
S	—	—	32

of the following parameters is used: SAR, adj SAR or adj RNa, RSC, and concentrations of HCO_3, CO_3, Ca, Na, Mg, pHc, ECw, or TDS.

- *Potential surface or subsurface soil sealing or layering by calcite formation*: Irrigation waters very high in ions that form calcites may result in calcite formation near the surface over time—this is less of a problem or concern than the effects of Na on soil physical problems. However, accumulation of Na-carbonate in arid regions hinders reclamation of sodic soils. Parameters of concern are Ca, Mg, HCO_3, CO_3, RSC, and pHc.
- *Precipitation of calcite (lime) in irrigation lines*: pHc and Saturation Index.
- *Toxic ion concentrations relative to soil accumulation or root toxicity and excess uptake within foliage*: Na, Cl, B, and metal ions.
- *Toxic ion concentrations relative to direct contact on foliage as irrigation spray*: Na and Cl.
- *Unsightly deposition on foliage, cart paths, and equipment*: HCO_3.

TABLE 6.4
Chemical Constituents in Typical Routine Irrigation Water Analyses for Turfgrass Situations

Water Analysis	Typical Units	Comments
General Water Characteristics		
pH	pH units, 1–14	Very high or low pH is a warning of possible problems.
Hardness	ppm, mg/L	Relates to Ca and Mg content and potential for scaling in pipes.
Alkalinity	ppm, mg/L	Measure of acid-neutralizing capacity of water. Reflects bicarbonate, carbonate, and OH (hydroxide) content.
Bicarbonate, HCO_3	ppm, mg/L, meq/L	Affects Ca and Mg precipitation from water.
Carbonate, CO_3	ppm, mg/L, meq/L	Affects Ca and Mg precipitation from water.
Assessment of Total Soluble Salts (Salinity)		
Electrical conductivity (ECw)	dS/m	Relates to potential for soluble salt stress on plant growth.
Total dissolved salts (TDS)	ppm, mg/L	Relates to potential for soluble salt stress on plant growth.
Impact on Soil Structure and Water Infiltration (Na Permeability Hazard)		
Sodium adsorption ratio (SAR, adj RNa, adj SAR)	meq/L	Measure of potential for adverse effects of Na on soil structure.
Residual sodium carbonate (RSC)	meq/L	Measure of potential for Ca and Mg to precipitate from irrigation water by reaction with bicarbonates or carbonates.
ECw and TDS	See above	These may also be listed in this section since they influence SAR.
Specific Ion Impact on Root Injury or Foliage Uptake and Injury		
Na	ppm, mg/L, meq/L	Na often injures plant roots.
Cl	ppm, mg/L	Cl often accumulates in foliage.
B	ppm, mg/L	Can cause root toxicity or shoot injury.
Specific Ion Impact on Foliage from Spray Contact		
Na	ppm, mg/L	Injury to sensitive plant foliage.
Cl	ppm, mg/L	Injury to sensitive plant foliage.
HCO_3	ppm, mg/L	Does not cause injury but can be unsightly on foliage.
Nutrients and Elements Normally Reported		
Nitrogen as total N, NO_3	ppm, mg/L	Nitrogen and the other nutrients contribute to plant nutritional needs. These should be considered as part of the "fertilizer" requirements
Phosphorus as total P, PO_4, P_2O_5	ppm, mg/L	
Potassium as total K, K_2O	ppm, mg/L	
Calcium as Ca	ppm, mg/L, meq/L	
Magnesium as Mg	ppm, mg/L, meq/L	
Sulfate as SO_4 or S	ppm, mg/L	Sulfate value also used to assess black layer potential.
Magnanese as Mn	ppm, mg/L	
Iron as Fe	ppm, mg/L	
Copper as Cu	ppm, mg/L	
Zinc as Zn	ppm, mg/L	
Boron as B	ppm, mg/L	
Sodium as Na	ppm, mg/L	
Chloride as Cl	ppm, mg/L	
Metal Ions		
Various metal ions are sometimes analyzed in irrigation water if a problem is expected or for a new irrigation source. Ions are normally reported in ppm or mg/L and include Al, As, Be, Cd, Co, Cr, Cu, F, Li, Mo, Ni, Pb, Se, Sn, Ti, W, V, and Zn.		
Miscellaneous		
Residual Cl_2	ppm, mg/L	Excessive chlorine from water treatment.
Total suspended solids (TSS)	ppm, mg/L	Suspended solids that are organic or inorganic in nature.
pH_c	pH units	Indicates potential for lime precipitation.

- *Nutrient content and influence*: The influence of irrigation water on soil fertility, plant nutrition, and potential nutrient imbalance problems. Macronutrients (N, P, K, Ca, Mg, and S), micronutrients (Fe, Mn, Cu, Zn, Mo, B, Ni, and Cl), Na, Si, metals, and ratios of various nutrients or elements.
- *Eutrophication impact on lakes*: P and N.
- *Potential for black layer formation*: SO_4, Na, Fe, and Mn.
- *Total suspended solids (TSS)*: Potential to add fines (suspended clay, silt, organic particles) to the soil surface and possibility of clogging irrigation system components.
- *Miscellaneous problems*: Residual Cl_2 or constituents are noted in Table 6.4. These are not commonly included by most lab analyses, but may be available by special request. For reclaimed water sources, the constituents in Table 6.4 may be determined at the treatment facility or sometimes onsite, but an agricultural assessment is needed for turfgrass ecosystems rather than the biological assessments required by most local, state, federal, and/or country rules and regulations, which are directed to human safety.

While all the above parameters are potential irrigation water quality problems that can be assessed with water quality data, four problem areas are considered the *big four salinity stresses.* These will be discussed in greater detail later in the chapter, but at this point it is essential to have a basic understanding of each problem. Each of these four critical problem categories must be assessed using water analysis data just as each is assessed in routine soil testing combined with special salt-related soil tests (see Chapters 4 and 5). Each category is a "salt problem" but differs from the other three problem areas in activating unique chemical effects on soil parameters that will eventually impact long-term turfgrass performance.

First, and most important, is *high total salinity (total soluble salts)* in irrigation water, which reflects the potential for a saline soil problem to develop, and for salt contamination of ground and surface waters, and affects plant selection on the site (see Chapter 2 for an overview of saline soils). Saline conditions inhibit water uptake by turfgrasses and cause salt-induced drought, which is called *physiological drought stress.* This is the most common and widespread salt-related water issue that occurs on turfgrasses, and it is the primary one that must be managed on most salt-affected turfgrass sites. Total salinity problems are site specific and must be assessed on that basis, and management strategies involving grass selection, cultivation (aeration), and irrigation scheduling for leaching and avoiding drought stress must be developed accordingly.

The second "salinity problem" is *sodium permeability hazard.* High sodium concentrations, especially in conjunction with high bicarbonates and relatively low Ca^{+2} and Mg^{+2} levels in irrigation water, can cause a sodium permeability hazard. The term "permeability hazard" is used because the most important effect of excess Na on soil CEC sites is to reduce water infiltration and percolation (see Chapter 3 for a review of sodic soils). Sodium induces soil structural deterioration (slaking, aggregate destruction, and clay and organic colloid dispersion), leading to subsequent water infiltration and percolation problems, low oxygen diffusion in the soil profile, and poor turfgrass rooting. When the soil is very high in sand content (>95% sand), Na does not cause structural deterioration in terms of breaking down aggregates but can cause any colloidal particles to be dispersed, often migrating downward in the soil profile with the wetting front during each irrigation cycle—thus, Ca amendment application rates can be reduced on these soils compared to the rates needed for fine-textured soils. Assessment and management strategies must (a) be based on site-specific soil and water conditions; and (b) be aggressively monitored and frequently adjusted to address specific constraints involving possible grass selection, amendments to the water and/or the soil, regular cultivation events, and careful irrigation scheduling (leaching).

The third salinity problem is the presence of *specific toxic or problem ions* that may (a) be direct root toxins as they accumulate or be taken up at high levels within plants to cause leaf injury and foliage salt stress (Na, Cl, and B), (b) include excess ions that can cause direct damage to foliage from spray contact (Na and Cl), (c) leave unsightly deposits (HCO_3), or (d) contribute to black layer development (SO_4). Turfgrass and landscape plant selection is critical for addressing the specific ion toxicity problems.

The fourth type of salinity stress is a combination of high-level *fertilizer additions and nutritional imbalances* that are induced by nutrients and elements in the irrigation water. Most of the imbalances are triggered by exceptionally high levels of one nutrient or element suppressing uptake of another or displacing a nutrient from the soil CEC sites. The impact of chemicals from water that are added to the soil from good-quality irrigation water is often ignored; but with a poorer quality source, chemical additions arise for the water itself, associated specific water treatments, and soil amendment additions. Coupled with the need for active leaching programs and often variable water quality over the years, these factors result in very dynamic changes over time in fertility and chemical stresses. Lack of attention and monitoring for these nutrient availability changes can result in the rapid onset of nutritional stresses and reduced turfgrass performance.

6.2 GENERAL WATER QUALITY CHARACTERISTICS

6.2.1 WATER PH

Water pH is a measure of water acidity or alkalinity with typical irrigation water within the pH range of 6.5 to 8.4 for natural waters. The pH scale goes from 0 to 14, where water at pH of 7.0 is neutral and the H^+ equals the OH^- ion concentration; while at pH < 7.0, water is acidic with H^+ dominant; and at pH > 7.0, OH^- ions dominate and the water is alkaline. Irrigation water pH is not usually a problem by itself, but abnormal pH can be an indicator of possible concerns (Ayers and Westcot, 1994; Yiasoumi et al., 2005). Irrigation pH can influence soil pH over time, cause corrosion of irrigation equipment and concrete channels, and alter pesticide efficacy and efficiency. Some possible problems are as follows:

- Low-salinity water (ECw < 0.2 dS/m or TDS < 128 ppm) can sometimes exhibit pH outside the normal range due to very low buffering, but not really be a problem. However, these ultrapure water sources can result in soil permeability problems, especially if calcium concentrations are <20 mm. For long-term use, ideally the irrigation water should have ECw ≥ 0.5 dS/m or TDS ≥ 320 ppm for best long-term soil infiltration and percolation.
- Irrigation water with pH > 8.0 often contains bicarbonates, while at pH > 9.0 carbonate and hydroxyl (OH^-) ions may be present.
- Because irrigation water that is alkaline often contains Ca, Mg, Na and, HCO_3, lime or Na-carbonate formation can occur, especially at high levels of these constituents in irrigation lines (scaling, especially in low-pressure drip systems) or in the soil. Soils with free lime from irrigation water can usually exhibit a pH between 7.3 and 8.2. Thus, alkaline pH water applied to an acid or neutral soil may cause pH to increase within this range. Precipitation of Ca and Mg from irrigation water by HCO_3 and CO_3 is especially critical if appreciable Na remains to accumulate and adversely affect soil structure.
- Water pH above 7.5 may adversely affect chlorine disinfection.
- When the irrigation water is used for chemical application, the effectiveness of some pesticides is reduced at pH outside the 6.0 to 8.5 range.
- Increasing acidity below pH 6.5 can be corrosive on metal components as well as concrete irrigation canal linings.

Water pH can also impact nutrient availability since it can influence soil and thatch or mat pH overtime, especially in the surface (Carrow, Waddington, et al., 2001). Extremely acid (pH < 5.0) or highly alkaline (particularly pH > 8.5) irrigation water can affect microbe populations, microbial breakdown of granular fertilizers and thatch or organic matter accumulation, and utilization of foliar-applied liquid sources when irrigation is activated shortly after spraying. Occasionally, very highly alkaline water, such as pH 9.0, being applied during or after fertilizer applications may tie up nutrients in microlayers near the soil surface where the turfgrass would have difficulty in uptake.

Sometimes groundwater pH can be very acidic if affected by acid mine tailings or acid-sulfate soil conditions. Acid mine drainage often has low pH, high Fe, Mn, Al, other metals, and SO_4, while acid-sulfate soil-affected waters will often exhibit low pH, high Fe, Mn, Al, other metals, SO_4, and Na. In Australia, rising saline water tables have resulted in widespread acid-sulfate soil formation in areas where the underlying minerals or sediments contain sulfide (Fitzpatrick, 1999). Thus, if these soil conditions have influenced the groundwater, caused acidic runoff, or come into contact with an irrigation lake, these constituents can be harmful to aquatic life as well as plants and soils that are irrigated with the water.

Water in lakes generally is sufficiently buffered to prevent major or rapid pH changes (see Section 6.2.2). However, small pH changes can occur daily, seasonally, and by depth in response to the balance of photosynthesis using dissolved CO_2 (which acts as carbonic acid, H_2CO_3) and respiration, which produces CO_2 (WOB, 2008). More dramatic pH changes can occur from pollution, as was previously noted. Also, if acid rain reduces the buffering capacity by removal of bicarbonate or carbonate, then more rapid pH changes can occur with only minor additions of more acidity. Acidity at pH < 6.0 can be toxic to fish and alter other aquatic relationships. One effect of more acid conditions is to enhance the dissolution of P and heavy metals in lake sediments.

6.2.2 Alkalinity, Bicarbonate, and Carbonate

Alkalinity is a measure of the water's ability to absorb H ions without significantly altering pH (i.e., the ability of the water to neutralize acids). The major constituents in water that provide buffer capacity against pH change are HCO_3^-, CO_3^{-2}, and OH^- ions, where each can react with excess H^+ ions (i.e., remove H^+ ions) to create CO_2 gas or water molecules. High alkalinity is an indication of the presence of these ions. Laboratories will express alkalinity as calcium carbonate equivalent (mg/L or ppm $CaCO_3$) or meq/L $CaCO_3$, where 1.0 meq/L $CaCO_3$ = 50.04 mg/L $CaCO_3$. Normal irrigation waters are between 20 to 300 mg/L of $CaCO_3$ equivalent.

Alkalinity is often included in irrigation water analysis reports but generally is not directly used in management decisions, except by the greenhouse or nursery industries to determine the need to acidify water. Instead, most turfgrass managers use residual sodium carbonate (RSC; see Section 6.4.2) data since they take into account the balance of HCO_3 and CO_3 versus Ca and Mg as a measure of the potential to precipitate out Ca and Mg from irrigation water either in lines or in the soil, leading to the domination of CEC sites by Na. Precipitation of Ca and Mg is important in terms of scale formation in irrigation lines and sodic soil formation.

In lakes or ponds, bicarbonate and carbonate can be replenished in water from Ca, Mg, K, and Na carbonates in the sediments, unless these become depleted via acid rain or other long-term acid additions to the site. Thus, alkalinity may be used to determine the buffering capacity of lakes or ponds where water with low alkalinity (soft water with <30 mg/L $CaCO_3$ equivalent) contains few basic ions, has a low buffering capacity to resist pH fluctuations, and is more susceptible to acidification. Hard waters usually have high alkalinities (>100 mg/L), many basic ions, and a high buffering capacity, and are less sensitive to acidification (Helfrich et al., 2001).

Concentrations of the individual ions HCO_3 and CO_3 are normally reported under the "General Water Quality Characteristics" section of reports, and this information is used for several types of problem assessments: (a) sodic soil potential—bicarbonate values can range widely, but as a general guideline HCO_3 > 120 mg/L (1.97 meq/L) and CO_3 > 15 mg/L (0.50 meq/L) start to become a concern, especially when Na concentrations are >100 mg/L (4.34 meq/L). The primary concern is that sufficient HCO_3 and CO_3 ions are present to react with Ca and Mg in the water and foster a sodic condition. These relationships are discussed more in Section 6.4, "Sodium Permeability Hazard." (b) Calcite formation in the soil—high concentrations of HCO_3 and CO_3 in combination with high Ca and Mg content can result in appreciable calcite or lime deposition in the soil.

See Section 6.4.2, "Residual Sodium Carbonate (RSC)." (c) Excess HCO_3 content may cause foliar deposition of lime on leaves impacted by irrigation water droplets. See Section 6.6, "Direct Foliage Injury and Miscellaneous Problems." And (d) irrigation water with high HCO_3 content may contribute to iron chlorosis by fostering iron carbonate formation in the soil and the removal of available Fe. There is no guideline for this problem, but values >500 mg/L (>8.2 meq/L) of HCO_3 may be a reasonable concentration to consider as a concern level for this problem.

6.2.3 Hardness

Hardness is often listed in irrigation water quality reports but, similar to alkalinity, is generally not used by turfgrass managers as much as RSC and the absolute values of specific ions. Hardness usually refers primarily to dissolved Ca and Mg, but other cations such as Fe, Mn, Al, and Zn can also contribute to hardness. Hardness is expressed in terms of $CaCO_3$ equivalent in units of mg/L of $CaCO_3$, and sometimes as grains per gallon (17 mg/L of free $CaCO_3$ = 1 grain $CaCO_3$ per gallon). Initially, hardness was used to determine the difficulty of producing soap suds and the tendency to form insoluble, greasy soap rings in wash basins. Also, it was used to assess the potential to leave scale deposits in pipes, especially when high P was present. For these purposes, as well as a general guide to lake water hardness and proneness to acidification, water hardness classes are as follows (Yiasoumi et al., 2005):

- *Soft*: <50 mg/L of $CaCO_3$ equivalent
- *Moderately soft to slightly hard*: 50–150 mg/L of $CaCO_3$
- *Hard*: 150–300 mg/L of $CaCO_3$
- *Very hard*: >300 mg/L of $CaCO_3$

6.3 TOTAL SOLUBLE SALTS (TOTAL SALINITY)

In Chapter 2, "Saline Soils" where total soluble salt problems were discussed, it was noted that the most common and important salt problem affecting turfgrass performance is soil accumulation of high total soluble salts leading to a *saline soil* condition. The most prevalent soluble salts are listed in Table 6.1, but the actual mix of salts can vary with the irrigation water source. Total soluble salinity in water is determined by *electrical conductivity of the water* (*ECw*) and expressed in dS/m or mmhos/cm (Tables 6.2 and 6.3). Also, total soluble salts may be reported using the designation *total soluble salts* (*TSS*) or the term *total dissolved salts* (*TDS*). However, the notation of TSS is also used to designate "total suspended solids." It is important to know which meaning is used in a water quality report because these factors are different—one measures dissolved salts, and the other suspended inorganic and organic matter. The authors will use TDS for total soluble or dissolved salts and TSS for total suspended solids.

TDS concentrations are reported in units of ppm or mg/L. The conversion of ECw to TDS is usually by the formula:

$$\text{TDS (in ppm or mg/L)} = \text{ECw (in dS/m or mmhos/cm)} \times 640$$

For example, if a groundwater has a TDS of 10,000 ppm, the ECw would be 13.3 dS/m. The *conversion factor* of 640 should be used when ECw is <5.0 dS/m, while at ECw > 5.0 dS/m the conversion should be 750.

All solutes do not have the same conductivity, and salt composition can differ from one source to another. Thus, some laboratories may use a different conversion factor, such as (a) 700 instead of 640, (b) 750 for waters with high-sulfate or highly saline water (i.e., ECw > 5.0 dS/m), or (c) 744 for seawater-affected waters. To avoid confusion, we would suggest the following.

Because it is ECw that is normally measured in the lab and then converted to TDS, the ECw value can be used instead of TDS. If a TDS value is desired, the lab or the turfgrass manager can use 640 as a conversion when ECw is <5.0 dS/m; and when ECw is >5.0 dS/m or the water source is a seawater-affected source or seawater blend, a conversion of 750 is appropriate.

Saline irrigation water can adversely affect plants in several manners (Carrow and Duncan, 1998) (see Chapter 9, "Selection of Turfgrass and Landscape Plants"). Seed germination and vegetative establishment are reduced by high salt levels. The most important adverse affect is by the osmotic influence, where salts in water and soil solution chemically attract water molecules, thereby reducing their availability for plant uptake. This effect is called salt-induced or *physiological drought.* Turfgrass symptoms to physiological drought include reduced growth rate, discoloration, wilting, leaf curling, and eventually leaf firing or desiccation that can lead to total leaf death. Initially, the discoloration may be the blue-green color associated with drought stress; but soon the combination of salt plus drought results in a general yellowing, followed by tissue desiccation, which can be yellow to dark brown depending on total salt impact severity and the salt tolerance level of the grass species and cultivar (see Chapter 9). Stand density and root volume gradually decrease over time.

Plant tolerance to the baseline salinity in irrigation water is very important for successful growth of turfgrasses and landscape plants under saline irrigation (see Table 6.5 and Figure 6.1).

TABLE 6.5
Irrigation Water Salinity Classification (i.e., Total Soluble Salts) Relative to Plant Salinity Tolerance, Infrastructure, and Management Requirements

Water Salinity	Lauchli and Luttge (2002)		AWA (2000)		
Rating	ECw dS/m	TDS mg/L	ECw dS/m	TDS mg/L	Description and Plant Tolerance
Very low	<0.78	<500	<0.65	<416	Fresh water. Sensitive plants. Few detrimental effects.
Low	0.78–1.56	500–1000	0.65–1.30	416–832	Slightly brackish. Moderately sensitive plants may show stress. Moderate leaching.
Medium	1.56–3.13	1000–2000	1.30–2.90	832–1856	Brackish. Moderately tolerant plants. Salinity will adversely affect most plants; requires selection for salt tolerance, careful irrigation, good drainage, and leaching.
High	3.13–7.81	2000–5000	2.90–5.20	1856–3328	Moderately saline. Moderately tolerant to tolerant crops must be used along with excellent drainage, leaching program, careful irrigation, and intensive cultivation and management.
Very high	7.81–15.6	5000–10,000	5.20–8.10	3328–5184	Saline. Tolerant crops must be used along with excellent drainage, leaching, irrigation, cultivation, fertility, and management programs.
Extremely high	15.6–50.0	10,000–32,000	>8.10	>5184	Highly saline. Very salt-tolerant halophytes. Excellent drainage, leaching, irrigation, cultivation, fertility, and management programs required.

Source: "Water Salinity Rating" section from Lauchli, A., and U. Luttge, *Salinity: Environment-Plants-Molecules*, Kluwer Academic, Boston, 2002; Australian Water Authority (AWA), Primary industries, in *Australian and New Zealand Guidelines for Fresh and Marine Water Quality* (Paper No. 4), Australian Water Authority, Artarmon, NSW, 2000.

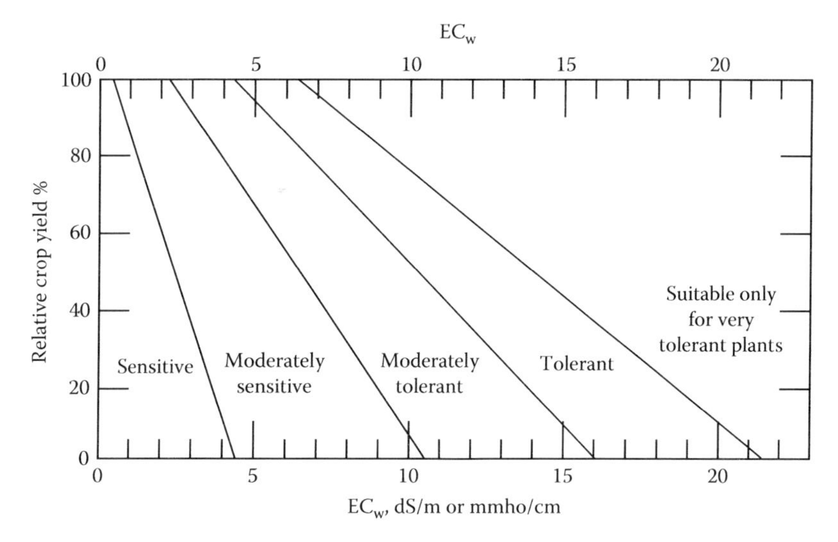

FIGURE 6.1 Irrigation water salinity (EC_w), relative salinity tolerance of plants, and relative plant growth under different salinity levels. Some very tolerant halophytes may exhibit 50% growth reduction at EC_w of >40 dS/m. (After Westcot, D. W., and R. S. Ayers, Irrigation water quality criteria, in G. S. Pettygrove and T. Asano (Eds.), *Irrigation with Reclaimed Municipal Wastewater: A Guidance Manual*, Lewis Publ., Chelsea, MI, 1985.)

When irrigation water contains appreciable total soluble salts, ECw is a major factor considered in selecting appropriate turfgrasses and landscape plants that can tolerate the base level of salinity. Classification criteria of irrigation water salinity relative to plant salinity tolerance have evolved as more salt-tolerant turfgrass plants, especially halophytes, are used. Traditional water classification schemes have considered irrigation water salinities of ECw > 3.0 dS/m to be highly saline, that is, generally unacceptable except for very salt-tolerant plants, excellent drainage, frequent leaching, and intensive management (Westcot and Ayers, 1985; Ayers and Westcot, 1994). However, some commonly used turfgrass species can often tolerate salinity levels well above ECw = 3.0 dS/m if adequate leaching is practiced. Thus, the authors prefer the water quality guidelines and descriptions reported by Lauchli and Luttge (2002) (Table 6.5). In Australia and New Zealand, irrigation water guidelines also reflect a broader salinity tolerance classification scheme (Table 6.4).

Species that are currently used as turfgrasses or are under development for turfgrass use on saline sites and that may exhibit threshold ECw (salinity at which plant growth starts to decline in response to salinity stress) values within the extremely tolerant range of ECw > 15.6 dS/m are some *Paspalum vaginatum* (seashore paspalum) cultivars, *Sporobolus virginicus* (saltwater couch and seashore dropseed), *Zoysia macrantha*, some *Stenotaphrum secundatum* (St. Augustinegrass) cultivars, some *Puccinella* spp. cultivars (alkaligrass), and some *Distichlis stricta* ecotypes (Brede, 2000; Carrow and Duncan, 1998; Barrett-Lennard, 2003). The cultivars or ecotypes of the above species that are most salt tolerant often exhibit ECw 50% growth reduction values in the 30 to 50 dS/m range (Carrow and Duncan, 1998). Refer to Chapter 9, "Selection of Turfgrass and Landscape Plants," for more detailed information.

While salt-tolerant turfgrasses have allowed more saline irrigation water to be used in some situations, practical implications related to using highly saline irrigation waters must be considered:

- As ECw increases above 3.0 dS/m, other salinity problems become "stacked" on top of total soluble salt stress because of the high concentrations and diversity of salts. These increased stresses include sodium permeability hazard, specific toxic ions, and nutrient challenges. Management becomes more demanding and costly. Unquestionably, proactive BMPs must be implemented for long-term turfgrass sustainability.

- As noted under the "Description and Plant Tolerance" column heading in Table 6.5, as ECw increases, so does the need for infrastructure improvements and enhanced management such as drainage systems, irrigation design uniformity and scheduling, leaching programs, cultivation equipment and programs, water treatment, water and soil amendments, increased application equipment needs, fertility programs that are much more dynamic in magnitude and frequency of changes, soil modifications and sand capping, and carefully and proactively managing the turfgrass to avoid additional biotic and abiotic stresses. Duncan et al. (2000) noted that many infrastructure and management decisions must be made prior to construction when using seawater, seawater blends, or other highly saline irrigation waters.
- Because of the above challenges, it is best to use as low ECw as available with the knowledge that just because a higher ECw water may be available as well as a salt-tolerant grass, a decision for using high-ECw water also is a decision for costly infrastructure changes, additional types of salinity stresses, and a greater degree of management expertise (Duncan and Carrow, 2005). The soil, including sand-based profiles, can accumulate 2–4 times higher concentrations of salts above the incoming water quality parameters even with good management (Duncan et al., 2009, Chapter 16 case study; Chen et al., 2010).
- Based on the authors' observations of sites using seashore paspalum, the highest ECw normally used for prolonged irrigation has been within the 15–20 dS/m range with excellent drainage, irrigation system and other infrastructure requirements, arid climate, and ample water quantity. At ECw 5–10 dS/m, long-term irrigation is not feasible on a salt-tolerant grass unless the salts can be effectively leached on a frequent schedule to avoid soil accumulation within the root system.

Salinity level of the irrigation water indicates the approximate soil salinity (expressed in terms of saturated paste extract electrical conductivity [ECe]) that can be achieved under "ideal" leaching conditions during periods when only irrigation water and not rainfall is applied. In reality, *soil ECe is often higher than ECw* unless the soil is very sandy with excellent leaching capabilities. Since ECw represents "the best" that the soil salinity will achieve without adding a better quality irrigation water or receiving rainfall, this value is important for the selection of turfgrasses that will tolerate this level of salinity on a specific site.

While there are turfgrasses that can tolerate very high salinities for periods of times, most of the species and cultivars used are within the 3 to 10 dS/m range. For example, many creeping bentgrasses exhibit a salinity tolerance of 3 to 6 dS/m, but there are some that can tolerate the 6 to 9 dS/m level, while all centipedegrass cultivars exhibit rapid growth reduction at 1.5 to 3.0 dS/m. Even for the very salt-tolerant species that can tolerate ECw well above 3.0 dS/m, irrigation water salinity as high as 3.0 dS/m can result in very rapid salt accumulation in soils that can stress even the salt-tolerant types unless a good leaching program is used. A more salt-tolerant grass allows more time to leach, take advantage of future rainfall, and avoid immediate salt stress; but salts cannot be allowed to continue to accumulate in the soil above the tolerance threshold of the specific grass cultivar.

In addition to influencing plant selection, irrigation water salinity level denotes the level of management expertise and intensity that will be required on a site. This is indicated in Table 6.5 in the comments about water salinity ratings. In general, as water quality declines, there must be increased emphasis on excellence in drainage systems, irrigation system uniformity, irrigation scheduling for proper water application, leaching programs, cultivation programs, fertility programs, soil modification, and overall proactive BMPs. Proper salt disposal will become increasingly important to avoid secondary salinization of soils, groundwater, or lakes. As irrigation water salinity increases, so does the potential for various toxic ion problems as well as nutrient and element imbalances.

With highly saline water, salts may impact irrigation system longevity and performance. Some irrigation companies have already made changes in system components to withstand more corrosive saline waters.

Water salinity also has biotic effects in irrigation lakes and freshwater ecosystems (NIWQP, 1998; Nielsen et al., 2003). Aquatic plants, invertebrates, fish, and birds can all be affected by water salinity and the stratified layering of various salinity levels in bodies of water. Hart et al. (1990) provide a good review of salinity effects on various aquatic organisms.

6.4 SODIUM PERMEABILITY HAZARD

In Chapter 3 ("Sodic, Saline-Sodic, and Alkaline Soils"), the problems associated with high Na levels in the soil were summarized; and it was noted that high Na content in irrigation water is a primary cause of sodic and saline-sodic soil conditions. The relative quantities of Ca, Mg, and Na in irrigation water are extremely important. Calcium is the primary ion that stabilizes soil structure (see Figure 3.1). Magnesium offers structural stability to a lesser degree. When excess Na (>100 to 200 ppm or 4.3 to 8.7 meq/L) is applied through irrigation water, the Na content builds up in the soil over time and, through the high volume of accumulated Na^+ ions, will eventually displace the Ca^{+2} ions that are the building blocks that enhance the structural integrity of the clay fraction in the soil profile. This "counterion" relationship, which is dominated by a larger hydrated radius Na ion (the nonhydrated Na ion is actually smaller than that of Ca, but it has a larger hydrated radius) with a weaker force or charge for holding clay particles together, eventually results in soil structural breakdown (deflocculation) (see Figure 3.3).

The result is a *sodic soil* (sometimes referred to as "black alkali" because the excess sodium solubilizes some of the organic matter fraction in the soil, which in turn rises to the soil surface and coats the soil particles). The deposit on the surface is black (with a slick, oily, or greasy appearance) or a *saline-sodic soil* (having both white salt deposits and black decomposed organic matter deposits on the surface; algae often colonizes these organic matter deposits during wet times), where excess Na^+ and high total salts are both present. Turfgrasses cannot thrive under these sodium permeability hazard conditions, since the soil structural breakdown results in a sealed soil with little or limited water permeability. Sodium carbonate will be present on these sites and contribute to soil sealing. Classic symptoms are heavily compacted areas, areas with constant puddles or very slow infiltration or percolation, sparse turfgrass density, and dead turfgrass. Secondary symptoms can include surface algae and black layer formation because of the constantly moist conditions in wet periods, an accumulation of sulfates, and the lack of oxygen getting to the turfgrass root system, since the soil structure has broken down with a sealed surface zone.

When Na predominates and is not countered by sufficient Ca and Mg, this ion will result in several adverse responses: (a) water infiltration into and percolation through the soil profile are greatly reduced by a decline in soil permeability (sealing) as soil aggregates degrade and colloidal particles are dispersed to form a more massive (and often collapsed) soil structure; (b) poor soil physical conditions increase the potential for low oxygen stress in the rhizosphere during wet periods; (c) the quantity of water required to leach excessive Na and other salts increases; (d) irrigation scheduling to apply routine irrigation volume as well as extra irrigation for leaching becomes much more challenging; (e) cultivation programs must be increased to maintain adequate downward water movement and oxygen flux; (f) water treatment by Ca addition or acidification may be required, depending on the balance of Na with Ca, Mg, HCO_3, CO_3, and ECw; (g) prescription soil amendment applications often are necessary; and (h) biotic problems increase (refer to Chapter 17, Section 17.6, "Biotic × Salinity Interactions").

The adverse effects of Na in irrigation water on water infiltration start at the soil surface and then continue downward in the soil profile so that eventually water percolation is reduced. Correction of sodic conditions requires the addition of Ca, displacement of Na from the soil CEC sites on clay particles, followed by leaching the Na displaced from CEC sites and present as Na-carbonate. Since the displaced Na has a tendency to go back onto CEC sites once it is displaced, the process of leaching is much slower than for the leaching of soluble salts that more easily stay in the soil solution.

While Na is the key ion in terms of adversely affecting soil structure, it is not the absolute concentration of Na that is important, but the balance and interactions of Na with Ca, Mg, HCO_3, and CO_3. These relationships were discussed in Chapter 2 ("Sodic, Saline-Sodic, and Alkaline Soils"). Sodium activity is also strongly influenced by ECw, soil texture, and clay type. The parameters of concern are (a) SAR or adjSAR; (b) RSC; (c) absolute concentrations of HCO_3, CO_3, Ca, Na, and Mg; (d) pH; (e) ECw or TDS; (f) soil texture; and (g) clay type.

6.4.1 SARw, adj SARw, and adj RNa

6.4.1.1 Understanding SARw, adj SAR, and adj RNa

The *sodium adsorption ratio* of the water (*SARw*, sometimes called *RNa*) is the traditional means used to assess the sodium status and permeability hazard (Ayers and Westcot, 1994). Na, Ca, and Mg concentrations (in meq L^{-1}, or $mmol_c/L$) are used to compute SARw, where

$$\text{SARw} = \text{RNa} = \frac{\text{Na}}{\sqrt{(\text{Ca} + \text{Mg})/2}}, \text{ in meq/L}$$

The traditional SARw is best used for irrigation waters that are low in HCO_3 and CO_3 at concentrations of <120 and 15 mg/L (<2 and <0.5 meq/L), respectively. In the SARw equation, HCO_3 and CO_3 concentrations are not included; therefore, it does not account for the effects of these ions. When HCO_3 and CO_3 ions are present in moderate to high levels, these ions can react with the Ca and Mg to form lime by precipitation in the soil or sometimes in the irrigation lines. Thereby, soluble and available Ca and Mg forms are depleted that are essential to displace Na from the CEC sites on clay surfaces where the Na causes breakdown by aggregation and dispersal of clay platelets. Ca reacts very easily with the bicarbonate or carbonate forms, while Mg is somewhat less reactive, but high Mg (usually resulting from exposure to ocean water that is high in Mg) can displace Ca from the CEC sites and then result in more precipitation of $CaCO_3$.

To account for the influence of HCO_3 and CO_3 ions in the irrigation water, an *"adjusted" SARw* (*adj SARw*) was developed by Bower et al. (1968) and used by Ayers and Westcot (1976). This initial adj SAR used the formula:

$$\text{adj SARw} = \text{SARw } [1 + 8.4 - \text{pHc}] = \text{SARw } [9.4 - \text{pHc}]$$

The *pHc value* is a theoretical, calculated pH value based on the irrigation water chemistry, which integrates the influence of Ca, Na, Mg, HCO_3, and CO_3 concentrations. The tables to make these adjustments are presented by Ayers and Westcot (1976). By itself, the pHc value has been used to indicate the tendency to dissolve lime by the water as it moves through the soil with a pHc > 8.4 indicative of dissolution, while a pHc < 8.4 indicates a greater tendency to precipitate lime in the soil or in irrigation lines if the pHc is very low.

Research of Suarez (1981) indicated that the initial adj SARw, as presented above, overestimated the sodium permeability hazard by as much as twofold since it did not adequately account for changes in Ca after irrigation water addition due to the potential for dissolution or precipitation of Ca (Stowell and Gelernter, 2010a). This preferred adjustment in the SARw was designated as adj RNa and uses a substitute Ca_x *value* in the SARw equation in place of Ca concentration. The adj RNa was calculated as

$$\text{adj RNa} = \frac{\text{Na}}{\sqrt{(\text{Ca}_x + \text{Mg})/2}}, \text{ in meq/L units}$$

The Ca_x factor comes from a table of HCO_3/Ca versus ECw where the table and example calculations are given by Hanson et al. (1999), Ayers and Westcot (1994), and Westcott and Ayers (1985).

Use of adj RNa is best for irrigation waters that are high in HCO_3 and CO_3, such as concentrations of >120 and >15 mg/L (>2 and >0.5 meq/L), respectively. In the published literature where SARw or adj SARw has been used in tables or figures, adj RNa can be substituted for these values.

Even though the adj RNa is superior to the initial adj SAR, a number of labs continue to use the initial adj SAR; some labs do report the adj RNa, while other labs have used the adj RNa calculation but report it as adj SAR. Due to this confusion, the authors are in agreement with Stowell and Gelernter (2010a, 2010b) about the following:

- Water test labs should use the adj RNa method (based on Ca_x adjustment) and not the older adj SAR (based on pHc adjustment) (see Table 5.1).
- Because the adj RNa terminology has been confusing and often not used, it is suggested that the term *adj SAR* be retained but based on the Ca_x method of Suarez (1981) to calculate it; and that the adjustment in the SARw can be reported as either adj RNa or adj SAR—with the laboratory clearly stating that it is based on the Ca_x adjustment of Suarez (1981).
- Stowell and Gelernter (2010a) have published in their *Pace Super Journal* the tools for estimating adj RNa from irrigation water quality reports.

6.4.1.2 Using SARw and adj SAR

As previously stated, when HCO_3 and CO_3 concentrations are low, SARw should be used; and when concentrations are higher, the adj SAR (Ca_x-adjusted SAR) is best. The top of Table 6.6 give the sodium permeability hazard classification if SARw or adj SAR is used alone. However, if either of these methods for determining sodium permeability hazard is used by itself, it can be misleading. For example, ultrapure irrigation water can sometimes exhibit a very high SARw because Ca and Mg levels are very low while Na is moderate. The authors have observed this on some coastal golf courses in the panhandle of Florida and Louisiana, where the high rainfall and sandy soils with low CEC have resulted in groundwater that is very low in Ca, Mg, and total soluble salts (i.e., ECw < 0.30 dS/m). In one situation, the irrigation water had the following parameters: 1 mg/L Ca, 0.5 mg/L Mg, and 140 mg/L Na. When the data are converted to meq/L, the SARw would be

$$\text{SARw} = \frac{6.09}{\sqrt{(0.091)/2}} = 28.6$$

Based on Table 6.6 (top), the SARw would suggest a high sodium permeability hazard, but under the situation, very little Na was present in the soil due to high leaching from rainfall and irrigation.

Clay type is very important to consider when determining whether particular irrigation water may present a sodium permeability hazard (Table 6.6, bottom). Clays can be classified as 2:1 and 1:1 types. Clays are crystalline in nature and are composed of layers such as tetrahedral or silica layers (Si-Al-O–rich layer) or octahedral layers (Al-Mg-Fe-O/OH–rich layer). The 1:1 clay types have one tetrahedral and one octahedral layer and are nonshrink and nonswell clay (i.e., nonexpanding and noncontracting), which means that after wetting and drying, they do not shrink, swell, or exhibit crack formation. Common 1:1 clays are kaolinite, allophanes, or any other clays rich in Fe and Al oxides. In contrast, the 2:1 clays have two Si tetrahedral layers and one Al-rich octahedral layer. These clays by nature shrink and swell, will demonstrate cracking on drying, and include smectites (montmorillonite), illites, and vermiculites. By viewing whether cracking occurs on drying of the soil, such as in a nonirrigated rough area, one can determine which clay type is present or dominates the soil profile.

As illustrated in Table 6.6 (bottom), 2:1 types, especially montmorillonites, are much more sensitive to structural deterioration by Na than are 1:1 clays. Even at an adj SAR of 6.0 for the irrigation water, when applied on a montmorillitic clay, the irrigation water can start to cause surface degradation of the soil structure. It should be noted that these relationships of adj SAR and clay type

TABLE 6.6
SAR_w and adj SAR Guidelines to Determine Sodium Permeability Hazard and Classification as Affected by Clay Type and ECw

Sodium Permeability Hazard	SAR_w or adj SAR (meq/L)	Comments
USSL (1954) Classification		
Low	<10	Can be used to irrigate almost all soils without structure deterioration. Salt-sensitive plants may be affected.
Medium	10–18	Appreciable Na permeability hazard on fine-textured soils with high CEC. Best used on coarse-textured soils with good drainage.
High	18–26	Harmful levels of Na accumulation on most soils. Will require intensive management, amendments, drainage, and leaching.
Very high	>26	Generally not suitable for irrigation at low to medium soil salinity levels.

Classification Considering Clay Type and ECw

	Degree of Problem		
	None	Increasing	Severe
EC_w (dS/m)(ultra pure water)	>0.50	0.50–0.20	<0.20
Adj RNa or SAR_w			
Montmorillonite (2:1)[a]	<6	6–9	>9
Illite, vermicultite (2:1)[a]	<8	8–16	>16
Kaolinite, Fe/Al oxides (1:1)[a]	<16	16–24	>24

[a] 2:1 clays are shrink or swell types; 1:1 are nonshrink or nonswell types.

Source: "Classification Considering Clay Type and EC_w" section from Ayers, R. S., and D. W. Westcot, *Water Quality for Agriculture* (FAO Irrigation and Drainage Paper 29), Food and Agricultural Organization, Rome, 1976.

Note: Use adjSAR when HCO_3 (>2.0 meq/L, 120 mg/L) or CO_3 (>0.5 meq/L, 15 mg/L) are high; and SAR_w for lower values. Also see Figure 6.2.

are even more pronounced as the percent clay content increases. Thus, a soil with 10% clay content would exhibit decreased Na effects at an adj SAR of 6.0 than would one containing 60% clay of the same clay type.

Figure 6.2 and Table 6.6 (bottom) illustrate that ECw influences SARw or adj SAR relationships. Essentially, there are two major ECw situations that should be considered: (a) ultrapure irrigation water (ECw < 0.50 and especially <0.20 dS/m) can exhibit a slight to severe reduction in water infiltration even at low adj SAR, especially when calcium concentrations are <20 ppm or 1 meq/L (this problem is discussed in detail by Duncan et al., 2009, Chapter 5, "Ultra Pure Irrigation Water"); and (b) at high ECw (>3.0 dS/m), high salt concentrations can aid in maintaining adequate permeability even when adj SAR is high (15 to 30). Appreciable Ca and Mg ions are usually present in high ECw water, so these act to counteract the effects of Na.

6.4.2 Residual Sodium Carbonate (RSC)

In the section on SARw and adj SAR, it was emphasized that determination of the sodium permeability hazard of irrigation water included consideration of SARw or adj SAR, ECw, clay type, and clay content. Another useful criterion is *residual sodium carbonate* (*RSC*), defined as

$$RSC = (CO_3 + HCO_3) - (Ca + Mg)$$

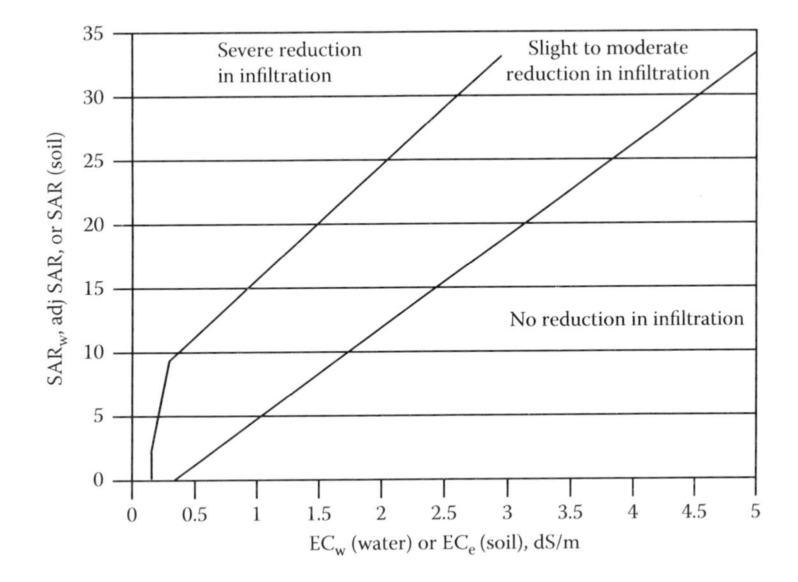

			Degree of Na Permeability Hazard		
			None	Slight to Moderate	Severe
Adj SAR or SAR$_w$	= 0	and EC$_w$ =	>0.7	0.7–0.2	<0.2
	= 3	=	>1.2	1.2–0.3	<0.3
	= 6	=	>1.9	1.9–0.5	<0.5
	= 12	=	>2.9	2.9–1.3	<1.3
	= 20	=	>5.0	5.0–2.9	<2.9

FIGURE 6.2 Sodium permeability hazard of irrigation water as influenced by EC$_w$ in graphic and table forms. As EC$_w$ increases, the adj SAR level at which water infiltration declines is greater. (After Oster, J. D., and F. W. Schroer, Infiltration as influenced by irrigation water quality, *Soil Sci. Soc. Amer. J.*, 43, 444–447, 1979.)

The ion concentrations are expressed in meq/L.

The RSC determines whether excess Ca and Mg remain in the irrigation water after reaction with the carbonates; or whether all Ca and Mg are precipitated from the irrigation water (Table 6.7) (Eaton, 1950; Wilcox et al., 1954). A negative RSC indicates that more Ca and Mg are in the water than carbonates, where the excess Ca and Mg can act as counterions to displace Na. In contrast, a positive RSC means that all Ca and Mg have been precipitated and excess bicarbonate and carbonate remain that could react with any Ca added, such as by gypsum application, once the Ca becomes soluble. The net effect is to reduce soluble Ca and Mg levels while leaving soluble Na to create sodic conditions.

One obvious problem in using RSC to assess sodium permeability hazard is that the concentration of Na is not included. Thus, it is possible to have a very high RSC, but not to create sodic conditions if little Na is in the irrigation water source. The adj SAR is a much better indicator of the sodium permeability hazard of the irrigation water, especially when ECw, clay type, and clay content are factored in the decision process. When Na is present along with a high (positive) RSC, sodium carbonate forms at the same time that Na displaces Ca and Mg from soil CEC sites; and the soil becomes increasingly sodic. In order to ameliorate sodic soils, sufficient Ca must be added to displace Na on the CEC sites as well as at least some of the sodium in the Na-carbonate. A good leaching program may leach the soluble Na carbonate easier than the Na on CEC sites, which must be first displaced by Ca. If Na-carbonate is not leached, the Na will dissolve and displace Ca on the CEC site from any Ca amendments, thereby negating the Ca amendment application. Table 6.7 lists guidelines for RSC relative to suitability for irrigation water use when sufficient Na is present to

TABLE 6.7
Residual Sodium Carbonate (RSC) Guides for Irrigation Water Suitability When Na Is Present at a Sufficient Level to Potentially Create a Sodic Soil Condition (Na > 100 ppm, 4.3 meq/L)

RSC Value (meq L^{-1})	Na Hazard
<0 (i.e., negative)	None. Ca and Mg will not be precipitated as carbonates from irrigation water; they remain active to prevent Na accumulation on CEC sites.
0–1.25	Low. Some removal of Ca and Mg from irrigation water.
1.25–2.50	Medium. Appreciable removal of Ca and Mg from irrigation water.
>2.50	High. All or most of Ca and Mg removed as carbonate precipitates, leaving Na to accumulate. How rapidly Na buildup occurs depends on Na content of the water.

Source: Data from Eaton, F. M., Significance of carbonates in irrigation water, *Soil Sci.,* 69, 123–133, 1976; Wilcox, L. V., G. Y. Blair, and C. A. Bower, Effect of bicarbonate on suitability of water for irrigation, *Soil Sci.,* 77, 259–266, 1954.

potentially cause sodic conditions (Na > 100 ppm or 4.3 meq/L can be used as a "red flag" to assess Na problems more closely) (Eaton, 1950; Wilcox et al., 1954).

RSC is very useful in defining irrigation water amendment and soil amendment needs. It is a valuable piece of information along with Na content to make informed BMP decisions. First, RSC determines how much (if any) *unreacted Ca and Mg* remain available in irrigation water to counteract any Na present. The unreacted Ca and Mg become part of the Ca "amendment" needs to supply sufficient Ca and Mg to counteract the Na.

Second, when Na is present and no residual Ca and Mg remains, but residual bicarbonate or carbonate is present, the RSC determines how much additional Ca amendment must be supplied to react with the excess or *unreacted bicarbonate and carbonate.* Only after sufficient Ca and Mg are available from the irrigation water and/or added Ca to react with all the carbonates will there be the opportunity to add additional Ca that will become soluble overtime and counteract the Na. Unless adequate Ca is added to react with the unreacted bicarbonate and carbonate (or the carbonates are removed by acidification), sodium carbonate will accumulate to create sodic conditions.

Third, RSC is used to determine the *quantity of acid in water acidification* to reduce the residual bicarbonate and carbonate so that they will not react with Ca or Mg in the water or soil solution. Sufficient acid is added to evolve off the carbonate ions (both $CO_3 + HCO_3$) as CO_2 gas and water. The formula is

$$\text{RSC} \times 133 = \text{pounds of 100\% } H_2SO_4 \text{ per acre-foot irrigation water}$$

In practice, sufficient acid is added to achieve 75% to 80% depletion of the residual carbonates, but to maintain pH > 6.0 and usually within the 6.5 to 6.8 range.

As noted, acidification of irrigation water is very important when it contains both excessive Na and carbonates, since an effective Ca amendment program cannot be achieved if most of the Ca reverts to lime or very insoluble forms. In Chapter 13, Section 13.3.4, "Case Study: Irrigation Water Calculation," a stepwise procedure will be presented to illustrate how RSC information can be used to determine water acidification and amendment needs. For sites with irrigation waters containing appreciable Na and carbonates, amendment needs are best determined by starting with the irrigation water quality conditions; and the RSC provides valuable information for these purposes (Carrow et. al., 1999; Carrow and Duncan, 1998). Water quality values that raise a red flag to look at the potential need for acidification are RSC > 1.25 meq/L, Na content > 100 ppm or 4.3 meq/L, and HCO_3 > 120 mg/L or 2.0 meq/L. These values do not mean that acidification is needed, but they do indicate that a closer investigation is warranted by considering all other relevant information.

A fourth use of RSC is to determine the quantity of gypsum to add per acre-foot of irrigation water to supply *Ca to react with the remaining bicarbonate and carbonate* to achieve a RSC = 0. The formulas used are

$$RSC \times 234 = \text{pounds of 100\% pure gypsum per acre-foot irrigation water}$$

$$RSC \times 86 = \text{kg of 100\% pure gypsum per 1000 } m^3 \text{ irrigation water}$$

Gypsum injection could be used to achieve the required gypsum. Alternatively, the water could be acidified to remove excess carbonates.

Finally, for situations where little or no Na is present, but Ca, Mg, and carbonates are all very high, the RSC aids in assessing the potential for calcite precipitation. In arid situations, there is sometimes concern that lime deposition on the soil from the irrigation water may cause sealing overtime (Carrow et al., 1999). Concentrations of these constituents in irrigation water usually are 100–400 mg/L HCO_3, 0–5 mg/L CO_3, 25–200 mg/L Ca, and 20–40 mg/L Mg. Assuming that irrigation water is very high in each of these aspects (at 811 mg/L HCO_3, 200 mg/L Ca, and 40 mg/L Mg in this example), there is sufficient Ca and Mg to react with all the 13.3 meq/L of HCO_3. This water would result in a combined total of 2,104 lb. of $CaCO_3$ + $MgCO_3$ per acre-foot of applied irrigation water or 48 lb. $CaCO_3$ + $MgCO_3$ per 1000 sq. ft. per 12 inches of irrigation water. As a comparison, a soil with 1% free lime in the surface 4-inch zone would contain about 230 lb. per 1000 sq. ft. of lime. In an arid climate where rainfall would not assist in dissolving the precipitated calcite and the irrigation program was consistent, accumulation could occur, especially at the bottom depth of routine irrigation water front penetration. However, with the use of acidifying fertilizers, periodic deeper leaching by irrigation or rains, and normal cultivation practices, all of these would assist in preventing a distinct calcite zone that could inhibit infiltration or percolation.

6.5 SPECIFIC ION IMPACT (ROOT INJURY AND SHOOT ACCUMULATION INJURY)

Irrigation water may contain excessive levels of certain ions that can (a) adversely affect plant root tissues due to soil accumulation, and (b) cause injury to shoot tissues due to uptake and accumulation in leaves. As total salinity increases in irrigation water, the potential for specific ion toxicity also increases. Germinating seed, young seedlings, and sprigs are especially vulnerable because of their juvenile developing root systems. The ions that most often cause toxicity problems include Na, Cl, and B. However, trace elements can also result in toxicity or ion competition problems over time in some situations (see Section 6.8, "Trace Elements"). The same guidelines are used for soil accumulation (root injury) and shoot accumulation (shoot injury) and are presented in Table 6.8.

In terms of foliar uptake, accumulation, and injury, any of the soluble salts in soil solution can be taken up and potentially accumulate in leaf tissues. However, as irrigation water salinity increases, two of the most common salt ions likely to be present are Na and Cl. Chloride is a very common anion in irrigation water and can easily be taken up by plants. As salts accumulate in the leaf tips of grasses or the outer margins of landscape plants in the topmost leaves, the salts can cause (a) osmotic stress by reducing water for cell uptake and inducing dehydration of cells, which can eventually lead to tissue desiccation and leaf firing; and (b) induce direct toxic effects, depending on the salt ion and level of plant genetic salinity tolerance.

Turfgrasses are generally less sensitive to Na and Cl uptake into leaves and foliar injury compared to other landscape plants, primarily because mowing removes accumulated Cl ions in the shoot tissues. When trees and shrubs accumulate excessive salts, the initial symptoms include leaf wilting, this is followed by firing on leaf margins (yellow to tan color as tissues die and desiccate), and finally leaves may fall from the plants. For tall trees, the upper leaves (active growing points) often exhibit leaf fall first.

TABLE 6.8
Specific Ion Toxicity (Na, Cl, and B) and Miscellaneous Chemical Constituent Problems in Sprinkler Irrigation Water for Sensitive Plants

		Degree of Restriction on Use for Sensitive Plants[a]		
Potential Toxicity Problem	**Units**	**None**	**Slight to Moderate**	**Severe**
Soil Accumulation and Root or Shoot Accumulation Injury				
Na	SAR (soil)	0–3	3–9	>9
	mg/L (ppm)	0–70	70–210	>210
	meq/L	0–3	3–9.1	>9.1
Cl	mg/L	0–70	70–355	>355
	meq/L	0–2	2–10	>10
Metal ions (trace elements)—see Table 6.12.				
Direct Foliage Injury (Sprinkler Irrigation)				
Na	mg/L	0–70	>70	
	meq/L	0–3	>3	
Cl	mg/L	0–100	>100	
	meq/L	0–3	>3	
Miscellaneous				
HCO_3^- [b] (unsightly deposits)	mg/L	0–90	90–500	>500
	meq/L	0–1.5	1.5–8.5	>8.5
Residual Cl_2 (chlorine)	mg/L	0–1	1–5	>5
SO_4 (black layer)[c]	mg/L	0–90	90–180	>180

Boron Toxicity Potential

B (mg/L)	Comments
<0.5	Very sensitive. Some ornamentals affected (see Hansen et al., 1999).
0.5–1.0	Sensitive. Some trees, shrubs, and ornamentals affected.
1.0–2.0	Moderately sensitive.
2.0–4.0	Moderately tolerant. Kentucky bluegrass.
4.0–6.0	Tolerant. Only B-tolerant plants should be used.
>6.0	Very tolerant plants should be used.

Source: After Ayers, R. S., and D. W. Westcot, *Water Quality for Agriculture* (FAO Irrigation and Drainage Paper, 29, Rev. 1), Reprinted 1994, Food and Agricultural Organization, Rome, 1994, http://www.fao.org/DOCREP/003/T0234E/T0234E00.htm#TOC; Hanson, B., S. R. Grattan, and A. Fulton, *Agricultural Salinity and Drainage* (Division of Agriculture and Natural Resources Pub. 3375), University of California, Davis, 1999; Australian Water Authority (AWA), Primary industries, in *Australian and New Zealand Guidelines for Fresh and Marine Water Quality* (Paper No. 4), Australian Water Authority, Artarmon, NSW, 2000, http://www.mincos.gov.au/publications/australian_and_new_zealand_guidelines_for_fresh_and_marine_water_quality/volume_3.

[a] Guidelines in this table are for "sensitive" landspace plants. Turfgrasses are generally more tolerant than many ornamental trees, shrubs, and flowers.

[b] HCO_3 deposition is not a toxicity problem but is unsightly on foliage or ornamentals.

[c] SO_4 (black layer). High SO_4 in the soil often arises from irrigation water, especially acidified water and some reclaimed water. Under anaerobic conditions, SO_4 can be reduced to sulfide forms and stimulate black layer formation. A good leaching program will leach SO_4 ions to prevent accumulation. Also, high Ca ion content can precipitate SO_4 as gypsum.

Root cells preferentially bind Ca to exchange sites on cell walls (similar to soil CEC sites on clays or organic matter), which is beneficial since cell walls and cell plasma membranes require high Ca content to maintain viability or function. High Na ion concentration in the soil results in Na competing with Ca ions on cell wall exchange sites, resulting in cell wall deterioration, especially near the root tips. The Na-induced Ca deficiency in root tissues is sometimes called "Na toxicity." Highly salt-tolerant turfgrass and landscape plants possess tolerance to Na toxicity and restrict the uptake of Na, but less salt-tolerant plants exhibit severe root injury from high soil and tissue Na accumulation.

Boron is often associated with saline hydrogeological conditions and is another element that can be a toxicity problem if too high in irrigation water. Surface water concentrations of B in natural ecosystems are normally <0.1 mg/L and rarely exceed 1.0 mg/L. Groundwater is usually <0.5 mg/L, but some groundwater can be as high as 5.0 mg/L, while agriculture drain water can be above this level if irrigation is from B-rich water (NIWQP, 1998).

Plant species vary in tolerance to B. Irrigation water quality–relative B tolerance guidelines are noted in Table 6.8 (bottom) (Ayers and Westcot, 1994; Hansen et al., 1999; AWA, 2000). The guidelines in Table 6.8 note the potential for B toxicity problems to occur over relatively short-term irrigation water use (i.e., <20 years) under conditions when soil B could accumulate (assuming no leaching losses or fixation losses). Kentucky bluegrass is the only turfgrass listed in these sources, and it is listed as moderately tolerant and able to tolerate 2.0 to 4.0 mg/L of B. Turner and Hummel (1992) report the B tolerance of grasses as creeping bentgrass (most tolerant) > perennial ryegrass > "Alta" tall fescue > Kentucky bluegrass > zoysiagrass > Bermuda grass.

Boron phytotoxicity on plants may be exhibited by yellowing followed by a dark necrosis on the margins of older leaves. As soil pH increases from 6.3 to 7.0, B is more tightly adsorbed on clays and Fe/Al oxides. Thus, at pH < 7.0, leaching may prevent B accumulation, while at pH > 7.0, light lime application (1–5 lb./1000 sq. ft.) to maintain high Ca levels can help fix the B is less available forms. Leaching is more effective on coarser-textured soils than on fine-textured ones.

The NIWQP (1998) reports that the B toxicity threshold in lakes for an aquatic plant is 10 mg/L. However, aquatic invertebrates, fish, amphibians, and wildfowl could tolerate higher levels.

6.6 DIRECT FOLIAGE INJURY AND MISCELLANEOUS PROBLEMS

Plant leaves of trees and shrubs directly impacted by overhead sprinkler irrigation water spray may exhibit foliage injury from *Na* or *Cl* in the water (Table 6.8). Leaves that are in the pathway of the irrigation water stream may die.

HCO_3 concentrations in irrigation water are not phytotoxic to leaves at >8.0 meq L^{-1} or 500 ppm. However, HCO_3 can cause unsightly whitish deposits on leaves of ornamentals, trees, and equipment (Table 6.8).

Residual chlorine (Cl_2) that is used to disinfect wastewater can be toxic at >5 ppm for many plants (Table 6.8). Damage is expressed as leaf tip injury. However, chlorine is normally not a problem on turfgrasses since most irrigation water is applied through overhead sprinkler systems, and this gas dissipates rapidly when aerated. Storage of high-chlorine water in lakes, ponds, or lagoons will accomplish the same dissipation phenomenon over time, especially if an aeration system is present in the storage area. Water from swimming pools may contain residual chlorine, but grasses seem to be tolerant. Other landscape plants may be less tolerant when irrigation with pool water is applied topically, but generally 100 ppm or more is required for injury. Observed chlorine injury on turfgrasses has occurred when treated reclaimed water was deposited into a closed storage tank and the chlorine levels spiked higher than normal during the disinfection process, resulting in elevated chlorine-laden droplets burning the grass. Opening the top vents in the storage tank reduced the problem. A second observation involved direct piping of reclaimed water into the wet well, and, again, the spike in chlorine treatment caused irrigation of elevated chlorine-laden droplets to burn the turfgrass. A significant corrosion problem occurred within the closed irrigation facility that resulted in rapid deterioration of pumping equipment. The recommendation is to deposit any chlorine-treated water into lakes or ponds with some type of aeration system to dissipate the chlorine prior to application on turfgrasses.

Total suspended solids (*TSS*) are inorganic particles (sand, silt, and clay), organic matter (plant parts, algae, and bacteria), and immiscible liquids (grease and oils) suspended in irrigation water. Duncan et al. (2010) reported that suspended solids are objectionable in irrigation water due to the following reasons: (a) clogging of screens, filters, or nozzles may occur; (b) inorganic particles can cause wear on pumps and equipment; (c) colloidal matter may decrease water infiltration into the turfgrass soil surface due to layering; (d) inorganic and organic colloids may protect microorganisms from the chemical action of chlorine or ultraviolet radiation (UV) disinfection (i.e., measured as turbidity at wastewater treatment plants); and (e) biodegradable organics can reduce oxygen status in the water (measured as biological oxygen demand [BOD] at wastewater treatment plants). Relative to the first three problems, suspended solids are measured as *total suspended solids* in units of mg/L or ppm and are often designated by the initials *TSS*. However, as discussed in this chapter, this can lead to confusion since TSS is sometimes used to also represent "total dissolved salts." Thus, the turfgrass manager must be careful in reading a water quality report to understand which water quality parameter is represented by the TSS notation. Reclaimed water used for irrigation with human access has stringent regulatory limits since TSS (reported at turbidity) influences disinfection treatment effectiveness and BOD. To insure effective disinfection, TSS is mandated by many states to be 5 to 30 mg/L in reclaimed water, depending on the state (USEPA, 2004).

Suspended solids are common in surface water sources, most often due to runoff of silt and clay or algae growth in stagnant waters. Reclaimed water is filtered to remove many of the TSS at the treatment facility because these materials would contribute to high turbidity levels, thereby reducing the effectiveness of disinfectant treatments. However, when reclaimed water is delivered to an irrigation pond and it contains ample N and P, algae bloom and eutrophication may result in increased levels of organic debris contributing to higher TSS levels. Groundwater, due to filtering by the soil, normally will have less TSS than surface water sources or reclaimed water.

Suspended solids arising from organic materials, such as algae, should be reduced by controlling the source since filtering is difficult to achieve for these materials. Control measures for organics are discussed by Duncan et al. (2009). The larger inorganic suspended solids, such as fine sand and some silt, may require use of filters and/or settling ponds. Sand does not stay suspended in irrigation water unless the water is moving or is agitated, so sand does not normally show up in laboratory TSS data. However, when sand is present in an irrigation source, it can cause severe wear on irrigation components. Norum (1999) provides a good discussion of sand problems and the use of sand separator systems to remove most sand contaminants. Silt and clay applied to sandy turfgrass soils can result in significant surface sealing and a reduction in infiltration, percolation, and oxygen diffusion.

TSS guidelines relative to the plugging of an irrigation system, especially drip systems, are present by Westcot and Ayers (1994) as follows:

- *None*: <50 mg L^{-1}
- *Slight to moderate*: 50 to 100 mg L^{-1}
- *Severe*: >100 mg L^{-1}

There are no guidelines published relative to potential to seal a surface of a turfgrass sod, especially on a sandy soil. If the particulate matter is organic in nature, such as live or dead algae that could contribute to an algae layer at the surface, control measures noted by Clark and Smajstria (1999), Yiasoumi et al. (2005), and Duncan et al. (2010) would be appropriate to maintain as low levels as possible. When abrasion is the problem from sand particles, a high degree of sand removal, usually by settling, would be desired. For silt and clay (often found in river or canal water sources) that may contribute to surface sealing of sandy soils, an irrigation water with a 50 mg/L (i.e., 50 ppm) TSS level would apply 3.1 lb. of TSS/1000 sq. ft. per acre-foot of irrigation applied (151 kg/ha per 30 cm of water applied). Wind deposition of silt and clay onto golf greens is often higher than this level. However, at 500 ppm TSS, accumulation would be at 30.1 lb. TSS/1000 sq. ft. per acre-foot of water, which over time may have a detrimental effect.

If a water quality report indicates a very high TSS value, such as 500 ppm, it is important to determine the following:

- Whether the lab is using the TSS notation to designate "total soluble salts" rather than total suspended solids, since total dissolved salts can be appreciably higher (up to several thousand ppm). If, on the water quality report, a value for electrical conductivity (EC) is given in dS/m units and multiplication of EC (in dS/m units) by 640 equals the TSS value in ppm or mg/L, then the TSS value reported is really total soluble salts and not total suspended solids.
- If the lab determined the TSS value by evaporating off the water and weighing the residue. In this case, the residue would be the total of both suspended solids and dissolved salts, and the conversion of EC to TSS noted above would not equal the same value.
- Whether the water source is truly high in suspended solids as noted by a high total suspended solids value that does not include dissolved salts. In such a case, the irrigation water should exhibit high turbidity, and settling of the solids should be apparent after a 24-hour period—sand particles settle in about 1 minute, but silt and clay will require 8 to 24 hours. In cases where TSS is high, a combination of management approaches may be necessary such as settling ponds, flocculation and settling using alum, filtration, pH adjustment, or measures to control algae (Duncan et al., 2009).

6.7 NUTRIENTS

All irrigation water will contain a certain level of nutrients in its composition, and reclaimed or saline irrigation waters may contain elevated levels of certain nutrients. Nutrients in irrigation water are a concern because of (a) contributions to the overall soil fertility and plant nutrient program for turfgrass and other landscape plants (all macro- and micronutrients), (b) promotion of algae and aquatic plant growth in irrigation lakes (P and N), (c) human health hazards (NO_3), and (d) the potential to contribute to black layer (SO_4). Because of the nutrient load in irrigation water, fertility programs must be adjusted to maximize turfgrass performance and to minimize environmental impact (King et al., 2000). In Chapter 16 ("Nutritional Practices on Saline and Sodic Sites"), nutritional aspects are discussed in detail.

Huck et al. (2000) reported nutrient content guidelines for irrigation water, especially reclaimed water, with low to very high value ranges (Table 6.9). Table 6.10 lists conversion factors to convert ppm or mg/L of an ion in irrigation water to quantity of ion applied per acre-foot of irrigation water applied. The ranges for all macronutrients except P are based on the possible effects on fertilization with continuous use of water. P recommendation is based on limiting eutrophication, since P is often the limiting nutrient that retards aquatic plant and algae growth.

High concentrations of a particular nutrient could lead to overfertilization or nutrient imbalance with counterions, especially for the cations. Because N responses by turfgrasses and aquatic plants can be more pronounced than for other nutrients, high N levels must be taken into account in the overall N fertilization program. For example, reclaimed water high in N and applied on cool-season grasses during hot periods can result in deterioration of the grass in response to excess N.

In the case of SO_4, higher concentrations could increase the potential for black layer formation if the SO_4 accumulates in the soil and anaerobic conditions occur. In Chapter 16 this issue is discussed, and Duncan et al. (2009) have presented measures to limit this potential problem.

TABLE 6.9
Guidelines for Nutrient Content of Irrigation Water to Be Used for Turfgrass Situations; and Quantities of Nutrients Applied per Acre-Foot on Irrigation Water

Nutrient or Element	Nutrient Content in Water in mg L^{-1} (or ppm)				Conversion to Lb. per 1,000 Sq. Ft. of Nutrient Added for Every 12 Inches of Irrigation Water Applied[c]
	Low	Normal	High	Very High	
N	<1.1	1.1–11.3	11.3–22.6	>22.6	11.3 ppm N = 0.71 lb. N per 1,000 sq. ft.
NO_3^-	<5	5–50	50–100	>100	50 ppm NO_3^- = 0.71 lb. N per 1,000 sq. ft.
P	<0.1	0.1–0.4	0.4–0.8	>0.8	0.4 ppm P = 0.057 lb. P_2O_5 per 1,000 sq. ft.
PO_4^-	<0.30	0.30–1.21	1.21–2.42	>2.42	1.21 ppm PO_4^- = 0.057 lb. P_2O_5 per 1,000 sq. ft.
P_2O_5	<0.23	0.23–0.92	0.92–1.83	>1.83	0.92 ppm P_2O_5 = 0.057 lb. P_2O_5 per 1,000 sq. ft.
K	<5	5–20	20–30	>30	20 ppm K = 1.5 lb. K_2O per 1,000 sq. ft.
K_2O	<6	6–24	24–36	>36	24 ppm K_2O = 1.5 lb. K_2O per 1,000 sq. ft.
Ca^{+2}	<20	20–60	60–80	>80	60 ppm Ca = 3.75 lb. Ca per 1,000 sq. ft.
Mg^{+2}	<10	10–25	25–35	>35	25 ppm Mg = 1.56 lb. Mg per 1,000 sq. ft.
S	<10	10–30	30–60	>60	30 ppm S = 1.87 lb. S per 1,000 sq. ft.
SO_4^{-2}	<30	30–90	90–180	>180	90 ppm SO_4^- = 1.87 lb. S per 1,000 sq. ft.
Mn	—	—	>0.2[b]	—	
Fe	—	—	>5.0[a]	—	
Cu	—	—	>0.2[a]	—	
Zn	—	—	>2.0[a]	—	
Mo	—	—	>0.01[b]	—	
Ni	—	—	>0.2[a]	—	

Source: After Huck, M., R. N. Carrow, and R. R. Duncan, Effluent water: Nightmare or dream come true? *USGA Green Section Record,* 38 (2), 15–29, 2000.

[a] These values are based on potential toxicity problems that may *arise over long-term use* of the irrigation water, especially for sensitive plants in the landscape—turfgrasses can often tolerate higher levels. For fertilization, higher rates than these can be applied as foliar treatment without problems.

[b] Based on Westcott and Ayers (1985).

[c] See Table 6.10 for conversion factors.

For assessing the potential to create a cation imbalance, ratios noted in Table 6.11 can be used as a general guideline, where the key ratios are Ca:Mg, Ca:K, and Mg:K. Note that these ratios are based on nutrient concentrations in meq/L (chemically equivalent basis) rather than mg/L. While K is seldom high in irrigation water, high Ca content is common. Some irrigation water, especially if influenced by seawater, is high in Mg. Another common situation is when excess Na in irrigation water inhibits K plant uptake and reduces K on soil CEC sites. High Na can also suppress Ca and Mg uptake and dominate the CEC status. Cation ratios that are extreme often require supplemental fertilization to obtain a balanced soil fertility status. Soil testing can assist in determining fertilizer and amendment needs.

6.8 TRACE ELEMENTS

Ayers and Westcot (1994) note that trace elements are almost always present in all water supplies, but at low concentrations. Some reclaimed water may exhibit higher levels of trace elements depending on manufacturing wastewater sources going into the treatment facility. Also, irrigation water affected by acid mine drainage or acid-sulfate soil conditions can have high trace element concentrations due to the extreme acidity that can dissolve soil trace elements (Gray, 1997; Sundstrom et al., 2002). Another irrigation source that may contain trace elements is untreated stormwater runoff from

TABLE 6.10
Conversion Factors for Calculating Pounds of Nutrient per Acre-Foot of Irrigation Water from Nutrient Content in the Water

Conversion Factors That Can Be Used

- 1 mg/L = 1 ppm
- 1 mg/L NO_3^- = 0.226 mg /L N
- 1 mg/L N = 4.42 mg/L NO_3
- lb. of N per acre-foot of water = mg/L N in water × 2.72
- lb. of P per acre-foot of water = mg/L P in water × 2.72
- lb. of K per acre-foot of water = mg /L K in water × 2.72
- lb. of P_2O_5 per acre-foot of water = mg/L P in water × 6.24
- lb. of K_2O per acre-foot of water = mg /L K in water × 3.25
- lb. of Ca per acre-foot of water = mg /L Ca in water × 2.72
- lb. of Mg per acre-foot of water = mg/L Mg in water × 2.72
- 1 mg/L SO_4^- = 0.33 mg/L S
- 1 mg/L PO_4^- = 0.33 mg/L P

Example: Irrigation water has 15 mg/L NO_3.

15 mg/L NO_3 = (15)(0.226 mg/L N)
= 3.39 mg/L as N

lb. N per acre-foot of water = (mg/L of N)(2.72)
= (3.39 mg/L of N)(2.72)
= 9.22 lb. N per acre-foot water

Or, 9.22/43.56 = 0.21 lb. N per 1,000 sq. ft. per 12 inches irrigation water

1 acre = 43,560 sq. ft.

previous surfaces (USEPA, 2003). While micronutrients may be included in a routine irrigation water quality analyses, other trace elements generally are not analyzed. However, if the irrigation water source is one of the above situations, it would be reasonable to obtain a full nutrient analysis.

The desired range for various micronutrients is at or less than that listed for LTV levels in Table 6.12. The values recommended by Westcot and Ayers (1985) are also included in Table 6.12, where it is suggested not to exceed these levels for long-term use. The STV limits used by AWA (2000) and listed in Table 6.12 can be used as maximum levels for normal use.

TABLE 6.11
General Guidelines for Cation Ratios in Irrigation Water

Ratio	Preferred Ratio Limits[a]	Comments
Ca:Mg	Below 3:1	Ca deficiency may occur.
	Above 8:1	Mg deficiency may occur.
Mg:K	Below 2:1	Mg deficiency may occur.
	Above 10:1	K deficiency may occur.
Ca:K	Below 10:1	Ca deficiency may occur.
	Above 30:1	K deficiency may occur.

Note: Ratios are based on cation content in meq/L basis.

[a] Irrigation water with ratios outside these guidelines can still be used, but supplemental fertilization may be required to maintain a soil nutrient balance.

TABLE 6.12
Recommended Maximum Concentrations of Trace Elements in Irrigation Water for Long-Term Values (LTV) and Short-Term Values (STV)

Element	AWA (2000) LTV[a] (mg/L)	AWA (2000) STV[b] (mg/L)	Westcot and Ayers (1985) LTV (mg L^{-1})	Remarks
Al (aluminum)	5.0	20	5.0	Can cause nonproductivity in acid soils (pH < 5.5), but a pH > 5.5 will precipitate the ion and eliminate any toxicity.
As (arsenic)	0.10	2.0	0.10	Toxicity to plants varies widely, ranging from 12 mg/L for Sudangrass to less than 0.05 mg/L for rice.
Be (beryllium)	0.10	0.50	0.10	Toxicity to plants varies widely, ranging from 5 mg/L for kale to 0.5 mg/L for bush beans.
Cd (cadmium)	0.01	0.05	0.01	Toxic to beans, beets, and turnips at concentrations as low as 0.1 mg/L in nutrient solutions. Conservative limits recommended because of its potential for accumulation in plants and soils to concentrations that may be harmful to humans.
Co (cobalt)	0.05	0.10	0.05	Toxic to tomato plants at 0.1 mg/L in nutrient solution. Tends to be inactivated by neutral and alkaline soils.
Cr (chromium)	0.10	1.0	0.1	Not generally recognized as an essential growth element. Conservative limits recommended because of lack of knowledge on toxicity to plants.
Cu (copper)	0.20	5.0	0.2	Toxic to a number of plants at 0.1 to 1.0 mg/L in nutrient solutions.
F (fluoride)	1.0	42	1.0	Inactivated by neutral and alkaline soils.
Fe (iron)	0.20	10	5.0	Not toxic to plants in aerated soils, but can contribute to soil (iron) acidification and loss of reduced availability of essential phosphorus and molybdenum. Overhead sprinkling may result in unsightly deposits on plants, equipment, and buildings.
Li (lithium)	2.5	2.5	2.5	Tolerated by most crops up to 5 mg/L; mobile in soil. Toxic to citrus at low levels (>0.975 mg/L). Acts similarly to boron.
Mn (manganese)	0.20	10.0	0.2	Toxic to a number of crops at a few tenths mg to a few mg/L, but usually only in acid soils.
Mo (molybdenum)	0.01	0.05	0.01	Not toxic to plants at normal concentrations in soil and water. Can be toxic to livestock if forage is grown in soils with high levels of available molybdenum.
Ni (nickel)	0.20	2.0	0.2	Toxic to a number of plants at 0.5 to 1.0 mg/L; reduced toxicity at neutral or alkaline pH.
Pb (lead)	2.0	5.0	5.0	Can inhibit plant cell growth at very high concentrations.
Se (selenium)	0.02	0.05	0.02	Toxic to plants at concentrations as low as 0.025 mg/L and toxic to livestock if forage is grown in soils with relatively high levels of added selenium. For animals an essential element but in very low concentrations.
Sn (tin)	—	—	—	Effectively excluded by plants; specific tolerance unknown.
V (vanadium)	0.10	0.50	0.1	Toxic to many plants at relative low concentrations.
Zn (zinc)	2.0	5.0	2.0	Toxic to many plants at widely varying concentrations; reduced toxicity at pH > 6.0 and in fine-textured or organic soils.

TABLE 6.12 (Continued)
Recommended Maximum Concentrations of Trace Elements in Irrigation Water for Long-Term Values (LTV) and Short-Term Values (STV)

Source: Based on Australian Water Authority (AWA), Primary industries, in *Australian and New Zealand Guidelines for Fresh and Marine Water Quality* (Paper No. 4), Australian Water Authority, Artarmon, NSW, 2000, http://www.mincos.gov.au/publications/australian_and_new_zealand_guidelines_for_fresh_and_marine_water_quality/volume_3; Westcot, D. W., and R. S. Ayers, Irrigation water quality criteria, in G. S. Pettygrove and T. Asano (Eds.), *Irrigation with Reclaimed Municipal Wastewater: A Guidance Manual*, Lewis Publ., Chelsea, MI, 1985.

[a] LTV: Guideline for when irrigation water is used for a long time (>20 years) on a site.

[b] STV: Guideline for when irrigation water is used for <20 years.

6.9 IRRIGATION WATER SELECTION

Some turfgrass sites may have more than one alternative irrigation water source with each source differing in quality. A feasibility study that analyzes water supply sources usually requires a qualified professional consultant to evaluate all potential sources with respect to supply adequacy, economic viability, engineering considerations, and environmental impacts. Duncan et al. (2009) listed the following general considerations that may apply to one or more of the sources that are particularly relevant to sites using saline irrigation water:

- Blending options and associated costs. Blending of irrigation water sources may be an option to minimize the salt load. This may be as simple as dilution of a saline source in rain-fed lakes; or more complex, such as blending desalinized water with a saline source. Volume of water for blending, storage capacity for each source, and flexibility in blending are all considerations.
- Quality variation over time is another consideration, and knowledge of spatial or temporal variations in water quality is essential to make appropriate best management decisions. Some saline sources may vary in quality over time, and knowledge of this is essential for good management. In other instances, quality variation may be by rainfall dilution within irrigation lakes that results in seasonal changes; but with multiple connected lakes, each lake may vary in overall salt load. Even within a lake receiving water from a saline stream, the quality near the inlet may be higher in salinity than across the lake. The lake may also vary by depth in quality (salinity stratification). The size of the lake, depth of the lake, need for possible aeration to minimize stratification, and location of pump intake are all considerations. In cases with spatial or temporal variation in irrigation water quality, irrigation water quality testing should be conducted at key times during the year so that the turfgrass manager can make adjustments in the management program.
- Pond and lake construction to avoid seepage into the pond of any saline groundwater or saline runoff. Lining and sealing the lake bottoms are considerations.
- Pond and lake (whether an irrigation pond or a water feature) location, design, depth, and inflow and outflow construction measures to avoid seepage into the soil and groundwater. The same would be true for any water conveyance features such as canals or pipes. Drainage catchments would be a consideration.
- On ponds or lakes where water withdrawal may exceed water recharge, especially in the summer, the influence that a drop in water level may have on fish, aquatic plants, and growth of undesirable plants along the exposed shore. When these water features are a part of a housing development, these issues are of concern to these home owners.
- Determination of water rights, competition for a source, permitting, and regulatory negotiations at local, state, and regional levels.

- Regulatory issues related to maintenance of in-stream flow for aquatic organisms, habitat, dilution needs, or the recreational needs of other users.
- When considering surface water collection into ponds, appropriate buffer zones should be used for water quality protection.
- Well water volume yield from aquifers and drawn-down determinations.
- Stream flow during dry periods versus irrigation demand.
- Reliability and water volume both in the long term and over the seasons of a year for all water sources. The anticipated effects of any water use restrictions that may apply to a water source and conservation regulations during drought periods should be included.
- Characterization of the underlying aquifer, which is the process of quantifying the physical and chemical features of an aquifer that may influence groundwater or the potential for contamination from an alternative saline irrigation water source. With more saline irrigation water that requires a leaching program, the potential for contamination of the existing aquifer must be determined. If this potential exists, very careful contouring and subsurface drainage with an appropriate outlet are necessary (Huck et al., 2000; Carrow and Duncan, 1998).
- A complete water quality test for any natural constituents in the water as well as any contaminants. Any permanent native grasses must be able to tolerate the salt levels in the water as well as any overseeding grasses.
- Potential to use an aquifer that is not used for potable purposes because of salinity, but may be available for plant ecosystem irrigation.
- Potential for interaction of water removal from a source on wetlands, streams, sinkhole problems, and so on.
- Energy costs to move water. This should be for well pumps and for transfer pumping costs—whether in pipelines or to pump from one pond to another.
- When more than one water source is used, consideration should be given to the potential loss of one or more of the sources due to drought, increased costs of maintenance due to corrosion, regulatory restrictions, or other reasons, and to the ramifications of losing a source.
- Costs associated with treatment of water prior to irrigation use. In recycling of storm and drainage waters for irrigation, treatment may encompass a typical water treatment facility. For use of desalinized water, the reverse osmosis (RO) or other treatment facility would be a significant cost. The most common water treatment is for irrigation water containing high sodium in conjunction with high bicarbonates that interfere with the use of calcium amendments to prevent formation of a sodic soil (Carrow et al., 1999).
- Total area to be irrigated on schedule, including turfgrass and landscape areas.
- Depth of the lake or pond, with shallow ponds accumulating increased heat during summer months, and tendency of the lake to turn over during rapid temperature changes in the spring or fall, leading to fish kills as oxygen depleted from bottom-stratified layers moves to the surface.

6.10 SUMMARY TABLE

Table 6.13 is a summary table with desired ranges of water parameters that may be found in a routine irrigation quality report. Also, listed is the usual ranges found in irrigation waters, average domestic water (i.e., potable water), and average reclaimed water (Ayers and Westcot, 1985; Stowell and Gelernter, 2001). Duncan et al. (2009) present eight different case studies with diverse irrigation water sources and their interpretation as practical experience in understanding routine irrigation water quality reports.

TABLE 6.13
Summary of Irrigation Water Quality Guidelines

Water Parameter	Units	Desired Range	Usual Range[a]	Average Domestic[b]	Average Reclaimed[b]
General Water Characteristics					
pH	1–14	6.5–8.4	6.0–8.5	7.7	7.1
Hardness	mg/L	<150	—	—	—
Alkalinity	mg/L	<150	—	—	—
HCO_3	mg/L	<120	<610	174	194
CO_3	mg/L	<15	<3	3.0	0
Total Salinity (Soluble Salts)					
EC_w	dS/m	0.40–1.20	<3.0	0.8	1.1
TDS	mg/L	256–832	<2000	617	729
Sodium Permeability Hazard					
SAR_w	meq/L	<6.0	<15	1.9	3.1
adj SAR	meq/L	<6.0	<15	1.9	3.1
RSC	meq/L	<1.25	—	−2.30	−1.88
EC_w	dS/m	>0.40	—	—	—
Specific Ion Impact on Root Injury or Foliage Uptake Injury					
Na	mg/L	<70	—	—	—
Cl	mg/L	<70	—	—	—
B	mg/L	<0.50	<2.0	0.17	0.44
Specific Ion Impact on Direct Foliage Injury from Sprinkler Irrigation					
Na	mg/L	<70	—	—	—
Cl	mg/L	<100	—	—	—
HCO_3 (unsightly on leaves)	mg/L	<90	—	—	—
Miscellaneous					
Residual CL_2	mg/L	<1	—	—	—
Total suspended solids	mg/L	<30	—	—	—
pH_c	1–14	>8.4	—	—	—
Selected Nutrients/Elements					
N	mg/L	<10	<2.2	—	—
P	mg/L	<0.1	<0.66	—	—
K	mg/L	<20	<2.0	4.0	26
Ca	mg/L	<100	<400	67	64
Mg	mg/L	<40	<60	24	23
SO_4 (black layer potential)	mg/L	<90	<960	171	196
Fe	mg/L	<1.00	—	0.16	0.20
Mn	mg/L	<0.20	—	0.01	0.03
Cu	mg/L	<0.20	—	0.04	0.03
Zn	mg/L	<1.0	—	0.12	0.08
Na	mg/L	<120	<920	70	114
Cl	mg/L	<70	<1062	82	130

[a] From Ayers, R. S., and D. W. Westcot. 1994. *Water Quality for Agriculture* (FAO Irrigation and Drainage Paper, 29, Rev. 1). Rome: Food and Agricultural Organization.

[b] From Stowell, L. J., and W. Gelernter. 2001. Negotiating reclaimed water contracts: Agronomic considerations. *Pace Insights*, 7(3): 1–4. San Diego: PACE Consulting.

7 Plant Analysis for Turfgrass

CONTENTS

7.1 THEORY AND PRACTICE OF PLANT ANALYSES

7.1.1 Basics of Plant Analysis

In addition to soil testing for salinity issues (Chapter 4) and for routine soil fertility status determinations (Chapter 5), the nutritional conditions of turfgrass sites can be evaluated by plant visual symptoms (deficiency symptoms) and plant analyses (tissue testing) (Carrow, Waddington, et al., 2001). Visual nutrient deficiency symptoms are best assessed when coupled with knowledge of site conditions that may favor deficiency of a particular nutrient. A detailed presentation of the relationships of visual symptoms and site conditions is provided by Carrow, Waddington, et al. (2001) in Chapter 2 ("Plant Nutrition") of their book *Turfgrass Soil Fertility and Chemical Problems: Assessment and Management*, which is also summarized by Baird (2007).

In this chapter, the focus is on plant analysis as a proactive tool for best management of saline sites. On salt-affected sites, plant analyses or tissue testing is more common than on nonsaline areas, similar to the use of soil and water testing, where each is a site-specific proactive monitoring practice to allow the turfgrass manager to make informed science-based decisions. In Chapter 16 ("Nutritional Practices on Saline and Sodic Sites"), the many soil fertility and plant nutrition challenges that are associated with salt-affected turfgrass areas are presented. The number of Salinity × Nutritional interaction issues coupled with the dynamic nature of soil and plant nutrition status result in an essential necessity for turfgrass managers to use all assessment tools available and on a more frequent basis.

While the terms *plant analysis* and *tissue testing* are often used interchangeably in the turfgrass industry, they do differ. Plant analysis is conducted in a laboratory using wet chemistry methodology or a combination of wet chemistry and combustion methodologies. Tissue testing is conducted in the field, and the process includes conducting colorimetric procedures on freshly extracted tissue sap using test papers, vials, and color charts; but these methods are not quantitative and are of limited use for quantitatively assessing nutritional status in turfgrasses (Plank and Kissel, 2010). We will use the term "plant analysis" in this book, but the reader should be aware that the turfgrass literature often does use the term "tissue testing." Good reviews of plant analysis basics and principles are provided by Westerman (1990), Marschner (1995), Kalra (1997), Reuter and Robinson (1997), Barker and Pilbeam (2007), and Plank and Kissel (2010). Review papers by Feigin (1985) and Grattan and Grieve (1999) discuss plant nutrient concentrations as affected by salt-affected soil and irrigation water situations.

Plants require specific amounts and balances of nutrients for optimum growth and reproduction. A deficiency of one essential nutrient or an imbalance between two nutrients during a critical growth stage can reduce the desired growth characteristics. The extent this effect may have on growth is related to the degree of nutrient deficiency or imbalance and sufficiency requirements of the specific grass cultivar. In some cases, a deficiency or a toxic level of a nutrient can cause total plant failure in a very short period of time.

Plank and Kissel (2010) discuss the method of plant analysis interpretation based on two "critical levels," with the lower one defining the *critical deficiency level* and the higher level defining the *critical toxicity level*, with the nutrient levels in between the two points being the *adequate range* or *sufficiency range* (Figures 7.1 and 7.2). The lower critical concentration was developed initially for row crops and forage crops where yield data were used and below which a 10% reduction in growth may occur. This does not imply that a severe deficiency exists since 90% of maximum yield is still

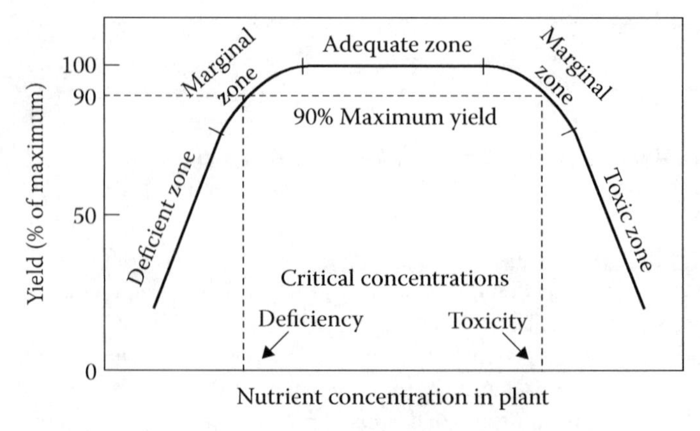

FIGURE 7.1 The relationship between macronutrient concentration in plants and growth or yield. (From Plank, C. O., and D. E. Kissel, *Plant Analysis Handbook for Georgia*, Agric. and Environ. Services Laboratories, Coll. of Agric. and Environ. Sci., University of Georgia, Athens, 2010, http://aesl.ces.uga.edu/publications/plant/.)

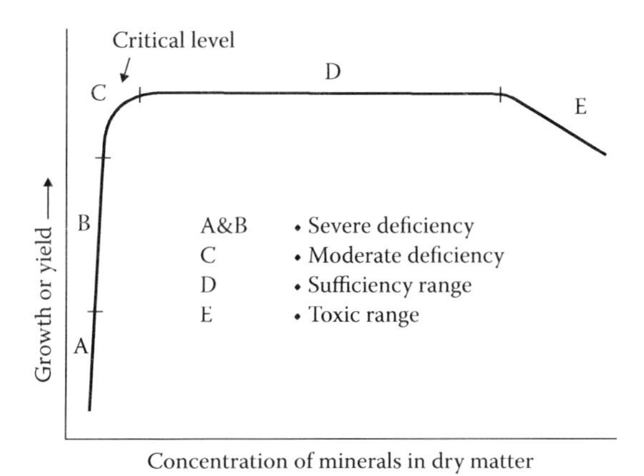

FIGURE 7.2 The relationship between micronutrient concentration in plants and growth or yield. (From Plank, C. O., and D. E. Kissel, *Plant Analysis Handbook for Georgia*, Agric. and Environ. Services Laboratories, Coll. of Agric. and Environ. Sci., University of Georgia, Athens, 2010, http://aesl.ces.uga.edu/publications/plant/.)

possible. This system of evaluation is currently in use in the University of Georgia Plant Analysis Laboratory and most other government and commercial laboratories.

Ideally, sufficiency ranges are developed by plotting yield or plant growth with nutrient concentration. Sufficiency ranges have been developed for turfgrasses by utilizing survey data from large populations of normal-appearing plants as well as some correlation studies, as is discussed in Section 7.5, "Interpretation and Recommendations of Test Results" (and see Table 7.1). Figures 7.1 and 7.2 illustrate two ways in which growth is related to nutrient concentration in plants. It is significant to note the differences in the slopes of the two curves on the left side; the slope in Figure 7.1 is more gradual, while the slope of Figure 7.2 is quite steep. The curve in *Figure 7.1 more nearly typifies one for macronutrients, and the one in Figure 7.2 typifies one for micronutrients* (Table 7.1). The graph in Figure 7.1 illustrates that for most macronutrients, a greater change in concentration occurs between sufficiency and deficiency. Figure 7.2 demonstrates the importance of accurate analyses and interpretations for micronutrients, especially at the low end of the sufficiency range where there is a small difference in nutrient concentration between sufficiency and a severe deficiency. On salt-affected turfgrass sites, where micronutrient deficiencies may occur more often compared to nonsaline areas, this sufficiency range demonstrates the importance of more frequent plant analysis and soil testing to avoid deficiencies.

7.1.2 Uses of Plant Analysis

Plant analysis can be used for two broad purposes (i.e., *diagnostic testing* and *nutrient monitoring*; Plank and Kissel, 2010). Diagnostic testing is to identify or verify a potential problem, while nutrient monitoring is to determine plant nutrient status regardless of the potential for any deficiency or problem. In either case, the objective is to guide fertilization needs. For salt-affected sites, nutritional deficiencies and nutrient or ion toxicities are much more common than on nonsaline sites due to the combination of irrigation water constituents, water treatment chemicals, soil amendments, and leaching programs on the movement of soil nutrients. Thus, plant analysis is used more on salt-affected sites for both diagnostic and monitoring purposes, and turfgrass managers of these areas should be familiar with the pros and cons of plant analysis. Plank and Kissel (2010), Reuter et al. (1997), and Reuter and Robinson (1997) provide excellent background discussions on plant analysis theory and practice, while Stowell and Gelernter (1998), Gelernter and Stowell (2002), and Plank and Carrow (2010) present practical insights for turfgrass situations.

TABLE 7.1
General (Generic) Turfgrass Plant Tissue "Sufficiency Ranges" and Phloem Mobility (Downward from Shoot to Roots from Foliar Application)

Nutrient	Common Sufficiency Range	Mobility in Plant
Macronutrients	%	
N	2.8–3.5	Mobile
P	0.20–0.55	Mobile
K	1.50–3.00	Mobile
Ca	0.50–1.25	Immobile
Mg	0.15–0.50	Mobile
S	0.20–0.50	Somewhat mobile
Micronutrients	ppm	
Fe	50–100	Immobile
Mn	20–100	Immobile
Zn	20–55	Somewhat mobile
Cu	5–20	Somewhat mobile
Mo	1–4	Somewhat mobile
B	5–60	Somewhat mobile
Cl	200–400	Mobile
Ni	<1	—
Other	%	
Si	1–3%	Immobile
Na	<0.50%	Mobile

Source: Carrow, R. N., L. J. Stowell, S. D. Davis, M. A. Fidanza, J. B. Unruh, and W. Wells, Developing regional soil and water baseline information for golf course turf, in *Agronomy Abstracts,* ASA, Madison, WI, 2001. Copyright Wiley-VCH Verlag GmbH & Co. KGaA. Reproduced with permission.
Note: Certain species and cultivars can vary from these general ranges.

A common sequence for plant analyses use on a new golf course where saline irrigation water is used requires the turfgrass manager to sample plant tissue from selected areas on a relatively frequent basis in the initial 2 to 3 years after planting, just as soil testing and water quality testing are more frequent. Once the grow-in phase is completed, it may require a couple of years to understand specific nutritional challenges on the site, the fluctuations in nutrient availability and plant uptake over seasons, and the reasons for these issues. Also, it will take some time to understand the more dynamic and seasonal nature of nutritional changes under saline irrigation. Thus, during these early years, tissue testing is often used both as a diagnostic tool and also for proactive monitoring to start establishing the baseline tissue concentrations on the site and how they vary during the year. The key areas to obtain tissue samples for long-term monitoring normally become clear from the seasonally accumulated data. Then, a monitoring program may be established on these key areas with diagnostic use of plant analyses at times as problems arise. However, the monitoring may be on fewer sites and with the timing adjusted as determined by the experience of the turfgrass manager.

Specific applications of plant analysis for turfgrass areas with emphasis on salt-affected turfgrass sites include the following:

- The most common reason for plant analysis on saline and sodic turfgrass sites is to monitor the nutrient status of plants throughout the growing season to determine whether each nutrient is present and available in sufficient concentration for optimum growth characteristics and physiological activity. The sufficiency range presented for tissue nutrient contents is based on growth where adequate growth is expected across the whole range. However, to maximize salinity tolerance, such as for Ca, K, Mn, and Zn, which are especially important nutrients for maximizing turfgrass salinity tolerance, the turfgrass manager may wish to maintain these elements at the upper end of the sufficiency range—see Chapter 16 for discussion on specific nutrients that require special attention on saline sites. For example, K tissue content sufficient for growth may not allow complete osmotic adjustment to salinity stress when K concentration is at the lower end of the sufficiency range. Just as soil tests are more often taken to monitor the dynamic nature of changes in soil nutritional status, so is plant analysis required to proactively view changes in plant nutrient status.

- For salt-affected sites, the next most common reason for plant analysis is to confirm suspected nutrient deficiency symptoms (i.e., diagnostic use of plant analyses). For example, saline irrigation water high in Na may cause Ca tissue concentrations to be lower than expected, even to the point of deficiency because of the dual impact from actual foliar adsorption as well as soil accumulation, on sites where the turfgrass manager is actually applying ample gypsum for soil Na control.
- Verify possible toxicities.
- Aid in evaluating the utilization efficiency of applied fertilizers.
- Assist in formulating and readjusting fertilization practices—granular products, liquid products, dual-product combinations, slow-release products, foliar feeding, highly soluble products, supplemental fertigation, and so on.

Plant analysis is a proven and effective means of predicting fertilizer needs for many crops. However, it does not completely replace a soil test (Gelernter and Stowell, 2002; Sartain, 2008; Plank and Carrow, 2010). Soil and plant analyses serve different purposes, and when properly used they complement each other in providing detailed information for maximizing the utilization efficiency of fertility programs. Soil testing is based on the concept that the concentration of a particular nutrient in a given volume of soil reflects whether or not the nutritional level of that soil is adequate for optimum crop growth or production. Plant analysis is based in part on the concept that the amount of a specific nutrient in the plant tissue is related to the plant availability of that element in the soil—*with the assumption that the plant has adequate roots and they are viable plus the nutrient is actually available for uptake.* Thus, plant analysis also reflects nutrient uptake conditions in the soil. Soil properties such as compaction, impervious layers, or poor drainage may inhibit the uptake of nutrients by plants. Or a low concentration of one nutrient in the plant may result from the excessive application of another nutrient. Conversely, favorable soil physical properties and optimum soil moisture may accentuate nutrient uptake even though the soil may not have an abundant supply of nutrients.

As a result of these soil–plant interactions, there are certain instances when contradictions occur between soil and plant analysis results. For example, assume turfgrass is growing on a soil in which the soil tests revealed a medium level of extractable Mg. A plant analysis from the area a few weeks later indicates that Mg is low. Immediately, the validity of the test results are questioned, which is an absolutely normal response. However, a closer examination of the plant analysis results revealed that the internal Ca and K concentrations of the turfgrass were high. After checking the information supplied on the history (basic initial and subsequent data) sheet accompanying the plant analysis results, it was noted that calcitic limestone had been used as the liming material and a high rate of K was applied in the fertilizer program. As a result of these two management practices, the levels of Ca and K in the soil were sufficient to reduce Mg uptake.

Another example of when plant analyses and soil test data may not be in agreement is when organic matter accumulation in the surface zone (1–2 inches, 2.5–5.0 cm) is high enough to result in excess moisture retention during frequent rain and high-humidity conditions, thereby causing low soil aeration, limited rooting depth, and reduced root viability. In these conditions, roots are confined to the surface zone and are not healthy or functionally adequate, which reduces soil nutrient uptake.

These examples illustrate how soil testing and plant analysis can be used together to make better nutrient management decisions. Plant analysis can also be used to supplement a soil-testing program. It is particularly useful in distinguishing between N and S deficiencies in turfgrasses since deficiency symptoms of the two elements are similar. Plant analysis offers an excellent means of delineating which element is deficient (which cannot be ascertained through soil testing). If this distinction is not made properly and the wrong corrective treatment is applied, plant growth can be affected appreciably.

Essential components of a good plant analysis program are:

- Obtaining a representative sample
- Timely collection of each sample
- Preparing the sample for analysis
- Accurate analysis of the sample for the target nutrient or nutrients
- Correct interpretation of the results
- Proper recommendations from the historical and present data

Each of these components is discussed in the following sections.

7.2 SAMPLING

7.2.1 Monitoring Sampling

Interest has increased over the years for plant analysis monitoring of turfgrass sites, especially on salt-affected sites where monitoring is for both growth and physiological functions. For monitoring, tissue samples are collected from the same location on a periodic basis. On a golf course, tissue monitoring could be from each green, tee, and fairway, but usually "representative" greens, tees, and fairways are selected that serve as a guide to the others. Sampling should be at the same location on a specific site—such as always at the front of the representative green. Frequency of sampling normally varies from monthly to weekly.

Sampling should be avoided immediately after granular or liquid fertilizer applications, during extreme climatic periods such as drought or water conservation–mandated times when the granular products might not be completely solubilized and integrated into the grass canopy, or immediately after exposure to other potential chemical contaminants (pesticides, plant growth regulators, etc.). Timely sampling will generally provide more accurate data for assessing any fertilizer adjustments in the best management practices (BMPs) for the salinity-challenged site.

Mowing clippings are used for turfgrass tissue analysis with mowing at the normal height used on the site. Clean mower baskets are acceptable if care is taken to wash off any soil that may contaminate the clippings. If clippings are transferred to a bucket, a plastic container is preferred. After mowing, randomly remove three to four handfuls of clippings from the basket of walking mowers or two to three handfuls from each basket of Triplex mowers and mix thoroughly. Take a subsample of about 1 cup (350 ml), and place in a brown paper bag or plastic bucket. Save the sample for washing and drying preparation prior to shipment or onsite analyses.

7.2.2 Diagnostic Sampling

The diagnostic role has been the traditional use of tissue testing for turfgrass situations to confirm a suspected nutrient deficiency (for growth or physiological function) or nutrient or element toxicity prior to applying corrective measures. Sample locations should be from (a) an area exhibiting the stress symptoms, and samples from turfgrass where the symptoms are just appearing are much better than from areas with severe symptoms or dead turf; and (b) an adjacent area that does not show any visual stress symptoms.

7.3 SAMPLE PREPARATION

Contamination on the leaves of turfgrasses or within the sample can make the test results invalid. Types of contamination include particulate matter, dust, dead or diseased plant tissue, and spray or other chemical residues on leaves (Arkley et al., 1960; McCrimmon, 1994; Plank and Kissel, 2010). Reuter et al. (1997) provide an excellent review of various washing procedures used on different crops. Plant leaf surface conditions differ with species and may affect contamination potential due to leaf morphology, wax types or load, and pubescent foliage (Reuter et al., 1997). On salt-affected sites, the frequent use of foliar nutrient applications and soil amendments (e.g., gypsum and others), as well as saline water and the application of any pesticides, all make timing of plant sampling somewhat more difficult. Sampling after a rain and when tissues are relatively dry is a good option if available.

Particulate matter in the sample may be sand or fertilizer, lime, or gypsum particles. If not removed, sand particles may be weighed with the ground sample and result in an incorrect percentage of nutrient in dry weight of "tissue." Obviously, any fertilizer, lime, or gypsum pellets may show up as tissue nutrient. Physical separation of particulate matter is necessary if these are present. Failure to remove sand from samples using proper washing techniques generally, with the exceptions of iron and aluminum, results in lower nutrient concentrations as compared to washed samples due to weight dilution. When samples are contaminated with sand, dust films, or soil particles, the analytical results for both iron and aluminum are high. To minimize the amount of sand in samples from golf greens, do not take samples immediately after topdressing. Wait until the sand is watered in and settled on or integrated into the surface canopy. This may require several days. Also, avoid taking samples on days when the grooming heads are attached to the mowers. The grooming heads result in increased sand and other foreign materials in the sample and prolong the washing procedure.

Diseased or dead plant material should not be included in a sample. Also, avoid sampling plants that have been damaged by insects and nematodes, or stressed extensively by cold, heat, moisture deficiencies, or excess moisture.

Turfgrass clippings often contain sufficient dust (soil or fertilizer or amendment dust) on the leaves to influence micronutrient values, especially Na, Fe, and Al, but usually not macronutrient values. If dust or soil contaminates are expected on the leaves, washing in a dilute soap solution is sufficient to remove this surface contamination (Arkley et al., 1960; McCrimmon, 1994). Clippings can be placed in a glass container of 1–2 quarts (1–2 liters) with a weak soap solution (5–6 drops of soap per quart), agitated for 30 to 60 seconds, poured into a colander; washed immediately, rinsed with good-quality tap water for 30 to 60 seconds, and laid out for leaves to air dry or dry in the sun.

Chemical contamination from foliar sprays (such as pesticides or fertilizer) can cause tissue deposits that are more difficult to remove than dust or soil. Another potential surface contaminate on salt-affected sites is the irrigation water constituents. Micronutrients are most affected. Nutrients from foliar sprays and at least some from the irrigation water may be taken up by the plant and be incorporated into plant tissue compounds—these would legitimately be part of the tissue nutrient content. However, surface deposits should not be. Arkley et al. (1960) demonstrated that a more vigorously applied washing solution was required to remove these materials such as a combination of weak soap (as described above) in a weak acid solution (0.1 N hydrochloric acid or, if the wash is conducted at the turfgrass facility, perhaps dilute acetic acid could be used since vinegar is easy to obtain). The wash procedure would be as outlined above, but using a wash time of at least 60 to 90 seconds.

The authors have observed that on sites with saline irrigation water, the micronutrient tissue levels often are unreasonably high. Washing clippings with water does not seem to be very effective. *Thus, we strongly suggest that on these sites, the longer washing cycle with a weak soap and weak acid solution, as noted above, be used followed by at least 60 seconds of washing* off the weak soap and weak acid with contaminates (see Table 5.1). Research has shown that internal nutrients are not leached from within the tissue unless the sample is washed for 1 hour or more (Arkley et al., 1960). If washing is not done onsite, the laboratory should be instructed to wash the samples in the above manner. This may cost more per sample, but is well worth the extra cost to obtain accurate plant analysis data.

If samples are washed onsite, the excess water should be shaken off and followed by partial air drying before shipment to the laboratory. Place samples in a paper bag or envelope, and allow the sample to set in a warm place for one half-day with the end of the bag or envelope open. This will generally remove excess moisture. Do not place the sample in a polyethylene bag or tightly sealed containers unless the samples are kept under refrigeration. When samples are shipped in polyethylene bags by routine mail, they have generally started to decompose by the time they are received for processing and the samples may not be suitable for analysis. If the sample is to be washed by the laboratory, air-dried samples should be kept in a fresh state by shipping immediately in a paper bag or envelope—if the sample is somewhat moist, air dry as noted above before shipping. Some international plant samples are being sent after sun drying and are generally sent by express shipment.

7.4 ANALYSIS OF THE SAMPLE

7.4.1 Conventional Wet Lab Analysis

Conventional plant analysis laboratories use *wet chemistry* to digest and analyze plant samples for the total quantity of 12 to 13 elements. Previously dried tissue samples, ground and weighed, are prepared for elemental analysis by destroying the organic matter using either wet chemical or thermal digestion procedures (Westerman, 1990; Kalra, 1997; Reuter and Robinson, 1997; Plank and Carrow, 2010). Wet chemical digestion involves the destruction of organic matter through the use of both heat and acids. Acids used in this procedure include sulfuric, nitric, and perchloric acids, either alone or in combination. This is a rather time-consuming procedure and may require up to 16 hours for complete digestion. A relatively new accelerated wet chemical microprocedure for organic matter destruction utilizes pressure and high temperature to shorten the digestion process to approximately 1 hour (see http://www.cropsoil.uga.edu/%7Eoplank/plantanalysisturf/Analysis/Ashing/ashing.html). Dry ashing is conducted in a muffle furnace at temperatures of 500°C to 550°C for 4 to 8 hours. Once the organic matter has been destroyed, the elements are dissolved in dilute nitric or hydrochloric acid, or a mixture of both. The whole sample is dissolved so there is not an "extractant" issue such as in routine soil testing and the total quantity of nutrient in the dissolved solution is analyzed, which results in high accuracy.

Recent developments in accelerated sample digestion coupled with new innovations in high-speed analytical equipment such as inductively coupled argon plasma emission spectrographs (ICAP) and combustion apparatus for N and S analyses make it possible for scientists to complete a plant analysis in 24 hours or less. ICAP instruments have the capability of simultaneously analyzing samples for 10 to 12 elements at the rate of one sample per minute, and combustion units analyze samples for N and S at the rate of one sample per 5 minutes. These rapid instruments achieve a very high degree of accuracy because they are calibrated against both known chemical and plant standards. Data are collected via computers and promptly returned electronically to clients. These advancements have made conventional laboratories very attractive to turfgrass managers in the last few years for obtaining rapid and accurate plant nutrient analyses.

7.4.2 Near-Infrared Reflectance Spectroscopy (NIRS)

Periodically in the turfgrass industry, the *near-infrared reflectance spectroscopy* (*NIRS*) procedure has been promoted by some as a means to obtain rapid tissue analysis information for diagnostic or monitoring purposes either by using an onsite NIRS unit or by shipping samples to a laboratory that utilizes NIRS (Plank and Carrow, 2010). Results can be obtained quickly because NIRS is a nondestructive procedure and precludes the digestion phase required with wet chemistry procedures. However, with the improvements noted in the previous paragraph, time in wet lab digestion and analysis procedures is less of an issue than in the past. Sample preparation usually involves only drying and grinding the sample. Once the sample has been prepared for analysis, scanning or analysis time typically is less than 3 minutes. Elements analyzed include N, P, K, Ca, Mg, S, Zn, Cu, Fe, Mn, B, and Na. Although this technology has been successfully used for several decades for determining N, total protein, carbohydrates, lipids, other organic chemicals, and moisture content in forages, grains, and oil crops (Clark et al., 1987; Foley et al., 1998; Masoni et al., 1996; Vazquez et al., 1995), its application in turfgrass plant analysis for nutrients other than N has not been nearly as accurate as wet lab analysis.

The basis of NIRS is to determine the reflectance of specific wavelengths (1 or 2 nm) over the infrared spectral range (750 to 2500 nm) and relate the degree of reflectance to a specific compound or element. Infrared wavelength energy is absorbed mainly by:

- C-H bonds, common in carbohydrates
- N-H bonds, common in proteins, amides, and amino acids
- O-H bonds, common in water

If the wavelength radiation matches the vibrational or rotational frequency of the chemical bond within a particular plant compound, it is absorbed and influences spectral reflectance. Statistical procedures correlate reflectance of one or more specific wavelengths to the true level of a compound or nutrient as measured by wet laboratory methods. A regression equation is developed to estimate the tissue content or quantity of a nutrient or compound based on the strength of reflectance from these wavelengths. This equation is then entered into the computer software for use by NIRS on future samples where wet laboratory analysis will not be conducted. However, achievement of statistically significant equations with high correlation (r^2 or correlation coefficient of determination; $r^2 > 0.90$; 1.0 is perfect) has not been demonstrated for nutrients except for N. Stowell and Gelernter (1998), Rodriquez and Miller (2000), Gelernter and Stowell (2002), Miller and Thomas (2003), and Plank and Carrow (2010) compared wet lab data to NIRS data for a number of macro- and micronutrients on grasses with limited success except for N. Plank and Carrow (2010) noted that except for N, none of the nutrient elements are directly involved in a C-H, N-H, or O-H bond. Reflected wavelengths are, therefore, always indirectly related to a nutrient rather than directly. This results in lower correlations of NIRS nutrient values (except for N) versus wet lab values than those achieved for organic components, which usually exhibit correlation coefficients of determination of $r^2 > 0.95$ because they are directly involved in C-H, N-H, or O-H bonds.

This does not preclude the use of NIRS in certain turf management programs because it can be effectively used for monitoring potential excessive or deficient levels of N. Nitrogen management is very important in turfgrasses because N influences numerous growth factors directly or indirectly. These factors include color, density, growth rate, organic matter accumulation, disease, and other factors.

7.4.3 Other Plant Analysis Approaches

Two other types of plant nutrient analyses are *sap nutrient content* and *ground-level spectral reflectance instruments* using visible (400–750 nm) and/or infrared (750–2500 nm) wavelengths (Bell and Xiong, 2008). Sap nutrient content is based on measuring the nutrient level in sap expressed from leaves or a grass stem. This measurement indicates nutrient levels in the xylem and phloem, which are not necessarily related to total tissue content. Interpretation of the data as to what the values mean have not been defined for turfgrasses; also, expressing sap from turfgrass tissues usually results in considerable tissue injury and mixing of cell contents with xylem and phloem sap.

Spectral reflectance handheld and mobile-sensing devices using certain wavelengths or combinations of wavelengths reflected off turfgrasses may be related to N content of leaves or other nutrients that may cause chlorophyll loss when deficient (Fe, S, Mg, and Mn) (Bell and Xiong, 2008). Generally these devices are measuring reflectance of wavelengths related to chlorophyll activity, but do not distinguish whether loss of chlorophyll is due to N, Fe, S, Mg, or Mn deficiencies or some other reasons that may affect chlorophyll content (high temperature, salinity, etc.). Also, visual observation of the turfgrass often reveals chlorosis as quickly as remote sensing.

7.5 INTERPRETATION AND RECOMMENDATIONS OF TEST RESULTS

In the first subsection (7.5.1, "Interpretation Based on Plant Analysis"), the focus will be on using plant analysis data to determine possible nutritional problems and to monitor plant nutritional status based on the tissue data—deficiency, sufficiency, and toxic ranges. However, the best approach to assessing nutritional status on turfgrass sites is by using all available means—irrigation water quality tests, soil tests, tissue tests, visual plant symptoms, and understanding site conditions that may favor a nutrient deficiency, toxicity, or excess. Thus, in the second subsection (7.5.2, "Interpretation of Tissue Tests in the Context of Site Conditions"), supplemental information will be summarized that aids in a more comprehensive approach to interpretation of plant analysis results (as well as interpretation of soil test and water quality test results), namely:

- Deficiency conditions: site conditions favoring deficiency of a nutrient
- Deficiency symptoms: plant visual deficiency symptoms
- Toxicity and excess conditions: site conditions favoring toxicity or excess levels of the nutrient

7.5.1 Interpretation Based on Plant Analysis

The basic approach for using plant tissue analysis data is to compare measured values to published "sufficiency ranges" for the grass (Table 7.1, Figure 7.1, and Figure 7.2). Nutrient levels below the sufficiency range imply a deficiency and rapid growth decreases, and visual nutrient deficiency symptoms would be expected if the low end value for a sufficiency is accurate. Thus, a measurement below the sufficiency range (i.e., in the "critical range") implies an existing deficiency, and immediate corrective action should be taken.

While this appears to be easy, interpretation of plant analyses for turfgrass can be more complex even without interactions from salinity. Turner and Hummel (1992), McCrimmon (1998, 2000, 2002, 2004), Campbell (2000), Carrow, Waddington, et al. (2001), Gelernter and Stowell (2002), Plank and Carrow (2010), and Plank and Kissel (2010) provide data on sufficiency ranges for different turfgrasses. Reliable interpretive data are lacking for a number of turfgrass species, particularly at different stages of growth and for nutrient concentrations near or at toxicity levels. Other factors that make the interpretative process complex are the effects of cultivar differences (which is common), unknown mixtures of species (creeping bentgrass and annual bluegrass), nematodes, and environmental factors affecting plant growth such as soil moisture, temperature, light quality, and the intensity they have on the relationship between nutrient concentration and plant response (Waddington et al., 1994; McCrimmon, 1998, 2000, 2004). If the laboratory is making recommendations based on the current test results, it is important to supply historical and other information requested on plant analysis forms when submitting samples to plant analysis laboratories.

Saline conditions also influence nutrient concentration versus growth response curves such as Figure 7.1. Grattan and Grieve (1999, 204) noted that

> [a] substantial body of information in the literature indicates that a plant may not exhibit the same response function under saline conditions as it does under non-saline conditions. In some cases, the optimal range may be widened, narrowed, or it may shift in one direction or the other depending upon the plant species or cultivar, the particular nutrient, the salinity level, or the environmental conditions.

For an initial "generic" approximation of turfgrass sufficiency ranges, Carrow, Waddington, et al. (2001) reported the values in Table 7.1. Also, the downward phloem mobility of the nutrient in the plant (i.e., from shoots to roots) is listed in this table. Nutrients such as Ca and Mn, which are immobile internally in plants, will not be transported as foliar-applied nutrient to roots. Thus, foliar-applied Mn would not be expected to increase soil Mn levels to suppress take-all disease, or Ca applied to the foliage would not be transported to roots to offset Na displacement of Ca in roots.

As noted, a defined sufficiency range may not apply to all situations or environments. In plants, nutrient concentrations are not absolute because nutrient uptake and internal mobility, nutrient ratios, as well as dry matter changes can affect the nutrient concentrations in plant tissues. Consequently, internal plant nutrient concentrations are not static; they change during the growing season in response to environmental and management conditions (Carrow, Waddington, et al., 2001) (Figure 7.3). Concentration and dilution occur due to the difference between plant growth rate and nutrient absorption as well as movement of the nutrients within and between plant parts—all in response to changes in growing conditions. Under normal growing conditions, nutrient absorption and plant growth closely parallel each other during most of the vegetative growth period. However, if the normal rate of growth is interrupted, nutrient accumulation (higher than expected nutrient values) or dilution (lower than expected nutrient values) can occur. Some of the factors that can result in nutrient accumulation include extremes in temperature for the grass species; heat or moisture stress; stress due to traffic, salinity stress, or other cultural practices; recent fertilizer applications; and stunting (reduced growth) due to a soil deficiency of a particular nutrient or nematode infestations. Factors that can result in dilution are growth factors that stimulate rapid growth, which may include highly favorable climatic conditions (moisture, temperature, light, etc.) and rapid growth response to N applications (Plank and Carrow, 2010).

When nutrient concentrations in turfgrass tissues change in response to enhanced growth or reduced growth conditions, as noted above, the levels would be expected to be within the sufficiency range as long as a soil nutrient is not truly deficient. Thus, a drop in a nutrient level within the sufficiency range may not mean that the plant needs fertilization; but it may simply be in response to growing conditions. However, if a soil nutrient is truly deficient, then the tissue content would be expected to remain in the deficient range and may continue to decrease.

Nematode activity on roots can produce nutrient deficiencies similar to those resulting from low soil levels. When elements such as Ca and P are deficient in the plant tissue, but soil pH and soil test P and Ca are adequate, this is a good indication that nematodes are the cause of the problem.

With the lack of information on sufficiency ranges for different species and cultivars within species plus the strong probability that saline conditions may alter sufficiency ranges determined under

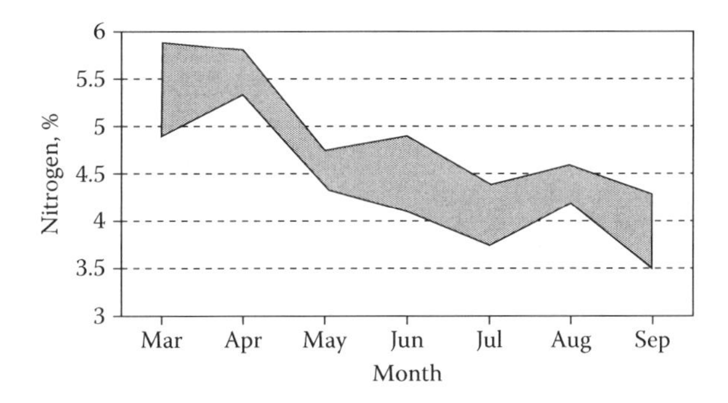

FIGURE 7.3 Seasonal changes of tissue N averaged over 18 bentgrasses in 1991 in northern Georgia. (After Plank, C. O., and R. N. Carrow, *Plant Analysis: An Important Tool in Turf Production,* Coll. of Agric. and Environ. Sci., University of Georgia, Athens, 2010, http://www.georgiaturf.com.)

nonsaline conditions, what can turfgrass managers do to determine appropriate sufficiency ranges for their grasses? Some suggestions that can assist in this process are as follows:

- Where published sufficiency ranges exist for a particular cultivar, start with these as a first baseline approximation.
- When published sufficiency ranges are not available for a particular cultivar, use data on similar cultivars for approximate ranges.
- A "generic" turfgrass sufficiency range is listed in Table 7.1 from Carrow, Waddington, et al. (2001) and can be used as a guide.
- Ultimately, the development of site-specific historical records provides the best turfgrass species and cultivar sufficiency ranges. These are based on maintaining records of tissue concentrations as they vary with season, management practices, and stress periods, including salinity stresses.

7.5.2 Interpretation of Tissue Tests in the Context of Site Conditions

On salt-affected sites, turfgrass managers should use all available tools to monitor soil and plant fertility and toxic ion issues: soil testing, plant tissue analysis, irrigation water quality tests, knowledge of site conditions relative to what nutrient or ion deficiencies or toxicities may occur, and visual deficiency and toxicity symptoms (Carrow, Waddington, et al., 2001; Duncan et al., 2009). We recommend starting with the water quality tests because when saline irrigation water is used, it has the most influence on soil fertility status and is the most important factor in understanding soil test results. Soil testing is next in importance, followed by plant tissue testing based on carefully selected indicator locations and taken in a timely fashion. Visual observation for nutritional deficiency or toxicity problems is an ongoing "check" on site conditions. Knowledge of what site conditions (soils, amendments applied, grass, etc.) may favor a nutrient deficiency or toxicity is essential in the assessment. Integrating all these sources of information is always better than reliance on any one technique or source of information. For any professional turfgrass manager, an understanding of these "tools" and expertise in interpretation are essential; but on salt-affected sites, the complexity and dynamic nature of soil fertility and plant nutrition in response to salinity additions and environmental challenges make such understanding and expertise even more critical. It is not unusual on salt-affected sites for turfgrass managers to use consultants to provide assistance; however, the site manager must be able to understand test results, interpret the data with knowledge of site-specific conditions, and assess any recommendations as far as implementation. Additionally, the site manager should be the most informed on site conditions that may favor deficiency, toxicity, or excess of a nutrient, as well as be able to rapidly observe possible deficiency symptoms. These aspects are summarized in the remainder of the chapter for each nutrient (Carrow, Waddington, et al., 2001; Sartain, 2008).

7.5.2.1 Nitrogen

Deficiency conditions. N deficiency is associated with sandy soils; high leaching conditions from rainfall or irrigation; ultrapure irrigation water; low organic matter content; clipping removal; loss by denitrification under anaerobic conditions, such as waterlogged and compacted soils; low pH < 4.8 inhibition of *Nitrosomonas* for nitrification; and infertile soils.

Deficiency symptoms. Symptoms include chlorosis or a gradual loss of green color that appears first on older (lower) leaves; reduced growth rate; senescence of lower leaves; loss of shoot density in response to loss of leaves, tillers, and whole plants; and increase in low-N diseases.

Toxicity and excess conditions. Direct toxicity as a salt and/or excessive growth and tissue succulence related to excessive application of N as fertilizer, the use of acidification products containing N for water bicarbonate treatment, or from effluent water with high N content. N is also a contributor to algae buildup in lakes and ponds.

7.5.2.2 Phosphorus

Deficiency conditions. P deficiency is more frequent on sandy, low-CEC irrigated soils; under low pH (<5.5) due to fixation with Fe, Al, and Mn; under high pH (7.5–8.5) from fixation with Ca; on soils containing hydrous oxides of Fe, Al, Mn or kaolinite fix P; on soils high in clay content fix P; during establishment due to high P demand and limited rooting; or due to reduced uptake in cold soils, site receiving little or no P over several years, and/or turfgrass grown on subsoil or infertile soil (decomposed granite, pulverized caliche, or volcanic pumice).

Deficiency symptoms. Symptoms include a reduced growth rate; dark green color, and lower leaves may turn reddish at the leaf tips and progress down the blade; and low density.

Toxicity and excess conditions. Excessive P application may induce Fe deficiency under some conditions. High soil P is a concern over P leaching into subsurface waters or runoff into surface waters and its contribution to algae buildup in ponds and lakes, contributing to eutrophication problems.

7.5.2.3 Potassium

Deficiency conditions. Deficiencies are most likely under high rainfall or leaching; sandy or low-CEC soils; acidic soils (pH < 5.5); clipping removal; sites receiving high Ca, Mg, or Na additions, often associated with saline sites, which can suppress K uptake or enhance leaching loss; high N fertilization, since more K is needed; and soils high in vermiculite, illite, or smectite at high pH.

Deficiency symptoms. Symptoms include interveinal yellowing of old leaves (lower) followed by dieback at the tip, and scorching or firing at leaf margins; eventually total yellowing of the leaf blades, including veins; weak, spindly turfgrass; and a greater tendency to exhibit wilting and wear injury.

Toxicity and excess conditions. Excessive K can contribute to total salinity stress; suppress Mg, Ca, or Mn uptake; and cause fertilizer burn.

7.5.2.4 Calcium

Deficiency conditions. Deficiency is most likely in root tissues causing poor root viability under acidic (pH < 5.5) conditions on low-CEC soils receiving high Na levels or with high Al, Mn, or H; high leaching; and irrigation water high in Na, which appears to reduce Ca content in shoot tissues over time and results in shoot deficiencies even when soil Ca appears adequate.

Deficiency symptoms. Symptoms include distorted appearance of new leaves; reddish brown to rose leaf blades; leaf tips and margins withering and dying; and stunted, discolored, brown to black roots.

Toxicity and excess conditions. Excess Ca can induce Mg, K, Mn, or Fe deficiencies in some cases.

7.5.2.5 Magnesium

Deficiency conditions. Mg deficiencies are enhanced by acidic (pH < 5.5) conditions, especially on sandy soils due to low CEC and high Al, Mn, and H; under high Na, Ca, or K additions from irrigation water or amendments; and from high leaching.

Deficiency symptoms. Symptoms include general loss of green color of older (lower) leaves progressing from pale green to cherry red, with leaf margins exhibiting blotchy areas of red. Leaf veins remain green, with some light yellowing striping between veins; and leaves start to die.

Toxicity and excess conditions. Excessive Mg can induce deficiencies of K, Mn, and Ca. Irrigation water influenced by seawater intrusion is high in Mg and can mimic Na permeability hazard conditions.

7.5.2.6 Sulfur

Deficiency conditions. Deficiencies of S are often associated with soils low in organic matter; sandy, low-CEC soils; high rainfall and leaching; areas not receiving atmospheric S additions; and high N with clipping removal.

Deficiency symptoms. Reduced shoot growth rate; yellowing of new leaves, with leaf tips and margins showing symptoms first; and older leaves may show symptoms later in the interveinal areas.

Toxicity and excess conditions. Excessive S applications can cause tissue burn; induce extreme acidity on soils not buffered by free $CaCO_3$; and contribute to black layer under anaerobic conditions. Some reclaimed water is high in S.

7.5.2.7 Iron

Deficiency conditions. Deficiencies are most likely at $pH > 7.5$; under poor rooting, excessive thatch, and/or cold and wet soils such as in the spring; under high soil P, especially at high pH; on high-pH calcareous soils in arid regions; in irrigation water high in HCO_3, Ca, Mn, Zn, P, or Cu; in low organic matter soils; and due to heavy metals from some sewage sludge.

Deficiency symptoms. Interveinal yellowing of new leaves (upper); leaves turn pale yellow to white; thin, spindly leaves; eventually older leaves exhibit chlorosis; and the turfgrass stand has a mottled appearance, with some areas exhibiting symptoms while others do not.

Toxicity and excess conditions. High foliar Fe can blacken leaves and, if sufficiently excessive, cause tissue injury; and it can induce Mn deficiency. Acidic, poorly drained soils can produce toxic levels of soluble Fe in roots.

7.5.2.8 Manganese

Deficiency conditions. Deficiencies are associated with high-pH, calcareous soils; peats and muck soils that are at $pH > 7.0$; dry, warm weather reduces Mn availability; high levels of Cu, Zn, Fe, and Na, especially on leached low-CEC soils.

Deficiency symptoms. Reduced shoot growth rate; small, distinct greenish-gray spots starting on youngest leaves; progresses to interveinal yellowing with veins green to light green; leaf tips may turn gray or white and exhibit drooping or withering. Turfgrass stand may appear mottled.

Toxicity and excess conditions. Toxicity to roots from Mn can occur in acid soils of $pH < 4.8$ and in anaerobic soils, especially if acidic; high Mn levels can induce Ca, Fe, and Mg deficiencies; and Si and high temperatures increase plant tolerance to Mn toxicity.

7.5.2.9 Zinc

Deficiency conditions. Deficiencies are more common on alkaline soils; with high levels of Fe, Cu, Mn, P, or N; with high soil moisture; in cool, wet weather and low light intensity; and in highly weathered acid soils.

Deficiency symptoms. Chlorotic leaves with some mottled, stunted leaves; leaf margins may roll or appear crinkled; and symptoms are most pronounced on young leaves.

Toxicity and excess conditions. Some municipal wastes may be high in Zn. High Zn may cause chlorosis by inducing Fe or Mg deficiencies.

7.5.2.10 Copper

Deficiency conditions. Deficiencies are more often observed on organic soils due to strong binding of Cu; heavily leached sands; high applications or soil levels of Fe, Mn, Zn, P, and N; and high pH.

Deficiency symptoms. Yellowing and chlorosis of leaf margins of youngest to middle leaves; white to bluish leaf tips that wither, turn yellow, and die; stunted growth; and leaves roll or twist.

Toxicity and excess conditions. Toxic levels can occur from some sewage sludges or pig or poultry manures; and use of high-Cu-content materials.

7.5.2.11 Boron

Deficiency conditions. High soil pH can induce deficiencies, especially on leached, calcareous sandy soils; high Ca can restrict B availability; dry soils; and high K may increase B deficiency on low-B soils.

Deficiency symptoms. Young leaves exhibit leaf tip chlorosis followed by interveinal chlorosis of young and older leaves; leaves curl; roots are stunted and thick; and plants may appear bushy or rosette in appearance.

Toxicity and excess conditions. B toxicity is much more likely than deficiencies due to irrigation water high in B; soils naturally high in B; overapplication of B; and the use of some compost amendments high in B.

7.5.2.12 Molybdenum

Deficiency conditions. Deficiencies are usually on acid, sandy soils; and acid soils high in Fe and Al oxides. High levels of Cu, Mn, Fe, and S suppress uptake.

Deficiency symptoms. Similar to N with chlorosis of older leaves and stunted growth. Some mottling and interveinal yellowing of leaves may appear.

Toxicity and excess conditions. Mo toxicities are important for grazing animals and are associated with high-pH soils that are wet.

7.5.2.13 Chloride

Deficiency conditions. Cl uptake is suppressed by high NO_3 and SO_4^2.

Deficiency symptoms. Deficiency symptoms are not described on turfgrasses, but other plants exhibit chlorosis of new leaves; wilting, especially at leaf margins; leaf curling; and eventually necrosis.

Toxicity and excess conditions. Cl as the chloride salt is a component of many soil salts that can be directly toxic to leaf tissues and roots; more often it reduces water availability by enhancing total soil salinity, and it can interfere with nitrogen nutrition. Many saline irrigation water sources are high in chloride ions.

7.5.2.14 Nickel

Deficiency conditions. Conditions associated with Ni deficiency are not clear due to the rare occurrence of Ni deficiency.

Deficiency symptoms. Deficiency symptoms are not described on turfgrasses. Barley shows partial chlorosis as interveinal yellowing followed by necrosis, and failure of leaf tip to unfold.

Toxicity and excess conditions. Ni toxicity can arise from the use of some high-Ni sewage sludges.

In summary, plant tissue samples are collected at a "specific point in time." Salinity management of perennial turfgrass ecosystems is not a short-term collection of implemented BMPs or a single data set, but a long-term science-based holistic strategy using quantitative tools such as combined water, soil, and plant tissue analyses to determine the sustainable grass management programs. As pointed out in this chapter, establishing baseline data and developing a historical collection of data over seasons and years are essential criteria when managing any salinity challenges on specific turfgrass sites.

8 Assessment for Salt Movement, Additions, and Retention

CONTENTS

8.1 ASSESSING SOIL PHYSICAL PROPERTIES

Site-specific assessment with respect to salt movement, addition, and retention as they relate to problems that may occur on the site is the subject of this chapter. However, attention should be given to preventing problems related to salt movement from the site. Duncan et al. (2008) addressed the latter aspect of offsite salt mobilization and transport; but the ultimate management options for reducing salt mobilization must be at the site-specific location. They present irrigation and drainage management options for agricultural situations, but many of the principles apply to turfgrass areas.

The constant dynamics of salt movement in the soil will challenge the best turfgrass managers. Salinity dynamics are the integrated results of irrigation water quality and quantity; precipitation; evapotranspiration (ET); soil physical, chemical, and biological properties; topography; drainage; and amendment additions. Every amendment that is added to the soil, whether it is fertilizer, wetting agents, or other organic or inorganic products or chemicals, will impact salt movement in the soil profile at some point and ultimately impart some level of positive or negative interaction on turfgrass growth, development, and sustainability. Understanding the dynamics of salt movement and their retention in the soil profile is critical to assessing the site-specific challenges needed for long-term management of turfgrass ecosystems.

On newly established sites or established turfgrass sites that have initiated irrigation with saline or sodic irrigation water, there is a pattern of salinity stress that is often observed. Recognition of these patterns is important in site assessment to identify specific salinity problems and their causes, and to develop best management options. The severity or rapidity of soil and plant salt stress symptoms will be strongly related to the combination of factors noted in the previous paragraph and to the degree that salinity management is proactive. Types of salinity problems and causes of a particular problem in a localized area are spatially diverse (i.e., salinity issues are not uniform over sites). The particular situations presented below will be discussed later in more detail, but an understanding of the common sequence is useful, namely:

- First, on any of the irrigated areas where slow infiltration (water movement into the soil surface), percolation (water movement through the rhizosphere), or drainage (water movement past the root system) occurs, total soluble salts may accumulate to the point of salt-induced drought stress in just a few days to weeks in the surface area of the rootzone. These stress symptoms may be expressed over a relatively large area regardless of topography.
- Second, areas where surface runoff causes salt accumulation may exhibit soluble salt stresses in the same time frame as the first stress symptoms previously noted, but in more random and localized areas related to surface (often gravity-induced) drainage patterns.
- Third, relative to soluble salts, sodic problems take longer to appear (longer period of sodium accumulation), and appear in a slower manner exhibited by reduced infiltration, percolation, and drainage with an increase in wet areas. This can be rather widespread over the site with symptoms first appearing in low areas (these can be small microdepressions on flat turf areas to larger areas) due to surface drainage patterns. The time frame for these stresses may be months to years.
- Fourth, subsurface movement of soluble salts, including Na, may be very slow and do not appear until usually 3 or more years after saline irrigation is started. Stress sites will appear as localized areas where subsurface movement has caused salts to come to the surface (upward capillary action, movement into root systems down slope, seepage from underground flow, etc.). These localized stress areas are rather difficult to predict ahead of time and are normally the most severe due to the very high salt loads resulting from several years of accumulation.

In developing a salinity management program on a site that has had long-term application of saline irrigation water, the high total dissolved salt accumulation issues noted in the first and second patterns may be resolved first, leaving the salt-challenged areas due to the third and fourth problem patterns (sodium accumulation). At this point, it is very important to assess each localized salt stress area in terms of potential for surface and subsurface salt movement, total salinity additions (such as with each irrigation cycle or fertilization or amendment application), and specific salt ion retention, since the corrective measures must be very specific to the particular stress area. This illustrates that on salt-challenged sites, a "one-size-fits-all" management program does not work—salinity issues must be dealt with by a best management practice (BMP) approach applied to the particular localized stress area. A "silver bullet" management practice or one magic product when applied over the whole site will not be effective.

Types of salt movement in soil profiles that contribute to the diversity of spatial and temporal patterns of salt stress on a site are summarized below. Where the salts emerge is called a *saline seep*. Brown et al. (1983), Seelig (2000), and Wentz (2000) provide insight and illustrations on saline seep formation in agricultural situations, but these seepage areas are very similar when dealing with turfgrass areas (Figures 8.1 and 8.2). Understanding underground salt movement is important for remediation purposes, including the location of drainage features (see Chapter 15, "Drainage and Sand Capping").

- Sloped areas and areas with rolling topography: primary salt movement is gravity-fed surface migration that is relatively rapid with free water at the surface; but secondary movement occurs by slow, gradual, but continuous subsurface lateral movement that can occur

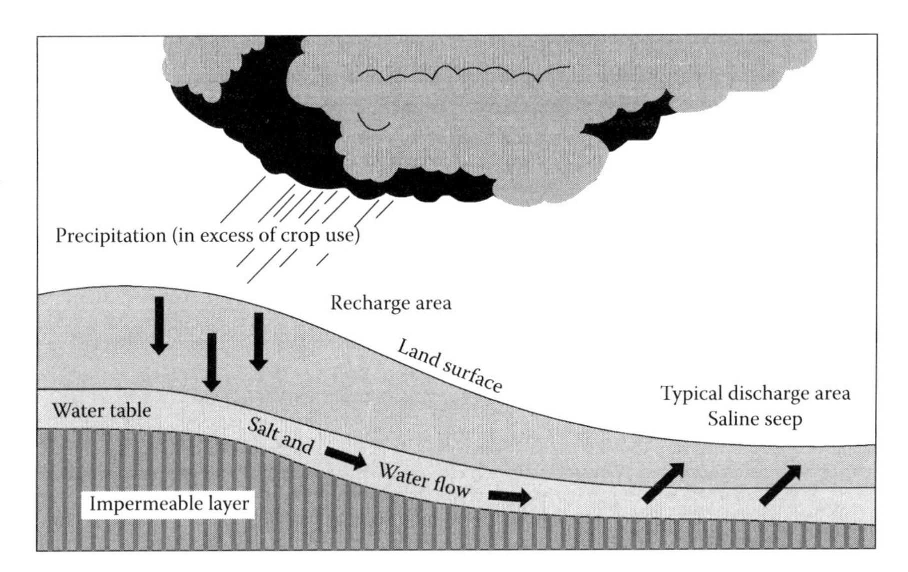

FIGURE 8.1 General diagram of saline seep. (After Seelig, B. D., *Salinity and Sodicity in North Dakota Soils*, ND State Univ. Extension Pub. EB 57, 2000, http://www.ag.ndsu.edu/pubs/plantsci/soilfert/eb57-2.htm#location.)

over weeks and months. Subsurface water will move by gravity and, if intercepted by a soil location with better macropores such as drain or irrigation trenches, will move along these features.

- Steep slopes leading to surface drains: primary salt movement is surface migration in conjunction with slower subsurface lateral movement; potential seepage sites often occur with lateral-migration subsurface salt emerging at unpredictable locations and not always at the lowest points but sometimes midslope.
- Capillary or uprising (upconing) of salts from an underlying water table, from underground water flow areas, or from localized salt accumulation zones in old arroyo (flat lake) areas; the salt emergence on the soil surface usually occurs in low-topography areas, but the salt can also emerge at the soil surface at other totally unpredictable sites, especially in arid and semiarid environments with minimal rainfall and high ET.
- On flat fairway, green, or any turfgrass area with little or no slope to provide surface drainage water movement, the primary salt movement generally occurs via movement into small depressions over time, where these small pockets develop into sodic scald sites with little or no grass density.
- Accumulation of migrating salts are caused by very high salt retention and subsequent nonmovement of salts when inorganic amendments (porous ceramics and diatomaceous earth products) that have high capillary microporosity and retain soluble salts in the (capillary) micropores are added to the soil profile. The retained soluble salts in these micropores are very difficult to remove via normal irrigation practices and generally require repeated pulse-cycle applications over time to leach the excess accumulated salts out of the capillary pores.

These salt movement situations result in considerable spatial variability of salt accumulation within the soil surface or subsurface profile, and these accumulation areas are often the most challenging—both to identify the underlying reason and for developing successful alleviation or remediation (in severe cases) programs. Salt accumulation in a localized soil area, especially if due to subsurface flow, is often the result of many years of slow movement, and salt loads are normally very

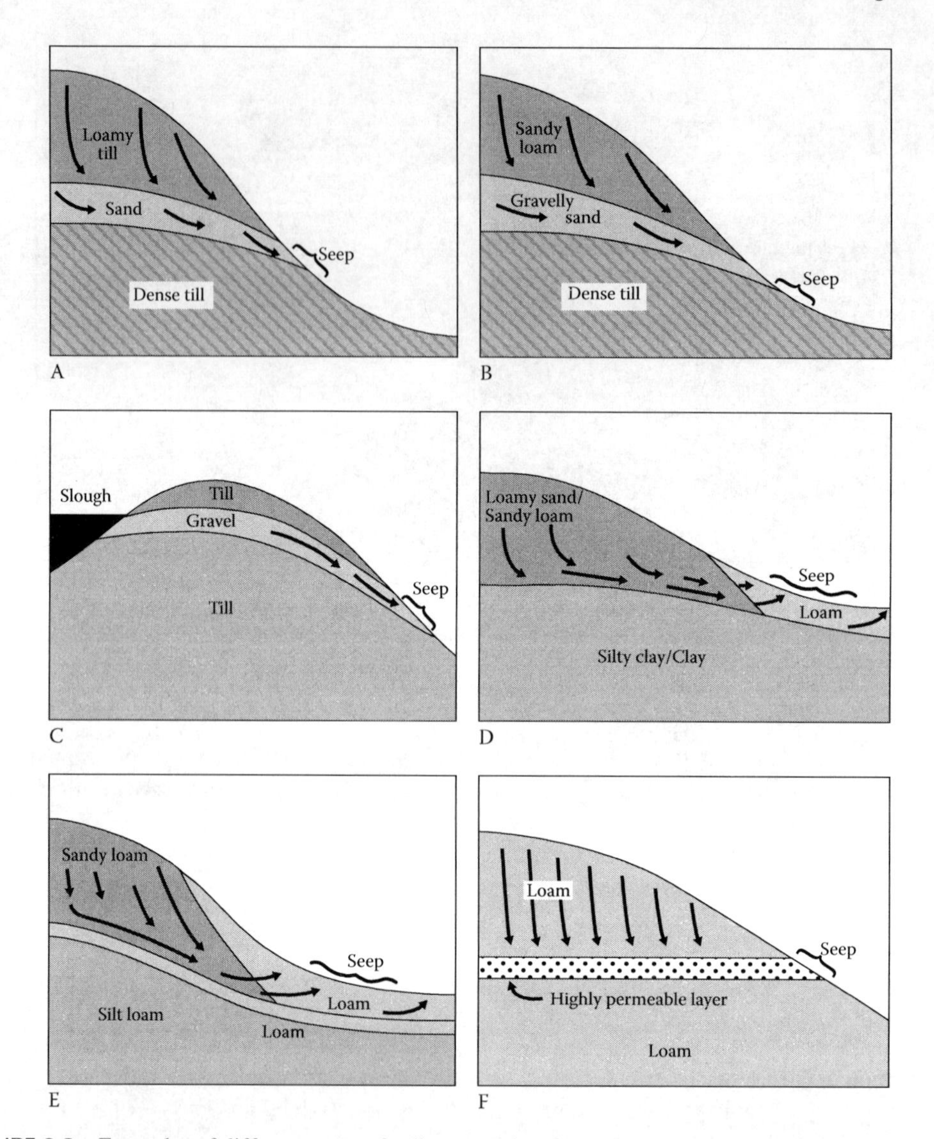

FIGURE 8.2 Examples of different types of saline seep situations. Any of these could occur on a turfgrass area. (After Seelig, B. D., *Salinity and Sodicity in North Dakota Soils*, ND State Univ. Extension Pub. EB 57, 2000, http://www.ag.ndsu.edu/pubs/plantsci/soilfert/eb57-2.htm#location.)

high. Seelig (2000) provides several illustrations in an online publication of different salt movement situations that contribute to saline seeps, and these would apply to turfgrass landscapes.

In Chapter 17, "Additional Cultural Practices," guidelines for the use of organic and inorganic amendments to aid in alleviation of salinity problems are addressed with additional BMP information. However, in the current chapter, organic and inorganic amendments are discussed from the viewpoint of affecting salt movement, total salinity additions, and specific salt ion retention as part of the site assessment process to determine the presence, severity, causes, and patterns of salt stresses.

8.1.1 Organic Amendments

Natural organic products such as peat moss (examples include sphagnum peat moss, hypnum moss peat, reed-sedge peat, sedimentary peats, and woody peats), biosolid materials produced by waste water treatment (sewage sludge) processes (an example is Milorganite®), earthworm castings, and animal- or plant-based products (e.g., hoof and horn meal; fish scraps and meal; seed meals such as cottonseed, linseed, or castor pumice; dried and composted manures; composted poultry waste; processed paper waste; processed tankage; or hydrolyzed feather meal) are often applied in various quantities to soil profiles (Carrow, Waddington, et al., 2001). These organic products have exchange sites that can sequester salt ions as well as moisture and vary in rate of decomposition and interaction with environmental conditions.

Organic matter can aid in reducing saline and sodic stresses when properly applied to enhance soil structure, especially soil macroporosity (Clark et al., 2009). In Chapter 12, Section 12.4, "Organic Amendments for Sodic Soil Reclamation," use of organic amendments for alleviation of salt-affected conditions is discussed. In terms of the focus of this section on site assessment, the application of organic matter types or quantities that would cause problems on nonsaline soils will cause the same problems on saline soils. Organic matter that retains moisture, which then dries, will also retain the soluble salts within the water and may result in a zone of salt accumulation in the organic matter zone, such as a surface thatch layer. If an organic layer is present in the soil profile that impedes water movement, this will become a zone of salt accumulation. Organic matter that exhibits hydrophobic properties, thereby hindering irrigation, will also do the same on salt-affected sites as well as normal sites.

Cautions when adding organic products to soil profiles include minimizing the volume applied during peak grass organic matter production cycles and commonsense applications when using saline irrigation water. Most of these products should be applied in conjunction with some cultivation event to integrate the specific organic amendment into the soil profile and to prevent surface layering. Salts also impact microbial activity and their normal cycles for decomposition of organic products, and these organic products have the capability to sequester excess salts prior to decomposition. Products containing fulvic acids are good choices when using saline irrigation water.

8.1.2 Inorganic Amendments

Inorganic amendments may impact soil salinity via influencing cation exchange capacity (CEC), macroporosity, microporosity, and moisture retention. Common inorganic products added to turfgrass soil profiles include zeolites (for increasing CEC and nutrient-holding capacity), porous ceramics (calcined clay products for moisture retention), and diatomaceous earth amendments (for moisture retention). Once these products are applied to a soil profile, they generally biodegrade very little over time but are extremely effective in doing their job—either nutrient retention or water and soluble salt retention.

- Zeolites are volcanic clays containing silica and aluminum oxides with negative charges. The CEC range for high-quality zeolites is 100–185 cmol/kg. Their primary use is for nutrient stabilization in the soil. These products generally have a preference for potassium and micronutrients such as calcium, and a nonpreference for sodium. Examples include, but are not limited to, Ecosand, Zeosand, Ecolite, ZeoPro, Clinolite, Z-Mendit, Zar-Min, Z-Ultra, and Z-Plus. (See Chapter 17, Section 17.4.3, "Zeolite," for additional information.)
- Porous ceramics are porous clay materials derived by heating to 1500–1800°F (i.e., *calcined*) to remove water and stabilize the sand substitutes both physically and chemically. Their primary use is for water retention via microporosity. They will retain soluble salts in the micropores (the range generally is 72%–83% capillary micropores depending on the individual product traits). Examples include, but are not limited to, Profile, Greenschoice, Permopore, Turface, Terragreen, and Zoneite Green. Interestingly, Murphy (2007) reported that many of these materials did not enhance plant water availability, which would indicate that the moisture is retained within the particles. If salts were present in the irrigation water, they could also be retained in the micropores.

- Diatomaceous earth products are deposits of fossil diatoms composed of silica dioxide. Examples include, but are not limited to, PSA, which is kiln fired but not calcined; Axis, which is calcined; and Lassenite (50% capillary porosity; 18% air porosity), which is calcined; and Isolite, which contains a noncalcined clay fraction. (See Chapter 17, Section 17.4.4, "Lassenite," for additional information.)

Leaching of salts in soils becomes more difficult in terms of time and quantity of water as the volume of soil pores <0.03 mm diameter increases, as is discussed in Chapter 11, "Irrigation Scheduling and Salinity Leaching" (and see Table 8.1). For example, the primary reason that it requires considerably more water to leach salts from a clay loam compared to a loamy sand, both at the same soil (saturated paste extract value of ECe) is that the clay loam contains more micropores or smaller pores. When an inorganic amendment with the majority of its internal porosity within the <0.03 mm diameter range is added to a sandy soil, these micropores enhance the chances of salt accumulation and make leaching of those retained salts much more difficult.

Thus, an important consideration when using inorganic amendments on salt-affected sites is the total quantity of pores below the macropore size range (i.e., pores <0.075 mm diameter) (Table 8.1). Of particular concern are the percentage of pores that is in the micropore (0.005 to 0.03 mm) and ultramicropore and cyptopore (<0.005 mm) diameter range. Micropores retain slowly available moisture, but most of the readily plant-available moisture is retained in the mesopores, while the moisture in the ultramicropores and cyptopores is not plant available. If the irrigation water is saline, salts may accumulate in the pores <0.03 mm with little leaching or flushing of salts by irrigation or rain. If salt concentrations are high in these pores, the salts can move by diffusion to achieve equilibrium with the bulk soil water over time.

The authors have observed situations where inorganic amendments with a high percentage of micropores associated with application rates of 10% to 20% by volume to the soil profile resulted in salt accumulation over time that was very difficult to alleviate. Some commonsense recommendations are as follows:

- If an inorganic amendment has a majority of its internal porosity smaller than the mesopore range (these pores provide readily plant-available water), then a considerable quantity of amendment on a volume basis would be required to enhance plant-available water since most of its pores are smaller. However, if saline irrigation water is used after the soil is amended, this site would have considerable microporosity that would require leaching to control soluble salt loads. Thus, a sand amended with 15% or 20% by volume compared to a sand amended at 5% to 10% by volume would be expected to exhibit more salt retention problems.
- An inorganic amendment with most of its pores in the mesopore range would not be expected to exhibit salt accumulation or leaching problems relative to another amendment with primarily micropores or smaller pores (<0.03 mm).
- Since inorganic amendments do not degrade or decompose, an error in amendment selection or quantity of application will result in persistent salinity management challenges. A caution when applying these sand substitute inorganic amendments includes commonsense "by volume or by weight" applications to sand-based profiles, especially when applying saline irrigation water because of its capability for sequestering soluble salt ions in water and retaining them in micropores. These products have excellent capabilities for water retention, but can create salt accumulation layers in soil profiles when applying saline irrigation water.
- Consider the previous comments when conducting a site assessment of salt problems on a turfgrass site that has had inorganic amendments applied in the past as to whether any salt accumulation patterns relate to inorganic amendment application areas.

TABLE 8.1
Soil Pore Size Classification and Relationship to Soil Moisture Retention

Pore Class	Comments
Total Porosity	All the pores contained within a soil.
Macropores	*Pores > 0.075 mm diameter (some use >0.10 mm for irrigated turfgrass).* Once the total pore space is saturated, the macropores would be the pores that drain within about 1 hour after a heavy rain. Macropores drain at –0.10 to –0.33 bar (–10 to –33 kPa) soil water tension, which is the soil field capacity. For sandy soils, the –10 bar tension is most appropriate, while the –0.33 bar is used for more fine-textured soils. These are often called macroporosity, aeration pores, or *aeration porosity*. When testing for greens mixes, laboratories using the USGA specifications use a soil moisture tension of –30 cm (i.e., a 12-inch depth of soil mix) or –0.03 bar as the "field capacity." All pores retaining water at this tension are considered capillary pores or *capillary porosity*.
Mesopores	*Pores of 0.03 to 0.075 mm diameter.* These contain readily available water for plants. These pores release water between –0.10 (field capacity) and about –4 to –5 bar (–400 to –500 kPa) soil water tension. Recreational turfgrasses are normally maintained in this range to avoid moisture stress. Since these pores release plant-available water readily and are rewet upon the next rain or irrigation event, salt accumulation would not be expected to be an issue within these pores.
Micropores	*Pores that are in the 0.005- to 0.03-mm diameter range.* These contain plant-available water, but the water is slowly available. On high-maintenance turfgrass sites, the water within these pores may not be replaced or pores "leached" on a regular basis during irrigation cycles so that salts could accumulate. The water within these pores is removed at soil water tensions of between –4 to –5 bar and –15 bar, where the permanent wilt point is defined as –15 bar.
Ultramicropores and Cyptopores	*Very small micropores of <0.005-mm diameter.* The water within these pores would be released at between –15 and –31 bar soil water potential. Salt accumulation is possible due to the lack of leaching or flushing during irrigation cycles.
Other Related Terms	
Hygroscopic Water	Water retained at more negative than –31 bar soil water potential. This water is tightly bound to soil particles.
Capillary Water	Water retained between –0.10 and –31 bar soil water potential. This is water retained in mesopores (easiest for capillary water movement), micropores, and ultramicropores or cyptopores (very slow capillary water movement).

Note: 1 mm = 1000 μm pore size diameter; and 1 bar = 100 kPa = 1000 cm soil water tension or potential.

8.1.3 Sand Particle Sizes and Salinity Accumulation and Movement

It is not unusual for sands to exhibit salinity challenges when receiving saline irrigation water. While the USGA specifications are appropriate, favoring the higher end of the sand size ranges is best. Sand sizing and the fractional volume of sands in each sizing category (coarse, medium, fine + very fine, and silt + clay) as revealed by a complete physical analysis are key determinants in assessing how effective the sand profile will be in either accumulating or moving soluble salts (via infiltration, percolation, and/or drainage) down through the soil profile. Especially for golf greens, but also for sands in general where saline irrigation water is applied, the fine + very fine (<0.25 mm in size) sands + silt and clay fractions should range in the 15%–18% total composition for effective salt movement over time at grass-planting time. Higher percentages can result in more salt retention as organics accumulate in the soil profile from clippings and root dieback.

As salinity increases >2000 ppm total dissolved salts (ECw or ECe = 3.125 dS/m), more coarse sands (>0.75 mm in size) may be needed to effectively move soluble salts down to drainage lines. The sand-based soil profile should contain primarily 0.25–1.00-mm-sized sand particles for acceptable saline water infiltration and percolation.

Soil fertility testing may reveal low CEC since the initial CEC of many sand-based profiles generally ranges below 1.0 cmol/kg; amendment with zeolite can aid in nutrient retention. Carrow, Waddington, et al. (2001) provided formula and tabular information on rates to amend the surface 4 inches (10 cm) of a sand to adjust CEC to target levels. For example, zeolite with a CEC of 120 cmol/kg added at 260 lb. per 1000 sq. ft. would increase CEC by 1.0 cmol/kg where a total CEC of 3.0 cmol/kg is acceptable. This rate is about 0.8% to 1.0% by weight of zeolite for the 4-inch (10-cm) upper soil zone. The primary soil profile positioning for zeolite should be in the top 3 to 4 inches (7.5 to 10 cm) of the upper soil profile (the primary area for root volume, crown, and rhizomes, and the major area for nutrient uptake from soil-applied fertilizers). Peat moss can be added in the 5%–10% by volume range, with the understanding that most of the peat moss will decompose during the grow-in program and that volumes >10% can at times retain excessive soluble salts prior to microbial breakdown.

8.2 DRAINAGE ASSESSMENT FOR DRAINAGE IMPEDIMENTS

Since the most critical BMP for long-term salinity management is salt leaching, sites should be carefully evaluated with respect to drainage problems. When drainage is limited on saline or sodic sites, numerous plant stress *symptoms and detrimental effects in association with soil salt accumulation* can occur, including:

Compaction	Excess growth
Rutting or persistent wheel marks	Wet wilt
Poor-quality playing surfaces	Interference with normal maintenance
Poor wear tolerance	Poor abiotic stress tolerance
Thatch accumulation	Enhanced disease activity
Poor turfgrass rooting and root dieback	Black layer problems
Weed encroachment	Algae or moss accumulation on surface
Leaching and volatilization of fertilizer nutrients	Denitrification alterations
Loss of microbial activity	Waste of resources (overwatering)
Increased root-borne pathogen problems	Cold soils
Soil oxygen depletion	Soil surface puddling

Leaching of a significant quantity of total soluble salts (saline soils) and sodium (sodic soils) requires water to enter the soil surface (infiltration), move through the rhizosphere (percolation), and move below the rootzone (drainage). Any barriers to water movement across the whole range of infiltration, percolation, and drainage should be identified and corrected to allow effective leaching programs.

Saline soils do not need amendments to displace sodium, and the total salt permeability management includes aeration and leaching with sufficient water volume to promote downward movement of soluble salts. Sodic and some saline-sodic soils traditionally have poor structure, especially in native soils since the structure has deteriorated due to slaking, colloid dispersion, and organic colloid migration induced by excess sodium accumulation (Carrow, Waddington, et al., 2001), often resulting in depressed "micropockets" at the soil surface, which are pre-scald or scald microsites. These soils require soil amendment applications (often granular gypsum or lime depending on the soil pH and sulfur load in the soil) to displace excess sodium ions that caused the soil degradation, as well as leveling or contouring of the depressed areas and cultivation.

Due to site-specific salinity interactions among the soil profile, the irrigation water, the turfgrass cultivar salt tolerance level, and unpredictable environmental extremes, several options should be evaluated to determine the combination of soil chemical amendments, irrigation water chemical amendments, cultivation types and frequency to promote continuous macropore continuity, and other practices that will enhance infiltration, percolation, and drainage. The management strategy is to "keep the salts moving away from the turfgrass root system" and hopefully down to the drainage lines.

Considerations for leaching salts include: the type of leaching (maintenance leaching, reclamation leaching, or pulse cycle leaching, discussed in Chapter 11); irrigation system distribution uniformity and efficiency; irrigation scheduling (e.g., long duration cycles, less frequent); the salt type (sodium propensity for dominating the CEC, total dissolved salt concentration or TDS, bicarbonate load, sulfate load, chloride concentration); and pore space balance or continuity between macropores and micropores. The capillary rise potential of the total salt complex once it has accumulated at some level in the soil profile and as influenced by evapotranspiration is another management consideration. Capillary rise of soluble salts of 1 to 3 feet (0.3–0.9 m) can occur on fine-textured soils during prolonged dry periods or dormant winter months if no rain or precipitation occurs to prevent upward salt migration. In sandy soils, the capillary fringe is usually 12 inches (30 cm) or less.

Infiltration barriers include upper soil profile bicarbonate complex layering and excess sodium accumulation that results in surface puddling, compacted or cemented zones, and lack of macropore continuity (aeration holes that rapidly collapse, such as occurs with expanding and contracting or 2:1 clay profiles).

Percolation barriers include calcitic layers at various levels in the soil profile, clay horizons, sand capping over different finer soil textures, hard cultivation pans, soil particle and colloidal migration, high organic matter zones, and porous ceramic or diatomaceous earth product layering. The higher the capillary porosity in these soil profiles, the greater the salt retention and the more difficult the management will be for those excess accumulated salts.

8.2.1 Surface Drainage

Surface drainage is important to prevent localized salinity accumulation and to control salt movement, total soluble salt additions, and specific salt ion retention on site-specific turfgrass areas. Options for surface drainage are listed in Table 8.2, with McIntyre and Jakobsen (2000) providing an excellent presentation of practical surface and subsurface drainage methods for turfgrass sites (see also Chapter 15, "Drainage and Sand Capping"). Drainage features should be carefully designed by drainage professionals.

Gravity flow in a controlled manner by appropriately designed surface contouring is the most important contributor to successful surface drainage. Most slopes ranging from 1% to 3% will provide adequate downslope water movement toward lower topography areas or drainage tiles. Steeper slopes can often result in reduced irrigation water infiltration and percolation as well as provide high-evapotranspiration areas that are conducive to rapid upward capillary pore movement of any accumulated salts to the soil surface. These areas often require solid-tine or core aeration events, followed by the application of zeolite for nutrient retention plus Lassenite (refer to Chapter 17, Section 17.4.4, "Lassenite," for more information), which has a favorable pore size distribution for plant-available water retention in order to stabilize turfgrass performance on the steeper slopes.

Turfgrass soils with very little slope are susceptible to *scald formation* if the irrigation water contains high Na levels favoring sodic soil development. The scald areas start with very small depressions that may not be noticeable to the eye by allowing water to collect in these micropockets and to start to concentrate Na. Over time, this area may start to noticeably depress at the surface (soil structure breakdown or deflocculaton) and expand in size, with the turfgrass deteriorating in the

TABLE 8.2
Surface Drainage Options to Consider in Site Assessment of Salt-Affected Sites to Control Salt Movement, Additions, and Retention

Basic contouring	Must be well designed on salt-affected sites to avoid improper surface water flow. When feasible, flat areas should be avoided on sites with fine-textured soils and 2:1 clays that receive sodic irrigation water to avoid scald formation.
Diversion channels or grassed surfaces	Designed to catch runoff water and divert it to an outlet or another conveyance channel. These channels can also include cement-lined or rock-lined ditches with or without solid pipe or tile.
Catch basins	Collect larger volumes of water in a low-topography area, draining into the subsoil or feeding directly to drain pipes, tiles, a sump (pump) pit, a dry well, or interception drains.
French drains	Trenches filled with pea gravel or very coarse sand, collecting surface water in flat or sloped areas; may or may not utilize tile.
Slit trenches	Generally smaller than French drains and may either extend into the subsoil or tie into tile. These trenches generally collect surface water in flat or sloped areas and may or may not utilize tile.
Interceptor drains	Modified slit trenches that can be installed on long sloped areas either in the middle of the slope or near the bottom of the slope to channel surface water to nearby French drains, catch basins, or diversion channels. These drains can also be installed just outside greens cavities to move surface water from bunker slopes or other surrounding elevated areas prior to the water encroaching onto greens surfaces. An additional option for these drains is to tie the drains into the herringbone drainage system in greens cavities or smile drains emerging from low-topography areas at the edge of greens cavities, carrying the excess surface water to catch basins or diversion channels.
Waste bunkers	Can also be utilized to catch excess surface water runoff and channel the water to catch basins, wetland areas, ponds, or lakes in the vicinity.

area as true sodic conditions occur. Eventually, the area may take on a black coloration due to high pH (pH of 9.8 to 10.4), which causes solubilizing of some of the organic matter that coats the soil particles with a black organic coating. Prevention is the best means to deal with these areas. They are most prone to develop on fine-textured soils with low infiltration rates. Practices to enhance infiltration are necessary and may include sand slitting with a Blec device, or similar cultivation and sand injection or drill-fill equipment. The injection sand can be mixed with 25% to 50% granular gypsum as well as a surface application of sand and gypsum to fill any microdepressions. Leaching will be necessary to leach soluble salts and Na.

Low-flow sprinkler heads may need to be installed in order to provide sufficient water volume during irrigation cycles to grow grass effectively on sloped areas and mounds. Contouring may be needed on certain areas to properly channel water to or from problem salt accumulation areas. Soil salt accumulation symptoms will often reveal the weaker turfgrass density areas or salt accumulation areas (scalds) as a result of surface salt movement.

Proper contouring must be implemented with adequate slope so that water can move to the target collection drainage site, but should avoid low, bowl-shaped deposit areas lacking effective surface drainage. On greens, three directions of surface water flow should be designed for proper drainage. On fairways and roughs, contouring should avoid large water flow volume channeling that could cause erosion, and the water flow should be spread uniformly over the sloped topography. Sloped areas directing surface flow of water to a site may also direct subsurface flow to the downslope turfgrass area, but the slow salt movement by subsurface flow on fine-textured soils in arid regions may not appear as stressed areas for years.

More frequent surface cultivation programs may be needed either to alleviate upper soil surface conditions that limit water infiltration or to actually improve or enhance the subsurface cultivation program. Sodic soils are especially a challenge since a cultivation operation often does not last long in terms of effectiveness (macropore spaces rapidly collapse) compared to a similar nonsodic soil. As a general rule, cultivation operations only last about half the time duration as on a nonsodic site of the same soil type. This enhancement program is often needed when fairway and rough soils are composed of fine-textured soils, soils with 2:1 clays, decomposed granite (DG), pulverized volcanic pumice, or pulverized caliche with little water and nutrient retention capabilities for growing turfgrass.

8.2.2 Subsurface Drainage and Challenges

With high salt deposition from each reoccurring saline irrigation water cycle, draining accumulated salts throughout the soil profile must be a priority management strategy. Various subsurface drainage methods are summarized in Table 8.3 (see also Chapter 15, "Drainage and Sand Capping"). Subsurface drainage removes gravitational water, controls water table depth, decreases seepage water problem areas, involves saturated flow for rapid removal of water via

TABLE 8.3
Subsurface Drainage Options to Control Salt Movement, Additions, and Retention

Surface slopes	Influence the subsurface flow of salts. When assessing subsurface drainage, consider the potential for subsurface flow to occur from highly sloped sites; interception drains may be required not just for surface flow but also for deeper unsaturated or saturated flow.
Interception drains	These can be installed midway up long slopes, at the bottom of those slopes, or just outside greens cavities; can also be used to collect subsurface water plus gravity-induced migrating soluble salts and help to channel that salinized water to other catchment facilities in order to minimize salt accumulation on stress-prone turfgrass sites.
Tile or perforated plastic pipe (of proper sizing)	Drainage utilizing shallow, closely spaced lines to remove water quickly. Infiltration and permeability (conductivity) are necessary. An example of use is the herringbone designs in USGA specification sand-based greens. Flat pipe is also being used in some of these subsoil drainage systems. Sometimes the authors have noted the need for drain lines to intercept subsurface water moving to the surface by capillary rise, such as water that seeped into drain or irrigation trench lines and then move by gravity flow in the trench line. Drain lines were installed along the top of the trench in lower areas leading to deep dry well drains.
USGA Greens Specifications	Require the installation of a designated sand mix (based on a complete sand physical analysis with or without organic or inorganic amendments) with drainage layer materials that are selected to (a) create a layered profile (perched water table) among sand, gravel, and subsoil drain lines; and (b) prevent rootzone particle migration into the drainage layer.
Subaeration/PAT systems	Can be utilized to either blow air or pull air through the subsoil drainage system. In addition, these subair systems have utility as a tool to assist in pulling upper sand-based and accumulated soil-soluble salts from the surface down to the drainage lines.
Mole drains	Temporary drainage channels that are formed with a mole plow. These drains work best in high clay content soils. On sodic soils, the slots or channels should be enriched with gypsum to counter the excess accumulated sodium in the subsoil and to keep the drains functional.

macropores (deep cultivation), includes unsaturated capillary flow via mesopores, percolates through the rootzone, and drains below the rootzone naturally or through tile or pipe. Subsurface drainage can also include surface water collection via some inlet and also manages water that infiltrates the soil surface.

Deep cultivation during renovation and prior to turfgrass installation can be beneficial when drainage is seriously impeded by heavy clay, sodic, or cemented soil zones. Deep ripping, paraplowing, and deep moldboard plowing are effective techniques, especially if combined with gypsum applications or slotting underlaid by a mole drain prior to grassing.

If turfgrass is already established, then deep cultivation can include individual probes using high-pressure air (Dryject) or water (Hydroject) to correct some of the subsoil salt accumulation drainage problems in localized areas. Drill and fill with sand and/or granular gypsum is an effective tool for filling holes, opening compacted or cemented layers, and maintaining macropores for effective channeling of soluble salts to subsurface drainage lines. On larger areas such as native soil fairways, slot draining and injection of sand and/or gypsum into the slots are alternative techniques for accumulated salt management to enhance subsurface drainage.

Some drainage limitations when using tile or perforated pipe (both normal-sized and flat drain pipes) include (a) if sand is too coarse, fines may move downward from the rootzone, clogging the drainage layer and preventing water movement into the tile where proper coarse sands drain at 80–120 inches (200–300 cm) per hour; (b) if roots and fines move into the drainage layer or into the pipe and tile, water movement can be restricted; (c) if deep cultivation is not implemented annually, fines can sometimes move down to the gravel–sand interface in greens cavities, layering with accumulation of soluble salts, colloidal material, and fines, and thereby restricting water movement into the drain lines; and (d) if accumulated Na or salts have sealed in a layer above the drain lines (this usually occurs on the low-topography areas in greens cavities), preventing water movement to those drainage lines (similar to a dam holding excess gravitational water and soluble salts), flat pipe can sometimes be vertically installed just inside the greens cavity to connect with French or smile drains to carry the water off the edge of the green. When installing drains in areas receiving irrigation water high in Na, application of high rates of gypsum in the trench, over the drain line, and in the backfill soil can help keep the soil more permeable.

8.2.2.1 Greens Cavities

When repeatedly applying saline water with each irrigation cycle, regardless of the schedule, initial salinization in sand-based soil profiles and eventually resalinization of the rootzone are grass management challenges. The constant movement of salts, vertically or horizontally and spatially over time, even in high-rate saturated hydraulic conductivity sands, will be influenced by cultivation and evapotranspiration issues. When assessing salinity problems on a turfgrass area, management practices such as cultivation and irrigation scheduling should be considered as to how they influence salinity movement, total soluble salt additions and subsequent accumulation, or specific salt ion retention.

Aeration (core aeration, solid tines, needle tines, spiking, and slicing) and other cultivation programs (verticutting, and grooming that penetrates the soil surface) coupled with windy environmental conditions can increase ET to the point of escalating the capillary upward movement of salts in the soil profile. Lack of aeration or other cultivation activities on a scheduled basis can lead to layering of salts that initially migrate with the wetting front down to a certain depth during each irrigation cycle. These challenges are not only site specific but also microsite specific to manage.

Short-duration, frequent irrigation cycles (including daily syringing) with saline water can rapidly increase upper soil surface salt accumulation and contribute to rapid capillary pore upward migration of salts near or onto the soil surface, especially with high-ET environmental conditions. The increased concentration of salts in the soil can rapidly accumulate at twice or more above the incoming irrigation water salinity concentration in as little as 2–3 weeks with frequent, short-duration applications.

Careful irrigation scheduling with pulse (repeat) cycles to apply sufficient water to promote net downward movement is recommended in order to move the wetting front containing the soluble salts as deep into the soil profile as possible, and hopefully below the root system or down to the drainage lines. The more effective the irrigation scheduling program, with uniform water distribution efficiency and deeper downward movement of soluble salts, the more effective the salt management program will be on these sand-based soil profiles. On USGA greens with a functioning perched water table, salts can accumulate within the lower part of the rootzone if no flushing occurs due to rain or sufficient irrigation water volume to allow periodic flushing. This is particularly a problem in dry periods where the turfgrass manager has switched to light, frequent irrigation on salt-sensitive annual bluegrass or bentgrass. On these sites, a flushing irrigation event should be applied as needed, often once every two weeks under these potential salt stress conditions depending on the salt load in the irrigation water.

Additional salinity management considerations in greens cavities are the quality of sands, the quantity of sand substitute amendments (whether organic or inorganic) by volume or by weight, and the functionality of the drainage system in actually removing salts away from the turfgrass root system. Particle migration—these particles include fine and very fine sand sizes <0.25 mm in size plus silt plus clay fractions (the so-called fines components), organic matter that is of colloidal size, and in some cases inorganic sand substitute amendments—occurs with regular irrigation cycles, and the migration will move with the wetting front during irrigation or rainfall events. Any inorganic amendments that are designed to sequester water will usually capture soluble salts in capillary pores, and removal of these accumulated salts is difficult and time consuming unless the pores are primarily in the mesopore ranges (Table 8.1).

When the irrigation water wetting front consistently moves to a certain depth in the soil profile and these fines subsequently layer near the bottom of that wetting front, salt accumulation will usually increase in these zones. Aeration, leaching, and drainage become critical management tools to minimize their impact on turfgrass density and overall performance. The strategy is to keep the salts moving downward in the soil profile and away from the turfgrass root system to the drainage lines.

A related problem associated with repeated, uniform irrigation patterns is that when the irrigation water contains appreciable Na and bicarbonates, a Na-bicarbonate layer may accumulate at the zone of normal irrigation water penetration. Normally, this is only observed in arid regions or during prolonged dry periods in other regions. Any fines moved in the water also will accumulate in this area. Acidic fertilizers, a better leaching program, and cultivation to penetrate through this zone will all assist in breaking up this layer, as would bicarbonate removal by water acidification.

An excellent monitoring program to determine the effectiveness in managing excess salt accumulation in sand-based greens cavities is to collect leached drainage water from the drainage system under the greens and analyze the salt load. This approach is a proactive approach to determining the efficiency of a leaching program and helps to quantify the total accumulated salts that can be removed from the sand-based soil profile with good management (an example is given in Duncan et al., 2009, Chapter 16). If no drainage water is noted in the greens or other drainage lines during dry periods, then salts are not leaching.

8.2.2.2 Fairway Topography

Regardless of the architectural design, rolling topography and elevation changes on fairways present unique site-specific challenges when applying saline irrigation water. Both initial surface runoff and slow subsurface migration of salts will move by gravity flow to accumulate at some low-topography area. Primary drain lines are often constructed to manage the surface flow of water and soluble salts, but these collection sites can also become major salt accumulation sites depending on the depth of the water table and the actual capability for moving salts from that catchment site to another area.

In addition, the slow but methodical subsurface migration of salts from elevated areas to lower-topography areas occurs over weeks, months, and years. Seepage areas often appear at unanticipated sites and require additional drainage installation for salt management. Microsite drainage installation (such as interception drains or mini–dry wells) and soil modification (excavation) may be required on persistently problematic soil salt accumulation areas when deep cultivation (Water Wick, Hydroject, Dryject, drill and fill, etc.) techniques have not resolved the problem on grassed fairways.

8.3 SALINITY DISPOSAL OPTIONS

Disposal options for excess salts are an environmental and a governmental issue (local, state, or federal), often requiring special permits. Possible options include desalinization (with an additional permit for safe disposal of the reject concentrate), injection into deep aquifers (usually ones impacted by ocean or other salinized water resources), dilution into streams or rivers, disposal into evaporation ponds, leaching fields and recapture of reduced saline water for reuse, or channeling to saltwater marshes or outlets to the ocean. In highly arid regions, salt may be allowed to sequester in the soil below the rootzone, but this requires an excellent irrigation program to move salts into this zone and not allow migration back to the surface. On areas with the potential for a rising water table over time, soil-sequestered salts would rise back to the surface and salinize the rhizosphere.

Disposal into dry wells (e.g., abandoned oil wells or abandoned irrigation water wells) has been reduced due to the potential for contamination of groundwater aquifers and underground rivers. Deep-ocean water disposal of reject concentrates from desalinization projects along coastal venues has been implemented in some countries, with the idea that the most saline ocean water can be found several miles deep and that disposal of highly saline concentrates into these areas will have minimal environmental impact to the overall marine ecosystem.

8.4 RESTRICTING SALT ADDITIONS

A common strategy at some golf course development sites (particularly island and coastal venues with limited topsoil availability) is to dredge bays, estuaries, ponds, and lakes for soil material to build fairway and rough soil profiles. These generally fine-textured soils are traditionally highly salinized and need to be completely flushed of excess salts prior to planting turfgrass. Proper and adequate drainage installation is essential for not only growing in the turfgrass but also achieving long-term sustainability and acceptable turfgrass performance.

Some native salinized soil sites can be capped with a sandy rootzone layer to facilitate leaching. However, the native soil should be deep ripped, and chemical amendments such as gypsum or lime should be applied prior to capping to avoid future drainage problems and the upconing of accumulated salts. Additional drainage may need to be installed prior to sand capping on selected sites.

On native soil sites that will not be sand capped, salinized subsoil horizons that are high in total dissolved salts or sodium should not be brought to the surface during construction. Additional chemical amendments may be required based on appropriate soil chemical and physical testing. This problem is especially critical in acid-sulfate–prone areas, where construction promotes the oxidation of iron pyrite by exposure to air, lowering the pH generally below 3.0 and causing significant soil ecosystem limitations, including aluminum and sulfur toxicity (Rosicky et al., 2006; Dent, 1992; Bush and Sullivan, 1999; Costigan et al., 1981) (see Chapter 3, Section 3.3, on acid-sulfate soils).

Many fertilizers have a salt index (or osmotic index: a measure of the osmotic pressure created in soil solution by the addition of a specific fertilizer product), and understanding the level of potential salinization that occurs with application(s) of these products on perennial turfgrass ecosystems is critical for successful salt management. Products listed in Table 8.4 provide a ranking of potential osmotic effects when applied to turfgrass sites.

Irrigation water salt load reductions can be accomplished by blending with lower salinity water to reduce undesirable total dissolved salts or high sodium concentrations and to dilute excess levels

TABLE 8.4
Potential Salt-Induced Injury Rankings for Various Fertilizer Products

Product	Salt or Osmotic Index Based on Equal Amounts of:	
	Product[a]	Plant Nutrient[b]
K sources		K_2O
Potassium chloride	116	1.94
Potassium nitrate	74	1.58
Potassium sulfate	46	0.85
Potassium magnesium sulfate	43	1.97
Monopotassium phosphate	8	0.24
N sources		N
Ammonium nitrate	105	2.99
Sodium nitrate	100*	6.06
Urea	75	1.62
Potassium nitrate	74	5.34
Ammonium sulfate	69	3.25
Ammonia	47	0.57
Diammonium phosphate (DAP)	34	1.61
Monoammonium phosphate (MAP)	30	2.45
Natural organic (5% N)	3.5	0.70
P sources		P_2O_5
Diammonium phosphate	34	0.64
Monoammonium phosphate	30	0.49
Concentrated superphosphate	10	0.22
Ordinary superphosphate	8	0.39
Monopotassium phosphate	8	0.16

Source: Adapted from Carrow, R. N., D. V. Waddington, and P. E. Rieke, *Turfgrass Soil Fertility and Chemical Problems: Assessment and Management*, John Wiley, Hoboken, NJ, 2001. With permission.

[a] Compared to sodium nitrate = 100.

[b] Value = divide first column by percent nutrient in the product.

of boron, bicarbonate, sulfate, and chloride. Blending with desalinized water (from ocean seawater, saline groundwater, sodic groundwater, or lagoon water sources) or reclaimed water (recycled and treated sewage water) is often used to reduce the total salt load. Desalinization (techniques include water generated by evaporation, distillation, and reverse osmosis) processing does not need to reduce the total salt load in the initial source down to drinking water status prior to blending due to cost and reject concentrate volume issues when planting and maintaining turfgrasses.

8.5 HYDROGEOLOGICAL ASSESSMENT FOR THE PROTECTION OF SURFACE WATERS AND GROUNDWATERS

A pre-construction hydrogeological site assessment should be conducted in order to determine (a) tidal influences in coastal venues and inland areas with low topography exposure to oceans,

(b) the depth of the water table and any associations with fluctuations in depth over seasons, (c) saline irrigation water applications over pristine drinking water aquifers to minimize or stop secondary salinization, (d) the potential for acid-sulfate soil formation that may influence water quality, and (e) the need for catchment basins or other containment facilities to minimize or stop the movement of saline drainage water into any surface or groundwater venues.

To protect groundwater, including aquifers, the site can be capped with 12–24 inches (30–60 cm) of heavy clay and compacted, drain lines can be installed, and then the site can be capped with appropriate native soil or sand prior to grassing. Containment facilities will need to be constructed to collect and channel any high-volume rainfall runoff to appropriate deposition sites with minimal discharge into the drinking water aquifer. TPC San Antonio (Texas, United States) has two golf courses constructed utilizing this technology.

8.5.1 Leaching and the Underlying Water Table

A high water table coupled with capillary rise can move excess subsoil salts into the turfgrass rootzone and be a primary limitation to maintaining functional turfgrass ecosystems. Potential salinity management strategies include (a) channeling excess salts using dikes, ditches, or sump pumps for shallow water table control; (b) planting deep-rooted trees and shrubs with capabilities to extract extra water beyond the turfgrass rootzone to help lower the water table; (c) removing soil drainage barriers below the turfgrass rootzone that may cause a perched water table using deep soil cultivation techniques or drilling small-diameter mini–dry wells to penetrate into more permeable subsoil zones; (d) raising inherently low-topography areas after removing any subsoil drainage hindrances and then properly contouring them to facilitate surface and subsurface water movement; and (e) possibly installing monitoring pipe outlets to proactively collect water table samples and track water table fluctuations.

8.5.2 Upconing Salt Migration Problems

Seepage sites or other areas that accumulate excess salts that are not migrating with gravity are subject to upward movement of those excess salts to the soil surface. Thinning out of grass canopy density is usually an initial symptom, and if the surface or subsurface deposition of soluble salts continues, the grass is usually killed completely in these scald areas.

Both surface and subsurface drainage management are required to minimize the problem. Capping with sand and raising the topography at the specific site with native soils are renovation options in conjunction with appropriate drainage installation.

8.5.3 Surface Runoff

A minimum 1%–3% slope is normally required to facilitate gravitation water movement for surface runoff. The water can then be channeled with contouring to primary drainage outlets. Erosion control programs will be needed with steeper slopes, mounds, and other similar topography changes.

When sloped bunker faces or elevated surrounds are located near greens cavities and surface runoff moves directly onto the greens surface, excess salts can accumulate to reduce grass density and long-term performance sustainability over time. Interception drains may be required either (a) outside the greens cavity to remove the surface runoff to designated catchment areas or primary drain lines; or (b) just inside the greens cavity, and the interception drains are connected with the normal greens drainage system.

8.5.4 Subsurface Salt Movement: Cascades and "Trains" versus Topography

The same technology as noted for surface runoff (as described in Section 8.5.3, "Surface Runoff") may be needed for subsurface migration of soil-accumulated salts. The subsurface salt migration speed is unpredictable and is dependent on a number of factors: the salt load in the irrigation water, the frequency and duration of irrigation scheduling, the efficiency of the irrigation system distribution on a specific site, the degree of slope and associated gravity flow dependency, soil profile characteristics, elevation changes, rainfall frequency and volume, and ET.

Soluble salts have a tendency to move through whatever flow channels may be available in a subsoil profile, often producing reoccurring cascade movements with gravity inducement from an elevated position. These "salt trains" have been found to accumulate in drainage line or mainline irrigation trenches and build up over years to create salt seepage areas in unpredictable locations on fairways and roughs, creating salt scald areas that kill turfgrasses.

Solute diffusion in undisturbed soil profiles controls many important processes: plant uptake of nutrients, sorption–desorption processes, and microbial degradation of organic matter (Laegdsmand et al., 2010). A new dynamic proactive monitoring method has been developed that effectively measures solute diffusivity as a function of water content with a relatively small effect from soil compaction in intact soil samples and across soil textures (Laegdsmand et al., 2010).

In summary, BMP salinity management assumes that total soluble salts as well as specific salt ions, regardless of source or concentrations, will continually move with normal irrigation and rainfall events. A second assumption is that generally at some point, these salts will accumulate in often unpredictable areas unless managed properly for the site; and when they accumulate, turfgrasses will eventually reveal salt stress symptoms. The overall management program must be holistic and proactive to minimize negative grass performance problems.

Section III

Management BMPs for Saline and Sodic Soil Sites

9 Selection of Turfgrass and Landscape Plants

CONTENTS

9.1 PLANT GENETIC AND PHYSIOLOGICAL RESPONSES TO SALINITY

9.1.1 INTRODUCTION

Turfgrass selection is a critical best management practice (BMP) for salt-affected sites. This chapter has sections targeted to different audiences, from the *turfgrass manager (e.g., Section 9.2)* to the *turfgrass breeder or geneticist (e.g., Sections 9.1.2, 9.3)*. Significant strides have been made in the past 10 years in providing turfgrass managers choices of more salt-tolerant grasses with improved cultivars and defining salinity tolerance of turfgrasses at the inter- and intraspecific levels. However, continued advancement will depend on careful attention to several challenging areas (Duncan and Carrow, 1999), namely, (a) appropriate salt-tolerance screening procedures capable of identifying the top salt-tolerant genotypes within a species (i.e., the top 1%) that are sustainable in field environments, (b) identifying the multiple mechanisms responsible at the species and cultivar levels for the superior tolerance of these genotypes, and (c) enhancement of these specific mechanisms in conjunction with appropriate whole-ecosystem salt management BMPs.

Halophytology is the study and development of plants in saline environments (Yensen, 2008). *Glycophytes* grow well on fresh or low-salinity (generally, total dissolved salts <1000 ppm or ECw < 1.56 dS/m) water and exhibit decreased productivity as salt levels accumulate in the soil or increase in irrigation water applied topically. *Halophytes* are plants possessing the capability of completing their growth cycles under highly saline conditions. *Oligohalophytes* or salt-resisting halophytes resist salty conditions, but do not grow well until exposed to low-salinity or freshwater cultivation. *Miohalophytes* are salt-tolerant plants that sustain productivity up to a genetically controlled specific salt threshold level and then decrease growth as salt levels continue to increase. *Euhalophytes* (*Distichlis palmeri*) are true salt-loving plants that have increased productivity with increasing salinity concentrations and actually grow better under salinized conditions compared to freshwater situations. *Crinohalophytes* excrete salts through salt glands (Yensen, 2008).

Halophytes manage salt loads in three basic ways (Yensen, 2008):

1. *Exclusion*: excluder plants (*Hordeum* spp., *Melilotus* spp., *Rhiozophora* spp., and most glycophytes) exclude high concentrations of salts from entering the root vascular system and generally very strictly regulate the uptake of specific salt ions. While some plants rely on exclusion as the primary means to manage high salt loads, all plants exhibit some degree of exclusion, that is, the plants that depend on excretion (noted below) also exhibit exclusion and accumulator mechanism(s) as part of their overall tolerance to protection from and exposure to high salt loads.
2. *Excretion*: excreter or crinohalophytic plants (*Distichlis* spp. and *Avicennia* spp.) excrete salts through salt glands or above-the-surface bladders and generally contain low tissue salt concentrations.
3. *Accumulation*: accumulator plants (*Atriplex* spp. and *Salicornia* spp.) sequester salts in cell vacuoles to avoid toxic effects and for osmoregulatory mechanisms, with up to 50% salt accumulation on a dry-weight basis. Some plants such as *Paspalum vaginatum* and *Puccinella* spp. contain salt exclusion, concentration-dependent salt ion uptake, and sequestration mechanisms without being actual phytoaccumulators.

9.1.2 Genetic and Physiological Responses of Grasses to Salinity

Among *abiotic stresses* (defined as an adverse, nonbiotic circumstance that disturbs the normal physiological functioning of a plant) imposed on perennial turfgrass and landscape plants, excess salinity stress is undoubtedly the most complex and unpredictable environmental stress. With a four-way interaction (Irrigation Water Quality × Soil Properties × Climatic Changes and Extremes × Turfgrass or Landscape Species and Specific Cultivar) constantly adding to the degree of stress, the challenge to sustain the ecosystem is dynamic and must constantly be monitored. The higher the genetic salt tolerance in a plant species and cultivar, the more flexibility you have in managing these plants with increasing exposure to soil salt accumulation and saline irrigation water.

Multiple plant stress-responsive genes, either upregulated or downregulated, are involved in multiple tolerance mechanisms that encompass the activation or deactivation of several cascades of molecular networks (ion pumps or channels; calcium signaling; salt overly sensitive [SOS] pathways; genes and transcription factors; mitogen-activated protein kinases; reactive oxygen species [ROS]; DEAD-box helicases; and hormones such as abscisic acid [ABA], gibberellins [GA], auxins, and CK-cytokinins) involved in stress sensing, stress-induced signal transduction pathways, and the expression of specific stress-related genes and metabolites (Hasegawa et al., 2000; Hussain et al., 2008; Turkan and Demiral, 2009).

Some of the upregulated plant stress-responsive genes encode osmolytes (glycine betaine and proline), receptors, ion channels, components of calcium signaling, and other regulatory enzymes or signaling factors that actually confer expression of salinity-tolerant phenotypes in coordinated fashion (Tuteja, 2007). Consequently, all turfgrass and landscape cultivars within a species are not created equal with regard to salinity tolerance, and each cultivar contains a unique compilation of salinity tolerance mechanisms that are sequentially programmed and functional in a series of cascading signaling-oriented responses to site-specific environmental stress challenges.

The major biochemical pathways for adaptive components of salt tolerance include synthesis of osmotically active methylated metabolites for osmoprotection; specific proteins; free radical scavenging enzymes; scavengers of oxygen radicals, or chaperones; photosynthetic adjustments (protecting chloroplast functions); adjustments in water relations (osmotic adjustments, retention and acquisition of water, and turgor pressure maintenance); stress-induced enhancement of photorespiration; adjustments in stomatal conductance; and maintenance of ion homeostasis (Serrano and Rodriquez-Navarro, 2001) at the whole-plant level, within specific organs (seed and vegetative nodal segments such as found on stolons and rhizomes), and at a specific cellular level (Ashraf, 2004; Parida and Das, 2005). As internal plant salt accumulation increases intraspecifically and interspecifically with each stage of plant development, and the tolerance mechanism activations cascade from the seed or vegetative nodal segment to the root and eventually to the shoots (whole plant), the complexity for salt acclimation and subsequent responses to stress-induced adaptation increases.

Halophilic enzymes require high salt concentrations (1–4 M) for stability and activity plus a high excess concentration of acidic over basic amino residues (Mevarech et al., 2000). The high negative surface charge of halophilic proteins, such as malate-dehydrogenase (hMDH) and 2Fe-2S protein ferredoxin, enhances solubility and flexibility at high salt concentrations. Nonhalophylic proteins generally tend to become nonfunctionally rigid and aggregate at those high salt concentrations.

Glycophytes, or salt-sensitive plants, primarily attempt to restrict salt uptake in an effort to minimize toxic tissue accumulations, and then they adjust their osmotic pressure by synthesizing organic solutes such as proline, glycine betaine, or sugars and their derivatives (Greenway and Munns, 1980). Halophytes, or salt-tolerant plants, regulate uptake (via exclusion mechanisms), then sequester and localize any absorbed salt ions in root and shoot cell vacuoles (away from photosynthetic machinery) to control cytosol salt accumulations while maintaining a high K^+/Na^+ ratio (Glenn et al., 1999).

Potassium uptake is an essential inorganic constituent in salt-tolerance adjustments in plants. Potassium is the most abundant cation in the cytosol (10% of total dry weight), is compatible with protein structure even at high concentrations, is involved in the neutralization of nondiffusible negative charges of anionic groups and the regulation of cellular osmotic potential (water potential and turgor pressure), and is important for both long- and short-term control of cell membrane polarization (Lebaudy et al., 2007; Gaymard et al., 1998).

Cellular membrane transport of potassium can be mediated by K^+ channels and secondary K^+ transporters (Gierth and Maser, 2007). Plants employ several transport systems for K^+ acquisition, catalyzing K^+ absorption across a wide spectrum of external concentrations, and mediating K^+ movement internally within the plant and subsequent efflux into the surrounding environment (Szczerba et al., 2009). Through Ca^{2+}-signaling cascades and regulatory proteins, K^+ movement (flux) reacts to variable external K^+ concentrations, to competition with other ions (especially Na^+) in the root environment, and when exposed to multiple plant biotic and abiotic stresses. Sodium can block the K^+-specific root cell transporters under salinity stress and compete with K^+ uptake through Na^+/K^+ co-transporters (Zhu, 2003). Three families of membrane proteins contain plant K^+ transporters (Gierth and Maser, 2007): K^+ uptake permeases (KT/HAK/KUP), K^+ transporters (Trk/HKT), and cation proton antiporters (CPA).

In the plant vascular system, control of K^+ transport in the phloem helps to build osmotic gradients that drive sugar sap flow from leaves to sink tissues (co-transport of sugars) during photosynthesis

(Lebaudy et al., 2007). In the xylem, root pressure drives K^+ secretion into the vessels and also drives crude sap sugar transport from roots to shoots in the absence of transpiration. Potassium also helps to regulate the aperture movement around stomatal pores on the leaf surface that is critical for transpiration.

Three classes of low-affinity K^+ channels have been identified (Szczerba et al., 2009). Gating of K^+ and other ion channels is essential in K^+ mineral nutrition, signaling, and both abiotic and biotic stress responses in plants (Dreyer and Blatt, 2009).

1. K^+-outward rectifying channels mediate Na^+ influx into plant cells through Shaker channel genes encoding for K^+ secretion, open during depolarization of the plasma membrane, and mediate K^+ efflux; SKOR is involved in K^+ release (loading) into the xylem sap and control of K^+ translocation (flow) toward the shoots (Gaymard et al., 1998). KCO1 is another outward rectifier with a steep Ca^{2+} dependency (Czempinski et al., 1997).
2. K^+-inward rectifying channels function in membrane hyperpolarization leading to K^+ uptake into cells; the AKT1 channel is involved in K^+ uptake from soil solutions (Lagarde et al., 1996), while KAT1 mediates K^+ influx in guard cells (Nakamura et al., 1995; Ichida et al., 1997). Exocitosis vesicles deliver clusters of K^+ channels to the plasma membrane during stomatal opening (Hurst et al., 2004).
3. Voltage-dependent cation channels conduct ions rapidly and filter selectively across cell membranes through gated pores (Gandhi et al., 2003; Sands et al., 2005). These channels are intrinsically sensitive to membrane potential changes, as well as open and close to generate complex electrical signals (action potentials) in neurons and other excitable cells (Yi and Jan, 2000). Fixed external surface charges of voltage-gated ion channels influence gating, conductance, and toxin binding (MacKinnon and Miller, 1989; Doyle et al., 1998; Elinder and Arhem, 1999).

A large set of transport systems, differing in K^+ affinity, selectivity, and energetic coupling, govern K^+ soil uptake, translocation, and compartmentalization with at least 35 genes coding for these transporters and channel systems (Lebaudy et al., 2007). Three families of K^+-selective channels (which are multimeric proteins) have been identified in plants: Shaker (KAT1, active at the plasmalemma, and voltage sensitive), TPK (tandem-pore K^+ channels active at the tonoplast), and Kir-like (K^+-inward rectifying channels active at the tonoplast). Two additional K^+-selective channels have been identified: TPC1, which is a K^+-permeable poorly selective cation channel (Peiter et al., 2005); and the cyclic nucleotide gated channel (CNGC; Talke et al., 2003).

In the tonoplast, three types of cation conductance have been identified: fast vacuolar (FV), slow vacuolar (SV), and vacuolar K^+ (VK) (Lebaudy et al., 2007). FV and SV conductances are not K^+ selective and mediate currents carried by different cations (K^+, Ca^{2+}, and Na^+). FV channels are inhibited by cytosolic Ca^{2+} concentrations higher than 0.1 μM (Tikhonova et al., 1997) and have a role in Ca^{2+}- independent ABA mediation of stomatal closure (Allen and Sanders, 1996; Pei et al., 1999). SV channels are inactive at low Ca^{2+} concentrations, but activated at Ca^{2+} levels higher than 0.5 μM (Hedrich and Neher, 1987; Roelfsema and Hedrich, 2005) and dominate tonoplast conductance. SV channels are involved in Ca^{2+}-signaling events, vacuolar Ca^{2+} release, cation homeostasis, and osmoregulation processes (vacuolar K^+ release during stomatal closure [Allen and Sanders, 1996; Peiter et al., 2005]). VK channels are highly selective for K^+ (Ward and Schroeder, 1994) and mediate K^+ release or uptake (Allen et al., 1998; Roelfsema and Hedrich, 2005). In summary, extracellular Ca^{2+} appears to alleviate sodium chloride damage by inducing H_2O_2 production; H_2O_2 activation increases cytosolic free Ca^{2+} concentrations, which in turn reduces Na^+ uptake through guard cell regulation of plasma membrane K^+ channels, leading to stomatal closure and reduction of water loss (Zhao et al., 2008). Thus, the synergism requirement for Ca^{2+} and K^+ in salinity tolerance responses is essential.

9.2 TURFGRASS SALINITY TOLERANCE

As illustrated in Section 9.1, "Plant Genetic and Physiological Responses to Salinity," multiple genetic mechanisms control the response of turfgrass and landscape plants when exposed to increasing salinity. Multiple salinity stresses affect turfgrasses, and often there is a mixed-stress response. The individual interactive deleterious effects of salinity and alkalinity have been studied (Shi and Sheng 2005; Li, Shi, and Fukuda, 2010a). Buffer capacity (CO32- and HCO3- concentrations), salinity (represented by Na^+ concentration), pH, and Cl^- reflected the reciprocal enhancement between salt stress and alkali stress responses, but salinity was the dominant factor followed by buffer capacity and pH as secondary factors (Li et al., 2010a). The deleterious effects of salinity or high pH alone were less than when combined as salt-alkaline mixed stress (Li et al., 2010b).

Also, multiple genetic mechanisms control the response of turfgrass and landscape plants when exposed to increasing salinity. Multiple phases of plant exposure to increasing salinity encompass reoccurring topical applications of total dissolved salts and specific salt ions (especially sodium and chlorides) from irrigation water that can be potentially foliarly absorbed into the surface canopy, which is then followed by soil accumulation over time that can approach toxic rhizospheric levels if the total salt complex is not properly managed.

As salinity stress increases over time, (a) many different genes are turned on or off as the plant is exposed to increasing salinity stress; (b) these multiple gene responses influence a number of specific salt tolerance mechanisms; and (c) unless the plant is exposed to sufficiently high salinity, important genetic responses controlling salinity tolerance mechanisms will not be triggered. Lee et al. (2005) demonstrated these points in determining salinity tolerance in seashore paspalum ecotypes and discussed the implications for breeder screening protocols. Recently, Tavakkoli et al. (2010, 632) noted similar conclusions for screening salt-tolerant barley:

> [S]everal processes are involved in salt tolerance and that the relative importance of these traits may differ with the severity of the salt stress. If the importance of different mechanisms to salinity tolerance differs by the severity of the stress, robust levels of salt tolerance may depend on more than one mechanism. Selection for improved salt tolerance therefore needs to be able to identify these.

Thus, appropriate, realistic salinity assessment protocols are essential to define the salinity tolerance and growth responses of halophytes (Subbarao and Johansen, 1994). While consistent salinity assessment criteria have been established for glycophytes, Lee et al. (2005) noted that no consistent criterion had been developed for halophytic grasses (Maas, 1994; Marcum, 2002). Using seashore paspalum as a model non-phytoaccumulator-type halophyte, they developed suggested criterion.

Halophytic plants *tolerate* salts and very strictly regulate the uptake of toxic salt ions such as sodium and chloride, thereby leaving the excess salts to accumulate in the soil. When both salt and alkali stress are combined, organic acid production and accumulation in the rhizosphere provide an additional mechanism to tolerate those simultaneous dual stresses (Chen et al., 2009).

Glycophytic plants do not possess the same number or activation level of tolerance mechanisms compared to halophytes. These plants will internally accumulate excess toxic salts over a certain salinity exposure time frame and, depending on the actual level of genetic salinity tolerance, will eventually experience reduced growth rates, nutritional imbalances, substantial loss of root volume, and significant loss of surface leaf area and canopy density.

All plants have a salinity tolerance threshold (usually expressed as the electrical conductivity of soil saturated paste extracts [ECe] = 1.5 electrical conductivity of water [ECw]) where growth rates, root growth, and topgrowth decrease below those under nonsaline conditions (Carrow and Duncan, 1998). The total growth reduction per unit increase in ECe determines the tolerance threshold and management options for a specific grass or landscape cultivar. Data in Table 9.1 provide SPE classification ranges for total soil salinity problems. Data in Table 9.2 provide general plant soil salinity tolerance threshold ranges based on nongypsiferous soils or gypsiferous soils.

TABLE 9.1
Classification of Soil Saturated Paste Extract (SPE) for Total Salinity Problems

Classification	ECe	TDS
Total salinity problems	dS/m	ppm or mg/L
Low	<1.5	<960
Moderate	1.6–3.9	961–2496
High	4.0–5.0	2497–3200
Very high	>5.0	>3200
Halophytes[a]	>20	>12,800

Source: Adapted from Carrow, R. N., and R. R. Duncan, *Salt-Affected Turfgrass Sites: Assessment and Management*, John Wiley, Hoboken, NJ, 1998.

[a] The primary limitation on sustainable management will be excess salt accumulation in the soil, and a very high degree of management expertise is required for sustainability. If salts are allowed to accumulate above the threshold salt tolerance level of the specific turfgrass or landscape cultivar, significant damage or even death of the plant can result.

Several parameters can be utilized to compare grass or landscape plant salinity tolerance thresholds (Carrow and Duncan, 1998). The most salinity-tolerant grasses or landscape plants would have the following:

- A high threshold ECe for both root and shoot growth. A higher ECe and the inherent capability to maintain 100% growth rates not only expand general tolerance to many saline sites but also provide more flexibility to manage those plants as salts continue to accumulate in the soil and the plant over time.
- A gradual or slow decrease in growth rate per unit ECe increase to sustain some functionality, resulting in prolonged growth rate reductions as the salinity impact increases.
- A high ECe for 50% growth reduction. Sustaining root growth and redevelopment as salts accumulate in the soil profile combined with maintaining adequate root and shoot growth to sustain root regeneration, canopy density, and wear and traffic tolerance (governed by sustainable growth rates) in turfgrasses is essential for long-term management.

TABLE 9.2
Plant Soil Salinity Tolerance Threshold Ranges Based on Soil Accumulation (ECe Thresholds for 0% Reduction in Growth and for 50% Growth Reduction)

Soil Salinity Tolerance	Threshold ECe 0% Reduction dS/m	$ECe_{50\%}$ Nongypsiferous Soils	$ECe_{50\%}$ Gypsiferous Soils
Very sensitive plants	<1.5	1.5–5.0	3.5–7.0
Moderately sensitive plants	1.6–3.0	5.1–10.0	7.1–12.0
Moderately tolerant plants	3.1–6.0	10.1–15.0	12.1–17.0
Tolerant plants	6.1–10.0	15.1–21.0	17.1–23.0
Very tolerant plants	>10.0	>21.0	>23.0

Source: Adapted from Carrow, R. N., and R. R. Duncan, *Salt-Affected Turfgrass Sites: Assessment and Management*, p. 34, John Wiley, Hoboken, NJ, 1998.

Note: Multiply ECe by 640 = TDS in ppm or mg/L.

- A turfgrass with high inherent growth rates under low salinity should be able to tolerate higher salt accumulation while still maintaining adequate growth as long as salt accumulation can be properly managed in both the soil and the plant (Marcum and Murdock, 1994).

Based on published research data prior to 1998 and summarized in Carrow and Duncan (1998), a general or relative salinity tolerance ranking of turfgrass species was assembled based on threshold ECe (Table 9.3). There are significant cultivar differences in salinity tolerance response within

TABLE 9.3
Relative Salinity Tolerance Rankings of Turfgrass Species Based on Threshold ECe Reported Prior to 1998

			Threshold ECe[a]		
Common Name	**Scientific Name**	**General Salinity Tolerance[a,b]**	**Average**	**$ECe_{50\%}$**	**Grass Type**
			dS/m		
Seashore paspalum	*Paspalum vaginatum* (Swartz)	VT-T	8.6	31	Warm
Alkaligrass	*Puccinella* spp.	VT-T	8.5	25	Cool
Saltgrass	*Distichlis stricta*	T	8.0		Warm
Kikuyu	*Pennisetum clandestinum*	T	8.0		Warm
Fairway wheatgrass	*Agropyron cristatum*	T	8.0		Cool
Western wheatgrass	*Agropyron smithii*	T	8.0		Cool
St. Augustinegrass	*Stenotaphrum secundatum*	T	6.5	29	Warm
Tall fescue	*Festuca arundinacea*	T	6.5	11	Cool
Perennial ryegrass	*Lolium perenne*	T	6.5	9	Cool
Slender creeping red fescue	*Festucs ruba* L. spp. *Trichopylla*	T	6.3	10	Cool
Buffalograss	*Buchloe dactyloides*	MT	5.3	13	Warm
Blue grama	*Bouteloua gracilis*	MT	5.2		Warm
Hard fescue	*Festuca longifolia*	MT	4.5		Cool
Creeping red fescue	*Festuca ruba* L. spp. *Ruba*	MT	4.5	10	Cool
Common Bermuda grass	*Cynodon dactylon*	MT	4.3	21	Warm
Hybrid Bermuda grass spp.	*Cynodon*	MT	3.7	22	Warm
Creeping bentgrass	*Agrostis palustris*	MT	3.7	8	Cool
Kentucky bluegrass	*Poa pratensis*	MS	3.0	14	Cool
Zoysiagrass	*Zoysia* spp.	MS	2.4	16	Warm
Carpetgrass	*Axonopus* spp.	VS	1.5		Warm
Centipedegrass	*Eremochloa ophiuroides*	VS	1.5	8	Warm
Annual bluegrass	*Poa annua*	VS	1.5		Cool
Colonial bentgrass	*Agrostis tenuis*	VS	1.5		Cool
Rough stalk bluegrass	*Poa trivialis*	VS	1.5		Cool

Source: Adapted from Carrow, R. N., and R. R. Duncan, *Salt-Affected Turfgrass Sites: Assessment and Management*, p. 34, John Wiley, Hoboken, NJ, 1998.

[a] VS (very sensitive) <1.5 dS/m ECe; MS (moderately sensitive) 1.6–3.0 ECe; MT (moderately tolerant) 3.1–6.0 ECe; T (tolerant) 6.1–10.0 ECe; and VT (very tolerant) >10.0 ECe.

[b] Cultivar dependent within a species.

species. In addition, those research studies prior to 1998 had no standardized screening protocols and were primarily conducted in nutrient solution water baths. The total salinity impact that was imposed on the turfgrasses in those studies varied considerably, hence the "relative ranking" designation for that table. Additionally, since most of those studies were conducted in water baths or other strictly controlled environmental conditions, those data do not always equate to absolute performance under actual field conditions, where other cyclic biotic and abiotic stresses may impact turfgrasses.

9.2.1 Salinity Tolerance at Seed Germination and with Vegetative or Immature Plants

Soil salt accumulation and the flux of these salts either upward (due to high evapotranspiration) or downward (via leaching) occur constantly regardless of temperatures or seasonal changes. Management to minimize soil salt accumulation is a primary strategy for long-term sustainability in perennial turfgrass and landscape ecosystems, but it must start with establishment. It is essential to realize that salinity tolerance at seed germination and for young seedlings, as well as for vegetative propagules, differs substantially from mature plants of the same species.

Turfgrass and landscape plants vary in their activation of salt tolerance mechanisms as plants progress from seed germination to seedling to established plant and from vegetative propagule (nodal segment) to plantlet to mature plant. Very few cultivars, including halophytes, will actually germinate in salinized water of ECw > 3.1 dS/m or 2000 ppm; and if soil accumulation of total salts and specific toxic salt ions has concentrated at even higher levels, initial germination, if it occurs, will be slow and sparse, and often results in rapid seedling death as that small plant attempts to initially root in the salinized rhizosphere.

9.2.1.1 Seeded Cultivars

Availability of water, irrigation scheduling, and water quality will govern seed germination and nodal segment plantlet initiation. Both germination rate and total germination are decreased in salt-affected soils due to the low water potential of saline water restricting actual water uptake by juvenile roots. Restricted saline water uptake alters morphology, physiology, and biochemistry functions in germinating seeds (Ashraf and Foolad, 2005). Seed priming is one technique that can be employed to improve the rate and synchrony of seed germination (Ashraf et al., 2008).

Osmo-priming or osmotic conditioning involves soaking seeds in solutions of sugars, glycerol, polyethylene glycol (PEG), mannitol, or sorbitol, followed by air drying prior to planting to enhance germination rate, percent germination, seedling growth, and biomass production under saline conditions; see Pill and Necker (2001) for Kentucky bluegrass (*Poa pratensis* L.), Hur (1991) for Italian ryegrass (*Lolium multiflorum* Lam.), and Al-Humaid (2002) for Bermuda grass (*Cynodon dactylon* L.).

Halo-priming soaks seeds in inorganic salt solutions (i.e., 0.1% $MnSO_4$ or 0.05% $ZnSO_4$) to promote seed germination and emergence, plus enhance subsequent growth (Babaeva et al., 1999) and biomass yield (Kadiri and Hussaini, 1999).

Hydro-priming involves cyclically soaking seeds prior to planting with or without air drying to "harden" the seeds in order to enhance germination and seedling emergence under saline conditions (Pill and Necker, 2001; Ashraf and Foolad, 2005).

Thermo-priming treats seed with either low or high temperatures to improve germination and subsequent seedling emergence under adverse environmental conditions (Watkinson and Pill, 1998; *Sorghastrum nutans* L. Nash; chilling treatment).

Hormone priming pretreats seeds with targeted concentrations of plant growth regulators (PGRs) to improve germination and biomass production under stress conditions (Lee et al., 1998). Growth regulators that have been used for seed priming include chlormequat chloride, auxins (IAA, IBA, and NAA), gibberellins and gibberellin antagonists, abscisic acid, ethylene, kinetin, salicylic acid (SA), polyamines (PAs such as spermine-Spm, spermidine-Spd, and putrescine-Put), ascorbic acid, triacontanol, brassinolide (i.e., 28-homobrassinolide), 2,4-D, thiamin, sodium salicylate, and pyridoxine (summarized in Ashraf et al., 2008).

9.2.1.2 Exogenous Plant Applications of Osmolytes, Osmoprotectants, and PGRs

Highly soluble, low-molecular-weight compounds can accumulate in high concentrations within cells to protect plants from stress via osmotic adjustments, detoxification of reactive oxygen species, protection of membrane structures, and stabilization of proteins (Ashraf and Foolad, 2007). These osmoprotectants protect cellular components from dehydration injury, and these compatible solutes include trehalose, proline, sucrose, polyols, and quaternary ammonium compounds (glycine betaine [GB], proline betaine, hydroxyproline betaine, pipecolate betaine, alanine betaine, and choline O-sulfate) (Ashraf and Harris, 2004).

Exogenous application of GB can ameliorate shoot accumulation of excess Na^+ and promote higher K^+ concentrations since GB-induced production of additional root cell vacuoles resulted in sequestration of Na^+ in those root vacuoles and reduced transport to the shoots (Rahman et al., 2002). Due to plant species and specific cultivar differences plus variable abiotic stress environmental conditions and plant developmental stages, the dose–response relationship still needs to be refined (Ashraf et al., 2008), especially in perennial turfgrass and landscape plants.

The same dose–response criteria exist for exogenous applications of proline under saline conditions (Ashraf et al., 2008; Ashraf and Foolad, 2007). Proline can enhance antioxidant activity to protect membranes from oxidative stress under salt stress (Yan et al., 2000), thereby providing osmoprotection and facilitating growth (Csonka and Hanson, 1991; Yancey, 1994). Exogenous application of proline has neutralized enhanced ethylene production induced by salinity stress (Chrominski et al., 1989) and decreased Na^+ and Cl^- in shoots (Lone et al., 1987) with proper application rates, but not in perennial turfgrass or landscape plants.

Exogenous application of kinetin (6-furfurylaminopurine) can reduce the uptake of Na^+ and Cl^- plus promote the uptake of K^+, thereby ameliorating the deleterious effects of salinity (Gadallah, 1999). Triadimefon partially mitigates salinity stress by enhancing antioxidant enzyme activity in roots, stems, and leaves (Jaleel et al., 2008). 28-Homobrassinolide ameliorates salinity stress by enhancing nitrate reductase, nitrogenase, leghaemoglobin, and carbonic anhydrase activities (Ali et al., 2007). Supplementary potassium nitrate and proline treatments significantly ameliorated adverse saline effects (Kaya et al., 2007).

Foliar applications of PGRs may be a more realistic approach for alleviating some of the deleterious effects of salinity than soil-applied products (Ashraf et al., 2008). But, most of the research has involved in vitro or annual crops and not perennials such as turfgrass and landscape plants. Additional potential hormonal products might include jasmonic acid and its methyl esters, thiamin (vitamin B1 antioxidant), 5-aminolevulinic acid (precursor in porphyrin biosynthesis such as in chlorophyll and heme), and salicylic acid (phenolic compound) (foliar applications are summarized in Ashraf et al., 2008).

At low concentrations, the reactive oxygen species (ROS) H_2O_2 pretreatment acts as a messenger molecule for abiotic stress acclimatory signaling as well as gene expression for antioxidative enzymes, and can trigger cross-tolerance to both biotic and abiotic stresses including salinity in shoots and roots (summarized in Azevedo Neto et al., 2005). Antioxidative enzymes (superoxide dismutase, ascorbate peroxidase, guaiacol peroxidase, glutathione reductase, and catalase) regulate H_2O_2 intracellular levels. Superoxide dismutase (SOD) is directly involved in salt tolerance enhancement (Gosset et al., 1994; Hernandez et al., 2000; Shalata et al., 2001), while Mn-SOD and Cu/Zn-SOD in chloroplasts enhance tolerance to salt stress (Tanaka et al., 1999; Badawi et al., 2004) during photosynthesis. Exogenously applied silicon can increase antioxidant enzyme activity (superoxide dismutase, peroxidase, catalase, and glutathione reductase) and reduce lipid peroxidation in roots under salt stressed conditions (Liang et al., 2003).

While the previous discussion demonstrates that a number of compounds have been researched for influence on salinity tolerance of immature plants, these have seldom involved turfgrasses or practical field situations. Thus, considerable research is required to determine the feasibility of using a particular material on turfgrasses. Additionally, since there have been too many examples of "proprietary" or "patented" materials marketed to the turfgrass industry without sound research and without full detailed label disclosure of the active ingredient, the authors would caution against use of any such materials unless these basic criteria are meet. Additionally, turfgrass managers must be aware of the magnitude of response versus economic input when using products. A response that is theoretically possible does not mean that it is of any importance unless the magnitude of response is sufficient to justify the economic input and the subsequent plant response.

9.2.2 Salinity Tolerance of Turfgrass Species and Cultivars

The criteria for assessing relative salinity tolerance of not only turfgrass species but also specific cultivars within a species for perennially grown plants have been a constant challenge, especially for halophytic ecotypes (Lee et al., 2005). Most initial assessments for plant salinity tolerance have encompassed nutrient solution screening of genetically variable accessions and, in more advanced screening programs, actual assessments in soil profiles based on specific ECe (electrical conductivity of saturated soil paste extracts). The creditability of these assessments in relating those specific grass performance data to actual long-term field exposure and reoccurring exposure to increasing soil salt accumulation and foliar exposure to saline irrigation water has led to turfgrass data summarizations based on "relative salinity tolerance" rankings of species (Table 9.3). The absolute salinity tolerance under field conditions depends on whether any additional climatic, abiotic, or biotic stresses are present that influence the ability of the plant to tolerate salinity stress.

The highest relative salt-tolerant ranked group includes seashore paspalum (*Paspalum vaginatum*), alkaligrass (*Puccinellia* spp.), saltgrasses (*Distichlis stricta* and *spicata*), kikuyugrass (*Pennisetum clandestinum*), fairway or crested wheatgrass (*Agropyron cristatum*), and Western wheatgrass (*Agropyron smithii*) (Table 9.3). The lowest ranked group includes Kentucky bluegrass (*Poa pratensis*), some zoysiagrass species, centipedegrass (*Eremochloa ophiuroides*), annual bluegrass (*Poa annua*), colonial bentgrass (*Agrostis tenuis*), and roughstalk bluegrass (*Poa trivalis*).

Data in Table 9.4 summarize some published salinity research on individual turfgrass species. The selected references provide the scope and progress over years for assessing and developing improved salt-tolerant turfgrass cultivars within each of the species.

Since considerable diversity exists among cultivars within each turfgrass species, data in Table 9.5 list the specific cultivars that have been marketed for their grass performance under saline conditions. Most private turfgrass companies as well as some universities have increased their cultivar breeding and development research programs over the past 10 years to improve overall salinity tolerance levels in both warm-season and cool-season turfgrasses.

When comparing grass species and their responses to salt spray (or foliar exposure to reoccurring exposure from saline irrigation water) versus actual tolerance to increasing soil salt accumulation, generalizations can be made. Alkaligrass (*Puccinellia airoides*), the saltgrasses (*Distichlis spicata* and *stricta*), and alkali sacaton (*Sporobolus airoides*) generally tolerate both salinity exposure situations quite well. Depending on specific cultivar and overall level of salinity tolerance, seashore paspalum (*Paspalum vaginatum*) has very high tolerance to salt spray due to the heavy wax load on the leaves and is quite tolerant of increasing salt accumulation in the soil; however, even this halophytic turfgrass can be overwhelmed if sodic or saline-sodic soil salts are not properly managed.

TABLE 9.4
Selected Salinity-Oriented Research References on Turfgrass Species, Including References after 1998 That Were Used for the Turfgrass Tolerance Ranking in Table 9.3

WARM-SEASON GRASSES

General Articles

R. R. Duncan & R. N. Carrow. 1999. Turfgrass molecular genetic improvement for abiotic/edaphic stress resistance. *Advances in Agronomy* 67: 233–305.

G. A. Rumman, E. Barrett-Lennard, & T. Colmer. 2009. Halophytic turfgrasses and their potential use on salt-affected areas. *Australian Turfgrass Management Journal* 11 (1): 46–47.

K. B. Marcum. 2008. Relative salinity tolerance of turfgrass species and cultivars. In M. Pessarakli (ed.), *Handbook of Turfgrass Management and Physiology*, pp. 308–406. New York: Marcel Dekker.

R. N. Carrow & R. R. Duncan. 2005. Just a grain of salt: As salinity increases, turfgrass management will need to increase, too. *TurfGrass TRENDS*, 70 (72), 74–75.

R. Munns. 2005. Genes and salt tolerance: Bringing them together. *New Phytologist* 167 (3): 645–663.

T. Yamaguchi & E. Blumwald. 2005. Developing salt-tolerant crop plants: Challenges and opportunities. *TRENDS in Plant Science* 10 (12): 615–620.

D. S. Loch, E. Barrett-Lennard, & P. Truong. 2003. Role of salt tolerant plants for production, prevention of salinity and amenity values. *Proceedings of the 9th National Conference of Productive Use of Saline Lands* (PURSL), 1–16.

K. B. Marcum. 2002. Growth and physiological adaptations of grasses to salinity stress. In M. Pessarakli (ed.), *Handbook of Plant and Crop Physiology* (pp. 623–636). New York: Marcel Dekker.

T. Colmer. 2000. Salt tolerance in plants. *Australian Turfgrass Management* 2 (5): 10–12.

S. Miele, M. Volterrani, & N. Grossi. 2000. Warm season turfgrasses: Results of a five-year study in Tuscany. *Agricoltura Mediterranea* 130: 196–202.

B. A. Smith. 1997. An investigation of the salt tolerance of turfgrasses. *The Grass Roots* 25 (4): 36–37, 39.

K. Marcum. 1994. Salt-tolerance mechanisms of turfgrasses: Scientists are slowly gaining an understanding of the ways in which turf responds to high-salinity growing conditions. *Golf Course Management* 62 (9): 55–59.

M. A. L. Smith, J. E. Meyer, S. L. Knight, & G. S. Chen. 1993. Gauging turfgrass salinity responses in whole-plant microculture and solution culture. *Crop Science* 33 (3): 566–572.

A. E. Dudeck & C. H. Peacock. 1993. Salinity effects on growth and nutrient uptake of selected warm-season turf. *International Turfgrass Society Research Journal* 7: 680–686.

Y-J. Kuo. 1992. Cell level selection for salt tolerance in some turfgrass species and confirmation of whole plant salt tolerance characteristics. Master's thesis, University of Illinois at Urbana-Champaign.

M. Ashraf, T. McNeilly, & A. D. Bradshaw. 1989. The potential for evolution of tolerance to sodium chloride, calcium chloride, and magnesium chloride and seawater in four grass species. *New Phytologist* 112 (2): 245–254.

A. E. Dudeck. 1988. Turfgrass salinity research. *TurfNews* (ASPA) 11 (3): 12, 19, 21.

G. L. Horst. 1986. Salt resistance and tolerance in turfgrasses. Proceedings of the 57th International Golf Course Conference and Show, p. 49050.

T. McNeilly, M. Ashraf, & A. D. Bradshaw. 1986. The potential for evolution of salt (NaCl) tolerance in seven grass species. *New Phytologist* 103 (2): 299–309.

M. Ashraf, T. McNeilly, & A. D. Bradshaw. 1986. The response of selected salt-tolerant and normal lines of four grass species to NaCl in sand culture. *New Phytologist* 104 (3): 453–461.

M. Ashraf, T. McNeilly, & A. D. Bradshaw. 1986. Tolerance of sodium chloride and its genetic basis in natural populations of four grass species. *New Phytologist* 103 (4): 725–734.

T. D. Hughes, J. D. Butler, & G. D. Sanks. 1975. Salt tolerance and suitability of various grasses for saline roadsides. *Journal of Environmental Quality* 4 (1): 65–68.

W. E. Cordukes. 1970. Turfgrass tolerance to road salt. *The Golf Superintendent* 38 (5): 44–48.

O. R. Lunt, V. B. Youngner, & J. J. Oertli. 1961. Salinity tolerance of five turfgrass varieties. *Agronomy Journal* 53 (4): 247–249.

V. T. Stoutemyer & F. B. Smith. 1936. The effects of sodium chloride on some turf plants and soils. *Journal of the American Society of Agronomy* 28 (1): 16–23.

H. L. Westover. 1928. Salt grass. *The Bulletin of the United States Golf Association Green Section* 8 (1): 14.

continued

TABLE 9.4 (Continued)
Selected Salinity-Oriented Research References on Turfgrass Species, Including References after 1998 That Were Used for the Turfgrass Tolerance Ranking in Table 9.3

***Paspalum vaginatum* Swartz**	**Seashore paspalum**

R. R. Duncan & R. N. Carrow. 1999. *Seashore Paspalum: The Environmental Turfgrass*. Hoboken, NJ: John Wiley.

R. R. Duncan. 2003. Seashore paspalum (*Paspalum vaginatum* Swartz). In M. D. Casler & R. R. Duncan (eds.), *Turfgrass Biology: Genetics and Breeding* (pp. 295–307). Hoboken, NJ: John Wiley.

R. N. Carrow & R. R. Duncan. 1998. *Salt-Affected Turfgrass Sites: Assessment and Management*. Chelsea, MI: Ann Arbor Press.

G-J. Lee. 2000. Comparative salinity tolerance and salt tolerance mechanisms of seashore paspalum ecotypes. PhD diss., University of Georgia.

M. Pessarakli & D. M. Kopec. 2008. Establishment of three warm-season grasses under saline conditions. *Acta Horticulturae* 783: 29–39.

G-J. Lee, R. N. Carrow, R. R. Duncan, M. A. Eiteman, & M. W. Rieger. 2008. Synthesis of organic osmolytes and salt tolerance mechanisms in *Paspalum vaginatum*. *Environmental and Experimental Botany* 63 (1–3): 19–27.

J. B. Unruh, B. J. Brecke, & D. E. Darcy. 2007. Seashore paspalum performance to potable water. *USGA Turfgrass and Environmental Research Online* 6 (23): 1–10.

G-J. Lee, R. R. Duncan, & R. N. Carrow. 2007. Nutrient uptake responses and inorganic ion contribution to solute potential under salinity stress in halophytic seashore paspalums. *Crop Science* 47 (6): 2504–2512.

P. L. Raymer. 2006. Salt tolerance in seashore paspalum: Not all varieties are created equal, research shows. *TurfGrass TRENDS* September: 62, 64, 66.

A. C. Hixson, T. Lowe, & W. T. Crow. 2005. Salts influence nematodes in seashore paspalum: Are seashore paspalum roots affected by plant-parasitic nematodes under high-salinity irrigation? *USGA Green Section Record* 43 (1): 9–13.

G-J. Lee, R. N. Carrow, & R. R. Duncan. 2005. Criteria for assessing salinity tolerance of the halophytic turfgrass seashore paspalum. *Crop Science* 45 (1): 251–258.

R. R. Duncan & R. N. Carrow. 2005. Managing seashore paspalum greens. *Golf Course Management* 73 (2): 114–118.

R. R. Duncan & R. N. Carrow. 2005. Preventing failure of seashore paspalum greens: Proper management techniques can prevent problems and failure in paspalum greens. *Golf Course Management* 73 (3): 99–102.

T. Carson. 2005. Seeded seashore paspalum. *Golf Course Management* 73(5): 28.

W. L. Berndt. 2005. Salinity alters growth habit of seashore paspalum. *Golf Course Management* 73 (5): 101–104.

N. B. Nicholas. 2005. Influence of saltwater on weed management in seashore paspalum. Master's thesis, University of Florida.

G-J. Lee, R. N. Carrow, & R. R. Duncan. 2005. Growth and water relation responses to salinity stress in halophytic seashore paspalum ecotypes. *Scientia Horticulturae* 104 (2): 221–236.

Z. Chen, W. Kim, M. Newman, M. Wang, & P. Raymer. 2005. Molecular characterization of genetic diversity in the USDA seashore paspalum germplasm collection. *International Turfgrass Society Research Journal* 10 (Pt. 1): 543–549.

G-J. Lee, R. R. Duncan, & R. N. Carrow. 2004. Salinity tolerance of seashore paspalum ecotypes: Shoot growth responses and criteria. *HortScience* 39 (5): 1138–1142.

G-J. Lee, R. N. Carrow, & R. R. Duncan. 2004. Salinity tolerance of seashore paspalums and bermudagrasses: Root and verdue responses and criteria. *HortScience* 39 (5): 1143–1147.

G-J. Lee, R. N. Carrow, & R. R. Duncan. 2004. Photosynthetic responses to salinity stress of halophytic seashore paspalum ecotypes. *Plant Science* 166 (6): 1417–1425.

G-J. Lee, R. R. Duncan, & R. N. Carrow. 2003. Initial selection of salt-tolerant seashore paspalum ecotypes. *Australian Turfgrass Management* 5 (3): 30, 32, 34–35. Also: 2002. *USGA Turfgrass and Environmental Research Online* 1 (11): 1–9.

R. R. Duncan. 2001. All seashore paspalums are not created equal: Research supports only a few of the species's new varieties. *Golf Course Management* 69 (6): 54–60.

Y-J. Kuo & T. W. Fermanian. 2001. Use seashore paspalum on phytoremediation of heavy-metal contaminated soil. *IXth International Turfgrass Research Conference* 9: 68–69.

A. E. Dudeck & C. H. Peacock. 1985. Effects of salinity on seashore paspalum turfgrasses. *Agronomy Journal* 77 (1): 47–50.

J. M. Henry, V. A. Gibeault, V. B. Youngner, & S. Spaulding. 1979. *Paspalum vaginatum* 'Adalayd' and 'Futurf'. *California Turfgrass Culture* 29 (2): 9–12.

J. F. Morton. 1973. Salt-tolerant silt grass (*Paspalum vaginatum* Sw.). *Proceedings of the Annual Meeting of the Florida State Horticultural Society* 86: 482–490.

TABLE 9.4 (Continued)
Selected Salinity-Oriented Research References on Turfgrass Species, Including References after 1998 That Were Used for the Turfgrass Tolerance Ranking in Table 9.3

***Cynodon* (L) Rich spp.** **Bermuda grass**

C. M. Taliaferro. 2003. Bermudagrass [*Cynodon* (L.) Rich]. In M. D. Casler & R. R. Duncan (eds.), *Turfgrass Biology: Genetics and Breeding* (pp. 235–256). Hoboken, NJ: John Wiley.

B. K. Bauer, R. E. Poulter, A. D. Troughton, & D. S. Loch. 2009. Salinity tolerance of twelve hybrid bermudagrass [*Cynodon dactylon* (L.) Pers. X *C. transvaalensis* Burtt Davy] genotypes. *International Turfgrass Society Research Journal* 11 (Pt. 2): 313–326.

M. Pessarakli & D. M. Kopec. 2008. Establishment of three warm-season grasses under salinity stress. *Acta Horticulturae* 783: 29–39.

C. M. Baldwin, H. Liu, L. B. McCarty, W. L. Bauerle, & J. E. Toler. 2006. Effects of trinexapac-ethyl on the salinity tolerance of two ultradwarf bermudagrass cultivars. *HortScience* 41 (3): 808–814.

K. B. Marcum & M. Pessarakli. 2006. Salinity tolerance and salt gland excretion efficiency of bermudagrass turf cultivars. *Crop Science* 46 (6): 2571–2574.

K. B. Marcum, M. Pessarakli, & D. M. Kopec. 2005. Relative salinity tolerance of 21 turftype desert saltgrasses compared to bermudagrass. *HortScience* 40 (3): 827–829.

J. M. Rutledge & C. H. Peacock. 2005. Turfgrass center report: What's new with bermudagrass? *North Carolina Turfgrass* March/April: 1–3.

G-L. Lee, R. N. Carrow, & R. R. Duncan. 2004. Salinity tolerance of selected seashore paspalums and bermudagrasses: Root and verdure responses and criteria. *HortScience* 39 (5): 1143–1147.

C. H. Peacock, D. J. Lee, W. C. Reynolds, J. P. Gregg, R. J. Cooper, & A. H. Bruneau. 2004. Effects of salinity on six bermudagrass turf cultivars. *Acta Horticulturae* 661: 193–197.

G. C. Munshaw, X. Zhang, & E. H. Ervin. 2004. Effect of salinity on bermudagrass cold hardiness. *HortScience* 39 (2): 420–423.

C. Rodgers. 2003. You've come a long way, Bermuda: The new seeded bermudagrasses offer high-quality turf and the obvious convenience of seed. *Golf Course Management* 71 (8): 91–95.

G. Zhang. 2003. Enhancing salt and drought tolerance of Poa and triploid Cynodon with BADH gene and somaclonal variation. PhD diss., Rutgers The State University of New Jersey—New Brunswick.

A. E. Dudeck, S. Singh, C. E. Giordano, T. A. Nell, & D. B. McConnell. 1983. Effects of sodium chloride on *Cynodon* turfgrasses. *Agronomy Journal* 75 (6): 927–930.

A. E. Dudeck. 1976. Salt tolerance of bermudagrass. *Proceedings of the 24th Annual Florida Turfgrass* 24: 95–97.

R. C. Ackerson & V. B. Youngner. 1975. Responses of bermudagrass to salinity. *Agronomy Journal* 67 (5): 678–681.

P. S. Ramakrishnan & R. Nagpal. 1973. Adaptation to excess salts in an alkaline soil population of *Cynodon dactylon* (L). Pers. *Journal of Ecology* 61 (2): 369–381.

***Buchloe dactyloides* (Nutt.)** **Engelm Buffalograss**

T. P. Riordan & S. J. Browning. 2003. Buffalograss [*Buchloe dactyloides* (Nutt.) Engelm]. In M. D. Casler & R. R. Duncan (eds.), *Turfgrass Biology: Genetics and Breeding* (pp. 257–270). Hoboken, NJ: John Wiley.

A. Van Dyke & P. G. Johnson. 2009. Buffalograss tolerance to post-emergence herbicides in the intermountain west. *Applied Turfgrass Science* 12: 1–2.

L. Wu. 2000. Buffalograss: This ancient American forage grass may have a future as turf. *Diversity* 16 (1/2): 42–43.

H. Lin & L. Wu. 1996. Effects of salt stress on root plasma membrane characteristics of salt-tolerant and salt-sensitive buffalograss clones. *Environmental and Experimental Botany* 36 (3): 239–254.

L. Wu & A. Harivandi. 1995. Buffalograss response to cold, shade, and salinity. *California Turfgrass Culture* 45 (1/2): 5–7.

***Zoysia* spp.** **Zoysiagrass**

M. C. Engelke & S. Anderson. 2003. Zoysiagrasses (*Zoysia* spp.). In M. D. Casler & R. R. Duncan (eds.), *Turfgrass Biology: Genetics and Breeding* (pp. 271–285). Hoboken, NJ: John Wiley.

D. S. Loch, B. K. Simon, & R. E. Poulter. 2005. Taxonomy, distribution and ecology of *Zoysia macrantha* Desv., an Australian native species with turf breeding potential. *International Turfgrass Society Research Journal* 10 (Pt. 1): 593–599.

continued

TABLE 9.4 (Continued)
Selected Salinity-Oriented Research References on Turfgrass Species, Including References after 1998 That Were Used for the Turfgrass Tolerance Ranking in Table 9.3

K. B. Marcum, G. Wess, D. T. Ray, & M. C. Engelke. 2003. Zoysiagrasses, salt glands, and salt tolerance. *USGA Turfgrass and Environmental Research Online* 2 (14): 1–8.

K. B. Marcum, G. Wess, D. T. Ray, & M. C. Engelke. 2003. Zoysiagrass, salt glands, and salt tolerance: Observing the density of salt glands may make selecting for salt-tolerant grasses a lot easier. *USGA Green Section Record* 41 (6): 20–21.

Y. L. Qian, M. C. Engelke, & M. J. V. Foster. 2000. Salinity effects on zoysiagrass cultivars and experimental lines. *Crop Science* 40 (2): 488–492.

K. B. Marcum. 1999. Salinity tolerance mechanisms of grasses in the subfamily Chloridoideae. *Crop Science* 39 (4): 1153–1160.

K. B. Marcum, S. J. Anderson, & M. C. Engelke. 1998. Salt gland ion secretion: A salinity tolerance mechanism among five zoysiagrass species. *Crop Science* 38 (3): 806–810.

K. B. Marcum & D. M. Kopec. 1997. Salinity tolerance of turfgrasses and alternative species in the subfamily Chloridoideae (Poaceae). *International Turfgrass Society Research Journal* 8 (Pt. 1): 735–742.

A. Patton. 2010. Selecting zoysiagrass cultivars: Turf quality and stress tolerance. *Golf Course Management* 78 (5):90–95.

***Eremochloa ophiuroides* (Munro) Hack.** **Centipedegrass**

W. W. Hanna & J. Liu. 2003. Centipedegrass (*Eremochloa ophiuroides*). In M. D. Casler & R. R. Duncan (eds.). *Turfgrass Biology: Genetics and Breeding* (pp. 287–293). Hoboken, NJ: John Wiley.

***Stenotaphrum secundatum* (Walt.) Kuntze** **St. Augustinegrass**

P. Busey. 2003. St. Augustinegrass [*Stenotaphrum secundatum* (Walt.) Kuntze]. In M. D. Casler & R. R. Duncan (eds.), *Turfgrass Biology: Genetics and Breeding* (pp. 309–330). Hoboken, NJ: John Wiley.

***Pennisetum clandestinum* Hochst. Ex Chiov.** **Kikuyu grass**

M. Radakrishnan, Y. Waisel, & M. Sternberg. 2006. Kikuyu grass: A valuable salt-tolerant fodder grass. *Communications in Soil Science and Plant Analysis* 37 (9/10): 1269–1279.

***Distichlis spicata* (L.) Green** **Inland saltgrass**

***Distichlis stricta* (Torr.) Rydb.** **Desert saltgrass**

M. A. Shahba, Y. Qian, & S. Wallner. 2009. Influence of proxy on saltgrass seed germination in saline soils. *International Turfgrass Society Research Journal* 11 (Pt. 2): 849–858.

M. A. Shahba, Y. L. Qian, & K. D. Lair. 2008. Improving seed germination of saltgrass under saline conditions. *Crop Science* 48 (2): 756–762.

M. Pessarakli & D. M. Kopec. 2008. Establishment of three warm-season grasses under salinity stress. *Acta Horticulturae* 783: 29–39.

M. Pessarakli, N. Gessler, & D. Kopec. 2008. Growth responses of saltgrass (*Distichlis spicata*) under sodium chloride (NaCl) salinity stress. *USGA Turfgrass and Environmental Research Online* 7 (20): 1–7.

Y. L. Qian, J. M. Fu, S. J. Wilhelm, D. Christensen, & A. J. Koski. 2007. Relative salinity tolerance of turf-type saltgrass selections. *HortScience* 42 (2): 205–209.

H. Rukavina, H. G. Hughes, & Y. Qian. 2007. Freezing tolerance of 27 saltgrass ecotypes from three cold hardiness zones. *HortScience* 42 (1): 157–160.

Y. L. Qian, J. A. Cosenza, S. J. Wilhelm, & D. Christensen. 2006. Techniques for enhancing saltgrass seed germination and establishment. *Crop Science* 46 (6): 2613–2616.

T. A. Aschenbach. 2006. Variation in growth rates under saline conditions of *Pascopyrum smithii* (Western wheatgrass) and *Distichlis spicata* (inland saltgrass) from different source populations in Kansas and Nebraska: Implications for the restoration of salt-affected plant communities. *Restoration Ecology* 14 (1): 21–27.

Y. L. Qian, S. Wilhelm, D. Christensen, T. Koski, & H. Hughes. 2006. Salt tolerance of inland saltgrass. *USGA Turfgrass and Environmental Research Online* 5 (24): 1–10.

M. Pessarakli & D. M. Kopec. 2005. Responses of twelve inland saltgrass accessions to salt stress. *USGA Turfgrass and Environmental Research Online* 4 (20): 1–5.

M. Pessarakli, K. B. Marcum, & D. M. Kopec. 2005. Growth responses and nitrogen-15 absorption of desert saltgrass under salt stress. *Journal of Plant Nutrition* 28 (8): 1441–1452.

TABLE 9.4 (Continued)
Selected Salinity-Oriented Research References on Turfgrass Species, Including References after 1998 That Were Used for the Turfgrass Tolerance Ranking in Table 9.3

H. Rukavina, H. Hughes, & Y. Qian. 2005. Freezing tolerance of saltgrass (*Distichlis spicata*) ecotypes. *HortScience* 40 (4): 1106.

K. B. Marcum, M. Pessarakli, & D. M. Kopec. 2005. Relative salinity tolerance of 21 turftype desert saltgrasses compared to bermudagrass. *HortScience* 40 (3): 827–829.

M. A. Shahba, Y. L. Qian, H. G. Hughes, A. J. Koski, & D. Christensen. 2003. Relationships of soluble carbohydrates and freeze tolerance in saltgrass. *Crop Science* 43 (6): 2148–2153.

H. Hughes, D. Christensen, T. Koski, & S. Reid. 2003. Desert saltgrass: A potential turfgrass? The need to conserve water has led researchers to attempt to develop native grasses into turfgrass. *Golf Course Management* 71 (4): 117–118.

M. A. Shahba. 2002. Environmental stress aspects of saltgrass. PhD diss., Colorado State University.

D. M. Kopec & K. Marcum. 2001. Desert saltgrass: A potential new turfgrass species. *USGA Green Section Record* 39 (1): 6–8.

R. Bonnart, A. Koski, & H. Hughes. 2000. Comparisons of mechanical scarification techniques for enhancing seed germination in two saltgrass (*Distichlis spicata*) seed lots. *HortScience* 35 (3): 466.

P. R. Kemp & G. L. Cunningham. 1981. Light, temperature and salinity effects on growth, leaf anatomy and photosynthesis of *Distichlis spicata* (L.) Green. *American Journal of Botany* 68 (4): 507–516.

COOL-SEASON GRASSES

General Articles

S. A. Bonos, M. Koch, J. A. Honig, T. Gianfagna, & B. Huang. 2009. Evaluating cool-season turfgrasses for salinity tolerance: Rutgers University scientists continue to unravel this important trait for future turfgrass cultivars. *USGA Green Section Record* 47 (6): 6–9.

P. D. Peterson, S. B. Martin, & J. J. Camberato. 2005. Tolerance of cool-season turfgrass to rapid blight disease. *Applied Turfgrass Science* 28: 1–8.

J. J. Camberato, P. D. Peterson, & S. B. Martin. 2005. Salinity alters rapid blight disease occurrence. *USGA Turfgrass and Environmental Research Online* 4 (16): 1–7.

S. F. Alshammary, Y. L. Qian, & S. J. Wallner. 2004. Growth response of four turfgrass species to salinity. *Agricultural Water Management* 66 (2): 97–111.

C. Rose-Fricker & J. K. Wipff. 2001. Breeding for salt tolerance in cool-season turfgrasses. *International Turfgrass Society Research Journal* 9 (Pt. 1): 206–212.

W. A. Meyer & S. A. Bonos. 2001. Trends in cool-season turfgrass breeding: Advances in biotechnology are helping scientists develop the turfgrass of the future. *Golf Course Management* 699: 61–64.

S. F. Alshammary. 2001. Salinity tolerance and associated salinity tolerance mechanisms of four turfgrasses. PhD diss., Colorado State University.

S. Alshammary, Y. L. Qian, & S. J. Wallner. 2000. Salinity tolerance of four turfgrasses. *HortScience* 35 (3): 414.

***Agrostis stolonifera* L.** **Creeping bentgrass**

S. Warnke. 2003. Creeping bentgrass (*Agrostis stolonifera* L.). In M. D. Casler & R. R. Duncan (eds.), *Turfgrass Biology: Genetics and Breeding* (pp. 175–185). Hoboken, NJ: John Wiley.

J. Neylan & A. Peart. 2009. Local ecotypes perform in bentgrass salinity trials. *Australian Turfgrass Management Journal* 11 (1): 34–37.

E. Zhang, J. Xing, T. Gianfagna, & B. Huang. 2006. Somaclonal variation in salinity tolerance for creeping bentgrass. *Proceedings of the Fifteenth Annual Rutgers Turfgrass Symposium*, p. 50.

J. Neylan, A. Peart, & D. R. Huff. 2005. A comparison of the effects of potable water versus saline effluent used for irrigation bentgrass (*Agrostis* spp. L.) and *Poa annua* L. cultivars. *International Turfgrass Society Research Journal* 10 (Pt. 1): 609–617.

Y. L. Qian & J. M. Fu. 2005. Response of creeping bentgrass to salinity and mowing management: Carbohydrate availability and ion accumulation. *HortScience* 40 (7): 2170–2174.

J. M. Fu, A. J. Koski, & Y. L. Qian. 2005. Responses of creeping bentgrass to salinity and mowing management: Growth and turf quality. *HortScience* 40 (2): 463–467.

continued

TABLE 9.4 (Continued)
Selected Salinity-Oriented Research References on Turfgrass Species, Including References after 1998 That Were Used for the Turfgrass Tolerance Ranking in Table 9.3

R. E. Koske & J. N. Gemma. 2005. Mycorrhizae and an organic amendment with bio-stimulants improve growth and salinity tolerance of creeping bentgrass during establishment. *Journal of Turfgrass and Sports Surface Science* 81:10–25.

S. O. Doak, R. E. Schmidt, & E. H. Ervin. 2005. Metabolic enhance impact on creeping bentgrass leaf sodium and physiology under salinity. *International Turfgrass Society Research Journal* 10(Pt. 2): 845–849.

C. Liu & R. J. Cooper. 2002. Humic acid application does not improve salt tolerance of hydroponically grown creeping bentgrass. *Journal of the American Society of Horticultural Science* 127 (2): 219–223.

K. B. Marcum. 2001. Salinity tolerance of 35 bentgrass cultivars. *HortScience* 36 (2): 374–376.

K. B. Marcum. 2000. Salt tolerance varies in modern creeping bentgrass varieties: Declining water quality requires planting of more-tolerant turfgrasses. *Golf Course Management* 68 (10): 54–58.

K. Marcum. 2000. Salt tolerance of modern creeping bentgrass cultivars. *Cactus Clippings* June: 16.

S. M. Redwine. 2000. Evaluation of drought and salinity tolerance in transgenic creeping bentgrass. Master's thesis, Michigan State University.

C. Kik. 1989. Ecological genetics of salt resistance in the clonal perennial, *Agrostis stolonifera* L. *New Phytologist* 113 (4): 453–458.

T. McNeilly, M. Ashraf, & C. Veltkamp. Leaf micromorphology of sea cliff and inland plants of *Agrostis stolonifera* L., *Dactylis glomerata* L., and *Holcus lanatus* L. *New Phytologist* 106 (2): 261–269.

L. Wu. 1981. The potential for evolution of salinity tolerance in *Agrostis stolonifera* L. and *Agrostis tenuis* sibth. *New Phytologist* 89 (3): 471–486.

I. Ahmad, S. J. Wainwright, & G. R. Stewart. 1981. The solute and water relations of *Agrostis stolonifera* ecotypes differing in their salt tolerance. *New Phytologist* 87 (3): 615–629.

I. Ahmad & S. J. Wainwright. 1977. Tolerance to salt, partial anaerobiosis, and osmotic stress in *Agrostis stolonifera*. *New Phytologist* 79 (3): 605–612.

V. B. Youngner, O. R. Lunt, & F. Nudge. 1967. Salinity tolerance of seven varieties of creeping bentgrass, *Agrostis palustris* Huds. *Agronomy Journal* 59 (4): 335–336.

***Poa pratensis* L.** **Kentucky bluegrass**

D. R. Huff. 2003. Kentucky bluegrass. In M. D. Casler & R. R. Duncan (eds.), *Turfgrass Biology: Genetics and Breeding* (pp. 27–38). Hoboken, NJ: John Wiley.

J. G. Robins, B. S. Bushman, B. L. Waldron, & P. G. Johnson. 2009. Variation within *Poa* germplasm for salinity tolerance. *HortScience* 44 (6): 1517–1521.

J. J. Camberato, P. D. Peterson, & S. B. Martin. 2006. Salinity and salinity tolerance alter rapid blight in Kentucky bluegrass, perennial ryegrass, and slender creeping red fescue. *Applied Turfgrass Science* 13: 1–4.

C. M. Grieve, S. A. Bonos, & J. A. Poss. 2006. Salt tolerance assessment of Kentucky bluegrass cultivars selected for drought tolerance. *HortScience* 41 (4): 1057.

J. A. Poss, C. M. Grieve, W. B. Russell, & S. A. Bonos. 2006. Assessment of Kentucky bluegrass salt tolerance with remote sensing. *HortScience* 41 (4): 999.

Y. L. Qian, R. F. Follett, S. Wilhelm, A. J. Koski, & M. A. Shahba. 2004. Carbon isotope discrimination of three Kentucky bluegrass cultivars with contrasting salinity tolerance. *Agronomy Journal* 96 (2): 571–575.

M. Pessarakli, K. B. Marcum, D. M. Kopec, & Y. L. Qian. 2004. Interactive effects of salinity and Primo on the growth of Kentucky bluegrass. *2004 Turfgrass, Landscape and Urban IPM Research Summary (Arizona)*, pp. 1–5.

G. Zhang. 2003. Enhancing salt and drought tolerance of *Poa* and triploid *Cynodon* with BADH gene and somaclonal variation. PhD diss., Rutgers The State University of New Jersey—New Brunswick.

Y. L. Qian. 2003. Salt tolerance should be considered when choosing Kentucky bluegrass varieties. *TurfGrass TRENDS* 59 (6): 60–62, 64.

M. R. Suplick-Ploense, Y. L. Qian, & J. C. Reed. 2002. Relative NaCl tolerance of Kentucky bluegrass, Texas bluegrass, and their hybrids. *Crop Science* 42 (6): 2025–2030.

Y. L. Qian, S. J. Wilhelm, & K. B. Marcum. 2001. Comparative responses of two Kentucky bluegrass cultivars to salinity stress. *Crop Science* 41 (6): 1895–1900.

TABLE 9.4 (Continued)
Selected Salinity-Oriented Research References on Turfgrass Species, Including References after 1998 That Were Used for the Turfgrass Tolerance Ranking in Table 9.3

G-Y. Zhang, S. Lu, S-Y. Chen, W. Meyer, R. Funk, and T. A. Chen. 2000. Transformation of Kentucky bluegrass (*Poa pratensis* L.) with betaine aldehyde dehydrogenase gene. *Proceedings of the Tenth Annual Rutgers Turfgrass Symposium*. p. 44.

G. L. Horst & R. M. Taylor. 1983. Germination and initial growth of Kentucky bluegrass in soluble salts. *Agronomy Journal* 75 (4): 679–681.

***Poa annua* L.** **Annual bluegrass**

D. R. Huff. 2003. Annual bluegrass (*Poa annua* L.). In M. D. Casler & R. R. Duncan (eds.), *Turfgrass Biology: Genetics and Breeding* (pp. 39–51). Hoboken, NJ: John Wiley.

J. G. Robins, B. S. Bushman, B. L. Waldron, & P. G. Johnson. 2009. Variation within *Poa* germplasm for salinity tolerance. *HortScience* 44 (6): 1517–1521.

J. Dai, D. R. Huff, & M. J. Schlossberg. 2009. Salinity effects on seed germination and vegetative growth of greens-type *Poa annua* relative to other cool-season turfgrass species. *Crop Science* 49 (2): 696–703.

J. Dai, M. J. Schlossberg, & D. R. Huff. 2008. Salinity tolerance of 33 greens-type *Poa annua* experimental lines. *Crop Science* 48 (3): 1187–1192.

J. Dai. 2006. Salinity tolerance of greens-type *Poa annua* L. Master's thesis, Pennsylvania State University.

J. Neylan, A. Peart, & D. R. Huff. 2005. A comparison of the effects of potable water versus saline effluent used for irrigation bentgrass (*Agrostis* spp. L.) and *Poa annua* L. cultivars. *International Turfgrass Society Research Journal* 10 (Pt. 1): 609–617.

***Poa trivialis* L.** **Roughstalk bluegrass**

R. Hurley. 2003. Rough bluegrass (*Poa trivialis* L.). In M. D. Casler & R. R. Duncan (eds.), *Turfgrass Biology: Genetics and Breeding* (pp. 67–73). Hoboken, NJ: John Wiley.

J. J. Camberato & S. B. Martin. 2004. Salinity slows germination of rough bluegrass. *HortScience* 39 (2): 394–397.

J. J. Camberato, S. B. Martin, & A. V. Turner. 2000. Salinity and seedlot affect rough bluegrass germination: The speed of emergence varies among cultivars and seedlots, but is generally slowed by higher salinity. *Golf Course Management* 68 (8): 55–58.

J. Camberato, S. B. Martin, & A. V. Turner. 2000. Cultivar differences in rough bluegrass germination occur with increased salinity. Clemson University Turfgrass Program, http://www.clemson.edu/extension/horticulture/turf/, pp. 1–4.

***Lolium perenne* L.** **Perennial ryegrass**

D. Thorogood. 2003. Perennial ryegrass (*Lolium perenne* L.). In M. D. Casler & R. R. Duncan (eds.), *Turfgrass Biology: Genetics and Breeding* (pp. 75–105). Hoboken, NJ: John Wiley.

S. Krishnan & R. N. Brown. 2009. Na^+ and K^+ accumulation in perennial ryegrass and red fescue accessions differing in salt tolerance. *International Turfgrass Society Research Journal* 11 (Pt. 2): 817–827.

J. J. Camberato, P. D. Peterson, & S. B. Martin. 2006. Salinity and salinity tolerance alter rapid blight in Kentucky bluegrass, perennial ryegrass, and slender creeping red fescue. *Applied Turfgrass Science* 13: 1–4.

***Festuca arundinacea* Schreb.** **Tall fescue**

W. A. Meyer & E. Watkins. 2003. Tall fescue (*Festuca arundinacea*). In M. D. Casler & R. R. Duncan (eds.). *Turfgrass Biology: Genetics and Breeding* (pp. 107–127). Hoboken, NJ: John Wiley.

D. C. Bowman, G. R. Cramer, & D. A. Devitt. Effect of nitrogen status on salinity tolerance of tall fescue turf. *Journal of Plant Nutrition* 29 (8): 1491–1497.

***Festuca* spp.** **Fine-leaf fescues**

C. J. Diedhiou, O. V. Popova, & D. Golldack. 2009. Transcript profiling of the salt-tolerant *Festuca rubra* ssp. *litoralis* reveals a regulatory network controlling salt acclimatization. *Journal of Plant Physiology* 166 (7): 697–711.

S. Krishnan & R. N. Brown. 2009. Na^+ and K^+ accumulation in perennial ryegrass and red fescue accessions differing in salt tolerance. *International Turfgrass Society Research Journal* 11 (Pt. 2): 817–827.

J. J. Camberato, P. D. Peterson, & S. B. Martin. 2006. Salinity and salinity tolerance alter rapid blight in Kentucky bluegrass, perennial ryegrass, and slender creeping red fescue. *Applied Turfgrass Science* 13: 1–4.

continued

TABLE 9.4 (Continued)
Selected Salinity-Oriented Research References on Turfgrass Species, Including References after 1998 That Were Used for the Turfgrass Tolerance Ranking in Table 9.3

B. A. Ruemmele, J. K. Wipff, L. Brilman, & K. W. Hignight. 2003. Fine-leaved *Festuca* species. In M. D. Casler & R. R. Duncan (eds.), *Turfgrass Biology: Genetics and Breeding* (pp. 129–174). Hoboken, NJ: John Wiley.

M. O. Humphreys, M. P. Kraus, & R. G. Wyn Jones. 1986. Leaf-surface properties in relation to tolerance of salt spray in *Festuca rubra* ssp. *litoralis* (G. F. W. Meyer) Auguier. *New Phytologist* 103 (4): 717–723.

M. O. Humphreys. 1982. The genetic basis of tolerance to salt spray in populations of *Festuca rubra* L. *New Phytologist* 91 (2): 287–296.

M. Harivandi, & K. N. Morris. 2010. Fineleaf fescue performance with recycled irrigation water. *Golf Course Industry* 22 (2): 54–56. (Also published in September/October 2009 issue of *Turf News*.)

***Deschampsia* spp.** **Hairgrasses**

L. A. Brilman & E. Watkins. 2003. Hairgrasses (*Deschampsia* spp.). In M. D. Casler & R. R. Duncan (eds.), *Turfgrass Biology: Genetics and Breeding* (pp. 225–231). Hoboken, NJ: John Wiley.

***Puccinellia distans* (L.) Parl.** **Alkaligrass**

A. Mintenko & R. Smith. 2001. Native grasses vary in salinity tolerance: Alkaligrass withstands salinity better than blue gramagrass, Idaho bentgrass and prairie junegrass. *Golf Course Management* 69 (4): 55–59.

MISCELLANEOUS GRASS SPECIES

Wheatgrasses with Some Salt Tolerance

Slender wheatgrass	*Elymus trachycaulus* (Link) Gould ex Shinners
Intermediate wheatgrass	*Elymus intermedia* (Host) Nevski
Tall wheatgrass	*Thinopyrum ponticum* (Podp.) Liu and Wang
Crested wheatgrass	*Agropyron cristatum* (L.) Gaertn.
Western wheatgrass	*Agropyron smithii* Rydb.

R. C. Johnson. 1991. Salinity resistance, water relations, and salt content of crested and tall wheatgrass accessions. *Crop Science* 31 (3): 730–734.

P. E. McGuire & J. Dvorak. 1981. High salt-tolerance potential in wheatgrasses. *Crop Science* 21 (5): 702–705.

M. C. Shannon. 1978. Testing salt tolerance variability among tall wheatgrass lines. *Agronomy Journal* 70 (5): 719–722.

O. J. Hunt. 1965. Salt tolerance in intermediate wheatgrass. *Crop Science* 5 (5): 407–409.

D. R. Dewey. 1962. Breeding crested wheatgrass for salt tolerance. *Crop Science* 2 (4): 403.

D. R. Dewey. 1962. Germination of crested wheatgrass in salinized soil. *Agronomy Journal* 54 (4): 353–355.

D. R. Dewey. 1960. Salt tolerance of twenty-five strains of *Agropyron*. *Agronomy Journal* 52 (11): 631–635.

***Chloris gayana* Kunth** **Rhodesgrass**

K. S. Deifel, P. M. Kopittke, & N. W. Menzies. 2006. Growth response of various perennial grasses to increasing salinity. *Journal of Plant Nutrition* 29 (9): 1573–1584.

***Sporobolus virginicus* (L.) Kunth** **Salt marsh grass**

K. B. Marcum & C. L. Murdoch. 1992. Salt tolerance of the coastal salt marsh grass, *Sporobolus virginicus* (L.) Kunth. *New Phytologist* 120 (2): 281–288.

S. Gulzar, M. A. Khan, & I. A. Irwin. 2003. Salt tolerance of a coastal salt marsh grass. *Communications in Soil Science and Plant Analysis* 34 (17/18): 2595–2605.

D. E. Aldous. 2004. Growth responses of four native seashore dropseed [*Sporobolus virginicus* (L.) Kunth] accessions to elevated salt concentrations. *Acta Horticulturae* 661: 199–205.

Source: Adapted from Carrow, R. N., and R. R. Duncan, *Salt-Affected Turfgrass Sites: Assessment and Management*, John Wiley, Hoboken, NJ, 1998.

TABLE 9.5
Specific Turfgrass Cultivars by Species That Have Been Marketed for Their Salt Tolerance Performance[a]

Species	Cultivars
Cool-Season Grasses	
Creeping bentgrass	Seaside, Seaside II, Cobra, SR1020, Penneagle Celebration, Mariner, Grand Prix, Providence
Weeping alkaligrass	Salty, Fults
Poa trivalis	Winterlinks, Laser
Tufted hairgrass	Barcampsia
Strong creeping red fescue	Inverness, Flyer, Ensylva, Shademaster II, Florentine GT, Sealink, Cardinal, Epic, Wendy Jean, Celestial, Garnet, Class One, Jasper II, Razor
Slender creeping red fescue	Dawson, Oasis, Seabreeze GT, Barcrown II, Shoreline
Blue hard fescue	Little Bighorn
Hard fescue	Reliant IV
Chewings fescue	Compass, J-5, Musica, Lacrosse, Cascade, Longfellow II, Culumbra II
Perennial ryegrass	Quickstart, Manhattan II & III, Manhattan 5GLR, Catalina, Catalina II, Charger II, Salinas, Barlennium, Gray Star, Gray Fox, Gray Goose, Silver Dollar, Brightstar SLT, Citation III, Citation Fore, Quicktrans, Paragon, Apple GL, Chaparral, Penguin II, Zoom, Pinnacle 2, Fiesta 3, Headstart, Palmer III
Kentucky bluegrass	Northstar, Moonlight, Moonlight SLT, Glade, Livingston, Blacksburg, Bariris, Apollo, Emblem Moonbeam, Eagleton, Liberator, Cabernet
Tall fescue	Wolfpack II, Dynamic II, Tomahawk RT, Alta, Apache II, Pure Gold, Tar Heel II, Barlexas II, Dynamic, Corona, Sidewinder, Ninja 3
Warm-Season Grasses	
Seashore paspalum	Salam, SeaIsle 1, SeaIsle 2000, Supreme, Platinum TE, Seaspray (seeded), Marina (seeded), Seadwarf, Seaway, Seagreen,Velvetene, Saltene, Salpas, SeaWolf, TX515, Aloha, Brazoria, Superdwarf, Durban CC, Seashore, Corrib, Neptune, Millennium, 223, Marimo, EE1, Boardwalk
Zoysiagrass	Diamond (highest); intermediate tolerance: Companion (seeded), DeAnza, Z-59, Crowne, El Toro, Emerald, JaMur, Marquis, Palisades, Royal, Victoria
St. Augustinegrass	Seville
Hybrid Bermuda grass	Tifway 419, Novatek™
Common Bermuda grass	Sahara, Riviera, Casina Royale, FloraTeX™, Savannah II, Yukon, Oz Tuff Green™

Note: The listing of a specific cultivar within a species does not endorse or verify actual turfgrass performance under long-term field conditions and cyclic salt stress soil accumulation and foliar exposure; it simply indicates some level of overall salinity tolerance as claimed by the company releasing the cultivar.

Some cultivars are being marketed as salt tolerant or having some degree of "improved" salt tolerance compared with other cultivars. There are no definitive designations on actual levels of salinity tolerance among the cultivars within a species and the relative salt-tolerance response under actual field conditions among species that are presented in this table. Salt challenges are site specific, and cultivar interactions with salinity are only one of four interactions on a particular turfgrass ecosystem. This table is provided as a source of information, and site-specific trials under saline environmental conditions are recommended to determine their suitability for turfgrass sustainability on that site.

[a] These cultivars are marketed to possess salinity tolerance greater than most other cultivars within the species.

Lower salinity-tolerant turfgrass species such as annual ryegrass (*Lolium multiflorum*), annual bluegrass (*Poa annua*), Kentucky bluegrass (*Poa pratensis*), and roughstalk bluegrass (*Poa trivalis*) are generally exceptionally sensitive to any irrigation water >200 ppm (>8.7 meq/L) sodium or chloride levels >355 ppm. Any accumulation of those excess salts in the soil above those concentrations can usually severely damage or even eliminate those grasses.

9.3 SALINITY TOLERANCE MECHANISMS

9.3.1 Phases of Salinity Stress in Plants

Physiological (biochemical) and molecular mechanisms of tolerance to osmotic and ionic components of salinity stress operate at the cellular, specific-organ, and whole-plant levels (Munns and Tester, 2008). Turfgrass response to initiation of salinity stress normally consists of two phases. The dual-phase plant growth response to salinity stress encompasses the following:

1. A rapid activation of the *osmotic control phase* (i.e., the *osmotic stress phase*) that slows and eventually inhibits growth rates governing root (root hairs, branch roots, and rhizomes) initiation, root system development and redevelopment, surface stolon development and extension, the crown region's functionality and sustainability, and subsequent leaf development. Inhibition of leaf development is exhibited by reduced leaf expansion in width and length, significantly reduced rate of new leaf emergence, lateral buds emerging slowly or not at all, reduced development of lateral shoots or stolons, and reduced branching in the surface canopy leading to less or decreasing surface canopy density. This osmotic phase activates immediately when the salt concentration around the root system increases to the specific cultivar tolerance threshold level: shoot growth (as a *defensive response* by the plant) is more sensitive than root growth—reducing leaf area development relative to root growth leads to reduced water demand and water use, thereby resulting in soil moisture conservation plus reduced escalation in potential soil salt accumulation (potential upconing of soil-accumulated salts via capillary pores as soil water potential is reduced and evapotranspiration increases the salt accumulation stress conditions in the upper soil profile).
2. A slower *ionic phase* (the *ionic cytotoxicity phase*), usually involving gradual accumulation and localization of specific salt ions internally in the plant, which either results in actual acclimation to the salt concentration (dependent on the actual inherent level of salinity tolerance in the plant) or subsequently reaches a concentration that accelerates sodium-induced mature (or older) leaf senescence and eventual whole-plant death. Chlorides will move rapidly to active growing points in the plant and to new or emerging leaves; tissue death often results. When the leaf death rate surpasses the capability to generate new leaves and the photosynthetic capacity to generate and supply carbohydrates to the young leaves (and the whole plant) is reduced, growth rates will eventually be significantly slowed down and eventually completely stopped. This ionic phase dominates only when the salinity levels are significantly higher than the overall cultivar tolerance levels and sodium and/or chlorides accumulate in the soil, and at toxic levels internally in the cells and whole plant, resulting in programmed death of tissue.

Plant adaptations to salinity stress internally include three distinct types (Munns and Tester, 2008): (a) osmotic stress tolerance and adjustments (salinity causes low soil water potential and physiological drought stress problems), (b) sodium and/or chloride exclusion, and (c) site-specific tolerance to both soil and plant-accumulated sodium and/or chloride concentrations. As salts accumulate in the soil, the ability of roots to extract water is altered due to the desiccation traits of salt ions on the root system and the creation of soil water potentials that prevent root extraction of water (physiological drought stress) even though soils may be at or near field capacity.

External salt exposure to roots will reduce cell growth rates and alter associated metabolic functions. As roots absorb these salt ions over time, toxic concentrations will gradually start to upwardly translocate in the vascular system (xylem), accumulating in certain tissues, and eventually affecting cellular and whole-plant functions.

Plant function as affected by constraints in saline environments includes the operation (usually upregulation) of genes in cells (although one salt-tolerant cell does not totally equate to whole-plant salt tolerance expression) and allied tissues in the alteration of plant growth rates. Sodium generally accumulates at toxic concentrations before chlorides, probably because chlorides are highly mobile in the plant, and sodium often translocates from roots to shoots at a slower rate and with the involvement of more regulatory mechanisms with accompanying energy expenditure by the plant.

To summarize the effects of salt ion stress on turfgrass and landscape plants (adapted from Munns and Tester, 2008):

Sodium stress	**Osmotic stress**	**Ionic stress—due to high leaf sodium**
Speed of onset	Rapid	Slow, must accumulate
Primary site of visible effects	Decreased new shoot growth	Increased senescence of older leaves; off color response
Chloride stress	**Osmotic stress**	**Ionic stress—due to high leaf chloride**
Speed of onset	Rapid	Moves rapidly to growing points
Primary site of visible effects	Decreased total growth	Increased senescence of newer leaves; leaf drop; nitrogen availability altered

9.3.2 Categories Governing Genetic Control of Salinity Tolerance

9.3.2.1 Tolerance to Osmotic Stress

Salinity-tolerant grasses must be able to tolerate osmotic stress as soil salinity increases either gradually (the usual case with saline irrigation water) or more rapidly, such as a saline water flooding event. For turfgrass plants to adapt to and tolerate the osmotic stress associated with increasing salinity, a number of plant responses are necessary. Plants must modify the long-distance sensing mechanisms in response to increasing salinity and send signals (via hormones, proteins, or their precursors) to the roots in response to the microenvironmental stress exposure. The plant must override the salt-induced inhibition of cell expansion and lateral bud (node) development. During photosynthesis, salt-induced stomatal closure must be decreased. As sodium and organic solutes accumulate in the shoots and eventually in the vacuoles, osmotic adjustments must increase to counteract the localization of this toxic salt ion. The actual cultivar genetic tolerance level will govern how these adjustments affect cell expansion in root tips and young leaves, actually changing flux in growth rates as well as initial stomatal conductance.

As water limitations occur with increasing salt loads in the irrigation water and subsequent secondary soil salinization, plants with higher actual functional plant salt tolerance levels have a greater chance for long-term sustainability. Plant roots undergo cyclic osmotic shock responses with exposure to various osmotica and as salinity levels gradually increase (such as with each exposure to repeated irrigation cycles with saline water), but recovery can be quite rapid (within an hour to a day depending on total salinity concentrations and the actual salt tolerance capability of the specific cultivar) as long as the osmotic shock does not result in cell plasmolysis and despite turgor not being fully restored; the turgor pressure response indicates that cell wall properties are changing (Frensch and Hsaio, 1994; Munns, 2002).

Leaves are much more sensitive to the cyclic osmotic shocks. As soil salinity increases, leaf growth rates decrease (which is a genetic downregulation response) primarily due to the osmotic effect of salts surrounding the root. Leaf cells lose water (decreasing cell water volume and turgor pressure), but osmotic adjustment counters this loss within hours; however, growth rates are reduced (a plant defensive response) in conjunction with this stress exposure adjustment. Leaves typically are reduced in area, are smaller, and become thicker (morphological changes) with progressive exposure to increasing salt concentrations. Lateral shoot development is inhibited. Older leaves will die as sodium concentrations increase and accumulate in critical tissues or organs. Young leaves will rapidly senesce with chloride accumulations, beginning with Cl accumulation initially at the growing points and then progressing to older tissues.

9.3.2.2 Sodium and Chloride Exclusion or Control in Tissues (Ionic Stress Tolerance)

Genetic exclusion of sodium (Moller et al., 2009) utilizes sensing and signaling mechanisms in the root system to not only control the net ion xylem transport to shoots, but also minimize or avoid toxic concentrations in the chloroplasts that will negatively affect photosynthesis (reduce growth rates). Long-distance transport of sodium to the shoots is reduced partially by sequestration into root vacuoles and partially through alteration of the actual transport processes in an effort to reduce sodium accumulation at toxic levels. Sodium ions that reach the shoots are also sequestered into shoot cell vacuoles (depending on actual salt tolerance levels) to minimize their effect on the photosynthetic operations of the cells.

Ionic tolerance at the tissue level involves control of vacuolar loading through the sensing and signaling mechanisms in the root. In the shoots, premature senescence of older leaves must be delayed. Ion toxicity in chloroplasts must also be delayed; otherwise, photosynthesis and growth rates are gradually reduced to nonsustainable carbohydrate production activity. If high concentrations of sodium accumulate in the shoots, significant energy is spent on sodium exclusion by the plant as a defensive response, and this energy expenditure can rapidly reach a level of diminishing returns (loss of stored carbohydrates) leading to eventual tissue and cell death. The key to countering this toxic accumulation is increased sequestration of sodium into leaf vacuoles. In conjunction with this vacuolar sequestration of sodium, high concentrations of compatible solutes need to accumulate in the cytoplasm as a protective (and defensive) adjustment by the plant. A good review of Na^+ transport in plants can be found in Apse and Blumwald (2007).

Similar control mechanisms operate for toxic levels of chlorides. Compartmentalization of both sodium and chloride ions at the cellular and intracellular levels is required to avoid toxic concentrations in the cytoplasm, such as mesophyll cells in leaves where active photosynthesis occurs. Sodium has a tendency to accumulate in older leaves with time as concentrations increase in the water, in the soil, and internally in the plant. Chlorides accumulate in growing points such as new leaves and root tips as increasing concentrations are exposed to and absorbed by the plant (Munns and Tester, 2008). Leaf blade Cl concentrations increase progressively (Prior et al., 2007). Sodium may be a more potentially toxic solute, but when Na^+ transport is genetically controlled better than Cl transport, Cl becomes an equally toxic salt since its influx is active and its efflux is passive (Munns and Tester, 2008).

Chloride loading from cells into the xylem is a passive mechanism controlled by anion channels (Gilliham and Tester, 2005). ABA downregulates these channels, which limits Cl translocation to the shoots when exposed to saline conditions due to reduced loading of Cl by anion channels or due to active retrieval of Cl from the root, stem, or petiole xylem transpiration stream (Munns and Tester, 2008). Chloride transport mechanisms are catalyzed by a $Cl^-/2H^+$ symporter (Sanders, 1980; Felle, 1994).

9.3.2.3 Whole-Plant Response to Salinity

Salt-induced osmotic effects outside the roots immediately affect stomatal conductance, decreasing stomatal aperture. Water relations are disturbed, and *in situ* synthesis of ABA occurs rapidly in photosynthetic tissues (Fricke et al., 2004, 2006). This root signal-induced stomatal regulation initially reduces transpiration rates, but the stomates reopen after a few hours and ABA levels also stabilize.

Photosynthetic rates per unit leaf area in salt-exposed plants may initially be unchanged even though stomatal conductance is reduced due to changes in cell anatomy (smaller leaves and reduced total leaf area, thicker leaves, and higher chloroplast density per unit leaf area) (James et al., 2002). On a whole-plant basis, any reduction in total leaf area will always equate to reduced photosynthesis and subsequent reduced overall growth rates. Feedback inhibition signals in response to reduced leaf growth plus accumulation of unused carbohydrates in growing tissues are thought to downregulate photosynthesis (Paul and Foyer, 2001). However, increased partitioning of carbon to sucrose may contribute to salt-stress tolerance (Udomchalothorn et al., 2009).

Cell-type specific processes contribute to plant salinity tolerance. Genes governing these processes include a Na^+ transporter (SOS1), the HKT family of transport proteins, sodium/myoinositol symporters and K^+ transporters, and particular H^+-ATPases (these genes are summarized in Duncan and Carrow, 1999; Yamaguchi and Blumwald, 2005; Hussain et al., 2008; Moller et al., 2009). Specifically, the HKT1:1 gene construct encodes a plasma membrane protein that mediates Na^+ influx into cells and regulates Na^+ distribution between roots and shoots through the retrieval of Na^+ from the xylem (Sunarpi et al., 2005). Small RNAs (microRNAs or miRNAs; and endogenous small interfering RNAs, or siRNAs) have a stress-regulated functional role in abiotic stress responses (Sunkar et al., 2007).

9.3.2.4 Oxidative Stress Acclimation

In addition to the salt-stress-induced acclimation adjustments in leaf morphology and chloroplast pigment composition, plants must acclimate their biochemical processes and activities so as to prevent oxidative damage to photosystems I and II. The sequence for salt-induced photoinhibition of PS II includes inhibition of CO_2 fixation and generation of reactive oxygen species (superoxide radicals, hydrogen peroxide-H_2O_2, and hydroxyl radicals—OH^-), which inhibits protein synthesis that is needed for repair of photodamage (Murata et al., 2007; Takahashi and Murata, 2008) and causes oxidative damage to membrane lipids, proteins, and nucleic acids (Halliwell and Gutteridge, 1986). One key process that avoids photoinhibition when exposed to stress conditions involves electron transfers to oxygen receptors other than water. This plant response requires the upregulation of enzymes (for the regulation of reactive oxygen species) such as superoxide dismutase (SOD), catalase (CAT), ascorbate peroxidase (APX: the H_2O_2 removal enzyme), dehydro-ascorbate reductase (DHAR), glutathione reductase (GR), glutathione-*S*-transferase (GST), and several peroxidases such as glutathione peroxidase (GPX) plus antioxidants such as reduced glutathione (GSH) to scavenge ROS (Apel and Hirt, 2004; Foyer and Noctor, 2005; Logan, 2005; Hussain et al., 2008; He et al., 2009).

Sodium chloride–induced oxidative stress on cellular oxidative metabolism involves accumulation of H_2O_2/O_{2*}, increased lipid peroxidation and carbonyl-group content, reduction of Fe^{3+} but not chelation of Fe^{2+}, increased plasmid DNA strand breakage induced by OH, increased accumulation of putrescine and spermidine, oxidation of ascorbate and glutathione redox pairs, and inhibition in activities of ROS-metabolizing enzymes (catalase, ascorbate peroxidase, and glutathione reductase) (Tanou et al., 2009). Sodium chloride–induced production of H_2O_2 appears to be linked to NAD(P)H-oxidase and amine oxidase regulation, with signaling from nitric oxide (Qiao et al., 2009), salicylic acid, protein kinase, and calcium (Ca^{2+}) channel activity (Tanou et al., 2009).

A reduction in the photosynthetic rate increases the formation of ROS, and also increases activity of enzymes that detoxify these species. The coordinated enzyme activity in specific cell compartments balances formation and removal rates for ROS while sustaining hydrogen peroxide (H_2O_2) levels that are required for cell signaling in response to stress exposure. The chloroplast is the regulatory point for all ROS detoxifying mechanisms that are needed for photosynthetic rate adjustments induced by saline soils (Logan, 2005).

Turfgrass and landscape cultivar genetic differences in salinity tolerance are not necessarily primarily due to differences in the capability to detoxify ROS. The biochemical defense mechanisms illustrate that ROS regulatory pathways exhibit plasticity, while being innately redundant, quite complex, and rapidly interactive (Munns and Tester, 2008). Genotypic differences in antioxidant activity probably encompass differential degrees of stomatal closure and responses that alter the rate of CO_2 fixation in conjunction with defensive photoinhibition protection.

9.3.2.5 Cellular and Whole-Plant Signaling of Salinity Stress

Root cells, when exposed to extracellular salinity (Na^+ and Cl^-), must sense both the ionic and osmotic components, and then respond rapidly to soil accumulation changes in salt ion external concentrations. Signaling within root cells is likely independent of ABA, but long-distance signaling to shoots is mediated at least partially by ABA with initial sensing of dehydration stress (Zhang et al., 2006; Munns and Tester, 2008). Extracellular exposure to salinity is sensed either at the plasma membrane or intracellularly once Na^+ crosses the plasma membrane (which means that a plasma membrane protein must be the sensor or be located immediately upstream of the sensor). The molecular basis for turgor sensing is not yet fully understood.

DELLA proteins, which are negative regulators of plant growth, may be the central coordinators that adapt growth rates to different stress environments (Achard et al., 2006), since they are mediators in the growth-promoting effects of gibberellins and integrate signals among hormones (probably including abscisic acid since it is directly involved in root-to-shoot plus cellular signaling and stomatal conductance; Zhu, 2002) when plants are exposed to abiotic stresses, such as salinity.

A likely signaling pathway specific to salinity stress involves increases in cytosolic free Ca^{2+} ($[Ca^{2+}]_{cyt}$) when root cells are exposed to extracellular salinity. Extracellular Na^+ exposure triggers several salinity-signaling components that apparently activate Ca^{2+} flux into the cytosol across the plasma membrane and eventually the tonoplast (Knight et al., 1997; Kiegle et al., 2000; Knight, 2000; Halfter et al., 2000). Examples include root-cell-type aequorin, and calcineurin B-like protein (CBL4/SOS3) interacting with CBL protein kinase (CIPK24/SOS2); SOS3, which is a calcium-binding protein, recruits SOS2, which is a Ser/Thr protein kinase, to the plasma membrane, where it activates the Na^+/H^+ antiporter, SOS1, by phosphorylation (Qui et al., 2002). The Ser/Thr protein kinase CIPK23 phosphorylates K^+ transporter AKT1 and enhances K^+ uptake under K^+-deficient conditions; two calcineurin B-like proteins (CBL1 and CBL9) are upstream regulators of CIPK23 (Xu et al., 2006).

Signaling pathways in salt-tolerant halophytic plant species are generally adaptive responses, while glycophytic species responses are thought to be more dysfunctional (Munns and Tester, 2008). Cellular response to extracellular Na^+ exposure ranges from biochemical and gene transcriptional functional changes to physiological, growth rate, and developmental alterations.

9.3.2.6 Ion Exclusion and Tissue Tolerance: Sodium Accumulation in Shoots

For perennial turfgrass and landscape plants, the main site for excessive sodium accumulation and associated toxic reactions is the leaf blade. The processes governing sodium sequestration into root vacuoles rather than deposition into the root xylem for eventual translocation to the leaves is a critical genetic tolerance mechanism (Munns, 2002). Most of the sodium that accumulates in the leaves will generally remain localized there and not be recirculated via the phloem back down to the roots. Therefore, sodium exclusion in the roots is essential. Seashore paspalum (*Paspalum vaginatum* Swartz) is one halophytic turfgrass species that genetically controls sodium accumulation via specific ion root vascular exclusion.

Four distinct components are involved in the net delivery of Na^+ to the xylem (Tester and Davenport, 2003):

1. Sodium passive influx into cells in the outer half of the root
2. Sodium reverse energy-dependent efflux from the above cells and directly into soil solution

3. Sodium energy-dependent efflux from cells in the inner half of the roots directly into the xylem
4. Sodium reverse passive influx from the xylem into the above inner root cells before delivery to the leaf blade

Sodium (influx) passively enters roots through voltage-independent or weakly voltage-dependent nonselective cation channels (Tester and Davenport, 2003) and possibly through other Na^+ transporters (high-affinity K^+ transporter [HKT] family) (Laurie et al., 2002; Haro et al., 2005; Moller et al., 2009). Candidates for gene-encoded nonselective cation channels include cyclic nucleotide-gated channels and ionotropic glutamate receptor-like channels (Demidchik and Davenport, 2002).

As water moves across the root cortex toward the stele, Na^+ ions are removed from the xylem stream into cells where they are potentially sequestered in the vacuoles depending on whether this tolerance mechanism is functioning (Munns and Tester, 2008). Sodium that enters cells in the outer part of the root is likely reverse pumped (with significant energy demand and gene-encoded efflux proteins) through plasma membrane Na^+/H^+ antiporters (Tester and Davenport, 2003) and/or Na^+-translocating ATPases (Cheeseman, 1982; Mennen, 1990).

Na^+ compartmentation in root vacuoles is achieved by tonoplast Na^+/H^+ antiporters (exchanger NHX family) (Pardo et al., 2006; Serrano and Rodriguez-Navarro, 2001). Na^+ must constantly be resequestered into root vacuoles since passive leakage via tonoplast nonselective cation channels of Na^+ occurs from those vacuoles back to the cytoplasm. More efficient root vacuolar compartmentalization may improve tissue tolerance by reducing cytosolic Na^+ concentrations (Apse et al., 1999; Gaxiola et al., 2001). Overexpression of AVP1 (gene-encoding vacuolar H^+-translocating pyrophosphatase) contributes to the electrochemical potential difference for H^+. Overexpression of NHX1 energizes Na^+ pumping into the root vacuole, increasing both Na^+ accumulation and Na^+ tolerance.

Sodium movement into and out of the xylem transpiration stream is quite complex. Na^+ moves in the root symplast across the endodermis (endodermal Casparian strip), is released from stellar cells into the stellar apoplast, and eventually moves into the xylem (Yeo et al., 1987; Munns and Tester, 2008). SOS1 is thought to be directly involved with loading of Na^+ from xylem parenchyma cells into xylem tracheids (Shi et al., 2002). The HKT gene family appears to be directly involved in Na^+ retrieval from the xylem. The members of this gene family function as Na^+/K^+ symporters and also as Na^+-selective transporters of both high and low affinity. Subfamily 1 genes contain low-affinity Na^+ uniporters, contain an amino acid serine residue, and are primarily Na^+ selective. Subfamily 2 genes contain an amino acid glycine residue, and catalyze both K^+ transport and high-affinity Na^+ influx (Horie et al., 2007).

9.3.2.7 Sodium Tissue Tolerance

Anatomical adaptations and intracellular partitioning are tolerance mechanisms in leaf cells exposed to elevated influx concentrations of Na^+ and Cl^-. Dicotyledonous halophytes implement two types of anatomical adaptations: succulence resulting from salt-induced cell size increase in vacuole volume, and excretion of Na^+ and Cl^- via salt glands (which are modified trichomes) or bladders (which are modified epidermal cells) (Flowers et al., 1986). Calcium concentration enhances diameter development and the salt secretion rates of salt glands (Ding et al., 2010).

Tissue tolerance of Na^+ requires the storage of Na^+ in vacuoles. This involves the Na^+/H^+ antiporters from the Na^+/H^+ exchanger gene family plus pyrophosphatase (Apse et al., 1999; Gaxiola et al., 1999, 2001).

In monocot turfgrasses, zoysiagrass (*Zoysia* spp.), Bermuda grasses (*Cynodon* spp.), buffalograss (*Buchloe dactyloides*), saltgrass (*Distichlis spicata* var. *stricta*), sand dropseed (*Sporobolus cryptandrus*), sideoats gramagrass (*Bouteloua curtipendula*), and curly mesquite (*Hilaria belangeri*) utilize salt glands in glycophytic tolerance responses to increasing salinity.

But, buffalograss and sideoats gramagrass are much more sensitive to salinity than the other species because their salt ion exclusion mechanisms are not as efficient as those of the other species (Marcum, 1999).

Compartmentalization of Na^+ and Cl^- are essential salinity tolerance mechanisms in all plants (Munns and Tester, 2008). Osmotic pressure of other subcellular compartments (including the cytosol) must be sequentially increased to maintain their volume as Na^+ and Cl^- concentrations increase in the vacuole; this osmotic pressure increase is adjusted by increasing the K^+ and compatible solute concentrations.

9.3.2.8 Compatible Organic Solutes

Gene control for synthesis and eventual metabolism of compatible organic solutes is essential for abiotic stress tolerance expression in plants (Hare et al., 1998; Chen and Murata, 2002; Rhodes et al., 2002). With soil salinity exposure, the plant must compensate for salt ion accumulation and potential toxicity plus balance turgor loss as it adapts osmotically to high Na^+ and Cl^- concentrations. Leaf tissue tolerance to high Na^+ and Cl^- concentrations is unquestionably an adaptive mechanism in both halophytes (Flowers et al., 1986) and glycophytes (James et al., 2006). Sodium exclusion provides the plant with the capability to avoid or at least postpone excessive accumulation and eventual toxicity; however, unless Na^+ exclusion is compensated for by adequate K^+ uptake, there is a greater demand for organic solutes to adjust the osmoticum (Munns and Tester, 2008).

With sequestration of Na^+ and Cl^- in the cell vacuoles, the organic solutes must be compatible with metabolic activity at high concentrations and must accumulate in the cytosol and other organelles at sufficient levels to balance the osmotic pressure of those accumulated ions in the vacuole (Flowers et al., 1977; Wyn Jones et al., 1977; Hasegawa et al., 2000; Munns, 2005). Sucrose, proline, and glycine betaine are the most common organic solutes that accumulate, but other compounds (trehalose, mannitol, polyols, sugar alcohols, and soluble sugars) can accumulate in high concentrations depending on the plant species and cultivar (Flowers et al., 1977; Hasegawa et al., 2000; Munns, 2005). Mannitol protects the chloroplasts against salt exposure, is a free radical scavenger, stabilizes subcellular structures such as proteins and membranes, and buffers cellular redox potential under stress (Sickler et al., 2007; Hussain et al., 2008).

In halophyte leaves, proline or glycine betaine can be found at relatively high concentrations (40+ mM on a tissue water basis) to contribute to osmotic pressure adjustment (over 0.1 MPa) after exposure to high salinity (Flowers et al., 1977). In glycophyte leaves, the solute concentrations are lower (~10 mM); if this osmolyte concentration is partitioned totally to the cytoplasm, significant osmotic functional pressure adjustments can occur. A secondary osmoprotectant function of low osmolyte concentrations is to stabilize the tertiary structure of proteins (Rhodes et al., 2002).

Excess localized concentrations or uncontrolled accumulation of organic solutes can alter metabolic pathways, diverting substrates from essential processes such as cell wall synthesis or protein synthesis (Munns and Tester, 2008). Additionally, the synthesis of organic solutes alters the energy balance in the plant at the expense of plant growth while at the same time providing survivability and recovery functions when exposed to high external salt concentrations (Raven, 1985). Assuming production of 0.5 mole of adenosine triphosphate (ATP) per photon and nitrate as the N source, about seven moles of ATP must accumulate 1 mole of NaCl as an osmoticum in leaf cells; an order of magnitude higher ATP is needed to synthesize 1 mole of an organic solute (Raven, 1985).

Concentrations of available Ca^{2+}, actual functional calcium activity, and stable Na^+:Ca^{2+} ratios in soil solution and internally in the plant are allied mechanisms involved in whole-plant and localized cellular salinity tolerance, especially with high sodium concentrations (Munns and Tester, 2008). Amelioration of sodium uptake by silicon is also involved in tissue tolerance (Epstein, 1994; Gong et al., 2006).

A general summary of plant salinity tolerance mechanisms would include the following (adapted from Carrow and Duncan, 1998):

1. Tissue or organ avoidance mechanisms to toxic ion accumulation (Na and/or Cl) in physiologically active areas
2. Toxic ion (Na, Cl) partitioning into specialized areas (vacuoles) or organs (older leaves) or through extrusion ports (salt glands)
3. Intracellular partitioning, sequestration, and compartmentalization of toxic ions
4. Synthesis and activity of compatible organic solutes
5. Cytoplasm, enzyme, protein, and chloroplast (photosystem) protection
   a. Amelioration of salts by soil bacteria (Yao et al., 2010) or arbuscular mycorrhiza (Sannazzaro et al., 2007)
6. Avoidance via translocation, uptake, excretion, or sequestration
   a. Retranslocation of salts from shoots to roots
   b. Restricted or excluded salt ion uptake (root absorption)
   c. Excretion and removal through specialized organs (glands or leaf drop)
7. Avoidance by controlling internal water deficits
   a. Osmotic and turgor pressure adjustments
   b. Greater succulence
   c. Oxidative stress acclimation
8. Key nutrient stabilization (Ca, K, Mn, and Zn) internally in the plant
9. Toxic ion amelioration (silicon)
10. Whole-plant and cellular signaling adjustments to stress
11. Ionic stress adjustments
12. Carbohydrate production sustainability and storage reserves
13. Reactivation of root regeneration (cytokinins); rhizome placement in the soil profile
14. Hormonal stabilization and functional physiological enhancement

9.4 LANDSCAPE PLANTS AND SALINITY TOLERANCE

A comprehensive treatise on salt- and boron-tolerant ornamental plants has been assembled (see Duncan et al. 2009, 367–441, with a broad-based reference listing on 442). Duncan et al.'s (2009) compilation of plants and their relative responses to varying salinity concentrations was summarized up to late 2008 and represents what was known at that time. Additional salt-tolerant plants that have been identified include the following:

1. L. Deeter, 2001. Researcher identifies perennials tolerant to salt. *Landscape Management*: 24. (No threshold data were presented.)
   Splendens sea thrift
   Karl Foerster feather reedgrass
   Helen Allwoood pinks
   Blue lymegrass
   Perennial fountaingrass
   Elijah blue fescue
2. J. Byron, 2002. Salt tolerance of landscape species evaluated. *California Agriculture* 56 (4): 121–122. (ECw = 2.34 dS/m)

Acacia redolens	Redolen acacia
Buxus japonica	Japanese boxwood
Cedrus deodara	Deodar cedar
Juniperus virginiana	'Skyrocket' juniper
Nerium oleander	Oleander

Olea europea	'Montra' dwarf olive
Pinus cembroides	Mexican piñon pine
Pittosporum tobria	Tobira pittosporum
Plumbago auriculata	Cape plumbago
Rhaphiolepis indica	Indian hawthorne
Sapium sebiferum	Chinese tallow tree
Washingtonia filifera	California fan palm

9.4.1 Saline and Alkaline Site Reclamation (Also See Chapter 20)

Some plant species can be planted and will survive on highly saline and alkaline soils, with thresholds in the ECe 10–20 dS/m range, and some tolerance levels approaching 30 dS/m (Table 9.6). *Chloris virgata* Swartz is a saline- and alkali-tolerant grass that has been used in northern China. But few turfgrasses can tolerate those salinity levels, except for perhaps halophytes such as seashore paspalum and alkaligrass, especially when other stresses (low mowing heights, traffic and wear, heat, cold, drought, insect infestation, and disease attack) are imposed on the perennial species.

Possible halophytic genera that could be planted for salt-affected site reclamation and stabilization include *Agropyron*, *Allenrolfea*, *Atriplex*, *Avicennia*, *Casuarina*, *Cenchrus*, *Diplachne*, *Distichlis*, *Eucalyptus*, *Juncus*, *Kochia*, *Kosteletzkya*, *Lycium*, *Maireana*, *Nypa*, *Pandanus*, *Panicum*, *Paspalum*, *Plantago*, *Puccinellia*, *Salicornia*, *Salsola*, *Spartina*, *Suaeda*, *Tamarix*, and *Zostera* (Yensen, 2008). A total of 1,861 halophytes and salt-tolerant species from 636 genera and 139 families (Chenopodiaceae and Poaceae have the highest numbers) have been estimated to exist in the plant kingdom. The highest potential number of halophytic and salt-tolerant species is found in Australia and New Guinea (Australasian area that is presently salinized comprises 357,330,000 ha) (Yensen, 2008). Specific halophytic and salt-tolerant species for several of the genera mentioned previously are summarized in Table 9.7.

Sporobolus ioclados is a saline halophytic desert grass that could be used (Khan and Gulzar, 2003). Other potential salt-tolerant grasses include (Brede, 2001): weeping lovegrass (*Eragrostis curvula* [Schrad.] Nees.) such as 'Consol', 'Ermelo', and 'Morpa'; Lehmann lovegrass (*Eragrostis lehmanniana* Nees.) such as 'A-68', 'Cochise', 'Kuivato', and 'Puhuima'; Wilman lovegrass (*Eragrostis superba* Peyr.) such as 'Palar'; Nuttal alkaligrass (*Puccinellia airoides* [Nutt.] Wats. & Coult.) such

TABLE 9.6
Prospective Plant Materials for Saline-Alkaline Soils

Scientific Name	Common Name	Threshold/Tolerance ECe (dS/m)
Sporobolus airoides	Alkali sacaton	14/26
Thinopyrum ponticum	Tall wheatgrass	13/24
Psathyrostachys juncea	Russian wildrye	13/24
Puccinellia airoides	Nuttall's alkaligrass	14/30
Leymus angustus	Altai wildrye	10/20
Elymus trachycaulus	Slender wheatgrass	10/22
Levmus multicaulus	Beardless wildrye	12/26
Elytrigia repens X *Pseudoroegneria spicata*	Hybrid wheatgrass	10/24
Poa juncifolia	Alkali bluegrass	12/24

Source: Adapted from USDA-NRCS, Plant materials for saline-alkaline soils, *Technical Notes Plant Materials*, 26, 1996; USDA-NRCS, Plant materials and techniques for brine site reclamation, *Technical Notes Plant Materials*, 26, 2001.

TABLE 9.7
Additional Plant Species That Can Be Used to Rehabilitate Salt-Laden Soils and/or Stabilize Sand Dunes

Scientific Name	Common Name	Reference
Lygeum spartum L.		Nedjimi, 2009
Portulaca oleracea L.*	Purslane	Kilic et al., 2008
Odyssea paucinervis (Staph) (Poaceae)+		Naidoo et al., 2008
Batis maritima+		El-Haddad and Noaman, 2001
Distichlis spicata+		
Juncus roemerianus+		
Salicornia bigelovii+		
Spartina alterniflora+		
Atriplex lentiformis+	Quailbush	Jordan et al., 2009
Atriplex prostrata+		Katembe et al., 1998
Atriplex patula+		
Panicum antidotale Retz.	Blue panicgrass	Aqeel Ahmad et al., 2009
*Suaeda fruticosa**		Khan et al., 2009
Cakile maritima+		Megdiche et al., 2009
Spondias tuberosa+	Umbu	Ciriaco da Silva et al., 2008
Spartina alterniflora Loisel+		Bradley & Morris, 1992
Festuca rubra ssp. *litoralis*+		Diedhiou et al., 2009
Hippophae rhamnoides	Seabuckthorn	Chen et al., 2009
Botanical name	**Common name**	**Cultivars[a]**
Aeluropus lagopoides (Linn.) Trin.	Ex Thw.[+]	
Agropyron cristatum (L.) Gaertn.	Crested wheatgrass	CD-II, Douglas, Ephraim, Fairway, Hycrest, Kirk, Nu-ARS AC2, Nordan, P-27, Parkway, Ruff, Roadcrest, Summit, Szarvasi, Vavilov
Agropyron junceum (L.) Beauv.	Sand couch	
Alopecurus arundinaceus Poir	Creeping foxtail	Garrison, Retain
Ammophila arenaria (L.) Link	European beachgrass	
Ammophila breviligulata Fern.	American beachgrass	Cape, Hatteras
Aristida adscensionis		
Arrhenatherum elatius (L.) J&K Presl	Tall oatgrass	9061649, Arel 41, Arone, Deal, Gala, Grano, Levocsky, Odenwalder, Roznovsky, SK-5, Sora, Tualatin, Wena, Wiwena
Bothriochloa saccaroides (Sw.) Rydb.	Silver beardgrass	
Bouteloua curtipendula (Michx.) Torr.	Sideoats grama	Butte, Coronado, El Reno, Haskell, Killdeer, Niner, Pierre, Premier, Trailway, Vaughn
Bouteloua gracilis (Willd. Ex Kunth)	Blue grama	Alma, Bad River, Hachita, Lag. Ex Griffiths Lovington
Bromus inermis Leyss.	Smooth brome	Achenbach, Alpha, Badger, Barton, Baylor, Beacon, Bravo, Bromar, Carlton, Elsberry, Fox, Homesteader, Jubilee, Lamont, Lancaster, Lincoln, Lyon, Magna, Manchar, Mandan 404, Martin, Polar, Radisson, Rebound, Sac, Saratoga, Signal, Southland, Tempo, York, 9005308
Calamagrostis canadensis (Michx.) Beauv.	Bluejoint reedgrass	Sourdoug, Saru 055

continued

TABLE 9.7 (Continued)
Additional Plant Species That Can Be Used to Rehabilitate Salt-Laden Soils and/or Stabilize Sand Dunes

Botanical Name	Common Name	Cultivars[a]
Cenchrus biflorus		
Cenchrus ciliaris		
Cenchrus pennesetiformis		
Cenchrus prieurii		
Chasmanthium latifolium (Michx.) Yates	Wild oats	
Chloris gayana Kunth	Rhodesgrass	Bell, Boma, Callide, Capital, Elmba, Finecut, Hatsunatsu, Karpedo, Kongwa, Katambora, Masaba, Mbarara, Mpwapwa, Nemkat, Nzoia, Pioneer, Pokot, Rongai, Samford, Topcut
Cynodon dactylon (L.) Person var. aridus	Giant Bermuda grass	NK-37
Dactylis glomerata L.	Orchardgrass	Able, Abserystwyth S.26, Akara, Albert, Amba, Ambassador, Artic, Baraula, Bartego, Bartyle, Barviva, Benchmark, Berber, Boone, Cambria, Chinook, Commet, Currie, Dactus, Dart, Dawn, DS8, Dorise, Duke, Frode, Hallmark, Haymate, Juno, Justus, K2-8, Kay, Latar, Mobite, Modac, Napier, Orbit, Orion, Paiute, Palestine, Pennlate, Pennmead, Peralvia, Persist, Piedmont, Pizza, Pomar, Porto, Potomac, Prairial, Prime, Rancho, Rapido, Saborto, Shinyo, Sterling, Sumas, Summer Green
Deschampsia caespitosa (L.) Beauv.	Tufted hairgrass	Bronzeschleier, Fairy's joke, Goldstaub, Holciformis, Nortran, Barcampsia
Digitaria eriantha Steud.	Pangolagrass	Leesburg No. 5, Mealani, Pangola, Slenderstem, Tiawan, Transvala
Distichlis spicata (L.)	Greene Inland saltgrass	Experimentals A138, DT16
Distichlis stricta (Torr.) Rydb.	Desert saltgrass	
Elymus canadensis L.	Canada wildrye	Mandan 419
Elymus cinereus (Scribn. & Merr.) A. Love	Great basin wildrye	
Elymus dahuricus Turcz. ex Griseb.	Dahurian wildrye	Arthur, Chabel, James
Elymus elymoides (Raf.)	Squirreltail	
Elymus glaucus Buckl.	Blue wildrye	Anderson, Arlington, Berkeley Hills, Mariposa, Ramaley, Stanislaus 2000 & 5000
Elymus junceus Fisch.	Russian wildrye	
Elymus lanceolatus (Schribn. & Sm.)	Thickspike wheatgrass	
Elymus salinus Jones	Salina wildrye	
Elymus smithii (Rydb.) Gould	Western wheatgrass	
Elymus trachycaulus (Link) Gould ex Shinnes	Slender wheatgrass	Adanac, Elbee, Highlander, Primar, Pryor, Revenue, San Luis, FirstStrike
Elymus triticoides	Buckley beardless or creeping wildrye	
Elytrigia intermedia	Intermediate wheatgrass	Amur, Chief, Clarke, Greenar, (Host) Nevski Greenleaf, Luna, Mandan 759, Manska, Nebraska 50, Oahe, Reliant, Rush, Slate, Tegmar, Topar, Trigo

TABLE 9.7 (Continued)
Additional Plant Species That Can Be Used to Rehabilitate Salt-Laden Soils and/or Stabilize Sand Dunes

Botanical Name	Common Name	Cultivars[a]
Eragrostis curvula	Weeping lovegrass	Consol, Ermelo, Morpa, Renner, A-(*Schrad.) Nees.* 67, Catalina
Eragrostis lehmanniana Nees.	Lehmann lovegrass	A-68, Cochis, TEM-SD, Kulvato, Puhulma
Festuca occidentalis Hook.	Western fescue	PI-518819
Festuca myuros L.	Myur fescue	
Festuca pratensis Huds.	Meadow fescue	Altesse, Barbarossa, Barkas, Barmondo, Bartran, Bartura, Belimo, Benfesta, Bundy, Comtessa, Darimo, Ensign, First, Mimer, Pegaso, Remko, Rossa, Stella, Swift, Trader, Wendelmoed
Fimbristylis spadicea (L.) Vahl		
Flaveria compestris J. B. Johnston	Marshweed	
Frankenia pulverulenta L.	Wisp-weed	
Glaux maritime L.	Sea milkwort	
Grayia spinosa (Hook.) Moq.	Hopsage or applebush	
Glyceria maxima (Hartm.) Holmb.	Manna grass	Pallida, Variegata
Glyceria striata (Lam.) Hitchc.	Fowl manna grass	
Halimodendron haledendron (Pallas) Schneid.		
Heliotropium curassavicum L.	Salt heliotrope	
var. *obovatum* DC		
var. *oculatum* Johnston	Pretty heliotrope	
Hierochloe odorata (L.) Beauv.	Sweetgrass	
Hilaria belangeri (Steud.) Nash	Curly mesquite	
Hilaria jamesii (Torr.) Benth.	Galleta grass	Viva
Hilaria rigida (Thurb.) Benth. Ex Schribn.	Big galleta	
Hordeum brachyantherum Nevski.	Meadow barley	
Hordeum jubatum L.	Squirrel's tailgrass or foxtail barley	
Hordeum marinum Hudson	Mediterranean barley or sea barleygrass	
Hordeum murinum L.	Rabbit barley	
Hordeum pusillum Nutt.	Little barley	
Imperata cylindrical (L.) Beauv.	Cogon	Red Baron
Juncus arcticus Willd.	Wiregrass	
Juncus gerardii Lois.	Blackgrass	
Juncus torreyi Cov.	Torrey's rush	
Juncus sweedyi Rydb.	Tweedy's rush	
Kochia prostrate (L.) Schrader	Prostrate Kochia	
Kochia scoparia (L.) Schrader	Summer-cypress	
Lasiurus scindicus		
Leersia oryzoides (L.) Swartz	Rice cutgrass	
Leptochloa uninervia (Presl) Hitche & Chase	Mexican sprangletop	
Leymus angustus (Trin.) Pilger	Altai wildrye	Eejay, Pearl, Prairieland, Mustang
Leymus arenarius (L.) Hochst.	Beach wildrye	Reeve

continued

TABLE 9.7 (Continued)
Additional Plant Species That Can Be Used to Rehabilitate Salt-Laden Soils and/or Stabilize Sand Dunes

Botanical Name	Common Name	Cultivars[a]
Leymus cinereus (Schribn. & Merr.) A. Love	Basin wildrye	Magnar, Trailhead
Leymus racemosus (Lam.) Tzvelev	Mammoth wildrye	Volga, Glaucus
Leymus triticoides (Buckl.) Pilger	Beardless wildrye	Gray Dawn, Rio, Shoshone
Ochthochloa compressa		
Panicum antidotale		
Panicum repens L.	Torpedograss	
Panicum turgidum		
Panicum virgatum L.	Switchgrass	Alamo, Blackwell, Bomaster, Caddo, Cave-n-rock, Dacotah, Forestburg, Haense Herms, Grenville, Heavy Metal, Kanlow, KY-1625, Nebraska-28, Pathfinder, Performer, Shelter, Summer, Sunburst, TEM-LoDorm, Trailblazer, Shawnee, Rehbraun, Squaw
Pascopyrum smithii (Rydb.) A. Love	Western wheatgrass	Arriba, Barton, Flintlock, Mandan 456, Rodan, Rosanna, Walsh, ND-WWG931, ND-WWG932
Paspalum dilatatum Poir.	Dallisgrass	Grasslands Raki, Prostrate, Sabine
Phalaris aquatica L.	Harding grass	Au Oasis, Australian, Castelar INTA, Estanzuela Urunday, Holdfast, Maru, Perla, Sirocco, Sirolan, Sirosa, Siro Seedmaster, Tresur Tala, Uneta, Wintergreen
Phalaris arundinacea L.	Reed canarygrass	Bellevue, Castor, Dwarf Garters, Feesey's Farm, Grove, Ioreed, Keszthelyi 52, Lara, Luteo-Picta, Motycha, Palaton, Peti, Rival, Szarvasi 50 & 60, Tricolor, Vantage, Venture
Poa bulbosa L.	Bulbous bluegrass	
Poa canbyi (Scribn.) Piper	Canby bluegrass	Canbar
Polypogon monspeliensis (L.) Desf.	Rabbitfoot grass	
Psanthyrostachys juncea	Russian wildrye	Bozoiski-Select, Bozoisky-II, Cabree, Idaho 100, Mankota, Mayak, Piper, Sawki, Swift, Tetracan, Vinall
Puccinellia airoides (Nutt.) Wats. & Coult.	Nuttal alkaligrass	Quill
Puccinellia fasciculate (Torre. L) Bicknell	Torrey's alkaligrass	
Puccinellia lemmoni (Vasey) Scribn.	Lemmons alkaligrass	
Puccinellia nuttalliana (Schult.) A. S. Hitchc.	Nuttal's alkaligrass	
Ruppia maritimaq L.	Ditchgrass	
Setaria palmifolia (Koen.) Stapf.	Palm grass	Variegata
Sorghastrum nutans (L.) Nash	Indiangrass	Cheyenne, Holt, Llano, Lometa, Nebraska 54, Osage, Oto, PI514673, Rumsey, Sioux Blue, Tejas, Tomahawk
Spartina alternifora Loisel.	Smooth cordgrass	Aureo-marginata, Bayshore, Vermillon
Spartina patens (Alt.) Muhl.	Saltmeadow cordgrass	Avalon, Flageo, PI415141
Spartina gracilis Trin.	Alkali cordgrass	
Spartina pectinata Link	Prairie cordgrass	Aureo-marginata

TABLE 9.7 (Continued)
Additional Plant Species That Can Be Used to Rehabilitate Salt-Laden Soils and/or Stabilize Sand Dunes

Botanical Name	Common Name	Cultivars[a]
Spergularia marina (L.) Griseb.	Salt sandspurry	
Spergularia rubra (L.) J.&C. Prsl.	Red sandspurry	
Sporobolus airoides (Torr.) Torr.	Alkali saccaton	Salado, Saltalk, Salty, Wilcox
var. *airoides*		
var. *wrightii*	Wright's saccaton	
Sporobolus cryptandrus (Torr.) Gray	Sand dropseed	PI255616 (W6 32700; W6 32704)
Sporobolus ioclados (Trin.) C. E. Hubbard[+]		
Sporobolus virginicus (L.) Kunth. Salt marsh grass		BT-1 (Saltfine®) or Seashore dropseed (Patent PP13652)
Stipagrostis plumosus		
Thinopyrum ponticum (Podp.)	Tall wheatgrass	Alkar, Jose, Largo, Orbit, Platte, Liu & Wang Tyrrell
Tripsacum dactyloides (L.) L.	Eastern gamagrass	Luka IV, K-24, Pete
Uniola paniculata L.	Sea oats	
Urochondra setulosa (Nees ex trin.) Nees[+]		
Vetiveria zizanoides (L.) Nash.	Vetiver grass	
Zoysia sinica Hance.	Seashore zoysiagrass	J-14
Zoysia machrostaycha[+]		
Zoysia matrella x *Z. tenuifolia*		A-1

Note: Some species have the potential to become weed problems in salt-affected ecosystems.

[a] This section summarized from D. Brede. 2000. *Turfgrass Maintenance Reduction Handbook*. Sleeping Bear Press, Chelsea, MI; USDA Grass Varieties in the United States. 1995. *USDA Agriculture Handbook No. 170*. CRC Press, Boca Raton, FL; http://www.ARS-GRIN.gov; and *Ecophysiology of High Salinity Tolerant Plants*, M. Ajmal Khan and Darrell J. Weber (Eds.), 2008, Springer, Dordrecht, the Netherlands.

*Salt phytoaccumulator.

+Haloyphyte.

as 'Quill'; weeping alkaligrass (*Puccinellia distans* [L.] Parl.) such as 'Fults', 'Salty', or 'Chaplin'; and Lemmons alkaligrass (*Puccinellia lemmoni* [Vasey] Schribn.). Sand dropseed (*Sporobolus cryptandrus*), alkali sacaton (*Sporobolus airoides*), and saltgrass (*Distichlis spicata* var. *stricta*) can also be included in this group (600 mM NaCl: Marcum, 1999). Inland saltgrass (*Distichlis spicata*) and Western wheatgrass (*Pascopyrum smithii*) can also be considered for salt-affected site restoration (Aschenbach, 2006). Kallar grass (*Leptochloa fusca*) can be used to improve degraded physical properties in saline-sodic soils (Akhter et al., 2004).

Most halophytes are not actual salt phytoaccumulators (also called phytoextraction), a trait that can be used to ameliorate and eventually rehabilitate highly saline ecosystems by scrubbing the salt out of the soil or water (Pilon-Smits, 2005). However, tolerant plants are available that can be used for the phytostabilization, reclamation, and rehabilitation of these salt-challenged sites (Table 9.7). Some of these species have the potential also to become weedy pest problems in some ecosystems (i.e., torpedograss). One halophyte (*Sesuvium portulacastrum* L.) has been used to phytodesalinize a salt-affected soil by phytoextracting levels of sodium that allowed growth of more salt tolerant glycophyte-type plants (Rabhi et al., 2010).

A critical point when reclaiming saline and/or alkaline sites is that some soil chemical and physical remediation may be needed prior to planting these plants, since specific salinity tolerance mechanisms sometimes require a transition period for activation, and root generation and regeneration involve juvenile roots that are quite sensitive to excess accumulated and localized salts. This establishment issue is critical for any seeded or vegetatively propagated plants from nodal segments unless salt primed (see Section 9.2.1), since germination and juvenile plantlet establishment or grow-in might not be successful if salt levels are excessive.

In summary, salt tolerance in turfgrass species and cultivars involves multiple genes, some upregulated and some downregulated, in response to total soluble salt and specific salt ion (sodium or chlorides) exposure in both the roots and the shoots. The plant tolerance or susceptibility responses include a cascade of chemical, physiological, and morphological sequences that are energy dependent and a regressive tax on subsequent photosynthetic activity. With continuing reduced photosynthesis and decreasing carbohydrate (energy) pools internally in the plant plus the accompanying nutritional disruptions, the plant will transition into a pattern of senescence and subsequent death if salinity is not properly managed. The complexity of this abiotic stress is unparalleled in the plant kingdom.

10 Irrigation System Design and Maintenance for Poor-Quality Water

CONTENTS

10.1 IRRIGATION SYSTEM DISTRIBUTION EFFICIENCY IS CRITICAL ON SALT-AFFECTED SITES

A poorly designed irrigation system (a) substantially hinders salinity control by leaching, (b) is a major contributor to spatial variability in salinity accumulation across the landscape, and (c) results in irrigation scheduling limitations that inhibit salinity management. In Section III, "Management BMPs for Saline and Sodic Soil Sites," a number of management practices or "strategies" for the turfgrass manager to use in salinity management are presented; but with the exception of turfgrass and landscape plant selection (Chapter 9), none are more important than the irrigation system—design and scheduling. *As irrigation water salinity increases, the essential requirements for excellence in irrigation system design, maintenance, and effective operation, utilization, and scheduling also increase.* In the current chapter (Chapter 10), irrigation system design and maintenance are the focus, while in Chapter 11 the topics are irrigation scheduling and salinity leaching. Huck (1997) and Duncan et al. (2009) present in-depth and practical treatment of how irrigation system design, maintenance, and operation and scheduling are profoundly affected by the presence of poor irrigation water quality and corrective suggestions, and the current chapter draws upon these sources.

To understand how irrigation design and scheduling are closely related with the spatial distribution of water and salt constituents over the landscape and within the soil rootzone, some illustrations are useful. Precision irrigation water management requires an irrigation design that optimizes both distribution uniformity and site-specific control (Tanji, 1996). With good-quality irrigation water (i.e., low salinity), turfgrass managers easily recognize that irrigation scheduling is much more of a challenge with a poorly designed system that does not exhibit good application uniformity. In order to apply sufficient water to areas receiving less coverage, other areas receive extra water to avoid drought stress, resulting in inefficient water use. Without excellent distribution uniformity, overwatered, saturated wet areas (with accompanying accumulation of excess salts) can develop where excess irrigation is applied that exceeds soil infiltration and permeability rates (Duncan and Carrow, 2000).

A system with high uniformity of distribution is critical in avoiding salinity accumulation patterns resulting from this design-induced scheduling problem as well as efficient salinity leaching (Carrow et al., 2000; Duncan and Carrow, 2000; Tanji, 1996). With saline irrigation water, the practice of overwatering some areas to achieve adequate irrigation on all areas greatly affects not only water distribution but also salt distribution and spatial accumulation, resulting in salinity stress responses by plants. Conversely, in the process of managing those wet areas and trimming back irrigation, drier areas can become salinized if a sufficient leaching fraction is not being applied. When salt accumulation is fostered in localized areas due to lack of irrigation uniformity and inadequate zoning to achieve site-specific irrigation, these limitations greatly complicate salinity leaching by (a) not allowing site-specific salinity leaching, and (b) requiring the same leaching fraction that is sufficient to leach the highest salinity areas but overirrigates all other areas. Reducing the number and magnitude of excessive salt accumulation areas will reduce water use for salinity leaching and allow more efficient leaching on those areas that require extra water to promote salt movement.

Not only does irrigation distribution uniformity impact salt distribution, but it also affects leaching fractions, leaching management, runoff, and total water consumption. Zoning must allow precision scheduling of water for good irrigation uniformity and for site-specific salinity leaching rather than using the same leaching fraction across the whole landscape. Control capabilities must allow matching water applications and precipitation rates to site-specific soil conditions (infiltration, percolation, and drainage rates) via pulse or repeat cycles as well as appropriate nozzles and zoning. Additionally, sites with multiple turfgrass cultivars, especially those cultivating both warm and overseeded cool-season turfgrasses, often require sophisticated sprinkler controls that allow precise management of leaching, such as subareas within tees, greens, and fairways on golf courses. The capacity of the hydraulic design (main and lateral line sizes) must account for increased water volumes needed for leaching fractions based on the irrigation water quality and on threshold salinity values of specific turfgrass cultivars.

An obvious implication from the above information is that successfully managing sites irrigated with poor-quality water may necessitate designing the irrigation system with capabilities and features not required when applying good-quality water. Depending on a site's water chemistry, specialized hardware to add amendments or remove odors may be needed. Higher salt concentrations and resulting corrosion with recycled and brackish waters may require alternative hardware components and/or other design considerations. Disinfectant chemicals used occasionally in effluent water have been known to degrade rubber gaskets and soft plastic components, a problem that occasionally surfaces after retrofitting older irrigation systems to apply recycled water.

As the use of halophytic turfgrass species such as seashore paspalum (*Paspalum vaginatum*) and inland saltgrass (*Distichlis spicata*) becomes more common, it is inevitable that more saline irrigation water sources will become increasingly common (Vermeulen, 1997; Duncan and Carrow, 2000; Duncan et al., 2009). Since highly saline waters are generally more corrosive, irrigation system components with increased corrosion resistance will be required as well as other special design considerations.

Soil conditions are another important consideration in irrigation system design and scheduling, but soil infrastructures play an even more important role with saline irrigation water, especially on more fine-textured soils. Irrigation decisions influenced by soil properties include the following; (a) the water-holding capacity of soils will dictate the length of time between irrigation events, (b) sprinkler precipitation and soil infiltration rates together will determine the maximum length of individual irrigation cycles, and (c) evapotranspiration (ET) and leaching fractions will determine the total amount of water and the number of cycles applied during each irrigation event (Oster et al., 1992; Tanji, 1996). When irrigation water is of good quality, these decisions are made with the primary attention to effective soil water distribution to meet plant needs. However, with saline irrigation water, irrigation decisions on total quantity of water to apply, length of irrigation pulse cycles, and number of pulse cycles must be made with the primary emphasis on how salt accumulation or leaching is affected and with secondary consideration to plant water requirements. Either lack of sufficient soil moisture or high soil salinity will induce drought stress on the plant, with high salts triggering physiological drought stress even at high soil water content.

Illustrations in the previous paragraphs demonstrate the critical nature of irrigation system design and scheduling interactions on salinity management. In the remainder of Chapters 10 and in Chapter 11, we will present more detailed treatment of irrigation system design, maintenance, and scheduling as related to soil salinity.

10.2 SYSTEM DESIGN CONSIDERATIONS FOR MANAGING POOR-QUALITY WATER

10.2.1 THREE CRITICAL IRRIGATION DESIGN CONSIDERATIONS

A number of irrigation system design considerations will be discussed in Section 10.2; but three design criteria are of special importance and are highlighted based on Huck (1997) and Duncan et al. (2009). The three critical design characteristics when using saline irrigation water are (a) sprinkler distribution uniformity, (b) site-specific application capability and capacity, and (c) control flexibility.

10.2.1.1 Sprinkler Distribution Uniformity

Sprinkler distribution uniformity, which refers to how evenly water is applied to the irrigated area, is of primary importance. If the system is not capable of a high degree of water application uniformity, then both soil moisture and soil salinity will be highly spatially diverse. The coupling of drought induced by lack of soil moisture and drought caused by excessive soil salt accumulation will result in a random patchwork of stressed sites across the turfgrass area. Irrigation uniformity is a "preventative" measure that greatly reduces site-specific problems from uneven distribution of irrigation water and its variable accumulation of soluble salt constituents.

A measure of how uniformly irrigation water is applied by the system is *distribution uniformity* (*DU*), which will be discussed in greater detail later in this chapter. If the amount of water applied is the same throughout the field, DU would be 100%, and the same amount of water would be applied at every location in the field—assuming that all water infiltrated into the soil at the point of application. While no irrigation system is capable of delivering 100% uniformity and different parts of the field will receive different amounts of water, distribution systems can be designed with very high (~80%) uniformity (Hanson et al., 1999).

The quantity of irrigation water required to leach soil salts in addition to replacement ET and adjustment for nonuniformity of the irrigation system is called the *leaching requirement* (*LR*), with an LR of 5% to 20% typical of saline sites. Where sprinkler DU is poor, there can be significant differences in amounts of water applied to the wettest versus driest locations; therefore, the poorer the DU, the more total water required to meet the salt-leaching requirement within the driest area while applying excess water and additional soluble salts to the wettest sites (Hanson et al., 1999; Skaggs and van Schilfgaarde, 1999). This is illustrated in Table 10.1, which shows the average leaching fraction required at different distribution uniformities needed to maintain a 5% leaching fraction in the area receiving the least total amount of water (Hanson et al., 1999). These differences in application rates result in salts being displaced lower in the profile where more water is applied and shallower salt displacement where less water is applied, thereby creating inconsistent turfgrass quality and playing conditions. Hence, salt movement and resulting turfgrass quality can be improved by more uniform total ecosystem water application (Hanson et al., 1999; Skaggs and van Schilfgaarde, 1999).

A high DU will assist in moving salts uniformly into the soil profile, reduce the total watering requirement when leaching, limit the subsequent development of random and unanticipated wet and dry areas, and potentially save water in the long term (Hanson et al., 1999; Skaggs and van Schilfgaarde, 1999; Zoldoske et al., 1994). The more uniform the irrigation application by the

TABLE 10.1
Average Leaching Fractions Needed to Maintain a Minimum 5% Leaching Fraction throughout the Field

Distribution Uniformity (DU%)	Average Leaching Fraction (LF%)
55 = poor	90
65	62
70	49
75	37
80	31
85	23
95 = good	14

Source: From Hanson, B., S. R. Grattan, and A. Fulton, *Agricultural Salinity and Drainage* (Division of Agriculture and Natural Resources Publication No. 3375), University of California, Davis, 1999.

sprinklers, the less need for excess irrigation to adequately irrigate dry or saline areas (Tanji, 1996). Supplemental drainage lines then become less important as less deep percolation and surface runoff will occur. Manual labor for spot irrigation with hand-directed hoses or portable sprinklers is also reduced. Short duration and frequent syringing with saline water have the potential to increase salt accumulation in shallow upper soil profile zones, and the necessity for these water applications must be included in the irrigation scheduling and leaching programs in concert with specific turfgrass species and cultivar water and salt tolerance requirements on the site.

10.2.1.2 Site-Specific Application Capability and Capacity

Maximizing sprinkler control down to the smallest manageable area is critical when working with poor-quality water. This allows maximum flexibility for varying site conditions (salinity level, soil types, slope, sun, wind exposure, etc.) that warrant special management when irrigating with poor-quality waters. The ultimate and ideal arrangement is one valve-in-head sprinkler per control station, commonly referred to as either "individual control" or "single-head control."

Even with high uniformity of water application, on salt-affected sites, still small stressed areas can still occur since surface and subsurface salt movement and spatial differences in ET on a site can also be causes of nonuniformity of salt distribution. Single-head zoning and control offer the best option to irrigate on a site-specific basis (a) to control water applications on small areas exhibiting any water stress—from runoff, localized dry spots, salt accumulation, or high ET in a localized area; and (b) to leach salts with the least quantity of water. Site-specific salinity leaching at the single-head level requires information on soil salinity distribution within the zone of influence of the head. However, recent advances in spatial mapping of soil salinity is progressing to the point where detailed information on salinity levels within different areas of a single irrigation head zone of influence can be accomplished (Krum et al., 2011; see Chapter 4, Section 4.3.2, "Approaches to Field Salinity Monitoring," for a discussion on mobile salinity-monitoring devices).

Computer central controllers combined with single- or individual head control (one station, one wire, and one sprinkler) provide the ultimate control system that allows leaching of specific areas along with micromanagement of wet and dry spots (Duncan et al., 2009). Single individual head control requires additional satellite controllers, but is well worth the investment on sites with poor water quality and/or where water use efficiency is a necessity. If the construction budget will not allow single-head control, then the following alternative can be used to effectively plan for future upgrading to this feature and within a reasonable water budget. Size the 110 VAC power system to accommodate enough satellite controllers to eventually provide single- or individual head control. Individual 24 VAC wires can be brought from each sprinkler back to the satellite controllers. The individual wires can be bundled as multiple sprinkler stations in the controller cabinet or in an underground valve box next to the satellite. This allows the option of adding satellites and splitting stations into single- or individual head control in the future as the budget permits. It also allows reassigning individual sprinklers to different stations without trenching in additional wire should management problems occur later. The cost of the additional wire is minimal when installing a new system and worth any additional long-term investment.

When saline water is applied by an overhead sprinkler irrigation system, vertical water flow results in salt movement (leaching) down into the soil profile with the infiltrating and percolating water (Hanson et al., 1999). Application of an LR of 5% to 20% additional water beyond the ET replacement requirement is needed to move the salts deep into the profile and to avoid toxic accumulations within the plant rootzone. Wherever ET replacement plus the LR have not been seen in the field, saline soils will develop (Pira, 1989; Tanji, 1996). Therefore, a relationship exists between deepwater percolation and subsequent salt migration because at lower soil profile depths, salt distribution and accumulation will depend on how uniformly water is applied and the quantity applied per irrigation event, or whether site-specific applications can be applied (Pira, 1989; Hanson et al., 1999). Since salinity distribution can be very variable over a site even when the system has high DU (see Figure 4.1), application of the LR only to the specific areas with high salt levels requires one- to two-head zoning. With irrigation zones consisting of three to six heads or more, the LR water will be applied on all areas not requiring extra water (i.e., an inefficient irrigation strategy for leaching). Sprinkler distribution patterns and irrigation scheduling programs, therefore, have a primary influence on salt distribution within the soil profile. It is not unusual for soluble salt accumulation in soil profiles, including sand-based profiles, to be two to four times higher than the incoming saline water when the irrigation and salt management strategies are not appropriate for the site.

10.2.1.3 Control Flexibility

Irrigation control flexibility entails any factor that provides the turfgrass manager with the most site-specific options to apply water as needed, when needed within a designated time frame, and at an acceptable volume to enhance salt movement for the specific site. The most important aspect of flexible control was noted in the previous paragraph, namely, single-head zoning and control. Other important aspects are (a) adequate total quantity of water to apply sufficient water in dry periods to meet the combined requirements of ET replacement and salt leaching; (b) piping and pumping capabilities to deliver the water volume needed; and (c) the ability to practice *pulse irrigation* (also called *cycle and soak* or *repeat cycles*) on all sites, where pulse irrigation is the practice of applying water just to the point of runoff, ceasing irrigation until the water infiltrates, then repeating cycles until the total ET replacement and leaching water quantity is applied (sometimes, it may take two or more evenings or early mornings to accomplish sufficient pulse cycles for effective leaching when the soils have a high clay content).

Pulse irrigation is very important for efficient salinity leaching (discussed in detail in Chapter 11), but it is also essential for water-use efficiency on all turfgrass sites with fine-textured soils. Soil infiltration, permeability, and internal drainage characteristics can vary considerably on complex sites such as golf courses where there is variation between native soil fairways, high sand content greens, and tees that may have been constructed with sand-based rootzones. Ideally, the precipitation rate of the sprinkler distribution system will not exceed the infiltration rate of the soils (Oster et al., 1992). However, in the real world, this is not always possible since sprinkler precipitation rates very often exceed soil infiltration rates. It is more often the case that computerized control systems must be used to deliver multiple irrigation cycles (i.e., pulse irrigation). This strategy provides time between each application cycle, allowing water to infiltrate and avoid surface runoff (Carrow et al., 2000). Additionally, pulse irrigation is effective in moving accumulated soluble salts through micropores in soil profiles and in inorganic amendments that contain high internal microporosity.

10.2.2 Additional System Design Considerations for Managing Poor Water Quality

In addition to sprinkler distribution uniformity, site-specific application capability, and control flexibility, the use of saline irrigation water warrants a number of other irrigation design considerations (Huck, 1997; Duncan et al., 2009). The overall performance of an irrigation system is often limited by the performance of an individual component or subsystem. It should be clear that ecosystem infrastructure requirements, such as irrigation system design, are essential to effectively manage accumulated soil salts, and are best installed during the preconstruction phase. Attempting to accomplish these critical infrastructure characteristics after construction and grass planting is very difficult, costly, and often impossible.

10.2.2.1 Irrigation Systems Hydraulic Design for LR

The leaching fraction must be accounted for when designing the irrigation system as this additional LR water volume must be considered when sizing pipe. If these flow requirements are not considered, either maximum flow velocities will be exceeded or the watering application window will need to be extended. Excessive flow velocities result in severe surge pressures and "water hammer" that lead to premature pipe and fitting failures and often shorten the useful life of the irrigation system. Extended water windows will interfere with players accessing the golf course, with maintenance operations, and/or with local regulations requiring all automatic irrigation to be completed within specific nighttime hours. These regulations are often mandated by municipalities when irrigating with reclaimed, recycled, or effluent water (Feil et al., 1997; Huck et al., 2000; Snow, 1994).

10.2.2.2 Sprinkler and Nozzle Selection and Spacing Evaluation

The sprinkler is the most important part of the irrigation system because it distributes the water over the land. How uniformly it accomplishes this task determines the effectiveness and efficiency of the irrigation system (Pair et al., 1983). Selection of the best performing sprinkler, nozzle, and spacing combination is therefore critical to the success in managing poor water quality with the irrigation system and in avoiding any soil salt accumulations along with development of excessively wet or dry areas.

Sprinkler and nozzle performance have been evaluated since the early 1940s with statistical calculations. The two most common measures of sprinkler uniformity are *Christensen's Coefficient of Uniformity* (*CU*) and *Low Quarter Distribution Uniformity* (DU_{LQ}), where each is expressed as percentages. The higher the percentage reported, the more uniform the water application, with 100% uniformity representing perfectly uniform coverage. However, this is unattainable because even rainfall does not fall with 100% uniformity.

A shortcoming of both CU and DU_{LQ} is that neither takes into account the location of the wetter and/or dryer application areas, such as whether drier or wetter values are concentrated in a localized area, or dispersed throughout the entire pattern where a surrounding high value may benefit an adjacent low value. Therefore, two different sprinklers that produce the same CU and DU_{LQ} results may deliver noticeably different performance in the field if the driest and/or wettest values are concentrated versus dispersed (Pair et al., 1983). Golf course irrigation designers have long recognized this problem with CU and DU_{LQ} and would, therefore, often rely on past field experience when selecting sprinklers, nozzles, spacing distances, and geometric configurations.

Personal computers changed this in the 1980s when the Center for Irrigation Technology (CIT) located at California State University, Fresno, developed its Sprinkler Profile and Coverage Evaluation (SPACE; 2007) software. Sprinklers can now be objectively evaluated before they are installed in the field through indoor testing and analysis of those data with the SPACE software (Solomon, 1988; Tanji, 1996; Barret et al., 2003).

For a small fee, the CIT staff tests a sprinkler and nozzle combination inside its laboratory to develop a profile. Duncan et al. (2009) present several sprinkler profiles from the SPACE software in both figure and densogram graphs, which represent water distribution by color intensity. Unfortunately, indoor evaluation has shortcomings. It cannot predict the influences of wind, elevation above sea level (air density), operating pressure changes due to land elevation changes (including slopes), air or water temperature, humidity, or other factors that may affect how water is distributed after leaving the nozzle. However, laboratory evaluation of individual sprinkler and nozzle combinations is the best method of sprinkler selection currently available, and is better than just installing sprinklers in the ground waiting for problems to surface later. If nothing else, indoor testing and computer modeling can allow for a better understanding of any given irrigation design limitations (Barret et al., 2003).

10.2.2.3 Pressure-Regulated Valve-in-Head Sprinklers and Remote Control Valves

Irrigation system design where poor irrigation water is present should specify that all valve-in-head sprinklers and remote control valves have internal pressure regulation devices. Depending on the manufacturer, pressure regulation devices may be optional. Pressure regulation maintains more consistent nozzle flow and precipitation rates, which both contribute to delivering the highest distribution uniformity possible.

10.2.2.4 Geometric Configurations (Square versus Triangular)

Irrigation texts stress the great importance of placing sprinklers on consistent geometric configurations to maximize distribution uniformity (Barret et al., 2003; Pair et al., 1983; Pira, 1989). Inground agricultural systems are most often established on square and rectangular spacing at least partially as a result of the need to move laterals with sprinklers on fixed spacing down the field from set to set (Pair et al., 1983). Turfgrass sprinkler systems are usually established on an equilateral triangular spacing (Pair et al., 1983). Field sprinkler research from the early 1940s demonstrated statistically that equilateral triangular spacing had the potential to deliver high uniformity without the need to overlap coverage beyond 50% to 60% of the diameter (Pair et al., 1983). Field experience at golf courses throughout the arid southwestern United States tends to support that equilateral triangular spacing is preferable over square spacing to optimize distribution uniformity. If proper head spacing is not used, even single-head control will not result in uniform application of water.

10.2.2.5 Combating Wind Effects on Distribution Uniformity

Wind of any speed distorts distribution profiles and the amount of distortion depends on the water droplet sizes created by the nozzle design (Figure 10.1) (Barret et al., 2003; Pair et al., 1983). As a rule of thumb when wind speed exceeds 5 to 6 mph, coverage uniformity can begin to suffer (Barret et al., 2003; Pair et al., 1983). Reducing the impact of wind on distribution has been attempted a number of ways such as reducing upwind or overall spacing distance and/or reducing operating pressures. Reducing operating pressure at the nozzle, while staying within manufacturer-recommended ranges, will produce larger water droplets that travel less distance in wind than fine

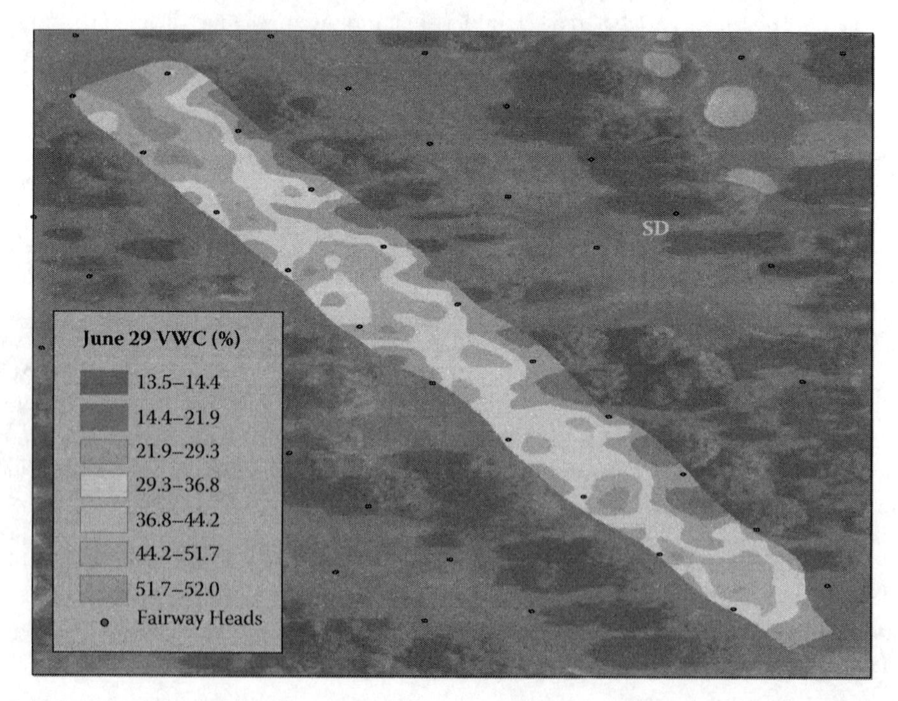

FIGURE 10.1 (See color insert.) Spatial variability of soil volumetric water content (% VWC) presented in a standard deviation map. Note the wind effects from right to left on spatial VWC patterns. Also, note the threefold difference in VWC on this fairway. Overlaying the theoretical water distribution pattern for these heads with the appropriate nozzles and system pressure would help determine system efficiency. (From Carrow, R. N., J. Krum, and C. Hartwiger, Precision Turfgrass Management: A new concept for efficient application of inputs, *USGA Turfgrass and Environmental Research Online* (TERO), 8 (13), 1–12, 2009a, http://usgatero.msu.edu/v08/n13.pdf. With permission.)

droplets produced at higher pressure (Barret et al., 2003; Pair et al., 1983). Improving water distribution performance in wind with spacing adjustments, especially where heads are offset to the upwind side, is risky. Unless wind speed and direction are very constant, especially during the nighttime hours when most irrigation is done, offsetting heads is not recommended (Barret et al., 2003).

10.2.2.6 Hydraulic Systems (Pump Station, Mainline, and Lateral Piping Network)

The pumping and piping systems need to deliver water throughout the site at an adequate, relatively uniform pressure to allow the sprinklers to operate efficiently. There is a direct correlation between sprinkler operating pressures and nozzle flow rates; thus, water pressure variations will affect the uniformity of in-field precipitation rates (Skaggs and van Schilfgaarde, 1999). Additionally, pressure will affect distribution uniformity if it is outside the sprinkler manufacturer's recommended range. Too low pressure will result in too large droplets and poor droplet distribution, while too high pressure creates excessive atomization and wind drift affecting droplet distribution (Zoldoske et al., 1987). Pressure regulating or boosting devices may be needed on sites with severe terrain or elevation changes to maintain uniform pressure.

The pumps and piping networks must also be sized to handle additional flow volumes for leaching requirements. The system should be able to complete a full irrigation cycle at maximum ET replacement plus the required leaching fraction within the time frame of any mandated watering window. For example, many recycled and effluent regulations state that all unattended automatic irrigation operations must be completed within a specific time frame such as between the hours of 9 p.m. and 6 a.m., a nine-hour window (Huck et al., 2000; Snow, 1994).

There may be additional reasons to shorten the watering window to less than what regulations may require. Special off-peak "time of use" energy rates are often available that can save considerably on electric pumping costs. Another potential reason for a shortened window of operation would be time for the irrigation application and leaching fraction to infiltrate the surface and provide a firmer playing surface for early morning golfers.

10.2.2.7 Parallel and Dual Mainline Distribution Systems

Recycled water use has become mandated for landscape irrigation in parts of the United States (California Assembly, 1991). However, many water providers and regulators will allow maintaining golf putting greens during nondrought conditions, and renovation or overseeding with potable-quality water while using a nonpotable source to irrigate the remainder of the golf course. They recognize that putting greens comprise a relatively small area of high-value, salt-sensitive turfgrass.

It is becoming more common for golf course irrigation systems to utilize parallel or dual (two) mainline loops to distribute the recycled or alternative water to the tees, fairways, and rough areas and potable water to the putting greens. Local regulations may also mandate dual water quality systems for setbacks from surrounding residences, environmentally sensitive areas, and/or potable-quality surface water and groundwater sources and their recharge zones (Barret et al., 2003). Dual systems require special equipment, including items such as air gaps and/or reduced pressure principal device (RPPD) backflow devices to protect the potable source from cross-connection contamination with the alternative or recycled water source. Minimum separation distances between the two mainline systems along with colored pipe (typically purple to signify nonpotable and either blue or white to signify potable) are also normally required. Local regulatory agencies should be consulted regarding specific design requirements for a parallel or dual-delivery system. Also, equipment must be included that is required to safely blend or deliver the alternative water to the greens in case of a drought where mandatory potable water use and/or water conservation restrictions may be imposed.

10.2.2.8 Dual Green Sprinklers and Leaching Sprinklers

It is becoming increasingly popular to install two sprinklers at the putting surface. The configuration between opposing part-circle sprinklers or one full and one part circle can vary based on the turfgrass manager's personal preference. Opposing part-circle sprinklers allow irrigating or leaching the putting surface with minimal effect on the surrounds. However, maintaining the arc adjustment and then managing areas where the overlap is insufficient or excessive can be challenging. Combining a full-circle (for regular irrigation applications) and opposing part-circle sprinkler for supplemental applications to surrounds or occasional leaching of the greens reduces the problems associated with exactly matching the overlap of opposing part-circle sprinklers.

10.2.2.9 Weather and Soil Moisture Monitoring Equipment

Various plant, climatic, and soil sensor tools for irrigation scheduling are discussed in greater detail in Chapter 11, "Irrigation Scheduling and Salinity Leaching," and summarized by Garbow (2008). Weather and soil moisture monitoring equipment can provide useful information to irrigation managers working with poor-quality water, but this information must be integrated into salinity management and irrigation scheduling in general. A leaching fraction can be added to evapotranspiration estimated by weather monitoring to allow scheduling irrigation events that meet both the ET replacement and leaching requirements. Soil moisture sensors can also be installed to monitor how deeply each irrigation application penetrates. By installing soil moisture monitoring sensors at various depths, the irrigator can determine when the entire rootzone has been adequately wetted. Refer to Section 10.3.4 on mobile sensor technology and Chapter 18 for additional information on salinity monitoring.

Equipment used to predict ET can range from simple evaporative gauges costing a few hundred dollars to sophisticated weather stations costing several thousands of dollars. The simplest of evaporative gauges use an exposed porous ceramic surface that evaporates water from a container to simulate ET. Manual readings are taken by the irrigator each day as to how much water has been lost from the calibrated container. The more sophisticated weather station systems monitor multiple parameters such as wind, relative humidity, temperature, and solar radiation and calculate ET with formulas that consider the influence of each parameter on evaporation and transpiration (crop water use).

The modified Penman-Monteith equation published in FAO Irrigation and Drainage Paper No. 56 (Allen, 1998; Campbell Scientific, 2010) is becoming the industry-accepted standard of the turfgrass scientific community and many state-operated weather system networks. However, recognize when purchasing a weather station that not all suppliers choose to adhere to this standard ET formula. Some weather stations, including those operated in conjunction with computerized irrigation control systems, use their own proprietary ET formulas (Campbell Scientific, 2010). This can cause great confusion when comparing data between weather stations operating under identical climatic conditions, and requires the crop coefficients to be adjusted for each different formula (Brown, 1999).

10.2.2.10 Setbacks and Buffer Zones

Recycled, brackish, and other saline irrigation waters can require setbacks or buffer zones from surrounding residences and property lines, environmentally sensitive areas, wetlands, potable surface, potable aquifer recharge zones, and potable groundwater well heads (Barret et al., 2003). Distances have been reported varying between 50 and 1000 feet depending on local regulations (Barret et al., 2003). Concerns addressed by setbacks and buffer zones range from overspray and wind drift of recycled water into designated areas, to concerns of leaching or runoff into potable surface water and groundwater sources. Wellheads are particularly susceptible to leaching immediately surrounding the area where soil has been disturbed when drilling the well. The concern is not always directly related to irrigation leaching, but to considering if a pipe were to break and flood the area immediately surrounding the wellhead. If irrigation is desired in the areas designated for setbacks, a dual-irrigation system delivering potable water as previously described is typically required.

10.2.2.11 Corrosion-Resistant Components

Salts in irrigation sources can range from under 100 ppm to amounts approaching that of seawater (i.e., in excess of 34,000 ppm). As total soluble salt content of the water increases, so does the need for the corrosion resistance of irrigation system components (pumps, mainline fittings, control cabinets, internal sprinkler components, wire splices, etc.). Exposure to both high-salinity water and salinized soils can cause corrosion of metal components (Clarke, 1980). Sprinkler manufacturers have responded with the increased use of plastics for control cabinets, sprinklers, and so on. High-quality, corrosion-resistant marine and naval grade paints, epoxy coatings, hardware, and metals commonly used in saltwater and various industrial applications (corrosive reverse-osmosis water and acid units) are options for refurbishing existing equipment (Micro Surface Corporation, 2010; Wink Fasteners, 2010). Plastic pumps as used in industrial handling of saltwater, corrosive acids, and deionized and reverse-osmosis ultrapure water have not yet entered the irrigation industry, but the potential exists as waters of lesser quality continue to be used for irrigation (Vanton Pump, 2010).

10.2.2.12 Chlorine and Chloramine Component Degradation Resistance

Potable and wastewater facilities have used disinfectants, typically chlorine, for the past 100 years to kill bacteria prior to distribution (Hudson, 2010). In arid regions, it is not unusual for treated wastewater to also contain appreciable salts due to concentration over time of salts within the water going into the treatment facility and open-air particulate settlement ponds. Though effective against waterborne microbes and pathogens, chlorine generates disinfection byproducts (DBPs) that with long-term exposure may cause cancer; therefore, the U.S. Environmental Protection Agency (EPA) has now set DBP limits (Hudson, 2010; Skipton and Dvorak, 2002). Consequently, many potable water treatment facilities have switched from using free chlorine to chloramine (a chlorine–ammonia mixture termed "combined chlorine" in water and wastewater reports) or a mixture of chlorine and chloramine (termed "total chlorine" in water and wastewater reports) (Hudson, 2010; Skipton and Dvorak, 2002). Chloramine has several advantages over free chlorine by producing less DBPs, and has less effect on the taste and odor of potable water. Chloramines react more slowly with the pathogens than chlorine, but since they do not evaporate rapidly from open-water bodies like chlorine, they remain active longer, performing the disinfection process and increasing the total contact time with the pathogen (Hudson, 2010; Skipton and Dvorak, 2002). Typical chloramine compounds used for disinfection are monochloramine (NH_2Cl), dichloramine ($NHCl_2$), and trichloramine (NCl_3) (Hudson, 2010; Skipton and Dvorak, 2002).

Chloramines are toxic to fish (both freshwater and saltwater), reptiles, and amphibians. Also, chloramines cannot be present in water used for kidney dialysis because they interfere with oxygen absorption if taken directly into a living organism's bloodstream. They are safe for consumption in drinking water by humans and animals because normal digestion neutralizes the ammonia before it can enter the bloodstream (Hudson, 2010; Skipton and Dvorak, 2002).

Although chlorine still can be, and is, used in both recycled and wastewater disinfection, it has been theorized that chloramines could be formed in wastewater as free chlorine reacts with organic nitrogen compounds. The significance of this is that chloramines have also been identified as potential problems with regard to the corrosion of metals and degradation of rubber and some specific irrigation system plastic components. Early recycled water users and irrigation equipment manufacturers often attributed premature plastic and rubber parts failures of sprinklers, pump seals, and gaskets contained in repair couplings and/or bell joint pipe to high residual chlorine resulting from the disinfection process. It now appears that the problem may actually be related to free chlorine that preferentially reacts with ammonia and organic amino acids contained in recycled water, and results in the forming of N-chloramine compounds. Research conducted at the University of Jordan supports this theory (Fayyad and al-Sheikh, 2001).

A study conducted by the American Water Works Association (AWWA) Research Foundation in 1993 determined that corrosion and degradation of certain metals and elastomers common to distribution plumbing parts and accessories were related to their exposure to chloramine compounds (Reiber, 1993). The study tested seven metal surfaces, seven common elastomers, and three thermoplastics. The results are reported in Table 10.2. Most irrigation components are now manufactured with components resistant to chloramine degradation, but occasionally an older system at a site retrofitted with recycled water may demonstrate problems. Fortunately, polyvinyl chloride (PVC) and chlorinated polyvinyl chloride (CPVC) compounds typically used to manufacture irrigation pipe, fittings, and lake liners appear to be resistant to chloramine degradation (Callery, 2003; Harvel Plastics, 2010; Vinidex Systems, 2010).

10.2.2.13 Water Treatment Systems

Evaluating water quality prior to designing a new irrigation system allows advance planning for water treatment systems that might be needed to improve infiltration or displace sodium (Carrow et al., 1999). If treatment equipment is included in the system, it is necessary to provide an area large

TABLE 10.2
Corrosion and Degradation Reactions of Common Distribution Plumbing and Appurtenance Materials to Free Chlorine and Chloramines

Material Type	Specific Product	Free Chlorine	Chloramine (Combined Chlorine)
Metals	Mild steel	Not specifically reported	Not specifically reported
	Copper	Moderately accelerated corrosion	Slightly accelerated corrosion
	Brass	Moderately accelerated corrosion	Slightly accelerated corrosion
	Bronze	Moderately accelerated corrosion	Slightly accelerated corrosion
	Pb-Sn solder	Slight pitting or corrosion	Slight pitting or corrosion
	Sn-Sb solder	No pitting or corrosion	No pitting or corrosion
	Sn-Ag solder	No pitting or corrosion	No pitting or corrosion
Elastomers	Natural rubber Acrylonitrilebutadiene Styrene-butadiene Chloroprene Silicone Ethylene-propylene Fluorocarbon	Hypochlorous acid disinfectants are significantly less damaging to elastomers as compared to chloramines	Both monochloramine and dichloromine solutions, with few exceptions, produced greater swelling, deeper and more dense surface cracking, a more rapid loss of elasticity, and greater loss of tensile strength as compared to equivalent concentrations of free chlorine. Chloramines are uniquely injurious to elastomers and produced conclusive results when compared to other forms of chlorine disinfectants. Natural isoprenes (rubber) and synthetic isoprenes are most susceptible to attack. Synthetic polymers developed for chemical resistance performed well in chloramine exposure.
Thermoplastics	Celcon®	Susceptible to chlorine attack	Susceptible to chloramine attack
	Delrin®	Susceptible to chlorine attack	Susceptible to chloramine attack
	Udel®	Impervious to chlorine attack	Impervious to chloramine attack

Source: From Walker, R., M. Lehmkuhl, G. Hah, and P. Corr, *Landscape Water Management Auditing*, Irrigation and Training Research Center, Cal Poly, San Luis Obispo, 1995.

enough to house and power equipment such as sulfur burners, acid injection, gypsum injection, or fertigation equipment near the pump station or irrigation reservoir. Also, consider adequate storage area for a supply of amendments for injection.

10.2.2.14 Miscellaneous Items

Sites using effluent and other nonpotable irrigation sources often are required to provide warning signs and/or tags at property lines, storage lakes, tags on remote control valves, and so on stating that the water is not suitable for human consumption. Readily visible signs indicating purple pipe or burial tape signifying nonpotable water lines are also commonly mandated (Feil et al., 1997). Occasionally devices to automatically cease irrigation operations at a particular wind speed are required (Snow, 1994). Onsite weather stations for ET replacement verification and minimum lake-lining thickness requirements (recycled effluent) have also been reported (Snow, 1994). These requirements can vary between states and local agencies; it is recommended that local regulators be consulted regarding codes in your area prior to irrigation system installation.

10.2.2.15 Recycled Water and Effluent Disposal

Occasionally effluent water delivery contracts require that a site use a set amount of water daily whether the turfgrass requires irrigation or not (Duncan et al., 2009). Under such circumstances, it may become necessary to irrigate out-of-play areas for the sole purpose of effluent disposal to avoid playing areas from becoming excessively wet. This process may be referred to as "slow rate land applied groundwater recharge." If out-of-play areas are not available for surface or sprinkler applications during the nonirrigation season, subsurface drip irrigation systems may be a viable alternative. These systems can apply small amounts of water 24 hours per day at rates matching soil percolation rates. The low application rate spread across a 24-hour period disposes of the effluent while maintaining the playing surface in a dry playable condition. Subsurface disposal may not be feasible in layered or extremely heavy clay soils where water may wick to the surface through capillary action. Another option is to install sand-based leaching fields with drainage and to apply recycled water, with the recaptured leachate being blended with other water sources for application on recreational turfgrass sites.

10.2.2.16 Potable Water–Recycling Equipment Wash Rack

It is common practice to wash mowing equipment in the rough at quick coupler locations connected directly to the irrigation system. However, when irrigating with higher salinity water sources, this strategy can accelerate corrosion and the ultimate deterioration of turfgrass maintenance equipment (Huck et al., 2000). Therefore, when irrigating with saline water sources, provisions should be made for equipment wash areas using lower salinity (potable) water to prolong equipment life and recycling wash racks that will conserve potable water resources. Additional equipment surface (such as mower understory) waxing may be needed to protect the equipment in the long term when applying increasingly saline irrigation water.

10.3 IRRIGATION SYSTEM DISTRIBUTION UNIFORMITY (WATER AUDITS)

Not all sites converting to poorer quality water will have the opportunity to upgrade or replace their irrigation system. Those sites that are fortunate enough to install state-of-the-art systems will also need to maintain their overall distribution efficiency performance over time. Therefore, regularly scheduled maintenance and occasional minor upgrades to optimize the irrigation system's performance and distribution uniformity are important for both new and old irrigation systems.

A formal *irrigation water audit* may be conducted on a site. The Irrigation Association (2003, 3) defined an irrigation audit as including the following: "Information about each area's technical characteristics and controller capabilities should be obtained. An irrigation audit involves collecting data, such as site maps, irrigation plans and water use records. Tuning of the irrigation system is accomplished during the inspection.... Field tests are then conducted to determine the system's uniformity and to calculate various zone precipitation rates."

Audits have two major components (Irrigation Association, 2003). First, there is a systematic evaluation of the components and scheduling of the system for maximum application uniformity and efficiency. This phase includes the following:

- Very detailed information is determined on the irrigation system design and performance. The main focus is on the irrigation system infrastructure on a specific site.
- Water source availability and needs are determined.
- Some soil information is determined as related to irrigation system design, zoning, and water application capability.
- Grass type and location are often noted.
- Climatic conditions may be available from a local or onsite weather station to assist in scheduling irrigation cycles.

Second, after the irrigation system is functioning at maximum capability, there is an evaluation of the irrigation water distribution uniformity, traditionally by the *catch-can method* (discussed later in this chapter).

In the next section, items included as a part of an ongoing maintenance program relating to the first phase of the water audit will be presented. Some turfgrass sites with professional turfgrass managers conduct these activities during their normal maintenance operations on a facility. For example, some golf courses have an irrigation assistant or specialist who may perform these tasks. However, on other turfgrass sites, system maintenance and scheduling may receive less attention.

10.3.1 Evaluating and "Tuning Up" Irrigation Systems (First Phase of Water Audit)

Without even considering the effects of wind and land slopes, distribution uniformity is often the weakest component of many irrigation systems new and old alike. An irrigation audit and catch-can test is a good method to evaluate and document water application efficiency. However, as noted, before performing the actual catch-can tests, a number of influencing factors should be inspected and their current condition documented. Developing an irrigation system "tune-up" checklist of the following items is suggested (Irrigation Association, 2003; Duncan et al., 2009).

10.3.1.1 Spacing and Geometric Configuration

The distance between sprinklers as well as geometric configuration (squares or equilateral triangles) should be uniform since both affect distribution and precipitation rates. Sprinkler spacing of a well-installed design will be within plus or minus 3 feet of the design specification. This provides for a reasonable margin of error and the accuracy that can be maintained in the field. Spacing within this range should be uniform throughout the primary turfgrass playing areas. Spacing adjustments to make the system "fit" the property should be made near perimeters in out-of-play rough areas (Barret et al., 2003). It is not practical to try to move hundreds of sprinklers to correct poor spacing; however, checking spacing in a few areas of questionable coverage and documenting specific problems can explain factors contributing to poor distribution uniformity and poor turfgrass performance conditions.

10.3.1.2 Lifting and Leveling Low Sprinklers

Low sprinklers must occasionally be lifted to compensate for thatch accumulation and soil settling to avoid surrounding turfgrass from disrupting spray patterns. Heads that are tilted to the turfgrass surface must be leveled so the trajectory of the nozzles will not be changed. Just a few degrees from level will change the radius wetted by the sprinkler.

Lifting and leveling sprinklers manually with nothing more than a shovel is very labor-intensive and difficult work. Each sprinkler raised and leveled can require 45 minutes or more time. A device sold as the Levelift mechanizes the process and greatly improves the efficiency of lifting and leveling low sprinklers (Levelift, 2010). This device uses the irrigation system's hydraulic pressure to place a safe amount of lifting force on the sprinkler canister. At the same time, water is bypassed to a probe (similar to devices used to supply deeply penetrating water and nutritionally feed tree roots) that is inserted into the ground to wet the soil surrounding the swing joint.

The injected water liquefies the soil surrounding the swing joint into a soup-like consistency, and the sprinkler is then automatically pulled to the proper grade. A small amount of dry soil or sand is then hand packed around the sprinkler body flange to fill the void created. A few irrigation flags are placed surrounding the head to discourage traffic from entering the area of softened soil. Within 24 hours, the liquefied soil will drain and become firm enough to support normal golf cart and maintenance equipment traffic. From start to finish, the time to lift a sprinkler with this device is typically between 5 and 15 minutes, depending on site-specific soil conditions (Levelift, 2010).

10.3.1.3 Sprinkler Brand, Model, and Nozzle Sizes

Assuming spacing is relatively uniform, sprinkler brands, models, and nozzle sizes should be uniform within the same area of coverage and control. Different brands, models, and nozzles have differing flow rates and distribution profiles. Therefore, mixing sprinklers and/or nozzle sizes will affect precipitation rates and distribution uniformity. Replacing worn nozzles or nozzles of varying sizes with new nozzles of the same size can provide a reasonable improvement in distribution uniformity and be cost effective.

Nozzle replacement can become necessary after 7–10 years of normal use. The time frame may be less if using water contaminated with sand and/or suspended solids that can accelerate wear on the irrigation components. Visual checks for damage can sometimes identify problems; however, the shank end of drill bits can be used as a gauge to more accurately assess wear between new and older, worn nozzles. If in need of nozzle replacement, compare high-efficiency, after-market nozzles with the original equipment manufacturer's replacements (Full Coverage Irrigation, 2010). The after-market nozzles have been reported to significantly improve distribution uniformity of older irrigation systems and systems with low-pressure problems.

10.3.1.4 Operating Pressure (Line Pressure and Sprinkler Nozzle Pressure)

Uniform coverage is compromised when operating pressures are not consistent and within manufacturers' specified ranges (Harvel Plastics, 2010; Barrett et al., 2003; Huck et al., 2000; Tanji, 1996; Oster et al., 1992). Pressure regulation valves in mainlines and at pumping stations should be regularly serviced and adjusted to deliver line pressures as specified in the original irrigation design and specifications. Depending on the severity of elevation changes throughout the site, pressure-reducing, -regulating, or -boosting devices may be a part of the system and will need occasional maintenance and adjustment.

To accommodate minor variations in line pressures across a golf course, valve-in-head sprinklers and remote control valves are available (sometimes as an optional feature) with internal pressure regulation devices. These devices also require periodic maintenance and repair. To test their performance, measure the operating pressure at the sprinkler nozzle with a "pitot tube" and pressure gauge. If nozzle pressures of valve-in-head sprinklers' internal pressure regulators vary by more than 5% of the manufacturer's specified ratings, readjustment, repair, or replacement of the regulator spring and/or regulator assembly may be needed. If the system is a block design, regulators at the remote control valve (if equipped) will require adjustment.

Operating pressure should be tested at various locations throughout the golf course during a normal watering cycle if high- or low-pressure problems are suspected. Portable pressure recording devices are the preferred method to collect data over a 24-hour period; however, a simple pressure gauge mounted on a quick coupler can be used to spot check problem areas in an emergency. Operating pressure data can help identify various problems, including too many sprinklers or satellite controllers operating simultaneously, and improperly operating pressure-boosting and/or -regulating devices and/or areas where pipe was not adequately sized to sustain the required pressure at a specific site.

10.3.1.5 Sprinkler Rotation Speed

Rotation speed should be checked and recorded as another diagnostic tool. Rotation speed will vary slightly depending on sprinkler brand, model, nozzle size, stator size, operating pressure, and condition of the gear drive or impact mechanism. Rotation speed should be reasonably consistent between like brand and model sprinklers for uniform water distribution. Impact rotors should complete one revolution in approximately 2 minutes (plus or minus 15 seconds), while gear drive rotors normally complete one revolution between 2.5 and 3.0 minutes.

The stream from a sprinkler rotating too rapidly will break into smaller droplets and be affected more so by wind (Barret et al., 2003; Zoldoske et al., 1987). Additionally, rapid rotation causes the main nozzle stream to curve due to a whip-like action resulting in a smaller wetted radius that affects distribution patterns. Sprinklers rotating too slowly may pause, stop rotating, rotate erratically, or not complete a full rotation during short irrigation cycles. This can create isolated puddles or localized wet and dry areas. Contact the manufacturer for exact specifications regarding proper rotational speeds for each model sprinkler.

When rotation speeds are found to be outside of the suggested ranges, they can often be corrected. Possible causes for improper rotation speed for impact sprinklers include improperly adjusted arm spring tension, worn or damaged nozzles, worn bearings and bushings, or misadjusted or bent drive spoons. The cause of improper rotation speed with gear drive sprinklers is typically related to mismatched nozzle and stator combinations (or stator settings with adjustable type stators) or debris partially plugging the stator passageways or bottom screen.

10.3.1.6 Control Systems

Upgrading to computer-driven solid-state control systems with flow management capabilities can help to maintain proper operating line pressures by not allowing too many sprinklers or controllers to operate simultaneously (Barret et al., 2003). Flow management options also compress the watering window by optimizing sprinkler and satellite controller operating sequences. Reducing the water window, while maintaining a proper operating pressure range and flow rate, improves both distribution coverage and energy use efficiency.

10.3.1.7 Other

When preparing to conduct an audit, record specific data from the areas selected to be tested for future reference, such as the hole number, location (fairway, tee, green, distance markers, etc.), satellite and station identification numbers, general conditions of the turfgrass, whether sprinklers are at proper grade, elevation, land slopes, tree interference, and so on (Barret et al., 2003; Irrigation Association, 2003; Walker, Lehmkuhl, Kah, and Corr, 1995).

10.3.2 Catch-Can Water Audit Approach for Uniformity Evaluations

As mentioned previously, poor irrigation water distribution uniformity results in salt distribution and soil salt accumulation problems, excessive water use when leaching, and the need for additional hand-watering labor to avoid both poor turfgrass density and poor playing conditions. The traditional approach to assessing sprinkler water distribution is by the "catch-can" method. In Section 10.3.4, a more recent "water audit" approach is discussed based on detailed determination of spatial variation in soil volumetric water content (VWC) using a grid of 8 to 10 feet

(2.5 to 3.2 m). Catch-can distribution uniformity testing is a necessary but frequently overlooked maintenance test for irrigation systems. Many turfgrass managers assume their systems operate at peak performance, but very few actually take the time to perform catch-can tests to measure actual performance. During the mid-1980s, the Irrigation Training and Research Center (ITRC) at Cal-Poly in San Luis Obispo, California, developed an irrigation system assessment and landscape water management program for the California Department of Water Resources. Several golf course irrigation systems were audited to determine their low quarter distribution uniformity (DU_{LQ}) during the project (see Section 10.3.3 for a discussion of DU_{LQ}). The golf courses DU_{LQ} results ranged from 50% to 90%, while most fell between 70% and 85% (Kah and Willig, 1993).

A similar study conducted in 2002 at five Florida golf courses, all with irrigation systems less than five years of age, produced average DU_{LQ} of 50% for fairways, 57% for tees, and 60% on greens (Miller et al., 2003). The variation of results between the California and Florida examples are assumed to be related to their specific irrigation system designs. Items such as larger versus smaller spacing distances and square as opposed to triangular configurations could all come into play. However, this cannot be confirmed because the articles summarizing the studies did not report these particular data.

The process of conducting catch-can tests is not difficult and with some training should be within the skill sets of assistant superintendents or irrigation technicians. Wind speed should be less than 8 miles per hour to collect meaningful catch-can data, but 5 miles per hour might actually be a more reasonable cutoff (Barret et al., 2003; Irrigation Association, 2003; Walker, Lehmkuhl, and Kah, 1995; Walker, Lehmkuhl, Kah, and Corr, 1995). The real question that must be asked is "What range of wind speed is typical during the normal irrigation time?" If a wind speed gauge (anemometer) is not available, the safe range of wind speed can be estimated with the "upwind:downwind" ratio test. Measure the distributed "throw" of water upwind and then downwind. Calculate the ratio of upwind divided by downwind; it should be less than 0.6 to proceed with the catch-can test (Walker, Lehmkuhl, Kah, and Corr, 1995).

Catchments are placed between at least two sprinkler rows; some auditors use symmetrical patterns, while others will use random arrangements. For landscapes where sprinklers are spaced less than 50 feet apart, catch-cans are typically placed near each sprinkler head in the area being tested and halfway in between each sprinkler (commonly referred to as "at the head and in between") (Walker, Lehmkuhl, Kah, and Corr, 1995). When data that are more precise are desired, additional catch-cans can be placed on 5- to 15 foot (1.6- to 4.7-m) centers.

The Irrigation Association's (IA) Certified Golf Irrigation Auditor (CGIA) Program currently recommends a minimum of 24 catch-cans be used in each area being audited. Spacing on greens and tees is suggested to be 15 feet on center. The minimum catchment spacing on fairways is suggested to be one catchment near each sprinkler and two catchments spaced uniformly between each sprinkler (Irrigation Association, 2003). Care is necessary so that the catchments placed nearest each sprinkler do not interfere with the spray pattern and trajectory, which could thereby deliver erroneous data. Sprinklers should operate long enough to collect a minimum of 25 mL in each catchment. A rule of thumb is five revolutions of each rotor or 15 minutes of operation will typically deliver this minimum volume. Cylindrical containers with straight walls allow direct measurement with a thin ruler; noncylindrical catchments require using a graduated cylinder and conversion of the data based on the throat-opening area of the catchment. Graduated calibrated containers are available that allow measurements to be directly read.

The amount of time the sprinklers operated and the volume of water collected in each individual catchment for analysis are recorded. The measured run time (as opposed to the programmed run time) is needed to calculate the field precipitation rate and to evaluate the accuracy of the control system. Normally, catch-can tests are conducted after completing repairs and adjustments identified in the "tune-up" process; however, to document the effects of the repairs and adjustments, performing before and after tests can be worthwhile.

TABLE 10.3
Estimated Distribution Uniformity (DU_{LQ}) by Sprinkler Type and System Quality

Sprinkler Type and Application	Excellent (Achievable)	Very Good	Good (Expected)	Fair	Poor (Needs Improvement)
Multiple-stream gear and impact rotors	85%	80%	75%	65%	60%
Single-stream gear rotors	80%	75%	70%	65%	55%
Single-stream impact rotors	75%	70%	65%	60%	50%
Fixed spray heads	75%	70%	65%	55%	50%

Source: Adapted from Miller, G., N. Pressler, and M. Dukes, How uniform is coverage from your irrigation system, *Golf Course Management,* 71 (8), 100–102, 2003; Oster, J. D., M. J. Singer, A. Fulton, W. Richardson, and T. Prichard, *Water Penetration Problems in California Soils: Diagnosis and Solutions*, University of California, Riverside, Kearney Foundation of Soil Science, Division of Agricultural and Natural Resources, 1992.

10.3.3 Evaluation of Catch-Can Test Data

Results of catch-can tests are used to calculate distribution uniformity. The Low Quarter Distribution Uniformity formula is most commonly used for turfgrass applications, while the previously mentioned Christensen's Coefficient of Uniformity is more often used in agriculture. DU_{LQ} is determined by sorting all catch-can data from the lowest to highest values. The average of the lowest 25 percent of values is divided by the average of all the values (Barret et al., 2003; Irrigation Association, 2003; Walker, Lehmkuhl, and Kah, 1995; Walker, Lehmkuhl, Kah, and Corr, 1995).

$$DU_{LQ} = (\text{Minimum}/\text{Average}) \times 100$$

where

DU_{LQ} = low quarter distribution uniformity
Minimum = Average of lower 25% of catchments
Average = Average of all catchments

Based on results of audits conducted in California by the Cal Poly ITRC while developing their water management program, Table 10.3 was developed as a guideline for ranking irrigation system performance by sprinkler type. The Irrigation Association's *Certified Golf Course Auditor Handbook* offers a similar table that has been modified to reflect a nationwide area influence and climate zones that regularly receive rainfall (Irrigation Association, 2003) (Table 10.4). Note that Tables 10.3 and 10.4 distinguish how performance varies between the various sprinkler types. Also,

TABLE 10.4
Estimated Distribution Uniformity (DU_{LQ}) by Sprinkler Type and System Quality

Sprinkler Type	Excellent (Achievable)	Good (Expected)	Poor (Needs Improvement)
Rotary sprinklers	80%	70%	55%
Fixed spray heads	75%	65%	50%

Source: From Ritchie, W. E., R. L. Green, and V. A. Gibeault, Using ET_O (reference evapotranspiration) for turfgrass irrigation efficiency, *California Turfgrass Culture,* 47 (3–4), 9–15, 1997.

sprinkler performance may also vary depending on brands, models, nozzles, and the age of the equipment (Zoldoske, 2003).

Fixed spray heads that are typically used in clubhouse and residential lawns, small landscapes, and flowerbeds that are spaced 10 to 18 feet (3.1 to 5.5 m) apart are least efficient. On the golf course, fixed spray heads are occasionally found in specially landscaped areas surrounding tees, snack bars, restrooms, and so on. Matched precipitation rate multistream rotor-type nozzle retrofits are now available that can be installed in major manufacturers' spray head bodies (Walla-Walla Sprinkler Company, 2010). The multistream rotor nozzles can improve DU_{LQ} significantly (often into the mid-70% to low-80% range) on systems spaced in the 10- to 30-foot (3.1- to 9.1-m) spacing range.

Single-stream impact and gear rotors are used in medium-sized areas, and generally are spaced from 20 feet to 50 feet apart (6.1 to 15.2 m). They are most commonly used for irrigated slopes, athletic fields, larger landscape beds, medium-sized lawn areas, and agricultural use. Single-stream rotors produce moderately good to high distribution uniformity.

Multiple-stream impact and gear rotors also come in various sizes and can be used at various spacing distances (typically 50 to 100 feet, or 15.2 to 30.4 m), depending on the make and model sprinkler, and irrigation system design. They are typically used for turfgrass irrigation of larger sites, such as golf courses or athletic field complexes. Multiple-stream (multiple-nozzle) style sprinklers typically produce the highest distribution uniformity when properly matched to their application. Golf course and large-area turfgrass rotors are typically designed with multiple-stream impact and gear drive rotors. Uniformity of 80% is achievable and is a realistic expectation with a properly designed, installed, and maintained multirow golf course irrigation system. Systems performing at less than the good ranking (70%) should be evaluated for areas of potential improvement such as nozzle replacement, head lifting, pressure adjustments, and so on. Results below the "poor" ranking (55% to 60%) following a system "tune-up" indicate major repairs or upgrades, or a complete system replacement may be warranted. Under these circumstances, a more extensive system evaluation by a qualified golf course irrigation designer is suggested.

10.3.4 Precision Turfgrass Management (PTM) Water Audit Approach

In Chapter 4, Section 4.3.2 ("Approaches to Field Salinity Monitoring"), it was noted that PTM, which is based on spatial mapping of key soil and plant properties, is a new approach with field applications related to soil sampling for soils and spatial salinity mapping. In Chapter 4, the emphasis was on spatial mapping of soil salinity, but another field application is offered as a new approach to a turfgrass irrigation system water audit (Carrow et al., 2009a, 2009b). The primary problem confronting agriculture and turfgrass and landscape irrigators in terms of achieving higher water-use efficiency and conservation is how to deal with site-specific soil water variability, both (a) spatially across the landscape and within the soil profile; and (b) temporally—over time (with seasons, climatic changes, and extremes). Site-specific turfgrass management, including irrigation management, requires site-specific information in order to determine when to irrigate, how much to apply, and where to apply irrigation water only on the specific sites needed within designated time frames.

Mobile sensor technology has been used in precision agriculture to move toward higher efficiency by dealing with site variability by detailed spatial mapping of areas (Corwin and Lesch, 2003, 2005; Yan et al., 2007). Mobile spatial mapping of site conditions also has potential for precision turfgrass management, especially with respect to water-use efficiency and conservation and salinity management on complex sites with a high degree of spatial and temporal variability. Carrow et al. (2007a, 2007b, 2009a, 2009b) and Krum et al. (2010) reported on research using mobile devices developed by the Toro Company (Bloomington, MN) that are compatible with GPS and GIS technology with the most advanced device (Toro Mobile Monitoring [TMM]) capable of rapid measurement of surface zone volumetric water content (where VWC data were used to map spatial evapotranspiration patterns), turfgrass stress (the Normalized Differential Vegetative Index [NDVI] to map plant stress), penetrometer resistance (PR to map soil compaction), and electrical conductivity of bulk soil conductivity (ECa) by soil depth in order to holistically map soil salinity.

Spatial mapping of VWC 4 inches (10 cm) in the surface in a dry period when the irrigation system is applying all the plant water needs would essentially be an alternative water audit method compared to the traditional catch-can method. In this case, just as it would be for a traditional water audit, it is essential before conducting the audit that the irrigation system be evaluated for maximum performance—all components are operating and adjusted properly, proper pressure is available for the heads, scheduling is appropriate, maximum volume is attainable, and so on. With these adjustments, the irrigation system distribution capabilities can then be assessed on a wall-to-wall basis with the results reflecting true system capabilities or deficiencies (such as improper head spacing) as well as any other factor that alters the spatial distribution of VWC—wind, slope, elevation, and so on—as noted by Carrow et al. (2009a, 2009b) (Figures 10.1 and 10 2).

Additionally, spatial mapping of VWC at soil field capacity provides valuable information as to areas with similar water-holding capacity due to similar soil texture and organic matter content—these are called "site-specific management units," or SSMUs (Carrow et al., 2009a, 2009b; Krum et al., 2010). This information can be used to develop more site-specific irrigation programs to foster water conservation. Each of these field applications (alternative irrigation water audit, salinity mapping, and better irrigation programming) are important for improving salinity management via the

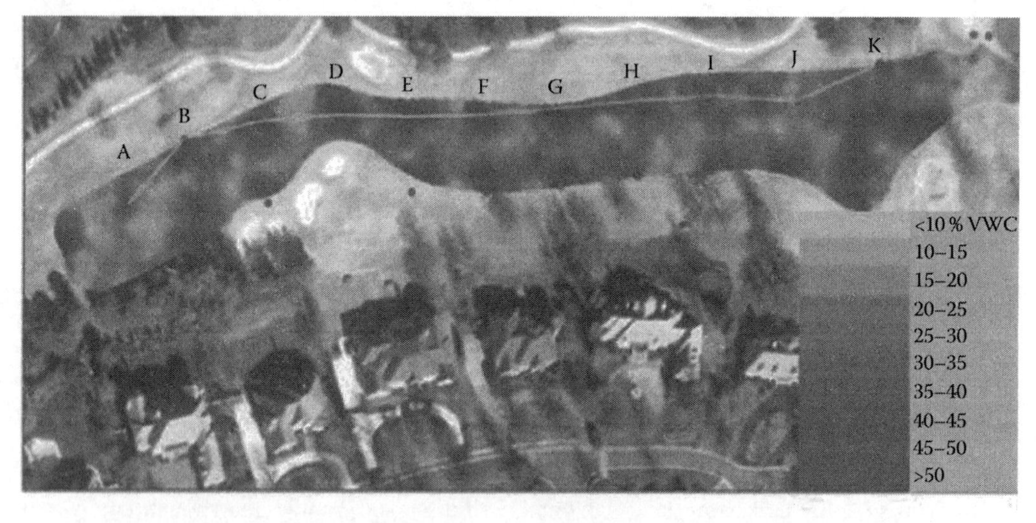

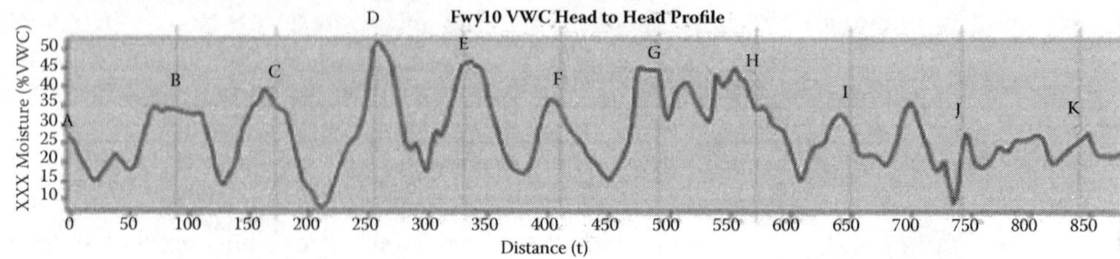

FIGURE 10.2 (See color insert.) Water audit by spatial mapping of soil volumetric water content (VWC) based on an 8- × 10-foot mapping grid using the Toro TMM mobile-mapping unit (Toro Company, Bloomington, MN). A fourfold difference in VWC is apparent. The diagram on the bottom illustrates VWC distribution along one irrigation main of this double row system with 90 feet head spacing. Spatial distribution illustrates nonuniformity of soil VWC due to lack of system uniformity of application of water. See Figure 10.3 for a different means to illustrate spatial variability in soil VWC of these data (Toro Company).

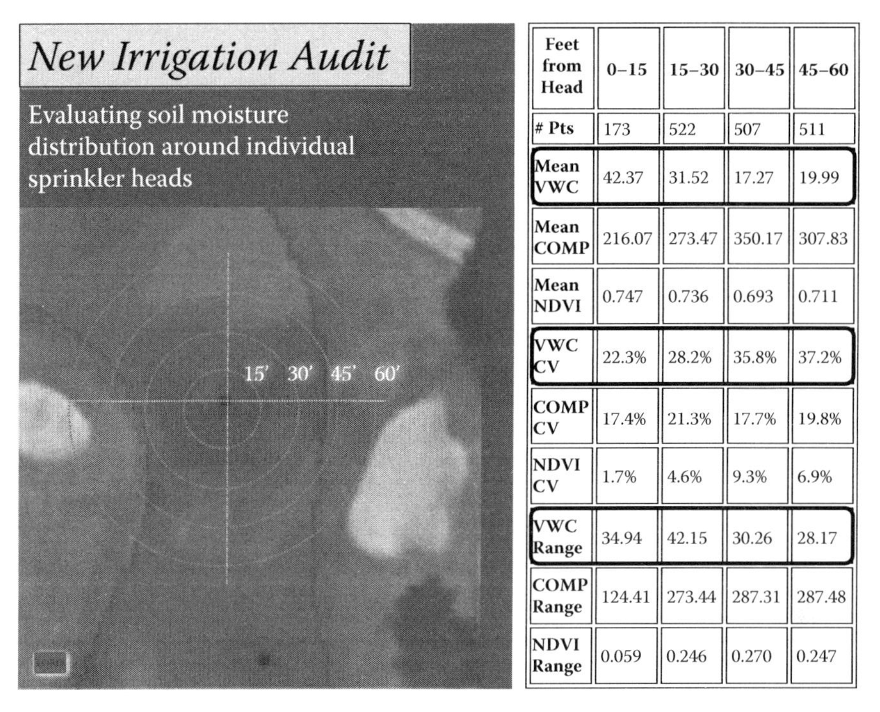

Feet from Head	0–15	15–30	30–45	45–60
# Pts	173	522	507	511
Mean VWC	42.37	31.52	17.27	19.99
Mean COMP	216.07	273.47	350.17	307.83
Mean NDVI	0.747	0.736	0.693	0.711
VWC CV	22.3%	28.2%	35.8%	37.2%
COMP CV	17.4%	21.3%	17.7%	19.8%
NDVI CV	1.7%	4.6%	9.3%	6.9%
VWC Range	34.94	42.15	30.26	28.17
COMP Range	124.41	273.44	287.31	287.48
NDVI Range	0.059	0.246	0.270	0.247

FIGURE 10.3 **(See color insert.)** Spatial mapping of soil volumetric water content (VWC) around a single irrigation head at various distances from the head as part of a new water audit approach. Map developed by spatial mapping with the Toro TMM mobile device (Toro Company, Bloomington, MN). CV = coefficient of variation of data at each distance, where a low CV is best. The same information can be developed with segments around each head. NDVI = normalized difference vegetative index, where 1.0 = ideal shoot density and turf color. COMP = penetrometer resistance data where higher values = harder soil surface.

irrigation system—both as preventative measures to limit salinity soil profile accumulation and as corrective leaching practices.

With the TMM device, detailed VWC distribution patterns can be observed within a single irrigation head zone of influence by using geostatistical approaches for interpolation of data between sample sites based on their mathematical relationships (Figure 10.3; Krum et al., 2010; Dr. Van Cline, chief agronomist, Center for Advanced Turfgrass Technology, Toro Company, Bloomington, MN, personal communication, April 2011). The irrigation head zone of influence can be subdivided into pie-shaped areas of the circle distribution to map both soil moisture and soil salinity patterns. The VWC patterns are used to determine head function and to make decisions of how to adjust for uniformity of application within the zone of influence of each head.

10.3.5 Developing Base Irrigation Schedules

The Irrigation Association's Certified Golf Course Auditing process suggests developing "base irrigation schedules" for programming the system (Irrigation Association, 2003). Although most turfgrass managers know best how to irrigate their particular site, calculating the base schedules can offer insight on developing a successful irrigation and leaching protocol intended to drive root systems to a greater depth. Base schedules are developed based on site-specific data collected during either the catch-can or spatial mapping audits. Items such as peak daily ET, soil moisture replacement, soil infiltration rates, soil moisture retention, sprinkler precipitation rates, and run time multipliers based on distribution uniformity are all considered when calculating the base schedule (Irrigation Association, 2003).

10.3.6 Other Miscellaneous Irrigation System Maintenance

Maintaining pump efficiency by optimum pump pressures and flow rates is critical to maintaining good water distribution efficiency. Regular pump testing allows comparing past and present performance to determine if operating conditions, energy use, and/or output of pressure or flow have changed due to wear of bowls, impellers, motors, and so on. Pump tests measure various operating aspects and estimate overall efficiency and power costs, while operating under the conditions of the test (Center for Irrigation Technology, 2010). Water flow rate, pump lift pressure, discharge pressure, and energy input are individually measured. Both well pumps and booster pumps should be tested every 1 to 3 years depending on annual usage and operating conditions. For example, a well pumping water that is contaminated with sand or suspended solids might be tested annually, while a booster pumping clean water might be tested only every 2 to 3 years. Public utilities, pump dealers, and independent pump-repair and -testing companies typically perform this service (Center for Irrigation Technology, 2010). In some cases, testing and repair costs may be shared between the pump owner and a local utility company because there is a mutual benefit to reduce energy use (Center for Irrigation Technology, 2010).

Maintaining air-release valves in proper operating conditions can reduce ruptured pipes and damage from water hammer. Annually or semiannually lubricating and "exercising" mainline and lateral valves by a closing and opening systems check can clean corrosion that forms on the threads of the actuator mechanism. Fabricating a long-stemmed oilcan that reaches the valve stem is suggested to treat frozen and stiff operating valves with a penetrating lubricant a few days prior to the exercising process. A qualified pump technician or electrician should annually tighten all the high-voltage (480 VAC) electrical connections, change oil in electric pump motors, and replace or repack pump shaft seals.

10.4 SITES WITH POORLY DESIGNED IRRIGATION SYSTEMS

10.4.1 Managing a Poor System Using Poor Water Quality

There will be cases where a new irrigation system is economically out of the question or will require considerable time to develop. Assuming that all reasonable measures possible have been taken to improve the distribution uniformity (DU_{LQ}) and less than desirable leaching and management capabilities still result, spot leaching with portable sprinkler equipment may be required (Kah and Willig, 1993). Low precipitation rate (so-called low flow) portable sprinklers (small nozzle, impact, or multistream rotor lawn models) mounted on portable bases can be used to leach areas of poor coverage, steep slopes, bunker surrounds, and/or native soil areas with low infiltration rates. In the most severe cases, agricultural (orchard) microspinner-type sprinklers with ultralow precipitation rates between 4.5 and 30.0 gallons per hour (approximately 0.08 to 0.50 gallons per minute) can be interconnected on lengths of flexible polyethylene tubing. These techniques are commonly used on push-up–constructed putting greens with poor internal and subsurface drainage to avoid saturating surrounds and green side bunkers with excess irrigation from full-circle green side sprinklers (Carrow et al., 2000; Gross, 1999).

Supplemental sprinkler systems, sometimes referred to as "cheater systems," composed of a few small lawn sprinklers can be permanently installed to leach and/or supplement irrigation to areas chronically lacking coverage. Installing these systems with their own manual or remote control valve can increase the flexibility of their use.

Porous pipe, also sold as "leaky pipe," and/or conventional "soaker" hoses are excellent tools to spot leach or irrigate small problem areas without disrupting golfers. Soaker hoses and porous pipe typically have low precipitation rates that are well matched for use on native soils. Recognize that low-precipitation-rate equipment (lawn sprinklers, microsprinklers, and soaker hoses) will require pressure and flow regulation if directly attached to the golf course irrigation system via quick couplers (Gross, 1999). These low-flow alternative water distribution systems are often run overnight for effective leaching of accumulated salts in persistently problem areas.

10.4.2 Economic Implications of Poor Irrigation System Designs

There is an old cliché that states, "There never seems to be enough time and money to do things right the first time, but there is always enough time and money to do things over." Too often, cutting corners to generate cost savings is applied to large-capital golf course construction projects, especially with irrigation systems. The original thinking is that a 5% to 10% savings on a $1 to $2 million expenditure is a significant amount. The decision makers often feel that stretching sprinkler-spacing distances or reducing pipe sizes to save a few percent on material costs cannot have a great impact on course conditions. Unfortunately, cutting corners on the irrigation system will be one of the greatest factors affecting the turfgrass manager's ability to succeed, especially when using poor water quality or attempting to achieve good water use efficiency and conservation. It is impossible to justify a $1 million plus irrigation renovation system on labor and cost savings alone, even when evaluated over the life of a new system. However, the long-term implication of cutting corners to save a few percent of the initial construction budget often results in additional maintenance costs that reach beyond those initial component savings. Consider the following example.

A $1.5 million irrigation system is expected to deliver a 30-year expected useful life. A 10% savings ($150,000) can be realized if smaller diameter mainline and lateral pipes are installed, and sprinkler spacing is expanded from 65 to 70 feet. However, additional hand-watering labor will be needed to manage dry areas and manually leach salinized areas that are lacking proper water distribution coverage while trying to maintain course conditions to the customer's expectation levels.

Compare those savings of $150,000 to hiring one additional $7.00 per hour laborer for 12 months. Over the 30-year useful life of the system, nearly three times those initial savings will be spent on hourly wages to compensate for system inefficiencies through hand watering and other irrigation system efficiency maintenance. This estimate does not include cost-of-living wage increases, taxes, benefits or additional materials, water, energy, and equipment associated with the hand-watering position—not to mention that any resulting decline in golf course conditions (such as those resulting in irrigation system downtime and repair) will affect revenues and harm the course's reputation among patrons.

Recognize that there is a close correlation between the number of sprinklers in the design, zoning (one- or two-head zones versus multiple-head zones), amounts and costs of materials needed, and number of labor hours needed to complete the installation. Typically, irrigation systems can be broken into three categories; each comprises approximately one third of the total initial water budget project cost.

- Labor for design, staking, trenching, pipe fitting, wire burial, wiring connections, and so on
- Pumps, pipe (mainline and laterals), fittings, swing joints, conduit, and wire (24 VAC direct burial and 110 VAC)
- Sprinklers and control systems (central computer, software, satellites controllers, and weather station)

Once installed, approximately two thirds of the total irrigation system's material costs and installation labor are both literally and figuratively buried. If undersized pipe is installed, the costs of both the pipe and installation labor have been wasted and cannot be recovered. A similar scenario occurs when sprinklers are spaced too far apart to be efficient in uniformly distributing water. The sprinkler can be recovered, but the wiring and pipe are often not worth the cost of labor involved to salvage them. This is partially why it can be less expensive to replace an entire irrigation system as opposed to salvaging portions, especially where sprinkler spacing and pipe sizing are the main problems. Experience teaches that it is more expensive to repair design flaws and problems after the fact than to install the system correctly in the first place. Or, as the previously mentioned cliché states, eventually enough time and money are found to do it over since turfgrass performance is the measuring stick!

In this chapter, we have outlined a number of irrigation system design and maintenance aspects that are affected by irrigation water quality, especially saline irrigation water. As more low-quality irrigation water is used on turfgrass sites, these considerations will become even more important, especially if the irrigation water is increasingly saline in nature. In Chapter 11, "Irrigation Scheduling and Salinity Leaching," routine irrigation scheduling on sites with saline irrigation water will be discussed as well as comprehensive strategies for effective salinity leaching; but efficiency in water use and effectiveness of salinity management start with irrigation system design and maintenance. The capability to effectively manage soluble salts in a perennial turfgrass ecosystem is only as good as the efficiency and capability of the irrigation system in concert with the ability of the turfgrass manager to optimize the utilization of that system to sustain turfgrass performance in the long term.

11 Irrigation Scheduling and Salinity Leaching

CONTENTS

11.1 ROUTINE IRRIGATION SCHEDULING: BASICS

Salinity leaching is the most important salinity management best management practice (BMP), since without excessive salt removal from the rootzone, all other turfgrass BMPs will be much less effective. Salinity leaching is very much affected by both irrigation system design and irrigation scheduling. Irrigation scheduling practices have a dual effect on salinity because poor scheduling practices can (a) increase salinity problems by fostering a greater magnitude of soil salinity accumulation and spatial variability, and (b) hinder salinity remediation by substantially decreasing leaching effectiveness and increasing quantity of water necessary for adequate leaching. Thus, in Section 11.1 ("Routine Irrigation Scheduling: Basics"), the focus is on good irrigation application practices that would apply to any site regardless of water quality, but are especially important with regular applications of saline water. In the remaining sections of Chapter 11, the emphasis is on irrigation-scheduling practices related to salinity-leaching programs.

This dual influence of irrigation scheduling is similar to what was emphasized in Chapter 10 ("Irrigation System Design and Maintenance for Poor-Quality Water"), where irrigation system design also has a dual effect on soil salinity by (a) markedly influencing the magnitude of soil salinity problems via the spatial distribution of salts in the irrigation water with greater spatial differences when the system is poorly designed and does not have good application uniformity; and (b) greatly affecting success of salinity remediation by leaching, where poor application uniformity results in hindering leaching effectiveness and increasing the quantity of water required for growing turfgrass (see also Section 11.5.1).

Good references related to irrigation scheduling include the book *Turfgrass Water Conservation* (2nd ed.) by Leinauer and Cockerham (2011), which is an excellent reference related to different turfgrass water conservation practices including irrigation system design and scheduling aspects as a supplement to the first edition (Gibeault and Cockerham, 1985). The Irrigation Association (2005) publication *Turf and Landscape Irrigation Best Management Practices* focuses primarily on the irrigation system for water conservation. Carrow et al. (2005) developed an online document outlining BMPs for water conservation on golf courses, and Carrow et al. (2009c) also presented online a template and guidelines document specifically for golf courses, which contains considerable information on irrigation-scheduling aspects.

11.1.1 THE IRRIGATION-SCHEDULING CHALLENGE: SPATIAL AND TEMPORAL VARIABILITY

Spatial and temporal variability in the numerous factors that influence turfgrass ET are the primary challenges in irrigation scheduling. Many turfgrass sites exhibit a number of *microclimates* where ET may vary across the landscape due to differences in (a) atmospheric conditions such as solar radiation, temperatures, humidity, and wind speed; (b) soil conditions such as slope, elevation changes, texture differences, organic matter content, compaction, waterlogging, presence of layers, soil depth, salinity, or other factors that may influence soil moisture retention or water infiltration; and (c) plant factors that alter ET requirements, such as grass type, canopy density, root volume, presence of trees or shrubs competing for water, management (mowing height, for example), pest injury, and others. For example, microclimate situations of many golf courses include the following:

- Greens, tees, fairways, and roughs may differ in grass type as well as canopy density, and those sites certainly differ in mowing and other management practices.
- Secluded greens with trees and shrubs on 2–3 sides and limited air drainage. Not only may the trees and shrubs block air movement, but also the turfgrass surface may be shaded at times. Often these sites have shade for part of the time each day, which further limits ET.

The problem on these sites is not excessive ET, but too little ET to dry the surface and allow better air movement into the soil as well as to limit disease activity.

- Fairways that have a consistent shade line—where shade is present from a period each day and often varies with seasonal changes. Shaded sites receive less solar radiation, and, therefore, shaded plants exhibit lower ET compared to the same plant in an adjacent full-sunlight area. Unfortunately, while shade is provided by the tree and reduces ET, the tree roots may extract any of the "saved soil moisture" in the adjacent soil profile under the turfgrass.
- Within a fairway, there may be soil texture and organic composition differences that influence soil water-holding capacity.
- Sloped areas on greens, fairways, and roughs also affect gravity-influenced surface and subsurface water runoff, thereby influencing effective rainfall and irrigation infiltration into the soil.
- Shaded tees; these often also have poor air drainage.
- South-facing slopes (in the northern hemisphere), which receive more direct solar radiation than an adjacent flat area.
- North-facing slopes (in the southern hemisphere), which receive less direct solar radiation than an adjacent flat area.

Water conservation and *water-use efficiency* require proper application of irrigation water in an efficient manner. Several decisions are involved in proper irrigation scheduling, and each of these decisions will influence the effectiveness of irrigation and how efficient water will be used—and inefficiency in water applications not only is a water conservation issue, but also will mean more salinity problems from the soil deposition of salts contained in the irrigation water. The major decisions are (a) when to irrigate—how long after the last irrigation or rainfall event before irrigation; (b) how much water should be applied to bring the rootzone moisture level up to field capacity; (c) how should water be applied to avoid leaching, runoff, or evaporation losses, where the run times and number of cycles play dominant roles in how to apply water most efficiently; (d) what time of day should irrigation be applied; and (e) whether supplemental irrigation, such as syringing or low-flow sprinklers, will be needed on specific sites due to climatic or other management restrictions.

Since site-specific irrigation requires that irrigation water requirement be adjusted to meet the demands of each microclimate, *what methods are needed to assist the turfgrass manager in site-specific irrigation?* Data on the plant, soil, and/or climatic conditions on the site (or near the site in the case of climatic-based irrigation) that influence irrigation requirements must be obtained. However, when monitoring the soil, plant, or climate to schedule irrigation, a problem arises—spatial variability. The best designed irrigation system will not apply water efficiently unless it is properly programmed or scheduled to deal with these issues within each microclimate or zone. Irrigation-scheduling tools and the budget concept of irrigation scheduling can assist in implementing the best BMP salt management program for the site.

11.1.2 Irrigation-Scheduling Tools: Climate, Soil, and Plant

11.1.2.1 Experience

Irrigation-scheduling practices are often based on the experience and site knowledge of the turfgrass manager, which are essential in BMP strategies. Indicator spots or problem areas where drought symptoms are first observed are used to aid in deciding when to irrigate, while at other times the decision may be made by an educated guess. The quantity of irrigation water applied may also be by experience where adjustments are made as needed to maintain adequate turfgrass performance. Weather conditions are normally taken into account in those management adjustments.

11.1.2.2 Climate-Based Irrigation Scheduling

Atmospheric-based irrigation-scheduling methods consist of three approaches. First, weather station meteorological models using weather station data to calculate the *estimated reference crop ET* (*ETo*), which is more precise than the older *potential evapotranspiration* (*ETp*) concept, is the most common climate-based method (Irrigation Association, 2005; Garbow, 2008; Carrow et al., 2005, 2009c). The ETo value is then adjusted to the *turfgrass or crop ET* (*ETc*) using a *crop coefficient* (*Kc*), or, more accurately, a *landscape coefficient* (K_L). The "standard" or best meteorological method is the FAO Penman-Monteith method for computing ETo and then adjusting it by a landscape coefficient to obtain ETc (Allen et al., 1998). Many golf courses and some other turfgrass sites have an onsite weather station where an estimated daily reference ETo can be calculated from climate data (Irrigation Association, 2005). Typical weather station locations are sites near the golf course or sports facility with full sunlight, flat topography, good air drainage, and no adjacent trees and shrubs.

Site-specific irrigation or precision irrigation requires that water application be adjusted from the weather station ETo to the microclimate conditions. The reference ET (ETo) must be adjusted for each microclimate site, since grass, soil type, radiation, wind, and other environmental or management conditions will differ from the weather station site. For example, for a golf course, each green must be viewed as an individual microclimate. Some states have statewide weather stations and provide ETo data for irrigation scheduling such as CIMIS (2010) in California or the AZMET (2010) system in Arizona. There are also a number of ET-based systems for onsite to local area irrigation scheduling that are determined by estimated ETo.

A second climate-based method is to use *weather pan* data to estimate ETc. With the advent of less expensive weather stations capable of connecting into irrigation controllers, weather pans are not widely used (Allen et al., 1998). Evaporation from a weather pan, *Epan*, integrates the climatic conditions at a site that influence ET. There are a number of pan types, and the Class A pan is the most common. The third climate-based method is by an *evaporometer* or *atmometer*, which is a cylindrical device with a ceramic evaporator on the top that is connected to a water reservoir (Bauder, 1999). A green canvas cap may be used to better simulate grass ET. As water is evaporated from the cylinder, the daily evaporation is measured. It can be calibrated against a weather station ETo or a weather pan Epan to provide a coefficient to make estimates of ETo.

11.1.2.3 Soil-Based Irrigation Scheduling

Interest in soil sensors for irrigation scheduling has been growing (Moller et al., 1996; Neylan, 1997; Charlesworth, 2005; Garbow, 2008; Munoz-Carpena, 2009). One reason has been the realization that real-time soil moisture status by soil depth is important site-specific information to determine when to irrigate and how much water to add. New developments in soil sensor technology now allow much more sophisticated sensor arrays and interfacing with control systems. Soil sensors are capable of monitoring soil moisture in 2–4-inch (50–100-mm) zones at multiple depths down to 1–3 feet in a real-time mode with remote transfer of the data for ease of use (Charlesworth, 2005; Munoz-Carpena, 2009).

A common question with soil moisture sensors is "What area does the sensor represent?" due to spatial variability across a landscape and within the soil (Schmitz and Sourell, 2000). Defining the boundaries of similar soil texture areas on complex sites with a turfgrass cover is difficult unless a considerable number of soil samples on a grid pattern are taken. Recently, Krum et al. (2010) and Carrow et al. (2009a, 2009b) reported on using a precision turfgrass management (PTM) approach where detailed spatial mapping with mobile devices was used to obtain soil water content at field capacity of a site to determine site-specific management units (SSMUs)—where these are areas of similar soil texture and organic matter content. Usually a complex site such as a golf course may have 5–6 SSMUs (distinct types) across its fairways with 1–3 SSMUs on a particular fairway. Once the boundaries of each SSMU type are known, soil sensors can be placed within representative

SSMU areas. Recent research by the Toro Company and Bob Carrow has carried this approach to the irrigation zone level for sites with one to three sprinkler head zones for further refinement and precision BMP measurements.

Closely associated with the question of "What area does a soil sensor represent?" is the question "What is a science-based means for sensor placement?" Mobile spatial mapping has been successfully used in precision agriculture (PA) for sensor placement. For example, Hedley and Yule (2009) used the SSMU approach with wireless time-domain reflectrometry soil sensors to irrigate a 35-ha maize field with diverse SSMUs. With this approach, the minimum number of in-place sensors can be determined; sensors can be placed based on a systematic, science-based approach; and the areas across the landscape that each sensor represents can be determined. Once the sites are located, it is important to follow good installation practices to insure accurate data. Installation must be according to guidelines to make sure good soil-to-sensor contact exists, and calibration must be accurate. Sensors that read only small soil volumes, or 1 to 3 cm away from a sensor body, require very careful installation.

The new generation of sensors and software can be made user-friendly for turfgrass situations while offering real-time data, multiple-depth moisture readings, translation of the information into useful visual formats, and the ability to electronically transfer data to remote sites by wireless means. These attributes can be useful for documenting the need to irrigate a site and for maintaining a history of soil moisture statue.

11.1.2.4 Plant-Based Irrigation Scheduling

Turfgrass managers often use plant symptoms to aid in irrigation scheduling with the "remote sensor" being the eye. On many sites, especially complex ones, observation of indicator areas where drought stress symptoms first occur is a common practice. As a plant is exposed to drying conditions and soil moisture is not replaced, *drought stress symptoms* will be exhibited, such as follows: (a) first, *growth rate decreases*, which is most noticeable in terms of shoot growth, but rhizome, stolon, and root growth and regrowth rates are all reduced; (b) *wilt symptoms may follow*, which is shown as leaf folding or rolling, a bluish-green color, and footprinting—the lack of turgidity by the shoot tissues results in plant tissues being pressed down and leaving an impression when stepped on or equipment is run over the surface; (c) *leaf firing follows*, which is chlorosis (yellowing) of the turfgrass as chlorophyll pigments in the plant decline and eventually tissue desiccation results, where desiccation may start as leaf tip injury and progress down the leaf, especially for young, actively growing tillers; and (d) *tissues desiccate*, where the turfgrass canopy starts to turn tan as dead leaves become dominant. As these symptoms progress, the grass normally will go into a *dormant state*, where the only living tissues are the crown, rhizomes, stolon stems, and some of the roots.

The ideal plant-based method would be to determine when the plant is going into a drought-stressed condition before it could be visually apparent. Throssell et al. (1987) were the first to use infrared thermometry to develop irrigation-scheduling indices on turfgrasses, based on measuring canopy temperature, air temperature, and relative humidity. Since then, others have investigated the potential for using canopy temperatures for turfgrass irrigation scheduling, but with limited success. However, this method is useful in identifying "hot spots" on greens or other turfgrass areas.

Multispectral reflectance has been used for plant-based monitoring of stress, including possible use for irrigation scheduling in precision agriculture (Frazier et al., 1999). Trenholm and colleagues (Trenholm, Carrow, and Duncan, 1999; Trenholm, Duncan, and Carrow, 1999) demonstrated that multispectral reflectance models could be used to assess stresses on turfgrasses, but not specifically moisture stress. To date, the focus has been on measuring light reflectance from the turfgrass canopy within the 350–1100-nm wavelength region, which includes the visible and photosynthetically active radiation (PAR) region of 400–750 nm (Geuertal and Shaw, 2004). Reflectance devices are available that can determine reflectance within a narrow-band (2–3 nm) or broadband (10–20 nm) mode. Loss of color and/or leaf area can increase reflectance within certain wavelengths that may be used in models to estimate overall plant stress, irrigation need, or perhaps a nutrient stress, such as N. The most common spectral reflectance approach has been to determine the Normalized Difference Vegetative Index (NDVI), which is a stress index that illustrates stress symptoms but not the actual cause (Krum et al., 2010).

Since turfgrass areas must be frequently mowed, it is possible to gather plant-based spectral data on a frequent basis, at the ground level, and incorporate information into GPS or GIS systems. Thus, plant-based sensing technology using GPS and portable sensors will likely become another tool for site-specific salt and drought stress information that may aid in more efficient, precision irrigation. For example, drought stress patterns after a good rainfall could help identify sites that exhibit moisture stress first—then field observation could be conducted to determine the cause, such as a sloped area, a south-facing slope, soil profile problems, elevation changes, an area with shallow rooted grass, and so on (Krum et al., 2010). Or, during a prolonged dry period where only irrigation water is applied, plant-based reflectance models may help to map irrigation distribution patterns. Then those sites that are underwatered or overwatered can be investigated with respect to specific reasons, and possible corrective measures can then be implemented.

11.1.3 Budget Concept of Irrigation Scheduling

Regardless of the irrigation-scheduling technology or tools used, the turfgrass manager should have a science-based irrigation concept that entails a "whole system" mind-set to irrigation. An easily understood irrigation approach is the *budget concept* to foster water conservation on a whole-systems basis (Figure 11.1) (Gibeault and Cockerham, 1985). The budget approach is a useful way to visualize turfgrass water management, similar to a bank checking account. Certain additions (inputs) of moisture are made, and there are losses (outputs) of moisture from the plant environment. At any point in time, the plant has available to it a certain reserve of available water in the soil within the plant's rootzone. The objectives of a wise turfgrass manager are to carefully review each component in order to maximize inputs, minimize outputs, and maintain a large soil moisture reserve. The budget approach can be a means to categorize BMPs for water conservation, but can include listing specific practices that enhance inputs, minimize outputs, and foster an extensive root system to capture rainfall more efficiently.

11.1.3.1 Inputs

Inputs of moisture are precipitation, overhead irrigation, dew, and capillary rise of moisture from below the root system. Precipitation and overhead irrigation are the major inputs. Normally, capillary movement to turfgrass roots from below the rootzone is minor except where a water table is within 2 to 4 feet of the roots. While the turfgrass manager cannot control natural precipitation, irrigation can normally be controlled with respect to when to apply water and the quantity applied during a specified time frame. Also, to insure effective *infiltration* of irrigation and rain events,

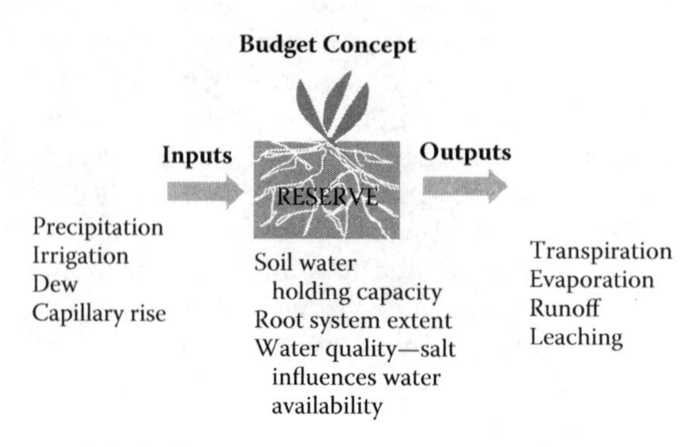

FIGURE 11.1 Components of the *budget concept* of turfgrass water management.

cultural practices to enhance soil infiltration are often necessary such as cultivation, topdressing, *pulse irrigation* (discussed in a later section), and proper water application rates to match the soil infiltration capacity with minimal runoff.

11.1.3.2 Outputs

Outputs or losses include runoff, leaching beyond the rootzone, evaporation, and transpiration. Reducing undesirable losses or outputs will retain more water for future use.

Runoff can be a problem on sloped sites and can be increased by fine-textured soils, thatched turfgrass zones, compacted soils, and applying water faster than the soil can receive it—thus, adequate infiltration and practices to enhance infiltration will reduce water loss by runoff. Runoff causes not only a dry site but also an excessively moist site in lower terrain areas. Reducing runoff requires correcting the above situations through cultivation, thatch control, or proper irrigation application rates.

Leaching or water movement beyond the root system is often an unrecognized water loss. Irrigators whose watering is based on the driest site often overirrigate other areas. Irrigating slightly beyond the existing root is acceptable because it provides a moist zone for further root extension. To reduce leaching losses, the irrigator must know the depth of rooting and depth of moisture penetration after applying a specific quantity of water. Well-designed irrigation systems that apply water uniformly reduce leaching losses. Also, proper zoning of irrigation sprinkler heads is important. Heads in similar areas should be zoned together. Poor zoning, with heads on slopes and low spots zoned together, results in poor uniformity of water application and minimal surface infiltration of waterlogged areas.

Evaporation is the vaporization of water from a surface. When moisture evaporates, it removes energy (heat) from the surface. Thus, evaporation helps cool the soil and plant if free water, such as dew, is on the leaf surface. Excessive evaporation is wasteful. Growers can control the quantity of water lost by evaporation. For example, immediately after irrigation, evaporation rates from the soil surface are high, but as the surface dries, evaporation dramatically decreases. Thus, light frequent irrigation results in high evaporative losses contrasted to heavier, less frequent applications. Other ways to reduce evaporation are to maintain good infiltration rates to get the moisture into the soil; maintain a dense turfgrass canopy to shade the soil surface; mow your turfgrass as high as feasible for your situation to insure further shading; avoid applying so much water that standing water occurs; and avoid afternoon irrigation, when ET is normally occurring at the highest rate.

Transpiration is the vaporization of water inside the plant leaf that diffuses through the cuticle and through the open stomata, which are pores on the surface of leaves where most transpiration occurs. During this process, heat is removed from the plant. In many situations, more than 90 percent of the moisture taken in by a turfgrass plant is utilized for cooling purposes. Transpiration is a desirable use of water, especially in hot environmental conditions. However, excessive transpiration can occur and thereby waste water. Overwatering turfgrass promotes excessive transpiration. The other transpiration variable in turfgrass situations is the genetically controlled transpiration rate that varies with species and cultivars.

11.1.3.3 The Reserve

The reserve of plant-available moisture at any point in time depends primarily on *soil water-holding capacity* and the *volume of soil occupied by the plant root system*. Obviously, over a period of time, irrigation and precipitation are the sources of the reserve moisture. Soil texture and organic matter influence "water-holding capacity," and, as a generalization, sands do not retain as much plant-available moisture compared to loam soils. Turfgrass growers can markedly improve the moisture reserve by managing to promote development and maintenance of a good, deep, extensive plant root system. This is especially important in humid and semiarid regions, where a deep root system aids in capturing rainfall and, thereby, can result in delaying irrigation events. Turfgrass breeders in some locations have placed considerable emphasis on the development of grasses that can tolerate the soil stresses that limit rooting and on grasses that can better maintain their roots during the hot, dry summer months. This approach has been developed and articulated by Duncan and Carrow (1999) and Carrow and Duncan (2003), where they identified key soil physical and chemical stresses that directly limit rooting development and longevity. Genotypic differences in root tolerance to each of these soil stresses exists and can be used to develop more drought-resistant cultivars within a species where the key root-limiting abiotic stresses are as follows:

- High soil strength
- Acid soil complex—a combination of acid soil pH (pH < 4.8) that results in Al/Mn toxicity to roots, nutrient deficiencies (Ca, Mg, and P), and usually high soil strength
- Low soil oxygen—from waterlogging, compaction, and soils with limited macropores
- Soil drought—different genotypes of a species can tolerate soil drying without root tissue loss better than others
- Excessive Na—causes root deterioration such as Na toxicity to roots by displacing Ca in root cells
- High soil temperatures—in combination with high air temperature, results in excessive carbohydrate loss in summer months for cool-season grasses

In arid climates with little rainfall to capture, deep rooting is less important (Brede, 2000). In these climates, the goal is to have sufficiently deep roots for a relatively deep and infrequent irrigation scheduling. However, since ET losses from the turfgrass rootzone start at the surface and progressively work downward, irrigation frequency and amount depend on the degree of surface drying and consequences on turfgrass quality; the creation of localized dry spots; excessively hard surface (soil strength); and the ability of the irrigation system to practically apply replacement ET losses back to the depth of water extraction. Maximum rooting may not be necessary, but reasonably deep rooting is critically important and essential for these roots to maintain viability. Thus, selecting adapted grasses with enhanced root development, redevelopment, and functional viability that are suitable to humid, semiarid, or arid situations, as noted above, is an important way to "increase the water reserve." A turfgrass with a 12-inch (30-cm) root system will have potential exposure to twice the quantity of plant-available moisture as one with only a 6-inch (15-cm) root system.

11.1.4 Pulse Irrigation

Pulse irrigation (also called *cycle and soak* and *repeat cycles*) is where water is applied in increments generally in the range of 0.20 to 0.33 inches (5 to 8 mm) with a time interval before the next pulse, and this cycle is repeated until the desired total quantity of water is applied. Each cycle limits the quantity of water to avoid runoff and saturated surface conditions. Instead, the surface soil moisture conditions result in unsaturated infiltration and percolation of the applied water, where water moves as a more uniform wetting front across both macropores and micropores. Runoff from the soil surface is minimized, and uniformity of application is maximized. The pulse irrigation method simulates a light, continuous rainfall that applies water at less than the soil's saturated infiltration rate. Such rainfall events are very effective in salt leaching.

Duncan et al. (2009) discussed this concept as related to salinity leaching when saline irrigation water is used. Pulse irrigation is essential for success in two major management challenges: (a) achievement of water conservation and water-use efficiency, and (b) achievement of effective salinity leaching. In Section 11.3 ("Maintenance Leaching and the Leaching Requirement"), pulse irrigation will be discussed in more detail. However, this brief introduction to the pulse irrigation concept demonstrates its importance for both of the above goals.

11.2 FACTORS AFFECTING SALINITY LEACHING

11.2.1 An Overview of Salinity Leaching

Salinity stresses result from the addition of excess total soluble salts and/or sodium in the irrigation water to the soil, and their subsequent accumulations are dominant salt stresses. Unless these are controlled, all other management practices not targeted to alleviating salinity stress, but used in an attempt to mask salinity, (a) cannot compensate for these stresses, and (b) will not result in the degree of grass performance response as would occur on a non-salt-affected site. Salinity management is essential, and the core of BMP salinity management is leaching. Leaching is the single most important management practice for alleviating or preventing soil salt accumulation stresses on turfgrass sites. A thorough understanding is essential concerning the following:

- Knowledge of factors that influence salinity leaching (Section 11.2)
- Types of leaching programs (i.e., maintenance vs. reclamation leaching programs) (Sections 11.3 and 11.4)
- Practices to enhance leaching effectiveness (Sections 11.5 and 11.6)

The total water requirement when a saline site is irrigated consists of the sum of three water needs: (a) sufficient water to replenish ET losses since the last rainfall or irrigation event in order to bring the soil back to field capacity status; (b) some additional water to compensate for nonuniformity of water application by the irrigation system; and (c) an additional quantity of water to prevent accumulation of excessive soluble salts to a level injurious to the grass, where this is called the *leaching requirement* (LR). The original definition of LR was the fraction of infiltrated water that must past through the rootzone to keep soil salinity from exceeding a level that would significantly reduce crop yield (USSL, 1954). This was modified by Rhoades (1974) into the form now considered as the traditional LR (Table 11.1). The LR concept will be discussed in greater detail in Section 11.3, "Maintenance Leaching and the Leaching Requirement," but for now, it is sufficient to note

TABLE 11.1
Determination of the Maintenance Leaching Requirement (LR)

Concept:

Once the soil salinity level in the turfgrass rootzone is at an acceptable or desirable level, the leaching requirement (LR) approach is used to maintain this level. The *leaching requirement* (LR) is the minimum amount of water that passes through the rootzone to control salts within an acceptable level. A traditional formula to determine LR is as follows:[a]

$$\mathbf{LR = EC_w / 5EC_e - EC_w}$$

where

EC_w = irrigation water salinity (dSm^{-1})

EC_e = threshold soil salinity at which growth starts to decline for the turfgrass on the site (see Chapter 9 or Duncan et al. [2009] and for an extensive listing)

Example:

Consider a turfgrass with a threshold EC_e of 6 dSm^{-1} and irrigation water with an EC_w = 2 dSm^{-1}, which means that the LR is 7.1% more irrigation water volume than to meet ET needs. Thus, if irrigation of 1.00 inch of irrigation water is required to replace soil moisture lost by ET, an additional 7.1% or (1.00 × 0.07) = 0.07 inches of water would be required for a total of 1.07 inches to maintain salinity conditions. An additional quantity of water would be required to compensate for nonuniformity of the irrigation water. It should be noted that a *more saline irrigation water* with higher EC_w or a *less salt-tolerant grass* would both *increase the LR*.

Source: Adapted from Rhoades, J. D., and J. Loveday, Salinity in irrigated agriculture, in B. A. Stewart and D. R. Nielson (Eds.), *Irrigation of Agricultural Crops* (Agronomy No. 30, pp. 1089–1142a), Amer. Soc. of Agron., Madison, WI, 1990.

that irrigation water quality (salinity level) and turfgrass salinity tolerance are the two major factors in determining LR. The traditional LR refers to a *maintenance-leaching program* to maintain salt levels below a critical level and not to a *reclamation-leaching program* to leach salts from an already salt-laden soil down to an acceptable level—these two situations are also discussed later in the chapter.

When the irrigation water contains appreciable salts, turfgrass managers must develop a mindset of *keep the salts moving downward*, and management decisions must be made in the context of how soil salt accumulation levels or movements are affected (Figure 11.2). For example, switching to a light, more frequent irrigation regime is a common practice on bentgrass or annual bluegrass putting greens in the summer months, but this practice results in salt accumulation at the surface and creates the potential for rapid capillary rise of salts from lower in the soil horizon. Or, during winter dormancy periods of warm-season grasses, managers may not realize that a dry winter can cause appreciable capillary rise of salts from salt-laden zones deeper in the soil, resulting in poor spring turfgrass greenup and overall performance. Thus, the salt management strategy should be one of managing salts before, during, and after managing the turfgrass. Short-term decisions based on convenience or speed can lead to future long-term headaches and poor turfgrass performance. This is one area of turfgrass management where cutting corners or expecting miracle cures is not going to work. Stay with the basics and base your management decisions on

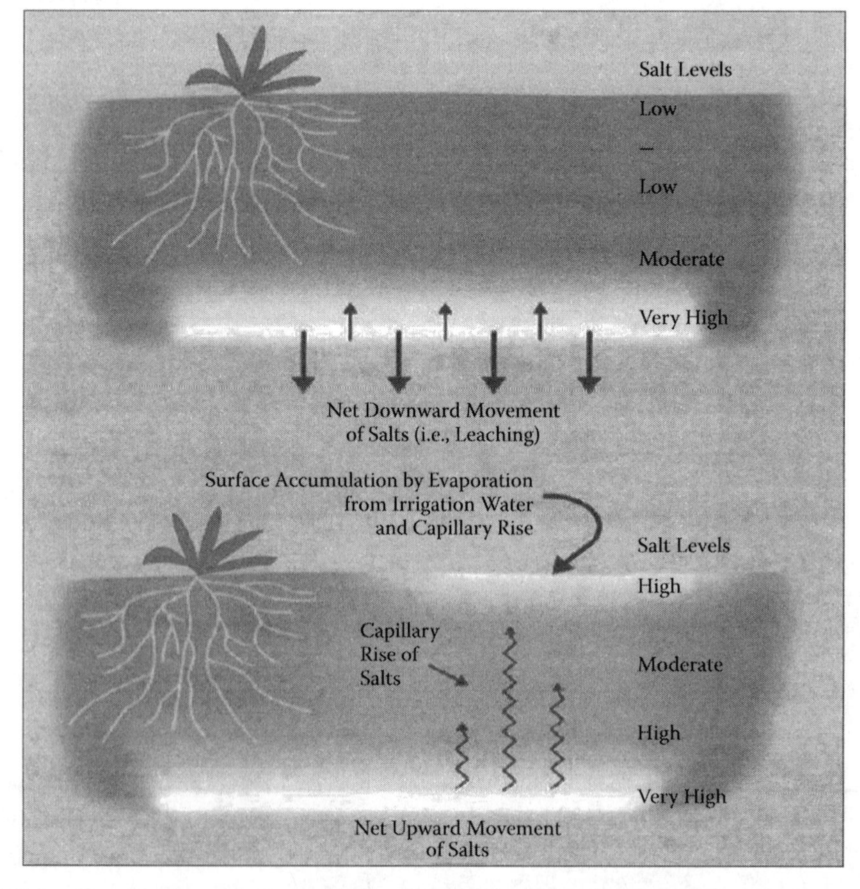

FIGURE 11.2 **(See color insert.)** Examples of salt levels throughout the soil profile. Top: Represents good leaching conditions with adequate leaching requirement (LR) applied. Bottom: Represents what happens when insufficient water is applied in midsummer with high evapotranspiration (ET) conditions. (From Carrow, R. N., M. Huck, and R. R. Duncan, Leaching for salinity management on turfgrass sites, *USGA Green Section Record,* 38 (6), 15–24, 2000. With permission.)

science and not testimonials, magic potions, or silver-bullet products or equipment. No single product or piece of equipment will solve or alleviate the multiple salt stresses that impact turfgrass ecosystems.

While the salinity management and -leaching principle is simple, achieving an effective leaching program that keeps salts moving past the rootzone is complex. Prior to discussing development of leaching programs, it is important to understand the various factors that influence the decision-making process. Salinity management is influenced by a number of factors, such as salt types and spatial distribution, soil properties, irrigation water quality, hydrological conditions, plant type, climatic conditions, irrigation system design, and irrigation-scheduling methods (Carrow and Duncan, 1998; Hanson et al., 1999; Yenny, 1994; Duncan et al., 2009).

11.2.2 Salt Type and Spatial Distribution

11.2.2.1 Salt Type: Soluble Salts versus Sodium

Development of an effective salinity management program starts with understanding which salt problems are present that may require leaching—the presence of high total soluble salt levels, high soil Na, potential for B accumulation, or a combination of any of these limitations to grass sustainability. Irrigation water quality tests will aid in determining the potential for developing each of these issues, while soil tests will show the current soil status with respect to each salinity problem. More detailed information on these specific salinity problems is presented in Chapter 4, "Salinity Soil Tests and Interpretation," and Chapter 6, "Irrigation Water Quality Test, Interpretation, and Selection."

Soluble salts: *high total soluble salt is the most common and injurious salt problem* (i.e., saline or saline-sodic soil) affecting turfgrass. The salts causing injury are soluble and result in high salt concentrations in the soil solution. When total soluble salt reaches excessive levels in the rootzone, turfgrass water uptake is reduced (osmotic potential inside the turfgrass root is greater than the osmotic potential outside the root), resulting in osmotic desiccation, or what is sometimes called *physiological drought* or *salt-induced drought*. As the stress continues, grasses often start to exhibit chlorosis, leaf tip necrosis, desiccation of lower leaves or individual tillers, and a decline in canopy density and quality (Carrow and Duncan, 1998; Yenny, 1994). Advanced canopy stress damage can include discoloration, such as yellowing, purpling, or browning depending on the individual turfgrass species and specific cultivar. These symptoms are often mistaken for disease problems, which may be false or, in some cases, predisposition to diseases stimulated by growth-rate-weakened turfgrass (Yenny, 1994). In other words, the primary problem is high total salinity and the secondary problem is the emergence of a visible disease challenge on the less aggressive turfgrass when environmental conditions are favorable for pathogen attack.

Because these salts are soluble, the bulk of the salts are in soil solution under well-irrigated soil moisture conditions; and removal of these salts only necessitates a sufficient volume of water and time to effectively promote downward movement and leaching to an acceptable level. Soil amendments to compensate for excessive Na (gypsum) or irrigation water treatment (acidification or gypsum) are not required when managing total soluble salts. Only sufficient irrigation water or rainwater (or a combination of the two) to meet ET needs plus irrigation to compensate for non-uniformity of the irrigation system plus the LR are necessary to leach rootzone soluble salts to an acceptable level (hopefully to the drainage lines) for the grass on the site. Thus, only sufficient water (quantity) applications are needed during this management stage, while soil or water amendments that are necessary for sodic situations will not improve total soluble salt movement. Cultivation (spiking, slicing, needle tine, core tine, Hydroject, Dryject, etc.) enhances this salt management strategy. However, a wetting agent that fosters a more uniform wetting front downward movement is an example of an amendment that could assist in more effective leaching and also helps to promote downward soluble salt movement between aeration holes.

With sufficient water moving through the soil, leaching may require from <1 to 4 weeks for reclamation purposes, depending on the soil texture, clay type, irrigation water quality, climatic conditions, total accumulated soluble salts in the soil profile, and other factors. In the case of maintenance leaching (see Section 11.3), the LR is to maintain adequate salt movement levels on a continuous basis with each irrigation cycle. However, accumulation of excessive soluble salts can also rapidly reappear due to elevated ET conditions, high salt additions from irrigation water without ample leaching, or soluble salts moving by capillary rise from below the rootzone upward into the root and crown area. Short duration, frequent irrigation, and syringing with saline water contribute to this phenomenon. Salt load in the soil profile can double with each cycle using this irrigation application strategy.

An especially noteworthy association with high soluble salt stress arising from irrigation water applications will be soil fertility and plant nutrient challenges (see Chapter 16, "Nutritional Practices on Saline and Sodic Sites"). The diverse chemical constituents in the irrigation water can easily create nutrient imbalances. When coupled with leaching programs that can leach desirable nutrients and variable water quality over the year (arising from changes in irrigation water quality or simply a wet and dry season where rain or lack of harvested rainfall influences leaching and irrigation lake quality), soil fertility and plant nutrition programs become much more dynamic—changes occur more frequently and to a greater magnitude compared to similar situations with good irrigation water quality. In general, if you are effectively leaching soluble salts, you are also leaching essential macro- and micronutrients through the rhizosphere.

In summary, relative to high soluble salts, this is the most serious and widespread salinity problem. Since the most common sources of soluble salts are the irrigation water and the frequency of irrigation cycles, the salt stress is distributed across the whole irrigated landscape. Soluble salts restrict water uptake and induce drought or desiccation stress on the plant; this process is the most easily controlled and rapidly controlled of the "salinity" problems. Soluble salts can accumulate or rapidly reappear under the right environmental and microenvironmental conditions, and dynamic nutritional challenges will be associated with this salinity stress. With these characteristics, it is easy to understand why total soluble salt stress is considered the *dominant salt stress* on sites where it is present and why it is a dominant stress since both water and nutrient uptake are adversely affected.

Excessive soil Na levels can lead to soil structural deterioration (i.e., sodic or saline-sodic soil) and to specific ion toxicity in shoot and root tissues (Carrow and Duncan, 1998; Duncan et al., 2009). The sodicity (sodium-rich) component, also termed *Na permeability hazard*, is measured by the soil sodium adsorption ratio of the saturated paste extract (SAR), the SAR_w (SAR or adjSAR of irrigation water), and the residual sodium carbonate (RSC) value of irrigation water (Duncan et al., 2009). It is not unusual for high total soluble salts and excessive Na to be present on a site, especially if the irrigation water contains Na as one of the dominant soluble salts. However, it is also possible for a sodic problem to arise from irrigation water with an unusually high ratio of Na to other salts, even when the total soluble salinity is modest; or, more commonly, when excessive bicarbonates cause precipitation of soluble Ca and Mg from the irrigation water as lime, thereby leaving Na as the dominant ion without a counterion (available and soluble Ca) to displace it from the soil CEC sites.

Repeated application of irrigation water containing excessive Na can result in increased Na on the CEC sites and formation of Na carbonate precipitates in the soil. Sodium can then cause structural breakdown by slaking, dispersion, and deflocculating processes (see Chapter 3, "Sodic, Saline-Sodic, and Alkaline Soils"). As clay content increases and/or the clay type is a 2:1 clay, these processes are more pronounced. The 2:1 clay types allow more moisture between clay plantlets as Na starts to dominate, resulting in more rapid dispersion. Soil structure deterioration from excess Na^+ on the soil colloid (clays, and colloidal organic matter) exchange sites causes a decline in water infiltration, percolation, and drainage (i.e., the reason for the term *Na permeability hazard*) as microporosity increases at the expense of macropores; low soil O_2, which further limits rooting;

waterlogged and poorly drained soils; sometimes black layer symptoms; negative nutritional availability problems; and sometimes surface moss or algae accumulation problems.

Leaching of Na requires additions of a relatively soluble Ca source to displace the Na from the soil cation exchange sites so that the Na comes into solution (usually as sodium sulfate) and can be leached with an adequate LR quantity of water (Carrow and Duncan, 1998). A soluble Ca source should be added whenever leaching with Na-laden irrigation water. If available Ca is not added, the Na problem can be compounded by the leaching of most of the remaining Ca in the soil profile, allowing replacement with Na supplied by Na-laden leaching water on the CEC sites, and therefore causing a complete sealing at the soil surface.

Compared to removal of high levels of soluble salts, a much longer time period will be required for remediation of sodic conditions, and considerably more water must move across the soil profile to move Na during the leaching process. The Na on CEC sites and in Na carbonate complexes must exchange or dissolve into solution over time and can reform if insufficient Ca and/or leaching is practiced. Generally, for the reclamation of a Na-affected (highly sodic) site, one or more years may be required to alleviate the Na-induced soil structural problems. The long time period is due to these ion exchange processes as well as poor physical conditions (layering and sealing) for leaching on an already Na-affected site. Additionally, the high Ca amendment rate and necessity of the Ca to be in contact with the Na-affected CEC sites and Na-carbonate throughout the rootzone contribute to the slow process of altering already sodic or pre-sodic sites. If an acre-foot of soil is considered to weigh 4 million pounds, 3400 lb./A (3808 kg/ha) of gypsum would be needed for each meq/L exchangeable Na to reclaim 1 foot of that soil. Obviously, proactive prevention of a sodic condition from forming is more important and less expensive than reclaiming a sodic soil. Highly sodic soils, especially on fine-textured soils, will not support a proper turfgrass canopy, even with halophytic turfgrass species. Do not expect highly salt-tolerant turfgrasses to remediate salt-laden soil profile infrastructure problems, especially sodic and saline-sodic stresses.

In summary, when irrigation water is a contributor to potential sodic soil formation, the key issues relative to a leaching and management program are as follows: sodic soil stresses do not form as rapidly as high-soluble salts; when formed, they require a considerably longer time frame for correction; leaching alone is not sufficient, but a relatively soluble and continuously available Ca source must be regularly applied to provide a displacement ion for Na; acidification of irrigation is often necessary if high water bicarbonates result in appreciable Ca and Mg precipitation as lime that leaves Na as the dominant ion; and nutritional challenges are even more dramatic and dynamic than for high-soluble salts since soil amendments, water treatments, and more limited roots influence soil fertility and plant nutrition along with the inherent irrigation water constituents and leaching programs.

Toxic soil levels of the salt B is another salt-related problem that requires leaching. The B often arises from the irrigation water source. Since B is absorbed to soil particles, two to three times the leaching water volume is necessary compared to the quantity needed for removal of total soluble salts. Boron leaching is generally more effective with acidic pH conditions. In conjunction with leaching, collection and offsite disposal of clippings can assist in reducing B since it is accumulated in turfgrass leaf tips. This strategy can also be used with total salt and sodium problems as a supplemental method of salt reduction with some turfgrass species, but landscape plants will generally need to be replaced when they accumulate high levels of Na and Cl. Halophytic turfgrass species, such as seashore paspalum, are not phytoaccumulators of excess salts such as sodium due to their exclusionary regulation of uptake of that salt ion (refer to Chapter 9).

11.2.2.2 Spatial Variability of Salts

As noted in Chapter 4 (Section 4.3, "Field Monitoring of Soil Salinity"), salts can exhibit great variability—horizontally across the landscape, vertically in the soil profile, and temporally over seasons. Identification of the specific stressed areas with total soluble salt accumulation at levels sufficient to cause concern for plants is critical in developing leaching programs that are effective (i.e., leaching programs that leach the salts, but with the least quantity of water and by avoiding overirrigation or underirrigation across the landscape).

One additional consideration for spatial salt accumulation, especially in fine-textured native soils, is the tendency for salts to accumulate in mainline and lateral irrigation line trenches. If those mainline trenches are positioned within fairways and are not paralleling the edge of the fairways or roughs, salt accumulation over months and years can lead to salt seepage areas (especially if there are any leaks in the mainline or lateral lines) and salt scald areas where the salts rise through capillary micropores to the surface with prolonged high ET events. Interception drains to collect subsurface (and surface) salt migration is one strategy; drilling mini–dry wells and filling the strategically positioned holes with gravel or coarse sand is another strategy.

11.2.3 Soil (Edaphic) and Hydrological Factors

Major differences in soil properties are especially apparent when comparing *sandy soils* (i.e., sands, sandy loams, and loamy sands) *versus fine-textured types* (i.e., those containing appreciable silt and clay) (Duncan et al., 2009). For example, on golf courses, sandy soils are typical of high-sand greens, while fine-textured types are representative of pushup greens (native soil greens), fairways, roughs, and many tees. Athletic fields may be either sand media or native soils. A number of soil characteristics that differ between coarse- and fine-textured soils profoundly influence salt and water movement plus retention and, therefore, effective leaching practices.

11.2.3.1 Cation Exchange Capacity

Cation exchange capacity (CEC), the ability of a soil to retain cations, is much higher for fine-textured soils compared to sands since CEC sites reside on clay particles and organic matter. As a result, fewer total soluble salts, Na, or B are required before CEC sites of sands are adversely affected compared to fine-textured soil CEC sites; and these salts start to accumulate in the soil solution where they are more active. While salts accumulate more rapidly to adverse levels in sands, removal by leaching is also generally more rapid due to normally high infiltration and percolation rates.

In terms of sodium accumulation, ESP on CEC sites can more rapidly increase on sandy soils, but the adverse effects on soil permeability may be less evident compared to a fine-textured soil since the latter are more prone to structural deterioration. On sandy soils, Na can cause colloidal matter (clay or organic colloids) to disperse and perhaps move downward with the potential to layer at some point in the soil profile (often at the bottom of the wetting front). This is primarily a problem in arid regions where the irrigation-scheduling programs tend to be more standard and the wetting front moves to the same soil zone with each irrigation application where colloids and any precipitated carbonates may become localized over time with associated salt accumulation. This process generally takes one or more years before any inhibition of water percolation may become evident. In contrast, on a fine-textured soil, Na may take a few weeks before adverse results are evident unless the Na load is very intensive, such as with seawater flooding or reoccurring salt spray on coastal sites.

11.2.3.2 Soil Pore Size Distribution

Pore size distributions within a soil are affected by texture, structure, inorganic amendments, and organic matter content, and have a major influence on water movement, water retention, salt movement, salt retention, and soil aeration. Data in Table 8.1 describe the various pore sizes. *Macropores* (aeration porosity) are soil pores with a diameter from >0.075 to 0.10 mm (depending on the reference), and these pore sizes are much more prevalent in sands than fine-textured soils, while in fine-textured soils *micropores* (<0.75 mm diameter, capillary pores, and moisture retention porosity) are more dominant. Macropores are critical for rapid water movement into the soil surface (*infiltration*), through the rootzone (*percolation*), and beyond the rootzone (*drainage*). When there is free water

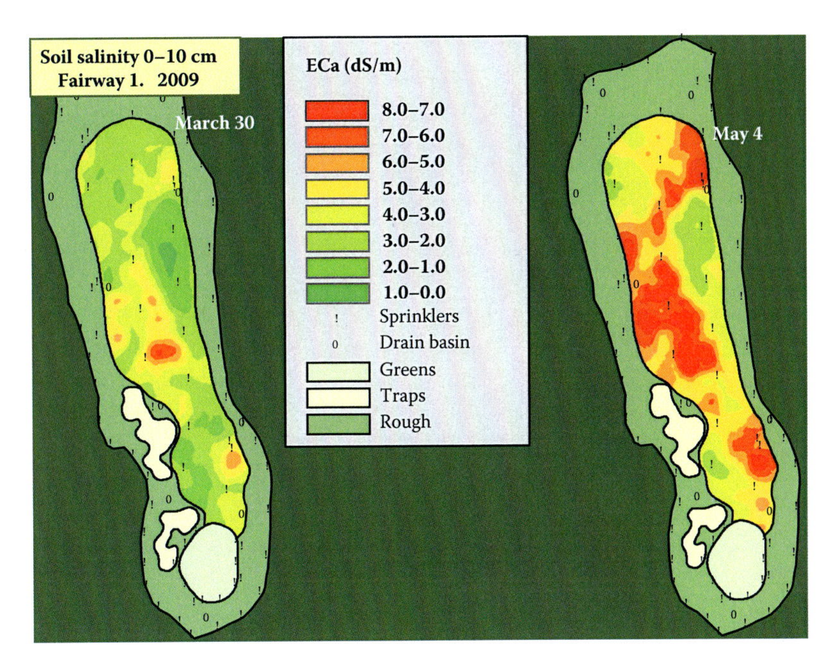

FIGURE 4.1 Soil salinity mapping with a mobile four-wenner array device on 30 March and 4 May 2009 during a period of very little rainfall and with irrigation water of 5.3 to 9.3 dS m^{-1} applied at Old Colliers Golf Club, Naples, Florida. The approximate conversion of EC_a to EC_e is $EC_e = 1.7\ EC_a + 6.07$. (From Carrow, R. N., J. Krum, and C. Hartwiger, Precision turfgrass management: A new concept for efficient application of inputs, *USGA Turfgrass and Environmental Research Online (TERO)*, 8 (13), 1–12, 2009, http://usgatero.msu.edu/v08/n13.pdf. With permission. Krum, J. M., I. Flitcroft, P. Gerber, and R. N. Carrow, Performance of mobile salinity monitoring device for turfgrass situations, *Agron. J.*, 103, 23–31, 2011.)

FIGURE 4.2 An experimental salinity-monitoring device (SMD) based on a four-wenner array for mapping surface and subsurface salinity, and the plant Normalized Difference Vegetative Index (NDVI). (From Carrow, R. N., J. Krum, and C. Hartwiger, Precision turfgrass management: A new concept for efficient application of inputs, *USGA Turfgrass and Environmental Research Online (TERO)*, 8 (13), 1–12, 2009, http://usgatero.msu.edu/v08/n13.pdf. With permission.)

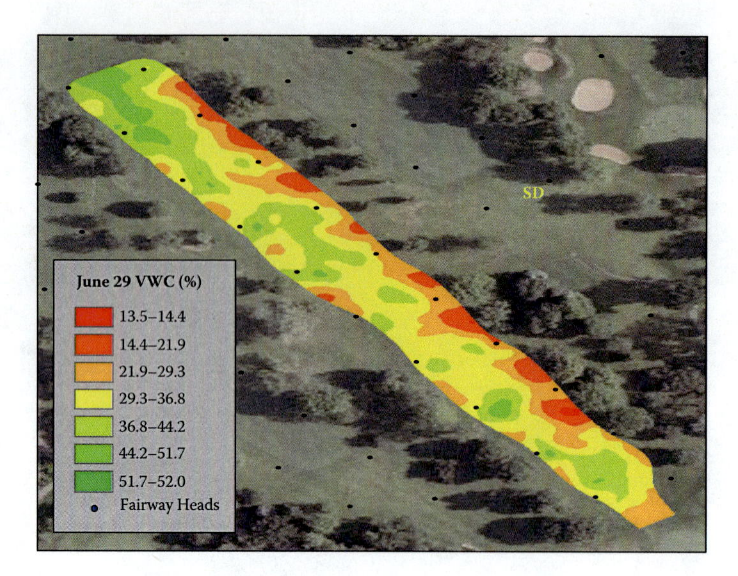

FIGURE 10.1 Spatial variability of soil volumetric water content (% VWC) presented in a standard deviation map. Note the wind effects from right to left on spatial VWC patterns. Also, note the threefold difference in VWC on this fairway. Overlaying the theoretical water distribution pattern for these heads with the appropriate nozzles and system pressure would help determine system efficiency. (From Carrow, R. N., J. Krum, and C. Hartwiger, Precision Turfgrass Management: A new concept for efficient application of inputs, *USGA Turfgrass and Environmental Research Online* (TERO), 8 (13), 1–12, 2009a, http://usgatero.msu.edu/v08/n13.pdf. With permission.)

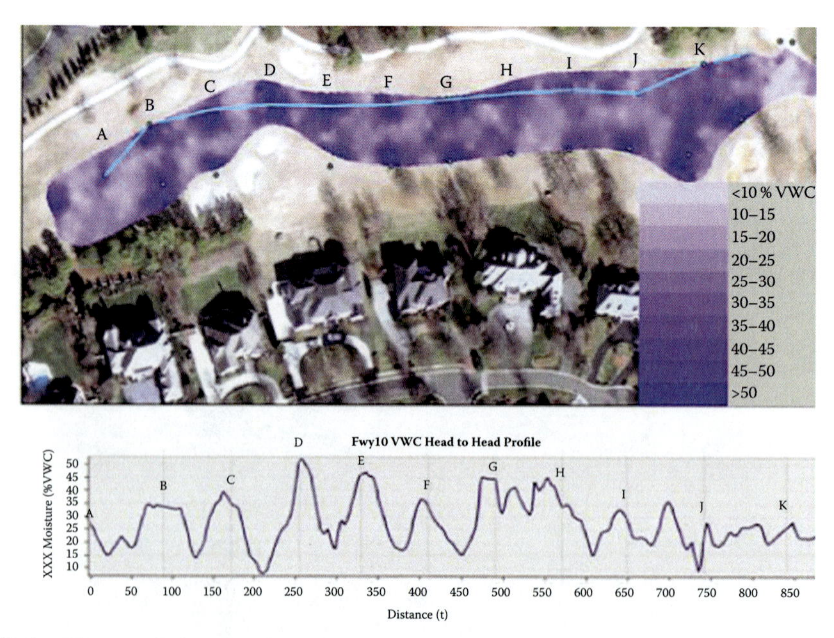

FIGURE 10.2 Water audit by spatial mapping of soil volumetric water content (VWC) based on an 8- × 10-foot mapping grid using the Toro TMM mobile-mapping unit (Toro Company, Bloomington, MN). A fourfold difference in VWC is apparent. The diagram on the bottom illustrates VWC distribution along one irrigation main of this double row system with 90 feet head spacing. Spatial distribution illustrates nonuniformity of soil VWC due to lack of system uniformity of application of water. See Figure 10.3 for a different means to illustrate spatial variability in soil VWC of these data (Toro Company).

New Irrigation Audit

Evaluating soil moisture distribution around individual sprinkler heads

Feet from Head	0–15	15–30	30–45	45–60
# Pts	173	522	507	511
Mean VWC	42.37	31.52	17.27	19.99
Mean COMP	216.07	273.47	350.17	307.83
Mean NDVI	0.747	0.736	0.693	0.711
VWC CV	22.3%	28.2%	35.8%	37.2%
COMP CV	17.4%	21.3%	17.7%	19.8%
NDVI CV	1.7%	4.6%	9.3%	6.9%
VWC Range	34.94	42.15	30.26	28.17
COMP Range	124.41	273.44	287.31	287.48
NDVI Range	0.059	0.246	0.270	0.247

FIGURE 10.3 Spatial mapping of soil volumetric water content (VWC) around a single irrigation head at various distances from the head as part of a new water audit approach. Map developed by spatial mapping with the Toro TMM mobile device (Toro Company, Bloomington, MN). CV = coefficient of variation of data at each distance, where a low CV is best. The same information can be developed with segments around each head. NDVI = normalized difference vegetative index, where 1.0 = ideal shoot density and turf color. COMP = penetrometer resistance data where higher values = harder soil surface.

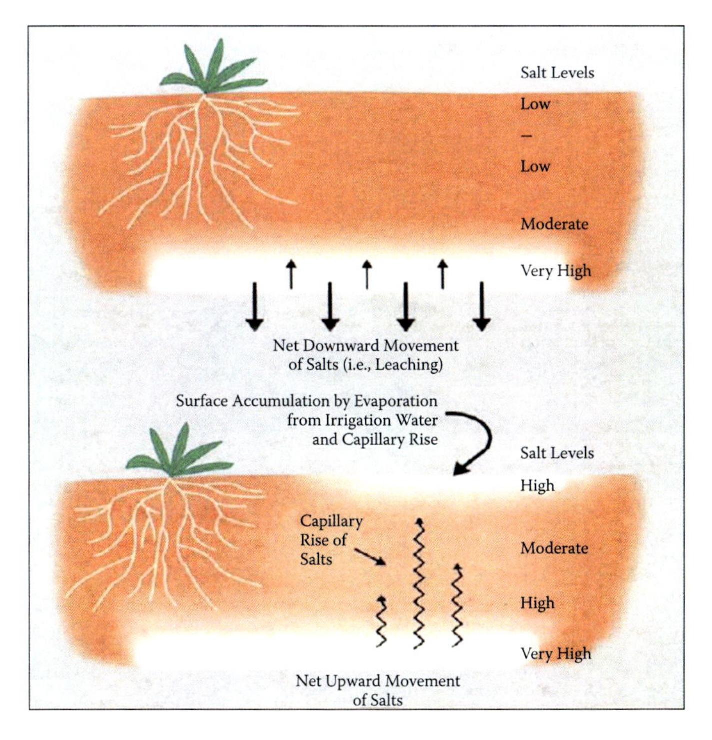

FIGURE 11.2 Examples of salt levels throughout the soil profile. Top: Represents good leaching conditions with adequate leaching requirement (LR) applied. Bottom: Represents what happens when insufficient water is applied in midsummer with high evapotranspiration (ET) conditions. (From Carrow, R. N., M. Huck, and R. R. Duncan, Leaching for salinity management on turfgrass sites, *USGA Green Section Record,* 38 (6), 15–24, 2000. With permission.)

PLATE 1. Saline soil resulting from application of saline irrigation water. Tan areas exhibit severe physiological stress (salt-induced drought stress). Plates 1–3 illustrate a progression of saline injury and symptoms.

PLATE 2. Saline soil caused by saline irrigation water. The whitish areas are surface accumulation of soluble salts.

PLATE 3. Saline soil with severe salt accumulation and injury to the turfgrass stand.

PLATE 4. Soluble salts move with water. In this case, irrigation water and salts are running off the irrigated rough onto the edge of a golf green and causing severe physiological (salt-induced) drought injury. Plates 4–7 illustrate different examples of soluble salt movement either via the surface or subsurface.

PLATE 5. Saline irrigation water is causing soluble salt injury on parts of the irrigation rough but the greatest injury is on the fairway where salts have moved from the rough and accumulated.

PLATE 6. Surface and subsurface movement of salts are causing excessive salt accumulation on the fairway, especially at the base of the hill on the left.

PLATE 7. Saline irrigation water and a fine-textured soil with low infiltration have caused salt accumulation in surface drainage areas with the grass exhibit soluble salt stress.

PLATE 8. Calcined clay material water added to the surface of this area by core aeration and topdressing resulting in excessive soluble salt retention and severe physiological (salt-induced) drought stress.

PLATE 9. Initial phase in development of a sodic scald area (top) and example of later development. Plates 9–11 illustrate increasing development and severity of sodic soil conditions, including scald areas.

PLATE 10. Saline-Sodic conditions caused by irrigation water that is both saline and has high Na concentrations. The predominately saline areas are whitish and the sodic areas are gray to black.

PLATE 11. Severe sodic scald conditions with typical black coloration due to extreme alkaline pH and solubilization of some of the organic matter to coat soil particles and cause the black color.

PLATE 12. Pre-acid sulfate conditions occurring on a parkland formed from dredged soil from the ocean. Soil is both extremely acid (pH < 3.8) and highly saline and sodic.

PLATE 13. Sand and gypsum slitting into a decomposed granite soil laid over an low percolation calcitic layer using a Blec SandMaster. Site receives saline irrigation water.

PLATE 14. Deep solid-time cultivation to enhance saline irrigation water infiltration and percolation for salt leaching.

such as after a heavy rain, most water movement into and through the soil profile is by macropore flow in response to gravity.

Total soluble salts do not adversely affect macroporosity, but excess Na does. The macropores are essential to remove excess water, provide soil gas exchange (oxygen flux and aeration), and provide root channels for good rooting extension. Macropores should be present across the entire soil profile depth, and there should be macropore continuity, which could be broken by a subsurface layer that does not contain many macropores. Since micropores retain water, they also retain salts. Effective leaching cannot be accomplished unless micropores are leached (generally accomplished with pulse irrigation cycles).

Sandy soils with >85% sand content exhibit sand particle-to-particle contact that opens up macropores between particles, and this arrangement resists compaction. Thus, salt leaching is much easier in sandy soils, especially when the sand content is >95%. If excessive "fines" are added to the soil or excessive organic matter accumulates, infiltration rates can decline, and salinity leaching becomes more difficult. Sometimes "sand substitutes" (calcined clay, diatomaceous earth materials, and zeolites) are added to sands, usually for moisture retention purposes (calcined clays or porous ceramics, and diatomaceous earth products) or sometimes for enhancing CEC (zeolites). Except for zeolites or Lassenite, materials normally contain predominantly micropores that are very fine and thus retain water with most of the water not plant available. Also, the total pore volume is increased (see Section 11.2.3.7, "Total Pore Space [Pore Volume (PV)]," later in this chapter). If saline irrigation water is used, these additional micropores will become filled with saline water, which will not easily leach (see Chapter 8, Section 8.1.2, "Inorganic Amendments"). The authors have observed several situations where >10% to 15% sand substitute (volume basis) amendments were added to the soil profile, and the irrigation water was saline. Salt leaching of these media was much slower and more challenging than with unamended sand profiles. If these inorganic amendments are not thoroughly mixed into the soil profile (i.e., greens sand mix), they sequester salts in the zone of their presence with each saline irrigation water application, and the salts tend to continue loading above this zone. Leaching of those accumulated salts has been very difficult, and grass performance has been significantly decreased.

In terms of salt leaching, as microporosity increases (regardless of sources such as clay, sand substitutes with considerable microporosity, soil compaction destroying structure, excessive organic matter, etc.), salinity leaching becomes more difficult and requires a greater volume of water. Salts leach relatively easy and rapidly in macropores, but, in micropores, most saturated flow (when sufficient water is added to create rapid, saturated flow conditions) bypasses the micropores and flows through the macropores. Under nonsaturated flow conditions, such as with pulse irrigation or during slow and prolonged rains, water can move through the micropores, but this is a much slower process to achieve sufficient flow to effectively leach the salts. However, considerably less water volume is required for effective leaching by pulse irrigation in order to create unsaturated flow conditions.

Even a thin zone or layer within a soil profile that has few macropores will not only limit water movement during normal irrigation and leaching events, but also result in salt accumulation above this layer. Any soil layer or horizon that inhibits water movement will be a major hindrance to effective leaching—whether it is at the surface (surface compaction) or subsurface (i.e., B horizon, cultivation pan, buried layer from flood deposition of fines, particle or colloidal migration, etc.). Cultivation operations to enhance infiltration and percolation (deep cultivation techniques) are essentially done to create temporary macropores. If the cultivation holes are filled with a medium to coarse sand (>0.25 mm), the macropores remain open for a longer period of time. Thus, turfgrass managers must be familiar with their complete vertical soil profile and should "visualize" (view speed of water infiltration) whether sufficient macropores exist for effective leaching downward into the deep subsoil or hopefully to the drain lines.

11.2.3.3 Clay Type

Clay type has a pronounced influence on water movement. Nonshrink and nonswell clays (kaolinite, allophanes, and Fe/Al oxides) are called *1:1 clay types*, and these do not crack when drying or seal by swelling when wet. Cultivation operations generally last longer on 1:1 clays than on the *2:1 clay types*, as discussed below. Also, a higher level of Na is required on 1:1 CEC sites before soil structure deterioration, usually at >24% Na saturation (ESP) compared to >3% to 6% Na for many 2:1 types (montmorillonite and illite). Generally, 1:1 clays are more resistant to soil compaction than 2:1 clays. Because 1:1 clays evolve in humid, high-rainfall areas, they often exhibit a B horizon where clay content is higher due to movement or migration over many years. For example, many southeastern U.S. Piedmont red clay soils (1:1 types) contain 40% to 50% clay in the B horizon versus 15% to 25% in the surface A horizon, and water movement is slower across the B horizon. Pulsing the water applications and using low-flow sprinklers plus soaker hoses are methods for effectively leaching these B horizon or higher clay composition soils when they become salt laden.

Most clay types in the United States, including arid and semiarid regions and many marine (coastal) clays, are 2:1 clays—these can be present in most climatic zones. When drying, 2:1 types are "self-cultivating" because cracks form, but on irrigated turfgrass sites, the soil usually does not dry sufficiently for cracking, except on nonirrigated locations. Unfortunately, under well-watered to saturated moisture contents, these clays swell and most macropores are lost. When total salinity problems develop on these soils, deep cultivation followed by filling the cultivation hole with sand or sand plus gypsum (sodic sites) is necessary to maintain a sufficient volume of macropores, to reach at least the depth of cultivation. In contrast, deep cultivation operations are effective for longer time periods on 1:1 clays even without filling holes with sand.

11.2.3.4 Soil Structure

Soil structure refers to the arrangement of sand, silt, clay particles, and organic matter into structural units or aggregates (see Figure 3.3). For example, a soil with appreciable silt and clay may have aggregates composed of sand, silt, and clay held together by organic matter aggregating agents arising from soil microbial activity. These aggregates are normally sand size or much larger and act as units that increase macroporosity for enhanced water movement and aeration. As aggregates are formed, macropores are developed between aggregates or structural units. Structure is very important as silt and clay content increases on fine-textured soils.

Soil compaction from recreational traffic destroys much of the structure and accompanying macropores in the surface 3-inch (7.5-cm) zone, but a well-structured soil will usually have some macropores deeper in the profile. The 2:1 clays are much more prone to soil compaction and structural deterioration than are 1:1 types. As noted previously, high Na causes structural deterioration of fine-textured soils. This structural breakdown is especially serious on 2:1 clays since they often have poor drainage even under low Na concentrations due to their swelling and sealing nature with sequential wet and dry cycles as well as the short duration in which aeration holes (macropores) remain open with regular irrigation cycles.

While high Na content does not cause "structural breakdown" of single-grain sand particles, it does cause any colloidal particles (clay or organic matter in nature) present to be dispersed and become susceptible to particle migration. Pond, lake, or river water with high turbidity can contribute these "fines" during irrigation. Often these fines accumulate at the normal depth of irrigation water (wetting front) penetration and can cause a layer that sequesters and concentrates excess salts like carbonates and sulfates, which could eventually induce black layer formation in conjunction with reduced oxygen movement through the soil profile. This sequence of events would then inhibit salt leaching and damage turfgrass roots.

11.2.3.5 Capillary Rise

Soluble salts move with the soil water. If there is net downward movement of water due to rain or irrigation to achieve an adequate leaching fraction, salt movement is downward away from the root system. But, these salts may accumulate below the rootzone in an area that can be very salt laden. If water moves upward (such as with high ET conditions), the salts also move upward. The most

TABLE 11.2
Evapotranspiration Averages by Environment for Turfgrasses under Well-Irrigated Conditions for Different Climate Conditions

Climate Situations	Average Evapotranspiration[a]
	Inches per day
Cool humid	0.10 to 0.15
Cool dry	0.15 to 0.25
Warm humid	0.15 to 0.20
Warm dry	0.20 to 0.25
Hot humid	0.20 to 0.25
Hot dry	0.25 to 0.35

[a] The actual ET varies with grass species and cultivar, wind speed, management level, and so on, but these values provide "ballpark" estimates. Also, as soil moisture level declines, ET decreases dramatically.

common upward movement is by *capillary rise* in the micropores and can result in major redistribution of salts within the soil profile (Jorenush and Sepaskhah, 2003) (Figure 11.2). Capillary rise of salts will be more rapid on fine-textured soils than sands because fine-textured soils contain more micropores.

Factors that enhance capillary rise of salts are low leaching rates, high ET conditions, and a high water table. Some water conservation regulations limit ET replacement to 70% to 85% potential ET (ETo) on turfgrasses in arid regions and do not take into account the need for an LR to prevent soil structural deterioration. Under high evapotranspiration (ET) conditions, salts may start to rise upward by capillary action driven by surface drying from ET losses (Table 11.2). If the replacement irrigation applications provide "no leaching fraction" and are equal to or less than ET, the net movement of salts will be upward into the rootzone (potentially desiccating roots) and eventually may result in surface salt accumulation (causing canopy thinning and reducing turfgrass growth rates) (Figure 11.2). When salt management and leaching programs are inadequate in moving accumulated salts down below the turfgrass root system, regular cultivations (core aeration, spiking, slicing, needle tining, etc.) coupled with high ET conditions can result in rapid loss of upper soil moisture, which can escalate the capillary rise of excess salts back to the surface.

11.2.3.6 Water Table

Water table location is another soil factor influencing salinity control. Sometimes, the natural water table level is near the surface. The *capillary fringe* of semisaturated water conditions above a free water table is usually 2 to 8 inches (5 to 20 cm) for sands and 8 to 12 inches (20 to 50 cm) for fine-textured soils. However, high-ET conditions and limited leaching can cause salts to rise well above these distances over time by long-term capillary action. Capillary rise on fine-textured soils is still strongly controlled by climatic conditions (i.e., ET) at a depth of 2.5 to 3.0 feet (0.8 to 0.94 m) and possibly down to about 5.0 feet (1.6 m). An example of long-term capillary rise is upward salt movement on a dormant turfgrass during dry, winter months from a salt-laden zone that may be well below the root system in the normal growing season when irrigation and rainfall keep the salts from rising.

A rising water table can bring salts that have accumulated above the water table into the rootzone area. In Western Australia, large areas of the landscape have become salinized by a slowly rising water table caused by the removal of deep-rooted trees and shrubs that prevented deep penetration of water and dissipated it as ET. This is a type of secondary salinization and is called *dryland salinity* (Barrett-Lennard, 2003).

Another water table issue is when the water table is near the surface and poor irrigation water quality requires a high LR. Over time, the water table may rise even higher and cause massive salinization of the rootzone. On sites where shallow water tables may rise, the turfgrass manager should investigate means to lower the water table when possible, such as additional drainage lines with possible sump pumps and distribution of the water into drainage canals, dry wells, or natural areas.

Turfgrass soils often contain layers in the soil that inhibit water percolation or drainage. This can create a temporary *perched water table* as water flow is slowed or stopped when the wetting front reaches this layer. Salts will then accumulate above the layer and can rise to the surface whenever low leaching rates or high ET occurs. A good concept to remember is "If a layer impedes water movement, it also impedes salt movement and, therefore, enhances salt accumulation."

Subsurface layers that are 1 to 3 feet below the surface are often overlooked in arid or semiarid regions where heavy rainfall events (that are sufficient to pond water up to the soil surface) are rare. But these "hidden layers" can contribute to major salt accumulation and layering so that when conditions favor capillary rise, the resulting water has very high salinity.

Another "perched water table" is found in many high sand content constructed profiles, such as the USGA green construction method where the interface between the rootzone media and a coarse sand layer creates a perched water table (USGA, 2010). In this case, ample macropores are present, but sufficient water is required to break the perched water tension and to initiate rapid drainage or "flushing" of the rootzone. During summer months when ET is high and saline irrigation water is used, salts above the perched water table zone may start to rise toward the roots and soil surface if extra leaching is not applied periodically (such as monthly). Prolonged drought, high temperatures, and dry, windy conditions can escalate or enhance the concentration of these salts and their capillary rise to the soil surface, especially if a lighter and more frequent irrigation regime is imposed. On older USGA-constructed greens with the perched water table, the authors have noticed that on rare occasions a sealed salt accumulation zone can develop at the interface between the coarse sand layer and the gravel layer with long-term use of saline water. If deep cultivation is not successful in opening up sufficient macropores for leaching down to the drain lines, a complete greens reconstruction may be needed to remedy the problem.

With the exception of the layer within the USGA greens, layers found in turfgrass soils that limit water percolation or drainage and enhance salt accumulation have few macropores due to excessive fines or compaction. Cultivation practices to break up these layers are important in salinity management. The cultivation depth must penetrate completely through the layer to allow for maintaining water flow when excess water application by irrigation or rainfall occurs and to prevent a zone of salt accumulation such as the bottom of the wetting front with each irrigation cycle, or for a normal LR in a maintenance-leaching program to result in net downward movement of the salts.

11.2.3.7 Total Pore Space (Pore Volume [PV])

The total pore space or pore volume of a soil also influences salt leaching (Rhoades and Loveday, 1990). Soils with higher PV require more water to leach the same quantity of salts, especially if the micropores are more prevalent. The PV range of sands, loams, and clays is about 35% to 40%, 40% to 50%, and 45% to 55%, respectively. For a soil depth of 12 inches (30.5 cm), 1 PV of applied water would represent 4.2 to 4.8 (sands), 4.8 to 6.0 (loams), and 5.4 to 6.6 (clays) inches (10.7 to 12.2, 12.2 to 15.2, and 13.7 to 16.8 cm, respectively) of irrigation water for effective leaching. Thus, more water is required to leach fine-textured soils than sands. When micropores are predominant, the leaching program must be much slower in order to allow time for the water to move by nonsaturated flow through these small pores and to minimize surface runoff.

11.2.4 Climatic Conditions

Climatic conditions impact leaching operations for salinity management by several means, such as rainfall quantity and patterns, temperature, relative humidity, wind speed, and time of year. Of prime importance is the climatic influence on ET and, therefore, the frequency and quantity of water to replace ET lost from the soil since the last irrigation or rainfall event (Table 11.2). High-ET conditions (high temperatures, low humidity, high solar radiation, and wind) require more frequent irrigation or higher quantities of water to replenish a deeper soil profile (i.e., deeper, less frequent irrigation). In arid climates or dry seasons, there is less rainfall to recharge the soil profile across a landscape, so irrigation is applied not just to replenish the ET losses but also to compensate for nonuniformity of the irrigation system in order to prevent dry areas. The net result of these factors is more water applied on a site.

When saline irrigation water necessitates a maintenance-leaching program with an LR, this furthers contributes to the total water needs—that is, as pointed out previously, the LR is normally only 5% to 15% additional water per irrigation cycle, assuming that pulse irrigation is used. With high ET conditions, the site manager is challenged to apply sufficient water within the irrigation cycle time frame to meet the total irrigation water requirement. Insufficient irrigation application to meet all needs results not because the LR is so high, but because the ET and/or quantity of water to compensate for nonuniformity are not estimated correctly or are not able to be applied due to time, upper management restrictions, or irrigation system constraints. The net result is that salts are deposited into the rootzone without adequate leaching to effectively remove them, and soil EC_e will increase. Salinity accumulation in the surface couple inches of the soil profile occurs where much of the crown, rhizome, root, and stolon tissues are located.

The second major influence on salinity management of a high ET climate is the impact on *capillary rise of salts* from below the surface. Without net downward movement of water or at least sufficient water to maintain equilibrium, salts will start to move upward by capillary action. If the zone of high salt concentration is dry, capillary action is minimal. However, this zone may contain sufficient moisture from past irrigation events or rain to allow for capillary rise. When the salt-laden zone is near the bottom of the existing root system, then it does not require much time before the lower rootzone is subjected to very high saline conditions in a short time frame. The double response of salt accumulation at the surface and salt increase within the lower rootzone can cause rapid and dramatic salinity stress problems on the plant.

Other influences of climate on salinity management are seasonal changes in rainfall distribution; effects of rain on water quality in irrigation lakes (for example, dilution of salinity in an irrigation lake that receives a saline water source); effect of prolonged high ET conditions on concentrating salts within an irrigation lake; and winter dormancy periods. Each of these climatic influences illustrates the necessity for the turfgrass manager to (a) always consider how climatic and surrounding area conditions (water harvesting of rainfall) are affecting salt accumulation in soil or lakes, and (b) monitor soil and irrigation water quality as needed to track changes before they become problems. Over time, due to the frequency of applications, the soils will eventually equilibrate to the irrigation water quality and accompanying salt load. If the salts are not properly managed, the soil salt accumulation can easily exceed the incoming salt concentrations in the irrigation water.

11.2.5 Irrigation System Design and Scheduling Capabilities

Irrigation system design considerations were the topic of Chapter 10, "Irrigation System Design and Maintenance for Poor-Quality Water," but are listed here as a reminder of the importance of the irrigation system design on salinity management and enhancement of leaching programs. Irrigation scheduling will be discussed later in this chapter as related to maintenance leaching (Section 11.3) and reclamation leaching (Section 11.4).

11.2.6 Grass Type and Salinity Management

Salinity tolerance of the turfgrasses on the site as well as any other landscape plants is especially important in determining appropriate salt-leaching programs. In fact, the initial LR used in maintenance leaching programs is based in part on the inherent salinity tolerance of the turfgrass (discussed in Section 11.3), while the soil salinity level that can be allowed before initiating a reclamation-leaching program is based on the turfgrass salinity tolerance.

11.2.6.1 Salinity Tolerance

Salinity tolerance of the grass is one of the most important factors affecting salinity management practices. The turfgrass *threshold* EC_e is used as a guide to acceptable soil salinity levels, where threshold EC_e is defined as the soil salinity at which growth starts to decline from a nonsaline condition (Carrow and Duncan, 1998; Duncan et al., 2009) (also see Chapter 9, "Selection of Turfgrass and Landscape Plants"). Salinity tolerance is used in determining LR where grasses with higher threshold EC_e have a lower LR compared to grasses with a lower threshold EC_e, since soil salinity can be maintained at a higher background level. Grasses with moderate to very high salinity tolerance may be irrigated to maintain the soil salinity at greater than the threshold EC_e, perhaps at EC_e of 25% or 50% growth reduction, as long as the grass has acceptable quality and is able to tolerate any stresses that are present. The grass vigor and ability to withstand wear from traffic are especially important on recreational sites when considering the appropriate maintenance EC_e.

One common misconception about highly salt-tolerant grasses (halophytes such as seashore paspalum) is that a high salinity tolerance, as exhibited by a high threshold EC_e, indicates that a grass should be maintained at that level. As salinity in the irrigation water increases, so do the potential adverse impacts on the environment, the soil infrastructure, the management costs, and the associated dynamic ecosystem challenges (Carrow and Duncan, 1998; Beltran, 1999; Duncan et al., 2009). The most important management benefit associated with using a more salt-tolerant grass compared to a less tolerant one when irrigation water is saline is that the grass response to salinity does not occur as rapidly, which allows time for the manager to implement corrective practices without significant visible turfgrass stress and loss of performance criteria.

11.2.6.2 Turfgrass Rooting and Salinity

When assessing salinity tolerance of turfgrasses, it is important to determine tolerance of the root system since salinity tolerance can vary with tissue type (Lee et al., 2005). Grasses exposed to highly saline environments must be able to develop and maintain a viable and extensive regenerating root system in order to exhibit adequate shoot salinity tolerance. For reclamation leaching, rooting depth determines the "rootzone" that must be leached. Thus, a deeper rooted plant will require more water for leaching during reclamation of a salt-challenged site.

Turfgrass rooting depth also impacts salinity management for routine maintenance leaching. Provided that adequate soil moisture is present in the lower one third of the root system to avoid salt accumulation (i.e., soil moisture is about "field capacity" in this zone), turfgrass growth is related to average rootzone EC_e regardless of the salt distribution within the rootzone. Thus, when monitoring soil EC_e by depth within the rootzone, the average EC_e is the value used to compare with the turfgrass salinity tolerance level selected, such as EC_e for 25% growth reduction. Especially with high-saline irrigation water, irrigation events should be scheduled to avoid depletion of soil moisture within the lower one third of the rootzone, which would result in increasing soil EC within this zone with serious salt stress occurring. Irrigation events are, therefore, scheduled more often than on a similar nonsaline site in order to maintain higher average soil moisture content and to avoid excessively high soil ECe. Also, a deep-rooted turfgrass will allow for more days between irrigation events than a shallow-rooted grass as long as the specific cultivar has adequate salinity tolerance to the water quality used for irrigation. While irrigation events may be scheduled more often on sites irrigated with saline irrigation water, the quantity of water should still be based on

ET replacement, compensation for nonuniformity of the irrigation system, and the LR. Remember that it is not unusual for soil ECe to range from two to five times higher salt accumulation above the incoming ECw irrigation water quality dependent on the effectiveness of the overall salt management program. Highly salt-tolerant cultivars provide time to make the necessary management adjustments (leaching, fertility, etc.) to sustain turfgrass performance during these fluctuating salt-challenging times.

11.2.7 Water Quality and Salinity Management

While many soil, climatic, and plant characteristics influence salinity management, one of the most important factors is the quality (salinity level) of the irrigation water. Not only is irrigation water often the source of the salts that must be managed, but also its quality has an essential influence on the quantity of water to achieve salinity control in both reclamation and maintenance-leaching programs. The influence of irrigation water quality on the LR in maintenance-leaching programs is discussed in detail in the following section (Section 11.3, "Maintenance Leaching and the Leaching Requirement"). When developing BMPs for salt-affected sites, it is wise to start with an irrigation water quality analysis.

11.3 MAINTENANCE LEACHING AND THE LEACHING REQUIREMENT

There are two different salinity-leaching approaches. First is *maintenance leaching*, where a small quantity of irrigation water is applied beyond ET replacement and water required to compensate for lack of irrigation system application uniformity with each irrigation cycle—maintenance leaching is the topic of Section 11.3. This approach is based on maintaining rootzone salinity at a level that will not hinder turfgrass performance and to insure slow downward movement of salts to a zone beyond the turfgrass root system. The second approach is called *reclamation leaching* (the topic of Section 11.4), where soil salinity is allowed to accumulate to a point approaching an unacceptable degree of salinity stress on the plant before leaching is started. Maintenance leaching is the best method for salinity control because it (a) requires the least quantity of water for salt control, (b) avoids salt stress, and (c) does not require a site to be closed due to waterlogging or excessive soil moisture. How irrigation water is applied has a profound influence on the effectiveness of salinity leaching and the total quantity of water needed to achieve salt leaching whether in a maintenance or reclamation mode. Thus, the information in Section 11.5, "Pulse Irrigation and Other Water Application Methods," is critical, especially to understand *pulse irrigation*, which achieves the greatest leaching with the least quantity of irrigation water.

11.3.1 Traditional Methods of Determining the LR

Maintenance leaching is where the term *leaching requirement* is used. As noted previously, the LR is the minimum amount of water that must pass through the rootzone to control salts (i.e., keep salts moving) in the rootzone within an acceptable level (Rhoades, 1974). In the traditional LR approach developed by Rhoades (1974), the whole area would be irrigated using the LR, as determined by the method discussed later in this section. For sites where soil salinity spatial differences are minimal, this method is effective; however, when significant spatial variability exists, this is not as "water efficient" as targeting each specific area as needed—which is the subject of Section 11.3.2. In the LR as determined by Rhoades (1974), irrigation water quality strongly influences the quantity of water necessary to leach salts (with more water required as salinity level in the water increases) as well as the actual level of salinity tolerance for the turfgrass (more tolerant grasses require a lower LR).

The LR is intended to be applied with every irrigation event for maximum effectiveness and to obtain salt leaching with the minimal quantity of water. This is especially important on all but the sandiest soils that have high infiltration and percolation rates where leaching is relatively rapid even when soil salinity is high due to delays in applying the LR. If the turfgrass has sufficient salinity tolerance to allow some soil accumulation of salts, then a turfgrass manager may delay leaching for a week or until the soil salinity is near the plant stress level. However, at that time, they must have the flexibility and time to apply sufficient water in order to leach to acceptable decreased salinity levels. This "hybrid" approach (between a true maintenance and a reclamation program) will still require more water in most cases than a continuous maintenance-leaching program. One situation where this approach is used is on golf courses that are closed for a day a week or on other turfgrass recreational sites with open dates.

On all other soils, if salts are allowed to accumulate, the resultant reclamation-leaching program most often requires too much water to avoid wet areas or the time frame is too long for high-use areas. This is the source of comments often made that applying additional water for salt leaching will result in wet, soggy conditions, or there just is not sufficient time to apply all the extra water. If a true maintenance LR program is used, however, these are not valid comments—but they are very valid when salt accumulation reaches a level requiring reclamation leaching. To illustrate, as a general rule for the total quantity of irrigation water required for a specific irrigation event, replacement of ET losses would account for 60% to 85%; 10% to 30% would be for nonuniformity of the irrigation system, and 5% to 15% for the LR, assuming pulse irrigation is used. For example, if ET replacement was 1.0 inch, the quantity of water for nonuniformity of the system 0.20 inch, and the LR of 7%, then the total quantity of irrigation water would be $1.00 + 0.20 + 0.08 = 1.28$ inches. The 0.08-inch LR is not the cause of undue wet conditions or too little scheduling time; rather, inaccurate estimates of replacement ET and nonuniformity of the irrigation system are the causes.

Several methods have been or could be used to determine the LR (Corwin et al., 2007; Carrow and Duncan, 1998). Corwin et al. (2007) recently reviewed and compared various steady-state models and more complex transient models that could be used for determination of the LR. These included several computer models that have been used for estimating LR for agricultural crops in California's Imperial Valley as well as the "traditional method" of Rhoades (1974), where two of the models resulted in an LR of 0.08 versus 0.13 for the traditional method. In the traditional method, Rhoades (1974) considered two very important factors, namely, (a) irrigation water salinity level (EC_w, dS/m), and (b) grass salinity tolerance using the threshold EC_e (the soil salinity, EC_e, at which growth declines compared to growth under nonsaline conditions) (see Table 11.1). The relationship used is as follows:

$$LR = EC_w / 5EC_e - EC_w$$

where

EC_w = electrical conductivity of irrigation water

EC_e = threshold EC = electrical conductivity of saturated soil paste at which turfgrass growth starts to decline by at least 10%

An example will illustrate the influence on LR of a change in the irrigation water quality. For a turfgrass with a threshold EC_e of 10 dS/m and the irrigation water quality is $ECw = 2.0$ dS/m, the LR by the above formula would be

$$LR = 2.0/5(10) - 2.0 = 0.042, \text{ or } 4.2\%$$

This LR would require 4.2% irrigation application above the replacement ET plus any for correction of nonuniformity of the irrigation system application. However, if the irrigation water quality has an EC_w of 4.0 dS/m, the LR then becomes 8.7%.

Likewise, a change in the turfgrass, such as an overseeded grass, that has a different threshold ECe for salinity tolerance can change the LR. In the above example, where the turfgrass threshold

EC_e is 10 dS/m and the irrigation water quality is ECw = 2.0 dS/m, the calculated LR is 4.2%. However, if the grass was less salt tolerant with a threshold ECe = 6.0 dS/m, then the LR would be 7.1%.

Some of the models proposed to estimate the LR that are discussed by Corwin et al. (2007) include factors beyond what are in the traditional LR, such as composition of irrigation water, salt precipitation processes, ET reduction under salinity, soil water content by rooting depth, preferential flow, and unsaturated flow effects. Some may ask, *"Which LR model should I use?"* In turfgrass situations with saline irrigation water and a perennial ground cover, salinity management by leaching is an ongoing process that must be monitored and adjusted. The "system" is not static or steady state, but common changes over a season are that irrigation water quality may change over the year, high rain periods alter soil salinity and salinity within irrigation lakes (especially when the rainfall is harvested and blended in the lake), turfgrass growth changes with dormancy periods, environmental stresses and traffic stresses may change that impact salinity tolerance, and other dynamic factors. What the steady state-based LR of Rhoades (1974) does is to estimate a reasonable LR by accounting for two of the most important factors, namely, irrigation water quality and plant tolerance to salinity. It is a "ballpark estimate" that provides a good starting place for determining an effective LR. From this starting point, a successful salinity-leaching program will require adjustment over time based on proactive monitoring of soil salinity with the following important considerations:

- Ongoing monitoring of changing irrigation water quality and soil salinity conditions by water quality and soil tests with adjustments made based on actual field conditions. If field monitoring demonstrates that leaching is not sufficient, then the LR should be increased; or, conversely, it should be decreased if leaching is adequate until an appropriate LR is defined.
- The actual LR does not change unless irrigation water quality changes or the grass salinity tolerance changes. The latter could occur when a cool-season grass is used to overseed a more salt-tolerant warm-season grass; or when, during renovation, more juvenile seedlings or vegetative plant parts are present, which have lower salinity tolerance than a mature plant.
- Sometimes the quality of the irrigation water source changes over time. This would necessitate a change in the LR based on changes in the EC_w.
- While the LR does not often change, the quantity of water to replace ET does change as weather changes. Thus, an underestimate of ET will result in salt accumulation, not because of an insufficient LR but because of insufficient ET, which accounts for the majority of the irrigation water replacement requirement.
- Lack of attention to irrigation system uniformity with respect to irrigation water application will result in nonuniformity of irrigation water application and, thereby, in salt leaching.
- A final assumption is that the entire soil profile is sufficiently hydrated to a soil water content level that will promote effective salt movement and leaching.

Assuming the initial rootzone salinity level is acceptable, when the LR is not sufficient to maintain salt leaching, two adverse salt responses occur: (a) first, salts applied in the irrigation water start to accumulate within the surface 2 to 3 inches (50–75 mm), and (b) capillary rise of salts from deeper in the soil and beyond the rootzone will bring concentrated salts back into the rootzone (Figure 11.1). Oftentimes, this zone of accumulated salts has a very high EC_e, and when it reaches the lower rootzone it can induce a rapid salinity stress (desiccation). To alleviate the salinity stress (physiological drought with reduced water uptake that is critical for plant transpirational cooling) now requires much more applied water than the LR amount because it is a reclamation problem.

This scenario is most often observed on high sand bentgrass/*Poa annua* golf greens irrigated with water of medium to high salinity where there has been a change to lighter irrigation events (including syringing) that may be applied more often. If conditions require shifting to a light, more frequent irrigation regime and, therefore, away from a true LR program for a period of time, then it is important to understand the implications. Even turfgrass managers with relatively low total salts in irrigation water (500 to 600 ppm) may experience this situation under extreme environmental conditions such as prolonged drought or arid climates, persistent windy conditions, temperatures >32°C (90°F), and ET rates consistently exceeding irrigation volume applications. The turfgrass manager may be achieving adequate leaching in the spring and early summer using ample irrigation or rainfall. However, by midsummer, three events can impede leaching: (a) first, with hot, dry weather, the ET increases the quantity of irrigation water needed simply to maintain soil moisture (Table 11.2); (b) turfgrass roots start to die back; and (c) turfgrass managers shift to light, more frequent irrigation, which does not supply sufficient water to leach and will concentrate salts near the soil surface due to inadequate percolation of the irrigation water deeper into the soil profile. The concentrated salts at the surface then exasperate water uptake through increased osmotic pressure, and potential salt damage to plant crowns and already shallow roots near the soil surface. If water conservation programs are mandated, salinity levels can increase rapidly in a short time period (in a matter of 1–2 weeks). The situation depicted in Figure 11.1 (bottom) is then initiated.

Light, frequent irrigation increases salt accumulation in the surface where most of the crown region's regenerated roots are located. And salts rise by capillary action into the rootzone from (a) a high-salt zone common for pushup greens, or (b) the perched water of a USGA green that is not adequately "flushed." Injury normally appears on the most elevated, open, and exposed greens or slopes where high ET conditions prevail due to high solar radiation and wind movement. Mounds, "berms," and slopes are normally among the first areas to show salinity and drought stress. Since the bentgrass/*Poa annua* is now under high-temperature stress, salt-induced drought is a serious additional stress symptom and disease outbreaks are commonly observed.

In this sequence of events, the basic problem is that the overall irrigation quantity is not sufficient to maintain leaching and a shift has been made from a maintenance-leaching program to one that will require reclamation leaching. The turfgrass manager must now apply an extra leaching irrigation every 1 to 4 weeks (depending on water quality and soil and environmental conditions) to avoid upper soil profile salt accumulation and attempt to reestablish an irrigation program that applies sufficient water to allow for adequate leaching until a maintenance-leaching program can be reestablished. The frequency between leaching events will vary depending on climatic conditions (solar radiation, wind, temperature, humidity, and resulting ET demand), water quality, rootzone depth and the threshold EC of individual turfgrass cultivars, and the depth of subsequent leaching and irrigation events. The leaching frequency and threshold EC can be accurately determined by use of an inexpensive portable EC meter (Vermeulen, 1997; see also Chapter 4, Section 4.3, "Field Monitoring of Soil Salinity"). Soil EC at the soil surface and throughout the profile can be monitored regularly (daily if necessary), and as the threshold EC is reached, leaching can be initiated to purge the perched water table.

A practical method to assure that the perched water table on a USGA green has been completely purged is to locate the outflow drain line exiting the green cavity and install an inspection port. Drainage flow can be observed and water samples collected and tested with the portable (EC) meter. Once the EC of the drainage water is at or near the EC of the incoming irrigation water, then acceptable initial leaching has been accomplished.

Although native soils may require 1 to 4 weeks to reclaim, well-drained and sand-constructed putting green rootzones with perched water tables can often be reclaimed in 1 to 3 days when high total salts are the primary salinity limitation. Between leaching events, additional irrigation may be needed on a light, frequent basis until turfgrass roots regenerate, which may not occur on bentgrass/*Poa annua* greens until cooler weather. If the irrigation water contains high levels of sodium and the sodium loads on the soil exchange sites, a calcium amendment such as gypsum must be used in conjunction with a proper leaching program to effectively move the excess sodium deeper into the soil profile.

11.3.2 Maintenance Leaching by Geospatial Variability in Soil Salinity

Earlier in this chapter and in Chapter 4 (Section 4.3, "Field Monitoring of Soil Salinity"), the nature of spatial variability and means of mapping geospatial patterns of soil salinity were discussed. Since field-mapping technology is rapidly developing, the ability to map soil salinity will change how maintenance- (and reclamation-) leaching programs are implemented. For maintenance leaching, an LR can be determined by the soil irrigation zone, especially with flexible irrigation systems with one to three head zones, based on the salinity level in that zone (Krum et al., 2011). To obtain an initial approximation of the LR, the soil salinity in ECe (saturated paste extract in dS/m units) can be substituted in the equation of Rhoades (1974) in place of the ECw. Then the irrigation within the zone or zones covering the area that exhibits the particular ECe can be adjusted. This is a site-specific salinity management approach that allows the LR to be adjusted by the actual soil ECe level, which can conserve water.

As noted in Section 4.3, "Field Monitoring of Soil Salinity," when salt accumulation sites are consistent, these become good areas to install *in situ* soil salinity sensors with at least two reading depths. This provides real-time data to make adjustments in the LR as needed as well as allow any capillary rise of salts to be observed.

11.4 RECLAMATION LEACHING

In the previous section, emphasis was on maintenance leaching and the LR concept for maintaining salinity levels at an existing acceptable reduced salinity level. In contrast, *reclamation leaching* is when salt accumulation is above the acceptable level (or very near) for the plant salinity tolerance, and salts must be leached to achieve an appropriate soil EC_e level. Since soil salinity levels are already excessive, reclamation leaching compared to maintenance leaching requires a higher quantity of water to decrease salinity within the rootzone to acceptable levels. Once this acceptable reduced salinity level is achieved, the LR irrigation approach (maintenance leaching) using less "extra" water can be used.

The reclamation approach is necessary in two primary situations in turfgrass management: (a) when a seriously salt-affected soil (e.g., highly saline and/or sodic condition) must be leached of excess salts before the grass can be established; and (b) when a turfgrass manager has not maintained an adequate LR, thereby the grass rootzone has increased in salinity to stress levels. This latter situation is most likely to occur during hot, dry summers when ET rates have increased, but the total water applied for ET plus the quantity of water to account for nonuniformity of the irrigation system and LR has not been adjusted to keep up with actual ET. Cool-season turfgrasses subjected to this sudden and intense salinity shock (a combination of drought, high temperature, and greater wear stresses from slower growth, all induced by salts) often do not survive. The "take-home lesson" for this situational stress is "proactive prevention" by adequate, continual application of sufficient LR water to keep salts moving downward and away from the turfgrass root system.

Reclamation leaching requirements can be estimated by the Rhoades and Loveday (1990) procedure (Table 11.3). This procedure takes into consideration the intended depth of salt leaching, desired final soil EC_e, current or initial soil EC_e, leaching water quality (EC_w), and soil type and irrigation method. Knowledge of the components in this equation and applying various situations are useful in understanding how different components change the water required to leach salts in a particular situation. The reclamation equation by Rhoades and Loveday (1990) is as follows:

$$D_w = k \times Ds \times EC_{eo} - EC_w/EC_e - EC_w$$

TABLE 11.3
Determining Reclamation Leaching Needs[a]

$D_w = k \times D_s \times EC_{eo} - EC_w / EC_e - ECw$

D_w = depth of water to apply for leaching (feet)

D_s = depth of soil to be reclaimed or leached (feet)

EC_e = final soil salinity desired in dS/m. This value is usually the threshold EC_e for the turfgrass being used or somewhat less than the threshold EC_e

EC_{eo} = initial or original soil salinity in dS/m

EC_w = salinity of irrigation water used for leaching in dS/m

k = factor that varies with soil type and water application method (efficiency of irrigation system)

For sprinkler irrigation applied by pulse irrigation that results in *unsaturated flow* conditions by allowing drainage for 1 to 2 hours (sands) to 2 to 8 hours (fine-textured soils) between a pulse irrigation event with repeated pulse events until the total quantity of water necessary for leaching is applied, use

k = 0.05 for high sand content each with >95% sand content (i.e., <5% silt and clay content)

k = 0.10 for all other soils

For continuous ponding or continuous sprinkler irrigation that results in *saturated flow* conditions with water applied to keep the soils saturated during leaching, use

k = 0.45 for organic soils

k = 0.30 for fine-textured soils

k = 0.10 for sandy soils

Source: Adapted from Rhoades, J. D., and J. Loveday, Salinity in irrigated agriculture, in B. A. Stewart and D. R. Nielson (Eds.), *Irrigation of Agricultural Crops* (Agronomy No. 30, pp. 1089–1142a), Amer. Soc. of Agron., Madison, WI, 1990.

[a] Adjustments in the k value for high sand content greens are based on the experience of Robert Carrow, Mike Huck, and Ron Duncan.

where

D_w = depth of water to apply (in feet)

D_s = depth of soil to be reclaimed in feet

EC_e = final soil salinity desired in dS/m

EC_{eo} = original or initial soil salinity in dS/m

EC_w = irrigation water salinity in dS/m

k = factor for soil type and irrigation method, where the "k" factor for pulse sprinkler application is 0.05 for 95% sands and 0.10 for all other soils; and the "k" factor for continuous ponding/flooding is 0.45 for organic soils, 0.30 for fine textured soils, and 0.10 for sandy soils

As an example of the use of this method to determine an estimate of the quantity of irrigation water to apply for reclamation leaching (i.e., D_w), assume (a) a high sand content golf green with an initial soil ECe = 12.0 dS/m (i.e., EC_{eo}); (b) the turfgrass being used has a salinity tolerance threshold EC_e of 5.0 dS/m, which is therefore the final desired soil EC_e; (c) the irrigation water used for leaching will be applied by pulse cycle means and has an EC_w of 2.5 dS/m; and (d) the desired leaching depth (D_s) is 16 inches to reach the drain tile, where 16 inches = 1.33 feet. Based on these conditions, the quantity of irrigation water to apply would be

$$D_w = 0.253 \text{ foot} = 3.0 \text{ inches of water}$$

To illustrate how the quantity of water for leaching can change dramatically with various situations, we assume that all other factors remain the same except for the following:

- The leaching water quality is EC_w = 1.0 dS/m rather than 2.5 dS/m. Then, D_w = 0.183 foot = 2.2 inches water.

- The leaching water quality is of lower quality at EC_w = 4.0 dS/m rather than 2.5 dS/m. Then, D_w = 0.532 foot = 6.4 inches water.
- The soil is not a high-sand but a pushup green of about 85% sand with a k factor of 0.10. Then, D_w = 0.506 foot = 6.1 inches water.
- The grass has a threshold salinity tolerance of ECe = 3.0 dS/m rather than a threshold ECe of 5.0 dS/m. Then, D_w = 1.26 feet = 15.2 inches of irrigation water.
- The grass has a threshold salinity tolerance of ECe = 3.0 dS/m rather than a threshold EC_e of 5.0 dS/m, and the soil is 85% sand with a k factor of 0.10 instead of 0.05. Then, the D_w = 2.53 feet = 30.3 inches of water.
- The grass is a silt loam fairway (k = 0.10) with a threshold EC_e of 3.0 dS/m and a leaching depth of Ds = 2.5 feet. Then, the D_w = 4.75 feet = 57 inches of irrigation water.

In each of these examples, the least quantity of water necessary for a reclamation leaching was 2.2 inches of irrigation water, and the most was 57 inches. If the site is one with an existing turfgrass, the 2.2 inches would be in addition to ET replacement and any irrigation to compensate for nonuniformity of the irrigation system. It is instructive to note that the LR for maintenance leaching is most often in the 5% to 15% range; for a 1.0-inch irrigation event to replace ET and apply sufficient water for nonuniformity of the irrigation system, the quantity is 0.05 to 0.15 inches of additional water. This illustrates why turfgrass managers with saline irrigation water should strive to achieve and maintain a maintenance-leaching program and not to get into a situation where reclamation leaching is necessary. Even to program 2.2 inches of extra water in addition to ET replacement and for nonuniformity of the system would be a substantial challenge on most sites, especially considering that the above calculations were all based on a pulse irrigation sequence (discussed in the next section)—that is, 0.15 inches of extra water does not create a soggy, waterlogged site or create a difficult scheduling problem, but 2.2-plus inches does. If a flood or high-volume irrigation regime was used to create saturated flow, the quantities of applied irrigation water would double or triple.

The *estimated influence of a quantity of rainfall* on salt leaching can be determined by substituting a low ECw value for the "irrigation water" when the other factors are known. For example, in the initial example situation, the quantity of water indicated was 3.0 inches (76 mm) irrigation water required with an EC_w of 2.5 dS/m. Assuming all other conditions are the same, we could use an EC_w of 0.10 dS/m for rainfall and estimate the quantity of rainfall to accomplish the desired leaching to the depth selected. In this case, the D_w (as rain) = 0.16 foot or 1.9 inches of rain. This quantity would be a good estimate if the rain came as a light continuous one that would maintain unsaturated flow. However, if the rain was heavier, then the k factor from Table 11.3 would become 0.10 rather than 0.05, and the quantity of effective rain would be double to 3.8 inches, with the assumption that all of the water infiltrated into the soil at the site of impact.

Assuming all other factors remaining the same in the initial example and then comparing the leaching needs using irrigation water with EC_w = 1.0, 2.5, and 4.0 dS/m resulted in D_w values of 2.2, 3.0, and 6.4 inches, respectively. This illustrates that water quality has a very important influence on reclamation leaching and on maintenance LR. A second implication is that turfgrass managers should use their rainfall periods to maximize leaching, especially when reclamation leaching is necessary. For example, irrigating just prior to the forecasted rain event to wet the soil profile to near-field capacity will maximize the leaching potential of whatever level of rain may fall on the site. Additionally, following a good rainfall period where substantial salt leaching has occurred, salts are now at an acceptable level, and the leached salts are below the rootzone, the rule should be to immediately initiate a maintenance LR program. Sometimes turfgrass managers do not initiate a maintenance-leaching program in order to "conserve water" with the result that salts again start to accumulate. Rather, the LR fraction should be implemented to prevent salts from rising back into the rootzone via capillary upward movement. If resalinization of the rootzone is allowed to occur, a reclamation-leaching strategy will necessitate substantially more water being applied than the volume needed for the maintenance strategy.

11.5 PULSE IRRIGATION AND OTHER WATER APPLICATION METHODS

A highly efficient irrigation system design with good zoning is a priority for effective leaching of salts. However, the method of water application, even with a well-designed system, strongly influences the quantity of water for effective leaching (Hanson et al., 1999; Rhoades and Loveday, 1990). A review in Table 11.3 of the "k" factor (determined by irrigation method and soil type) in the reclamation-leaching formula illustrates the importance of specific water application regimes with pulse irrigation the most efficient. Potential means to apply water for reclamation or maintenance LR needs are (a) via heavier applications that favor saturated flow of water in the soil—flow primarily via macropores; or (b) by lighter applications, especially in repeated cycles or pulses that result in unsaturated flow—flow through both macropores and micropores for greater effective salt leaching.

Heavy, continuous water application by sprinklers where the soil is essentially saturated or near saturated throughout the leaching period would be similar to soil conditions that may occur from heavy rainfall or continuous ponding or flooding of water above the soil surface. Water application by any of these methods requires the most water quantity to achieve leaching, especially on fine-textured soils. Under *saturated flow* or near-saturated soil conditions, water flow is primarily through the larger macropores and water does not effectively leach between the macropores (i.e., within soil aggregates or micropore areas). On high sand content soils, which do not form aggregates but have more single-grain sands in the structure, saturated flow works better than on fine-textured soils for salt leaching. However, if the high sand content soils have a high organic content in the surface zone or contain appreciable volume of inorganic sand substitutes (calcined clays, diatomaceous materials, some zeolites, etc.) that significantly increase the total and micropores space volumes, these soils will also require higher quantities of water for leaching—that is, the k factor in Table 11.3 should reflect the estimated microporosity of the media and not just whether it is a "sand" when used in the formula calculations.

Heavier applications of irrigation water that result in saturated surface soil moisture conditions will foster greater runoff and uneven distribution of water over the landscape with more excessively wet and dry spots. Also, saturated flow favors the development of "finger flow" or "preferential flow" conditions within areas with more macropore channels. Both runoff and preferential flow within a soil would obviously adversely affect salinity leaching across the landscape, but also result in very inefficient use of water from a water-use efficiency or conservation standpoint. Under conditions of water applied to foster saturated flow, it is likely that wetting agents would not have much effect in creating a more uniform wetting profile.

Pulse irrigation (also called *cycle and soak*) is where water is applied in increments generally in the range of 0.20 to 0.33 inches (5 to 8 mm) with a time interval before the next pulse, and this cycle is repeated until the desired total quantity of water is applied. Each cycle limits the quantity of water to avoid runoff and saturated surface conditions. Instead, the surface soil moisture conditions result in unsaturated infiltration and percolation of the applied water, where water moves as a more uniform wetting front across both macropores and micropores. Runoff from the soil surface is minimized, and uniformity of application is maximized. The pulse irrigation method simulates a light, continuous rainfall that applies water at less than the soil's saturated infiltration rate. Such rainfall events are very effective in salt leaching. Pulse regimes of applying water are very effective and efficient in leaching salts. Normally only one quarter to one half the water is required for pulse irrigation versus heavy continuous irrigation—in other words, it takes 2 to 4 times as much water to leach to the same degree using heavy application as would be needed when using the pulse method.

Pulse irrigation is also a very water-use-efficient means for water conservation purposes for irrigation even on sites without salt issues since it maximizes irrigation water infiltration (i.e., increases effective irrigation). A deeper, less frequent irrigation regime by pulse irrigation for enhancing water-use efficiency and salt leaching can do the following:

- Reduce the number of localized dry areas.
- Allow the maintenance of viable roots deeper in the profile to take advantage of any water that moves deeper into the soil from precipitation, while a shallower zone of irrigation water penetration by lighter applications can result in root pruning. Root pruning may appear to allow salts to be leaching rapidly below the lower rootzone, but these salts would accumulate in a relatively shallow zone that could quickly rise by capillary action back into the rootzone with high-ET conditions, causing root desiccation.
- Result in salt movement deeper into the soil compared to light, frequent irrigation, or heavy events that result in primarily saturated flow.
- Allow for more opportunities to take advantage of precipitation events and, thus, allow canceling or delaying irrigation events.
- Reduce runoff and eliminate many excessively wet and dry areas.

Thus, pulse irrigation is both a good water-use-efficiency and -conservation strategy and a good salt-leaching protocol. A question that often arises in arid regions is "How deep and how infrequent?" As an example, assume that the full rootzone is recharged to field capacity on a Bermuda grass fairway with a 2.0-feet-deep root system and a fine-textured soil. During dry-down, the water is extracted from the surface zone first and then progressively moves downward. By the time that water is extracted to <50% field capacity in the surface 1.0 foot of soil, the turfgrass may be exhibiting a too slow growth rate for recovery from wear or may even be showing some drought stress symptoms; yet the lower rootzone may be at 75% to 95% field capacity. Thus, the controlling factor for irrigation is when the surface soil and turfgrass conditions require irrigation for the use requirements of the grass on the site. In this case, sufficient water would be applied by pulse regime to recharge the surface 1.0 foot. During the next one or two dry-down cycles, the deeper 1.0 foot of soil would gradually dry to a point that it would require recharging. This particular irrigation cycle would require more water and perhaps two nights of pulse irrigation to achieve the quantity of water for stabilizing the moisture in the soil profile and rhizosphere. With the above example, if the irrigation water was saline, the LR fraction would cause a net downward flow of water, and the lower soil profile should not be allowed to dry too much—perhaps to about 75% for field capacity before recharging. *The point of this discussion is that neither efficient water conservation nor salt leaching can be accomplished without a much wider adaptation of pulse irrigation regimes.*

Generally, the time interval between pulse cycles is 0.5 to 1 hour for sands, 1 to 2 hours for loamy sands and sandy loams, 2 to 4 hours for loams, and 3 to 6 hours for clays. Sometimes when the soil has low infiltration or percolation rates and the quantity of total water applied is relatively high, the pulsing strategy can be accomplished over several nights—it does not have to be limited to a single-night application. In reclamation-leaching situations where pulsing may be conducted over several consecutive nights, temporary cart and equipment traffic control may become necessary to minimize potential rutting and compaction of saturated native soils, particularly in fairways and roughs and when golf carts are not restricted to cart paths. A good surface cultivation program to maintain adequate water infiltration without runoff will reduce the time interval between irrigation pulses and may allow a higher quantity of water per pulse application.

As noted for pulse irrigation, water flow within the soil is primarily as unsaturated flow, which moves as a more uniform wetting front downward through the soil profile. Water movement occurs more in the micropores than in the macropores; therefore, leaching is more effective since soluble salts in the micropores will move with the water. Wetting agents often aid in maintaining a more uniform wetting front for leaching under a pulse irrigation regime or unsaturated flow conditions. Also, wetting agents reduce localized dry spots associated with water-repellant (hydrophobic) sands.

11.6 ADDITIONAL PRACTICES TO ENHANCE LEACHING EFFECTIVENESS

Whether the leaching program is designed for maintenance or reclamation leaching, it was noted that pulse irrigation would enhance the effectiveness of salt leaching with the least quantity of irrigation water application. There are also other factors that will enhance salinity leaching and, as the salinity level increases in irrigation water, the importance of these practices also increases. Some of these factors have been discussed in previous chapters, and others will be noted in subsequent chapters. They are listed here to emphasize their importance in an overall BMP approach to salinity management. The most important considerations when saline or sodic irrigation water is present are as follows:

- Use salt-tolerant turfgrasses and landscape plants. More salt-tolerant plant species allow more time to make management adjustments, including irrigation changes for leaching. Chapter 9, "Selection of Turfgrass and Landscape Plants."
- Target for a well-designed irrigation system (high uniformity of application and flexibility in control) to avoid spatial salt accumulation and to more effectively leach salts. Chapter 10, "Irrigation System Design and Maintenance for Poor-Quality Water."
- Practice good irrigation-scheduling protocols, including pulse irrigation discussed in Section 11.5 as well as items noted in this chapter. Accurate determination of ET, irrigation requirement adjustment of nonuniformity of the system distribution, and LR are all important.
- Use proper soil amendments, especially for sodic soils. Chapter 12, "Remediation Approaches and Amendments."
- Practice good application protocols for amendments. Chapter 13, "Amendment Application Options and Guidelines."
- Insure adequate infiltration, percolation, and drainage. Chapter 14, "Cultivation, Topdressing, and Soil Modification"; and Chapter 15, "Drainage and Sand Capping."
- Adjust fertilization to compensate for irrigation water quality nutrients and ions, soil and water amendments, and leaching losses of nutrients. Chapter 16, "Nutritional Practices on Saline and Sodic Sites."
- Adjust cultural practices to compensate for salinity effects on environmental, traffic, and biotic stresses as well as changes in greens-grooming protocols. Chapter 17, "Additional Cultural Practices." One cultural practice is the use of wetting agents (see Section 17.4.1, "Wetting Agents").
- Monitor via soil, tissue, and water quality tests as well as site observations and use of sensor technology. Chapter 18, "Proactive Monitoring of Progress."

In summary, development of a salinity management leaching program requires a holistic approach with consideration of a number of soil, water, and grass factors. *Leaching of salts is the most important component of any salinity management program.* Unless salts are consistently leached from the rootzone, resalinization will occur from continued irrigation water salt additions and from capillary movement from below the rootzone. The "peak time of year" for massive resalinization and the accompanying decline of turfgrass performance is often mid- to late summer. This is the least favorable time for salinity stress and the most difficult time to institute reclamation leaching. The best option for managing salinity is a continuous, routine *maintenance*-leaching program using an adequate LR. The most common reason for not applying sufficient irrigation water volume for leaching of salts is to underestimate the daily ET requirement for replacement of soil moisture lost by ET as well as to not account for nonuniformity in the irrigation system in terms of quantity of irrigation water to apply, rather than to underestimate the LR fraction of total irrigation water needed to move salts.

Also, it is important to understand that many of the scientific principles outlined in the current chapter were initially developed for annual agricultural crop production situations where daily equipment and pedestrian traffic, maintenance of a perennial grass playing surface, and continuous need for low mowing heights are not an area of concern. Therefore, the situation may occur where the outlined procedures cannot be implemented without compromising playing conditions or where implementation of other management programs (deep cultivation, traffic restrictions and control, and course closure) might be limited, required, or not completely effective.

11.7 SALINITY MANAGEMENT DURING ESTABLISHMENT

Turfgrasses, both glycophytes and halophytes, are quite vulnerable to saline soil and water conditions during early development. Genetically controlled plant salinity tolerance during establishment is much lower than for mature turfgrasses, including very salt-tolerant halophytes. Most of the salt-tolerance mechanisms are not genetically activated in juvenile plantlets early in the growth cycle. Thus, none of these salt-tolerant grasses can be established from sprigs or seed using highly saline irrigation water initially after planting and during early growth due to the severe growth regulatory and nutrient imbalance effects that are caused from salts and soil accumulation. Root system development would be severely or completely suppressed. Even sodding would be a problem because the root system would have difficulty in developing root hairs, in branching, and in developing stolons and rhizomes. Besides the high salt impact on the root system, turfgrass growth rates will be reduced, which can prolong the grow-in period by 1 to 4 months depending on salinity tolerance of the grass, irrigation water salinity levels, irrigation system efficiency for leaching, soil infrastructure, fertility program adjustments, and salt management capabilities of the grass manager.

As a guideline, the salinity of irrigation water for juvenile plantlets should be <20% of what the mature grass would tolerate without exhibiting salinity stress (refer to Chapter 9). Wind, high temperatures, plus exposed sandy surfaces during establishment and early grow-in can place very high evaporative demands on the overall turfgrass system. Under these frequent irrigation cycle conditions, salts can easily accumulate in the surface during establishment. After the initial establishment phase, grow-in (achieving full canopy coverage and transition to a mature stand with mature plants) may allow gradual transitioning to somewhat higher irrigation water salinity levels, but more rapid canopy coverage will occur with better quality water. If highly saline irrigation water is to be used on a routine basis, then application of this water should be delayed until full grow-in if possible, and definitely not be applied during the initial germination (seeds), plantlet initial rooting, and subsequent establishment phases. Proper management techniques can minimize the need for an expensive replanting and aid in rapid establishment. The key BMP management consideration when dealing with salinity and juvenile plants is to alleviate or minimize any high-soil and irrigation water salinity issues prior to planting; even halophytes will not remediate highly salinized soils or tolerate high-salinity water during the early growth stages.

11.7.1 Alleviation of Na-Induced Soil Physical Problems in the Surface Zone

Aggressive deep and shallow cultivation, gypsum application, medium to coarse sand topdressing, and leaching are key management options during the critical establishment period. If the soil is already sodic from the influence of seawater or salt spray, or for other reasons, the best time to apply the quantity of gypsum required to aid in reclaiming the soil is prior to establishment since large quantities can be applied and deep tilled. This is especially important for soils containing even modest quantities of silt or clay. The quantity of soil profile–integrated gypsum depends on the soil texture and how sodic the soil is (i.e., percent Na saturation on CEC or exchangeable Na percentage = ESP) (Carrow and Duncan, 1998; also see Chapter 13, "Amendment Application Options and Guidelines").

Even if large quantities of gypsum have been added, it is important to apply a surface application of 30 to 60 lb. gypsum per 1000 sq. ft. so that the soil surface does not seal from Na dispersion, which would make continuing irrigation difficult. As the grass matures and poorer quality water is applied, gypsum applications should be made periodically to maintain surface permeability and always be made (integrated into the soil profile with cultivation) to avoid creating an Na-affected layer deeper in the soil profile from the frequent irrigation applications during establishment where water penetration tends to be in a uniform zone, often at 3 to 4 inches of depth. Applications to more mature turfgrass stands are often done in conjunction with core aeration to properly integrate the gypsum into the upper soil profile. The soil-applied granular gypsum is also an excellent slow-release Ca source for plant nutritional needs, which is required in these situations. Acid-sulfate sites, because of pH level <4.0 and high Al, Mn, and S levels, may require not only gypsum but also lime applications mixed into the rootzone and some surface-applied lime during the grow-in to the full canopy density period to form gypsum and stabilize the excess sulfates. Dolomite applications should be avoided since additional Mg is added with this product and ocean water already has high Mg concentrations.

11.7.2 Reduction of Total Salts for Establishment

On any soil that has high total salts, these should be leached prior to turfgrass establishment to a level where the grass will not be exposed to salinity stress. Leaching should be sufficiently deep to prevent capillary rise of salts between irrigation events. An extreme case of a salinization of the soil surface is sites exposed to seawater by flooding. Seawater has a total salinity level of $EC_w = 46$ dS/m (using a 750 conversion, $EC_w = 54$ dS/m when a 640 conversion is used; the most appropriate is 740 for seawater). Total salts will only be reduced below 46 dSm^{-1} or 34,560 ppm TDS under the following conditions: after a heavy rainfall (cyclone, typhoon, and hurricane) or prolonged rainy period, by use of lower salinity water resources or by blending with lower salt-containing water resources. Thus, use of as high-quality water as is available is strongly recommended, and even with halophytic grasses, seed germination and establishment by vegetative means will be greatly reduced as salinity increases and specific salt ions accumulate in the soil. For seashore paspalum, the safe limits seem to be irrigation water of TDS < 2500 ppm regardless of means of establishment to achieve a high percentage of plant survival and reasonable rate of establishment (some cultivars with higher salinity tolerance can be managed at higher TDS with proper management). Vegetative propagation by sprigging does not appear to offer any higher initial tolerance to salinity than by seeding (only one seeded cultivar is available at this time, ‘SeaSpray’) for seashore paspalum. Paspalum sprigs seem to be more sensitive to salinity as temperatures increase with salinity tolerance of 2500 to 8000 ppm when temperatures are in the 75°F to 90°F range, but at above 90°F, overall salinity tolerance may be reduced by at least half. While establishment may be possible at somewhat higher salt levels, the process will be very slow (possibly extending grow-in to full canopy density by 1 to 6 additional months depending on the season) and budgeting issues will be costly. For other halophytic species, the limits will likely be different. If the soil was initially saline, preestablishment irrigation must be applied to reduce the salinity levels to an acceptable level. Preestablishment leaching would be applied after adding gypsum treatments. Leaching should be ample enough to leach salts to at least 8 inches deep in the soil profile, and then the site should not be allowed to dry or the salts will rise by capillary action to resalinize the surface. For highly saline sites, leaching beyond 8 inches deep would be preferable.

11.7.3 Maintenance of a Uniformly Moist Soil Profile

Assuming no rain, the soil salinity will be no lower than the irrigation water salinity used during establishment if an excellent leaching program is maintained. Drying concentrates salts in the soil solution (often twofold or higher than the incoming water salinity level), thereby increasing the potential salinity stress on the turfgrass. During the initial establishment phase, irrigation should be frequent enough to avoid drying of the surface; but if the soil is more fine textured, care should be taken to avoid waterlogging. By maintaining a moist soil, salinity stress is reduced. As noted, on sites where the soil was highly saline, initial leaching is required to reduce surface soil salinity to acceptable levels. Thereafter, irrigation should be frequent enough to prevent excessive drying of the surface and to prevent the capillary rise of salts from below (i.e., leaching salts deeper into the soil).

11.7.4 Adequate Initial Fertilization and Careful Monitoring of Micronutrients

A good starter fertilizer with adequate phosphorus (2–3 lb. P_2O_5 per 1000 sq. ft.) should be applied to the surface and worked into the top inch at planting to promote establishment. Soil test analysis will reveal the need for additional fertilizer nutrients. After initial establishment, spoon-feeding nutrients should be practiced, but as the transition to poorer water quality is made, it becomes even more critical (while higher annual rates of fertilizer are required, the rate per application is similar to non-salt-affected sites—but the frequency is greater). Use of highly soluble fertilizers and fertigation through a well-designed irrigation system would be very beneficial during establishment, grow-in, and progression to a mature stand. High-leaching events can deplete micronutrient (Fe and Mn) levels, and careful soil and plant monitoring is necessary on a continuous basis. Calcium and Mg are subject to leaching and should be monitored closely. Potassium should be applied frequently (often weekly with prescription application strategies) in a 1:2 ratio ($N:K_2O$) with N. After initial establishment on sites using seawater blends or highly saline irrigation water, K nutrition may be even higher in some cases and still must be frequently applied. The key BMP consideration is "Do the plants have nutrient sufficiency levels internally to sustain growth and development with the salinity exposure on the site?"

12 Remediation Approaches and Amendments

CONTENTS

12.1 OVERVIEW OF REMEDIATION APPROACHES IN SODIC SITUATIONS

In the best management practice (BMP) approach for salt-affected turfgrass sites, a number of chemical and physical amendments have been discussed for use on salt-affected soils in chapters on individual BMP strategies. These amendment applications were presented in the context of improving soil physical conditions for better leaching or as routine maintenance operations to improve turfgrass performance under salinity stress; that is, fertilization, sand capping, soil modification by sand, topdressing, wetting agents, cytokinins, science-based organic and inorganic amendment additions, and others. Saline soils, as noted in Chapter 2, "Saline Soils," do not require Ca amendment treatment to displace Na since Na is not the dominant salt stress in these soils, but total soluble salts is the primary salinity management limitation issue. However, sodic and saline-sodic soils, as described in Chapter 3, "Sodic, Saline-Sodic, and Alkaline Soils," do require vigorous chemical reclamation efforts, especially the addition of chemical amendments that provide available Ca as a displacement ion for Na on the soil cation exchange capacity (CEC) sites (Table 12.1). In the vast majority of reclamation cases for turfgrass sites, Ca arises from either direct application of an Ca amendment (normally gypsum) or application of an S form (acidifying agent) + lime (native

TABLE 12.1
Soil and Irrigation Water Situations That Require Ca Amendment as a Displacement Ion for Na in Soil Reclamation

Initial Soil Salinity Classification	Requirement for a Ca Amendment to Provide Ca to Replace Na
Irrigation Water High in Na (i.e., High SARw)	
Sodic[a]	Yes, continuous basis to remediate and to prevent future sodic conditions
Saline-Sodic[a]	Yes, continuous basis to remediate and to prevent future sodic conditions
Saline[a]	Yes, to prevent future sodic condition
Non–salt affected[a]	Yes, to prevent future sodic condition
Irrigation Water Low in Na (i.e., Low SARw)	
Sodic[b]	Yes, but only until the soil is remediated to a nonsodic condition
Saline-Sodic[b,c]	Yes, but only until the soil is remediated to a non–saline-sodic condition
Saline[b]	No calcium amendment needed
Non–salt affected	No calcium amendment needed

[a] All of these situations require leaching to remove Na on a long-term basis.
[b] These require leaching until the initial Na or high total salts are leached.
[c] If a saline-sodic soil has appreciable Ca in the surface horizon, a Ca source may not always be needed when a low SAR_w water is used.

in the soil or applied to the soil) to chemically form gypsum in the soil. *In Chapter 12, the focus is on remediation approaches (options) for sodic and saline-sodic soils involving amendments that are applicable to turfgrass ecosystems.* When the irrigation water is high in Na, reclamation (repeated and carefully scheduled Ca amendment applications, cultivation, and leaching) is an ongoing process.

Except for the book by Carrow and Duncan (1998), most of the extensive publications on sodic and saline-sodic soil management has focused on agriculture situations (USSL, 1954; Abrol et al., 1988; Rhoades and Lovejoy, 1990; Rengasamy and Olsson, 1991; Sumner, 1993; Jayawardane and Chan, 1994; Naidu et al., 1995; Chhabra, 1996; Oster and Jayawardane, 1998; Levy, 2000; Suarez, 2001; Rengasamy, 2002; Qadir and Oster, 2004; Qadir, Noble, et al., 2006; Qadir, Oster, et al., 2006; Qadir et al., 2007). The USSL publication (1954) and the FAO publication by Abrol et al. (1988) are considered foundational publications.

Amendment remediation (reclamation) approaches for sodic and saline soils entails using one or more of four methods:

- Direct *application of one or more soluble Ca amendments* with gypsum being the most widely used Ca source.
- Direct *application of a sulfur source plus lime* (lime either applied or already contained in the soil at sufficient concentrations to activate the necessary chemical reactions) to create gypsum in the soil.
- *Application of organic matter* as a secondary amendment to assist in stabilizing aeration and moisture dynamics in the soil and to provide nutrients for microbial activity. Various organic materials are often used during preestablishment, during establishment, or on mature turfgrass stands to improve soil physical conditions under sodic stress. Organic amendments are considered secondary remediation approaches since they are best used in conjunction with a Ca source and are much less effective if used as the sole amendment in salt-challenged turfgrass systems.
- *Phytoremediation* is the use of live, higher order plants for the reclamation of sodic and saline-sodic soils. Synonymous terms are *vegetative bioremediation*, *phytoamelioration*,

and *biological reclamation*, but the latter term would also include microbial-based amelioration products (Qadir, Oster, et al., 2006; Qadir et al., 2007). Note that this strategy differs from the application of organic matter amendments in that a live plant is used in phytoremediation. In agricultural situations, specific salt-tolerant plants with a crop value may be used initially to help in reclamation of a new site, and then such plants are included in future crop rotations.

- *Microbial-based remediation* for salt-affected soils has been a very minor component within the "phytoremediation" approach, but Singh and Dhar (2010) recently discussed the use of cyanobacteria for reclamation of salt-affected sites. Additional microbes have been investigated for their functional effectiveness in phytostabilization of saline soils (Sannazzaro et al., 2007; Yuan et al., 2007; Yao et al., 2010; Grandlic et al., 2009; Azam and Ifzal, 2006; Tang et al., 2009). In Chapter 20, "Reclamation, Drainage Water Reuse Schemes, and Halophytic Forage Sites," salt reclamation topics related to nonturfgrass situations are discussed and phytoremediation is addressed in more depth.

In the vast majority of amendment-based reclamation cases for turfgrass sites, either of the first two approaches (Ca amendment application, or S + lime to form gypsum) is the basic remediation method. It is not uncommon for some organic amendment applications to be used, at least during establishment, sometimes on mature turfgrass ecosystems. Phytoremediation in the same sense that salt-tolerant crops are used in agriculture situations is not applicable to turfgrass systems. For example, in agricultural systems, prior to using the site for crop plants, a halophytic plant may be used for 1–2 years in order to add organic matter, create root channels, and possibly remove salts if the crop is harvested and shoot material or total biomass containing sequestered salts is removed. Instead, in turfgrass systems, appropriate, adapted salt-tolerant perennial grasses are selected as permanent grasses on the site, and these plants then contribute to the overall remediation of a site (such as *Paspalum vaginatum* on dredge spoils; see Onwugbuta-Enyi and Onuegbu, 2008)—this will be discussed in greater detail later in this chapter.

When sodic or saline-sodic soils are present due to irrigation water additions of Na, but the soils are very *sandy* in nature with little or no clay content, then aggressive "soil reclamation" actions are less important because there are no "soil structural aggregates" to break down. Sufficient available Ca will still be necessary to deal with stabilization of the limited quantities of clay, silt, or organic colloids and to provide available Ca for the plant root system as protection from high Na levels, but the quantity of Ca will be much less than when applied to fine-textured soils. In *fine-textured soils*, as clay content increases, especially 2:1 clay types, reclamation is extremely important and an ongoing process if the irrigation water contains very much (>125 ppm or 5.4 meq/L) Na. Even a limited quantity of clay (3%–5%) can cause problems by dispersing and migrating to the zone of irrigation penetration (bottom of the wetting front), where it may contribute to layering along with Na carbonates and bicarbonates and organic colloids. Percolation rates decrease, and oxygen flux is reduced through this layer in the soil profile.

Excessive Na in sodic and saline-sodic soils markedly reduces soil permeability for water and gas exchange (see Chapter 3, "Sodic, Saline-Sodic, and Alkaline Soils") where the Na may already exist in the soil or be frequently added via the irrigation water. Thus, *the basic chemical amendment is Ca* in order to displace Na on the CEC sites and with the displaced Na then leached from the rootzone (Table 12.1). Without the Ca to improve soil structure that has deteriorated due to excess Na, the soil will exhibit poor soil physical properties (e.g., poor aeration; low infiltration, percolation, and drainage; inability to leach soluble salts effectively; waterlogging; and poor rooting).

As noted, the two major approaches for providing Ca to the soil are (a) by applying a Ca source capable of supplying relatively high levels of moderately soluble Ca (ideally, releasing Ca with each irrigation or rainfall cycle)—usually gypsum supplemented with other Ca sources; or (b) use of a sulfur source in conjunction with added lime or native calcite already present in the soil to create gypsum within the soil (Table 12.2). As a secondary amendment, various organic materials are

TABLE 12.2
Equivalent Rates of Calcium Compounds, Acids, and Acid-Forming Amendments Required to Replace 1 meq/kg of Exchangeable Na in the Soil or to Increase the Ca Content in the Irrigation Water by 1 meq/L Either by Direct Addition of Ca or by Neutralizing HCO_3 + CO_3 in the Water to Allow Ca in the Water to Be Available

Chemical Name[a]	Chemical Formula	Pounds Equivalent to 1 lb. Gypsum[h]	Soil or Water Amendment	[b]Pounds/ Acre—6″ Soil to Replace 1 meq ex. Sodium	[b]Pounds/ Acre—Ft. of Water to Obtain 1 meq/L Free Ca
Gypsum (23% Ca; 19% S)	$CaSO_4 \cdot 2H_2O$ (100%)	1.00	S,W	1720	234
Elemental S	S (100%)	0.19	S	326	43.6
Sulfuric Acid[f,g] (33% S)	H_2SO_4 (100%)	0.61	S,W	978	133
Anhydrite (29% Ca; 24% S)	$CaSO_4$	0.79	S	1358	—
Lime Sulfur (Ca Polysulfide) (9% Ca + 24% S)	CaS_x	0.78	S,W	1410	191
Calcium Chloride (27% Ca)	$CaCl_z \cdot 2H_2O$	0.85	S,W	1538	200
Calcium Chloride Hi-Cal (13% Ca)	$CaCl_z \cdot 2H_2O$	1.76	S,W	3076	400
Calcium Nitrate (15.5% N; 20% Ca)	$Ca(NO_3)_2 \cdot H_2O$	1.15	S,W	1978	269
Calcium Thiosulfate (6% Ca, 10% S)	$Ca(S_2O_3)_3 \cdot$ CATS	3.83[e]	S,W	6387[e]	869[e]
Potassium Thiosulfate (17% S; 25% K_2O)	KTS	1.12[c]	S,W	1883[c]	256[c]
Ammonium Thiosulfate (26% S; 12% N)	THIO-SUL	0.73[c]	S,W	1233[c]	167[c]
Ammonium Polysulfide (40% S; 20% N)	NITRO-SUL	0.48[c]	S,W	800[c]	109[c,d]
Monocarbamide dihydrogen Sulfate (MCDS) (US-10. 10% N; 18% S)	N-Phuric. Urea + Sulfuric Acid	1.06[c]	S,W	1779[c]	242[c,d]
Dicarbamide dihydrogen Sulfate (DCDS) (28% N; 9% S)	N-Control. Urea Sulfate	2.11[c]	S,W	3557[c]	484[c]
Monocarbamide dihydrogen Sulfate (MCDS) (15% N; 15% S)	pHairway. Urea + Sulfuric Acid	1.26[c]	S,W	2139[c]	291[c]
Phosphogysum of Flue Gas Desulfur. Gypsum	24% to 29% Ca	0.79–0.96	S	1359–1651	—
Aluminum Sulfate (14.4% S)	$Al_2(SO_4)_3 \cdot 18\ H_2O$	1.30	S	2219	—
Iron Sulfate (11.5% S)	$FeSO_4 \cdot 7H_2O$	1.65	S	2786	—
Ferric Sulfate (17% S)	$Fe(SO_4)_3 \cdot 9H_2O$	1.12	S	1875	—

TABLE 12.2 (Continued)
Equivalent Rates of Calcium Compounds, Acids, and Acid-Forming Amendments Required to Replace 1 meq/kg of Exchangeable Na in the Soil or to Increase the Ca Content in the Irrigation Water by 1 meq/L Either by Direct Addition of Ca or by Neutralizing HCO_3 + CO_3 in the Water to Allow Ca in the Water to Be Available

Chemical Name[a]	Chemical Formula	Pounds Equivalent to 1 lb. Gypsum[h]	Soil or Water Amendment	[b]Pounds/ Acre—6″ Soil to Replace 1 meq ex. Sodium	[b]Pounds/ Acre—Ft. of Water to Obtain 1 meq/L Free Ca
Iron Pyrite (30% S)	FeS_2	0.35	S	602	—
Langbeinite (Sul-Po-Mag) 21% K, 11% Mg, 22% S	$2MgSO_4 \cdot K_2SO_4$	1.24	S, W	2133	290

Source: Adapted from Sanden, B., A. Fulton, and L. Ferguson, Managing salinity, soil and water amendments, University of California, Kern County, 2010, http://cekern.ucdavis.edu/Irrigation_Management/MANAGING_SALINITY,_SOIL_AND_WATER_AMENDMENTS.htm.

[a] Adapted from Sanden et al. (2010).

[b] Salts bound to the soil are replaced on an equal ionic charge basis and not equal weight basis. Laboratory data show that an extra 14% to 31%, depending on initial and final ESP or SAR, of the amendment is needed to complete the reaction.

[c] Based on acidification only from the sulfate component contribution and not N or K contributions that may occur in the soil.

[d] See Table 13.2 for Ca equivalent based on including ammonium-based acidity.

[e] Based on the soluble Ca content and not the sulfate component contribution.

[f] Sulfuric acid can be purchased as an acid or generated by sulfur dioxide generators from elemental S. Elemental S is oxidized into SO_2, then SO_2 combines with water to generate H_2SO_3.

[g] *Sulfur* is also spelled as *sulphur*, *sulfate* as *sulphate*, and *sulfuric* as *sulphuric*.

[h] Purity. Listed at 100% purity. For amendments of less than 100% purity, the following formula can be used to calculate an amount equivalent to 1 pound of 100% gypsum:

$$Equivalent\ amount = \frac{100\%}{\%\,purity} \times Pounds\ Equivalent$$

For example, assuming 93% sulfuric acid is to be used:

$$Equivalent\ amount = \frac{100\%}{93\%} \times 0.61 = 0.66\ lb.$$

often used during preestablishment or establishment, or on mature turfgrass stands, to improve soil physical conditions under sodic stress. Organic amendments are considered secondary amendments since they are best used in conjunction with a Ca source and not as the sole amendment. The use of sand as an amendment to physically improve soil conditions for better salinity leaching is addressed in Chapter 14, "Cultivation, Topdressing, and Soil Modification," and Chapter 15, "Drainage and Sand Capping." In Chapter 15, it is emphasized that on highly sodic soils, with substantial Na additions from irrigation or subsoil capillary rise, and high 2:1 clay content, the best initial amendment approach is sand capping with a Ca amendment at the sand-to-soil interface, followed by a soil-applied Ca source on an ongoing basis as needed. These excess Na conditions are typical, for example, in turfgrass sites established on marine clays.

Reclamation of sodic or saline-sodic soils for a new turfgrass area to be irrigated with good water would be primarily targeted as a preestablishment effort until the site is suitable for a permanent grass. However, in cases where the irrigation water is high in Na and this water source is the major contributor to the sodic soil condition, reclamation is both a preestablishment and an ongoing sustainable process that does not end. When reclaiming a salt-affected soil or preventing a salt problem due to poor irrigation water quality, several factors must be considered:

- The site-specific conditions and especially the particular combination of soil chemical and physical properties and irrigation water quality—addressed in Section II of this book, "Site Assessment BMPs for Saline and Sodic Soil Sites" (Chapters 4–8)
- Choice of a remediation approach or combination of approaches where this decision influences what amendment or combination of amendments to supply Ca since there are different options, which are addressed in this chapter
- Rates of amendments and application method options, which are discussed in Chapter 13, "Amendment Application Options and Guidelines"

The most common chemicals added to the soil to increase the ratio of Ca to Na are gypsum, elemental S + lime source, or sulfuric acid + lime source (Table 12.2). Some materials can only be soil applied as a granular, while others can be water applied (liquid or suspension). Common irrigation water amendments are gypsum (soluble or suspended) and sulfuric acid. Other soil or water amendments are sometimes used due to local availability. Basically, amendments to treat sodic or saline-sodic soils are Ca compounds or S compounds. When S sources are used, lime (or native calcite or free calcium carbonate) must be present in the soil to generate a Ca source when the S reacts with lime to create gypsum. Amendment options are discussed in detail in the following sections.

12.2 CALCIUM SOURCES

Since many sodic and saline-sodic soils occur under alkaline pH where lime ($CaCO_3$ or calcite) is present in the soil often in relatively high content, it would appear that ample Ca should be available for Na displacement. However, lime alone has a very low dissolution rate that is not sufficient to provide the level of soluble Ca necessary in the soil solution to adequately displace Na on CEC sites (Table 2.1 and Table 16.2). Only when the lime dissolution rate is increased by addition of a sufficient quantity of acid-forming materials will it release sufficient Ca on a sustained basis by forming gypsum—this is discussed in Section 12.3, "Acid-Forming Materials + Lime Source." Thus, more soluble forms of Ca than lime are necessary; but due to the relatively high quantity of Ca necessary to displace the Na, it is not desirable to have all the Ca "water soluble" at one time. If this was the case, total soluble salts would be high and the Ca would be very susceptible to leaching loss. Rather, a moderately slow release form of Ca is desirable, such as granular gypsum ($CaSO_4 \cdot 2H_2O$) (refer to Table 2.1 for calcium product dissolution rates).

For Ca or Mg amendments directly applied to the soil (i.e., not as gypsum formed over time by S source + lime), there are two means for improvement of soil structure degradation caused by high Na, namely, (a) the most important is by the displacement of Na from the clay CEC by Ca that leads to structure formation and stabilization; and (b) if the soil ECe is increased at the surface, then the negative effects of Na are decreased—otherwise, high soil ECe results in less Na permeability hazard even when the soil sodium adsorption ratio (SAR) (soil exchangeable sodium percentage [ESP]) is relatively high. This is illustrated in Figure 3.1, where at high soil ECe (or high irrigation water ECw), infiltration remains good even at relatively high SAR. To obtain the latter salt effect, soluble Ca salts and other salts such as langbeinite (a soluble Mg and K sulfate) can produce similar results. During earlier phases of reclamation of a highly sodic soil before a grass is established, water-soluble forms may provide more rapid initial response in improving water infiltration due to the high soil ECe and the decrease in Na permeability hazard that is created. However, since these materials can more easily leach and produce soil ECe higher than most grasses can tolerate under the high rates used, the shift should be toward more slow-release granular Ca forms like gypsum, which solubilizes with each irrigation cycle or rainfall event (Aydemir and Najjar, 2005).

Various Ca amendments are discussed in the remainder of this section as well as langbeinite, an Mg and K sulfate–based amendment (Table 12.2).

12.2.1 Gypsum

There are two calcium sulfate minerals used for sodic and saline-sodic remediation: gypsum ($CaSO_4 \cdot 2H_2O$; dihydrate) and anhydrite ($CaSO_4$) (Table 12.2). Gypsum is by far the most commonly used form, it is widely available, and it is moderately soluble, which allows available Ca to be released over time in sufficient solubilized quantities to displace Na on the CEC sites (Table 2.1). Gypsum purity depends on the source; but products used for sodic soil reclamation should be >70% purity; and on sandy sites, impurities of silt or clay should be avoided from mined products. Some mined gypsum can be as low as 20% purity, and shipping costs would increase as purity decreases. Gypsum requirement for a soil is based on the Na on the CEC sites rather than Na in the Na-carbonate or Na-bicarbonate forms, where the reaction of gypsum targeted to displace Na on the clay CEC sites is

$$2Na^{+}CLAY + CaSO_4 \cdot 2H_2O \leftrightarrow Ca^{+2}CLAY + Na_2SO_4 \text{ (leachable)} + 2H_2O$$

The Na_2SO_4 created in this reaction is leachable as long as an effective leaching program is practiced on a continuous basis. If no leaching program is in effect, the Na will accumulate and Na on the CEC sites will increase again, while the Ca in the gypsum will tend to form lime.

As indicated, in addition to Na on CEC sites, Na may also be present in sodic and saline-sodic soils as Na carbonate and Na bicarbonate where the carbonate forms precipitate due to lack of adequate leaching, ongoing Na additions, and poor soil structure (see Chapter 3, Section 3.1.2, "Process of Physical Degradation of Soil Structure by Na"). But, as stated, Ca from the gypsum is not targeted to the carbonate and bicarbonate pool of Na, but to the Na on the clay CEC sites since it is this Na that adversely affects clay structure and subsequent breakdown (deflocculation). The carbonate and bicarbonate forms, however, often contribute to reduced infiltration and percolation by precipitating along with lime and sometimes partially sealing soils, especially by causing a layer at the depth of irrigation water penetration (bottom of the wetting front) in arid regions that results in a layer of reduced water flow. Accumulation of Na carbonate and Na bicarbonate is much less apt to occur in semiarid or humid regions except in prolonged dry periods due to leaching with precipitation events. The carbonate and bicarbonate forms are leachable and will leach along with the Na sulfate—see the solubility for these salts in Table 2.1. As the Ca displaces Na on the clay micelles, soil structure should improve and allow easier leaching of Na sulfates (i.e., Na from the CEC sites that binds with the sulfate component from the gypsum) and carbonate and bicarbonate forms. It is essential that a leaching program be in operation, or all of these Na forms will continue to accumulate and layer in the soil.

Gypsum can be applied as (a) granular material by gypsum, (b) gypsum injector units that inject gypsum-saturated water into the irrigation system, (c) gypsum injection into the irrigation system of very fine gypsum suspension, and (d) some turfgrass managers have deposited very fine grade gypsum into a small irrigation lake near the intake while irrigating the suspended gypsum. Gypsum dissolution rate depends on particle size and degree of crystallization with mined forms often more highly crystallized. Higher solubility is related to smaller particle sizes and less crystallization (Abrol et al., 1988). High Na levels can increase gypsum solubility to some extent, while lime may decrease dissolution rate. Oster and Frenkel (1980) suggested increasing the gypsum requirement in highly calcareous, sodic soils by 10% to 30% to compensate for the lime effect. Lime has been reported to sometimes coat part of the gypsum particles and reduce dissolution.

A gypsum source with fine particle size may be preferred (a) during reclamation conditions, when Na levels are already high and the goal is to decrease the Na ESP on the CEC as quickly as possible; (b) when the irrigation water is high in Na so that Na additions are high (i.e., >125 ppm or

5.4 meq/L) on a continuous basis; or (c) if applied by a gypsum injector as a suspension and saturated solution into irrigation water. Otherwise, a particle size range from 0.25 to 2.0 mm diameter is a good choice to provide a wider time frame for dissolution (slow pulse irrigation cycle release). Also, in highly calcareous, sodic soils, a wider particle size range is preferred since fine particles are most prone to being coated by lime while particles of 1 to 2 mm diameter are not coated.

During reclamation of a site already highly sodic and prior to turfgrass establishment, gypsum should be mixed to as deep as the expected rootzone or to at least 12 inches (30 cm) with a final surface application prior to establishment. For established turfgrass sites that are highly sodic and/or receive appreciable Na via irrigation, maintenance of a continuous supply of gypsum at the surface is important; but deeper injection of gypsum by a cultivation (such as drill and fill), sand-silt, or slit drainage device is also very beneficial for maintaining better soil structure throughout as much of the rootzone depth as possible. Units such as the Blec Sand Master, Drill & Fill, and WaterWick are examples of devices capable of injection of a combination of sand and gypsum or straight gypsum into soil profiles to help create and maintain vertical macropore channels in sodic soils.

The influence of gypsum on soil pH is sometimes not well understood. In Chapter 3, Section 3.1.3, "Plant and Soil Symptoms of Sodic Conditions," it was noted how high ESP and Na_2CO_3 levels result in pH > 8.5 and up to pH 10.4. When Ca displaces Na on the exchange sites and Na_2CO_3 is leached, the soil pH will decrease to about pH 8.3, and then it does not decrease any further (due to the alkaline soil conditions) unless an acidifying agent is applied in sufficient quantities to dissolve any free $CaCO_3$. Since many of these soils are calcareous, pH will remain at about pH 8.3; and at this pH, the toxic OH^- ion levels (which generally occur at pH > 9.0) are not of concern. Thus, gypsum can reduce pH, but only in highly sodic soils and in the range stated above. The focus of this chapter is primarily on soil chemical changes that occur during remediation of sodic soils, but one very important change in highly sodic soils is the reduction in OH^- ions that are toxic to roots as well as excess Na^+ that also cause root deterioration as the reclamation process proceeds.

12.2.2 Anhydrite

Anhydrite is a calcium sulfate mineral with the formula $CaSO_4$. It has lower solubility than gypsum ($CaSO_4 \cdot 2H_2O$; dihydrate), but solubility comparisons are greatly influenced by particle size, the presence of other solution ions, temperature, and other factors; so there is not a straightforward relationship among the factors (Stowell and Gelernter, 2009). The higher gypsum equivalent of anhydrite versus gypsum (1.26 vs. 1.0, respectively) in Table 12.2 reflects less water of hydration; however, this does not imply higher solubility or Ca availability, but reflects only percent Ca content. Anhydite is a mined material and is normally mixed with other chemicals such as gypsum and calcite. Since it has lower solubility, anhydrite should not be used as the primary Ca source in cases where relatively rapid Ca availability is required, but this product is an option for situations where gradual, long-term dissolution is acceptable—the alternative strategy being more frequent application of gypsum or using gypsum with larger particle sizes. As with fine particle sizes of gypsum in the presence of appreciable calcite or lime, anhydrite particles may be subject to some lime formation on the surfaces that can reduce dissolution rates, and application should involve an aeration event to integrate it into the soil profile.

12.2.3 Gypsum Byproducts (PG and FGDG)

In some locations, gypsum byproducts from manufacturing processes may be available that are lower in cost than gypsum (Branson and Fireman, 1980; Abrol et al., 1988; Amezket et al., 2005; DeSutter and Cihacek, 2009; Gharaibeh et al., 2010). Phosphogypsum (PG) and flue gas desulfurization gypsum (FGDG) are the two most common forms, and these products contain from 24% to 29% Ca content. These materials are normally 5 to 10 times more soluble than mined gypsum due to fine particle size. Soil application is the standard practice that is used for both PG and FGDG. Application procedures and situations where soil-applied PG and FGDG can be used are similar to those noted for gypsum. For sodic soil reclamation prior to turfgrass establishment where high Ca amendment rates are required, these products should be seriously considered as alternatives to gypsum.

12.2.4 Other Ca Sources

Other Ca sources that can be directly applied to the soil or into the water source are much more soluble in nature than gypsum (Table 2.1); therefore, they increase irrigation ECw if applied via the irrigation system or soil ECe when directly applied. These materials are best used as *supplemental Ca additions* in conjunction with the primary Ca coming from gypsum or an S source + lime.

One example is *calcium chloride, $CaCl_2$*, which is expensive unless it is available as a byproduct material. It has not been widely used in agriculture due to concern over Cl absorption on agronomic and horticultural crops, but turfgrass plants are not very sensitive to high Cl except as it contributes to total soluble salts. Calcium chloride has very high solubility and easily increases the ECw of irrigation or ECe surface soil solution water, while adding available Ca for Na displacement. Calcium chloride should be considered in the following situations:

- For sodic soils with SAR above 20 in the surface horizon, since high soil ECe or water ECw reduces the adverse effects of Na on permeability as shown in Figure 3.1, where high ECw decreases Na permeability problems even when SAR is high. Because $CaCl_2$ is much more soluble than gypsum, the reduction of Na permeability at the surface is more pronounced. As with gypsum, the Ca in $CaCl_2$ also improves soil structure by displacing Na from the clay CEC.
- Any soil with a highly sodic B or C horizon of SAR above 20 because the soluble Ca can move more rapidly into the subsoil with the Cl and provide the counterion to Na as long as there is adequate leaching water applied. Thus, this highly soluble Ca source would help remediate the subsoil sodic conditions more rapidly than granular gypsum applied on the surface.
- On saline-sodic or sodic soils where the irrigation water has very low EC_w (<0.50 d S/m or <320 ppm TDS) in order to enhance water infiltration by injection of $CaCl_2$ into the irrigation water. Duncan et al. (2009) discuss ultrapure water effects on water infiltration into soils and note that increasing the ECw > 0.50 dS.m with $CaCl_2$ or a gypsum injector to inject Ca-saturated water greatly benefits water infiltration in these soil situations—whether the soil is salt affected or not.
- During initial reclamation of sodic or saline-sodic soils prior to grass establishment in order to reduce reclamation time.

On established turfgrass, the quantity of $CaCl_2$ is controlled by (a) avoiding salt burn similar to salt fertilizers, where approximately 4 lb. salt (granular) material per 1000 sq. ft. (1.95 kg/100 m^2) is within the range for turfgrass foliar burn under the right conditions such as moist surfaces and high humidity; or (b) concern over soil ECe becoming too high for the specific grass salt tolerance level. The soil reaction is

$$2Na^{+}CLAY + CaCl_2 \cdot 2H_2O \leftrightarrow Ca^{+2}CLAY + 2NaCl \text{ (leachable)} + 2H_2O$$

As with all Na displacement amendments, there must be an effective, long-term leaching program to remove the displaced Na.

Calcium nitrate, $Ca(NO_3)_2 \cdot H_2O$, can be a granular-applied fertilizer or added in the irrigation water by a fertigation system. The total quantity of calcium nitrate will be determined by the turfgrass N requirement, and this Ca product source will provide applicably less Ca than a typical gypsum application; thus, it should be considered as a *minor supplemental Ca source* to be used with other sources. However, because of its high solubility, Ca in calcium nitrate is very active in replacement of Na. The halophytic grass seashore paspalum readily responds to Ca nitrate as a fertilizer; so it is often used for this species. In the liquid form, this Ca source can be foliar absorbed also and can aid in sustaining shoot tissue Ca when high Na irrigation water is applied on a consistent basis.

Lime sulfur (CaS_x) (Ca polysulfide) is a liquid containing S in the sulfide form as well as providing a source of Ca (Table 12.2). This material is readily soluble. The Ca is directly available, and the S can react with lime over time in the soil to create gypsum, with the S reaction discussed later in this chapter.

Calcium thiosulfate, $Ca(S_2O_3)$, is an oxyanion of sulfur produced by reaction of sulfite ion with elemental sulfur in boiling water. This product (also called *calcium hyposulfite*) is stable only in neutral to basic (alkaline) solutions. A liquid product that has the formula 0-0-0-6Ca+10S contains 0.63 lb. Ca and 1.04 lb. S per gallon. Calcium thiosulfate is not generally used to acidify alkaline conditions, but can be used to supply soluble Ca in the irrigation water, and when applied to the soil the S will aid in the creation of gypsum when combined with lime.

12.2.5 Soluble Mg Amendments

Langbeinite (soluble potassium-magnesium sulfate mineral: $K_2SO_4 \cdot 2MgSO_4$; Sul-po-mag is one source) has been used to reclaim fine-textured (clay loam) sodic and saline-sodic soils (Artiola et al., 2000; Aydemir and Najjar, 2005). The average chemical composition of langbeinite is 22% K_2O (18.2% K^+), 18.5% MgO (11.1% Mg^{2+}), and 67% SO_4^{2-} (22% S) (Hagstrom, 1986). Langbeinite requires 50% less irrigation water than gypsum to displace and leach exchangeable sodium from the soil profile. Artiola et al. (2000) reported that langbeinite improved hydraulic conductivity rates of the saline-sodic soil, but not as effectively as gypsum. However, this product is more expensive than gypsum or other amendments and, as a result, is often used as a fertilizer nutrient rather than a sodium remediation product.

Aydemir and Najjar (2005) found that, initially, langbeinite improved infiltration better than gypsum due to higher solubility increasing soil ECe and suppressing the Na permeability hazard, but "[o]verall the study indicated that gypsum application to reduce Ksat (infiltration) was more efficient than K-Mag application for the duration" (p. 6). They indicated that the selectivity of K, Mg, and Ca over Na on the exchange sites was 64%, 82%, and 88%, respectively. Sumner and Naidu (1998) reported that the relative flocculating power for Na^+, K^+, Mg^{+2}, and Ca^{+2} was 1.0, 1.7, 27.0 and 43.0, respectively. The divalent ions are more effective flocculating agents because they can be attracted to two clay micelles and also have greater bonding strength to displace Na due to the divalent versus single charge. One reason for Mg to be somewhat less effective in Na displacement and clay flocculation than Ca is the larger hydrated Mg ion radius that would physically keep clay micelles further apart, where the hydrated radii for Mg and Ca, respectively, are 1.08 and 0.96 nm (Sumner and Naidu, 1998).

12.3 ACID-FORMING MATERIALS + LIME SOURCE

The most commonly used acid-forming materials are S based, and these are applied in conjunction with a lime (calcite) source to form gypsum in the soil. The "lime" source may be added as lime, or it may be present in the soil as free calcium carbonate or calcite. Since many sodic and saline-sodic soils are also calcareous soils, the native calcite in the soil can be used as a Ca source, especially during initial reclamation when the grass has not yet been established and soil mixing (sulfur integration) is easier.

12.3.1 Elemental S

Elemental S is insoluble and can only be applied as a soil treatment. It does not directly supply Ca, but furnishes Ca by reaction of sulfuric acid formed from the S, which reacts with lime ($CaCO_3$) in the soil, resulting in gypsum formation. Usually a high-purity grade of 85% to 99% S is used on turfgrasses; but for reclamation purposes on bare soils prior to establishment, grades from 50% to 99% have been used. Finely ground particles (passing at least an 80 mesh screen, with 100 mesh preferred) provide the fastest solubilization and chemical response for gypsum formation. Pelletized granules of fine particles and porous granules of popcorn S (formed from molten S and water) are also available. Other formulations include S plus 10% bentonite as a conditioner, S-slurry suspension of very fine S in water with 2% to 3% attapulgite clay conditioner, and agriculture grade S that is a mixture of particle sizes.

When S is applied to a soil containing $CaCO_3$, a series of reactions must occur before gypsum is formed. The reactions are

$$2\ S + 3\ O_2 + 2\ H_2O \rightarrow 2\ SO_3 \text{ (biochemical reaction by } \textit{Thiobacillus}\text{)}$$

$$SO_3 + H_2O \rightarrow H_2SO_4 \text{ (sulfuric acid)}$$

The first reaction is by biochemical oxidation of S by the bacteria genus *Thiobacillus*. In climates where wintertime soil temperatures are <55°F (13°C), this reaction does not occur due to low bacteria activity under cold soil temperatures. Thus, in cool coastal climates where the turfgrass receives high Na inputs from irrigation, salt spray, or flooding in winter months, soil-applied S (as well as elemental S from other sources discussed below) may not be transformed into the active H_2SO_4 (sulfuric acid) form. Once sulfuric starts to be formed in the presence of lime, the next chemical reaction occurs rapidly and is

$$H_2SO_4 + CaCO_3 \text{ (added lime or native calcite)} \rightarrow CaSO_4 \text{ (gypsum)} + H_2O + CO_2$$

The gypsum then reacts as discussed in the gypsum section to replace Na on the cation exchange sites. Elemental S usually cannot be applied to established turfgrass at >5 lb. S per 1000 sq. ft. without the potential for crown injury (Carrow, Waddington, et al., 2001). Thus, for highly sodic soils where high S rates are required, this granular product would need to be applied prior to turfgrass establishment with sufficient time for the reactions to occur and for the H_2SO_4 to be depleted to levels that will not damage the grass root system.

If free $CaCO_3$ is not in the soil or is not added as lime, elemental S will not produce Ca for Na replacement. Instead the reaction becomes

$$4\ Na^+CLAY + H_2SO_4 \leftrightarrow 4\ H^+CLAY + 2\ Na_2SO_4$$

Under these conditions, Na^+ can potentially be replaced by H^+ and leached, but soil structure does not readily improve (since no Ca is available for flocculation) except when soil pH declines to pH < 6.0 and appreciable $Al(OH)_3$ (soluble Al^{+3}) is created that allows clay micelle Al-induced flocculation, since H^+ alone is not a good flocculation ion. Soil pH would decline in this situation where no lime is present. This situation could occur in humid regions on soils that are acidic or neutral, where the source of Na for these sites is often from coastal flooding, salt spray, or groundwater high in Na from saltwater intrusion. Divalent Ca^{+2} is very effective in flocculating clay colloids together to create structural units, thereby improving soil permeability so that leaching can occur. However, H^+ ions do not cause flocculation, even though swelling and shrinking may decline, except when the trivalent Al^{+3} ion becomes prevalent and increasingly available at pH < 6.0.

If a turfgrass manager wishes to use elemental S on acidic or neutral sites, lime should be periodically applied to the soil surface to provide a Ca source, and pH regularly monitored in the surface 1-inch (2.5-cm) and 4-inch (10-cm) zone to avoid excessively low surface pH (Carrow, Waddington, et al., 2001). The surface inch may become highly acidic, but a soil test obtained from a 4-inch depth soil sample will not exhibit much change if the S is concentrated at the top—where crown damage can occur. At pH < 5.0, the potential for decreased turfgrass vigor is greater due to nutritional imbalances, element toxicities (Al and Mn), and/or nutrient deficiencies (Ca, Mg, and K).

When irrigation water high in Na requires continuous use of elemental S and the turfgrass manager is depending on native free $CaCO_3$ near the soil surface or has not applied lime in a timely fashion, the lime may become depleted in the surface zone over time and result in reduced infiltration of water as Na dominates that zone—additionally, the potential for excessive acidity can occur as well in this zone, as noted in the previous paragraph. Yet, sufficient free $CaCO_3$ may be present at lower depths in the profile to allow for gypsum formation and eventual Na leaching into the lower zones. To maintain surface soil permeability, additional lime should be added to the surface because this is the critical area for infiltration.

Gypsum formed by H_2SO_4 reaction with lime has very fine particle size, is readily solubilized, and is thus very active in replacing Na. While it requires 2 to 4 weeks of soil temperature above 55°F (13°C) for some of the elemental S to be transformed into H_2SO_4 and then for gypsum to start being produced, once the gypsum is created, it is fast acting in countering the Na on the CEC.

One point of caution: when greens sands are calcareous and not of silica or quartz origin, these softer sand-sized particles are subject to particle size alterations when granular sulfur products or water acidification products are applied at high rates and on a continuous basis as they react with the carbonates and bicarbonates in those sand particles. The result can be a change or alteration to finer particle sizes, additional "fines" particle migration, and the potential for subsurface layering that can reduce percolation through the profile. Topdressing on greens, preferably with silica or quartz sand, will help to maintain desired particle sizes at the surface.

12.3.2 Sulfuric Acid and Sulfurous Acid Generators (SAGs)

Prior to turfgrass establishment on sodic soils, sulfuric acid (H_2SO_4) can be applied directly to the soil with special safety equipment because it is highly corrosive and dangerous (Abrol et al., 1988). This allows high rates of acid for sodic soil amelioration and mixing into the soil for reaction with the Ca source being relatively fast. This option is especially good if the soil contains appreciable calcite and is already highly sodic. In some cases, the soil may have a strong calcite layer that would impede long-term salinity leaching. Mechanical breaking of this layer plus sulfuric acid additions can create cracks through the calcite layer and weaken it for better water percolation in the future. In established turfgrass situations, sulfuric acid is injected into irrigation water to adjust water pH to 6.0 to 7.0, with lower pH (<6.0) resulting in corrosion of metal irrigation system parts and release of elevated concentrations of Al and Mn. When injected into irrigation water, equipment to handle high-purity liquid sulfuric acid is necessary.

On established turfgrass sites, sulfuric acid injection or a sulfurous acid generator (SAG) to create sulfuric acid is used on sodic and saline-sodic soils for two major reasons: (a) *water acidification* for the purpose of removing CO_3^{-2} and HCO_3^{-1} from the irrigation water by gaseous evolution as CO_2, and (b) as an S-based acid to be used with added lime or native calcite to create gypsum for remediation of sodic soils (Miyamoto and Stroehlein, 1975a, 1975b, 1986; Bailey and Bilderback, 1998; Stowell and Gelernter, 1998b; Gale et al., 2001; Goss, 2003; Zia et al., 2006; Duncan et al., 2009; Kidder and Hanlon, 2009; Ezlit et al., 2010; Whitlark, 2010). Another possible secondary purpose for using these materials is to lower the soil pH to suppress several diseases, such as summer patch and take-all patch (Duda, 2008) and others (Chapter 17.6.2, "Predisposition to Diseases"). Acid pH enhances manganese soil levels and can promote the growth of antagonistic microbial populations that aid in suppressing some pathogens. The effectiveness of S amendments in the presence of lime or native calcite is discussed further in Chapter 13, and pH reduction will not occur unless the lime is totally dissolved from the surface—which is counterproductive on a sodic soil with irrigation water containing Na.

Concerning these two main reasons for S-based materials to be used, the creation of Ca ions for displacement of Na ions is most important for sodic soil reclamation. Irrigation acidification is a primary concern for preventing sodic conditions due to high CO_3^{-2} and HCO_3^{-1} in the irrigation water reacting with soluble Ca and Mg (creating insoluble and unavailable compounds) in the water (or soil), and allowing Na to accumulate and become dominant on the soil CEC with reduced competition from Ca ions. However, as is discussed later, when the acidifying agent in water results in SO_4 formation, this can be used to generate gypsum for sodic soil reclamation (i.e., providing a dual benefit of reducing water bicarbonates and carbonates while forming gypsum).

Water acidification for the purpose of removing CO_3^{-2} and HCO_3^{-1} from the irrigation water is the primary reason turfgrass managers usually consider sulfuric acid injection or SAGs to create sulfuric acid with injection into the irrigation water (Lubin, 1995; Stowell and Gelernter, 1998b; Goss, 2003; Duncan et al., 2009; Whitlark, 2010). Articles by Bailey and Bilderwick (1998) and Kidder and Hanlon (2009) provide a good review of neutralizing excess bicarbonates in irrigation water. Acidification reduces the residual sodium carbonate (RSC) level and allows more Ca and Mg in irrigation water to remain soluble and available, rather than precipitating as low-solubility lime forms and accumulating in the soil. This is an important treatment under certain conditions, such as when the following occur:

- High CO_3^{-2} and HCO_3^{-1} content in the irrigation water is in conjunction with significant Na to cause sodic soil conditions; plus high enough concentrations of Ca and Mg ions to become available for Na displacement on the soil CEC; plus excess carbonates and bicarbonates remaining after reacting with the Ca and Mg (i.e., water with a high RSC). This is the most critical situation to consider for irrigation water acidification since any Ca amendment program will be very ineffective under these conditions of excessive CO_3 and HCO_3 ions, with a high percentage of the Ca ending up as lime.
- High CO_3^{-2} and HCO_3^{-1} content in the irrigation water in conjunction with significant Na to cause sodic soil conditions; but the Ca and Mg concentrations may not be very high. In this situation, it may be beneficial to reduce the carbonate and bicarbonate levels so that they do not react with applied Ca (such as from gypsum) in the soil and thereby reduce the Na counterion effectiveness of Ca amendment application. Also, less Na-carbonate forms will be precipitated.
- High CO_3^{-2} and HCO_3^{-1} content in the irrigation water along with high concentrations of Ca and Mg, but Na is low. In this case, Na is not the concern, but whether calcite precipitation occurs to the extent to cause soil sealing is the issue. As noted earlier, carbonate accumulation would be most likely in an arid environment over several years with the zone of accumulation at the depth of routine irrigation water penetration (bottom of the wetting front). Use of acidifying fertilizers and deep cultivation through the sealed zone would be additional options to water acidification for managing this problem and may be more cost-effective than acid injection.

Other liquid materials containing sulfate-S have been injected via the irrigation system primarily to acidify the irrigation water and decrease CO_3^{-2} and HCO_3^{-1} content in the irrigation water—N-Phuric acid, pHairway, and N-Control—as well as the use of "acid substitutes." These are discussed in Section 12.3.3, "Other Sulfur-Based Amendments." Also, in Chapter 13, "Amendment Application Options and Guidelines," the protocols to determine how much acid will be required to remove residual carbonates from irrigation water using the RSC method or titration procedure are discussed.

A second reason to use these S-based acid systems (beyond the primary reason of water acidification to remove excess CO_3^{-2} and HCO_3^{-1} from the irrigation water) is to create gypsum if there is sufficient soil lime. The authors recommend that for every pound of H_2SO_4 applied (whether by sulfuric acid injection or SAGs) per 12 inches of irrigation water, 1 lb. of lime should be applied to the soil surface. This will result in formation of 1.36 lb. of gypsum per 1000 sq. ft. For example, if a system added 11 lb. of H_2SO_4 per 1000 sq. ft. over the time period that 12 inches of irrigation water was applied and the turfgrass manager applied 11 lb. of lime at the beginning of the period, this would create about 15 lb. gypsum per 1000 sq. ft.

This procedure is very effective on sites with high Na irrigation water because it has a threefold benefit: it (a) acidifies the irrigation water to remove CO_3^{-2} and HCO_3^{-1} ions; (b) allows the S to be used to create gypsum, which contributes to the total gypsum requirement of the site; and (c) "scrubs" out excess SO_4 that could contribute to black layer if subsoil anaerobic conditions occur. Since the lime must be at the surface, it is recommended that lime be applied periodically during the year based on estimated water use and H_2SO_4 added via the water. If the irrigation water already has appreciable SO_4^{-2} present before acidification (as a nutrient), such as some saline reclaimed water sources do, the lime addition should be based on total pounds of H_2SO_4 plus SO_4 applied per 12 inches of irrigation water—the SO_4 does not need to be calculated as H_2SO_4 since the formula weight is very similar.

Sometimes the question arises about not using S-based water acidification systems on sodic soils since these have poor soil structure and any soluble SO_4 could contribute to black layer, which can also deteriorate soil physical conditions. This issue is discussed in Chapter 16, Section 16.1.1, "Factors Contributing to Nutritional Challenges: Reclaimed Water."

Sulfuric acid directly injected or injected in a SAG after it is generated by oxidizing elemental S into SO_3 gas, which combined with water creates H_2SO_4 (H_2SO_3 is formed, but reacts the same as H_2SO_4), does not require biological transformation as does elemental S. The S is already in the H_2SO_4 form and since *Thiobacillus* bacteria are not needed for oxidization, the sulfuric acid is not dependent on soil temperatures for activity. Outside of these differences, sulfuric acid reacts in the soil in the same manner as elemental S and requires free $CaCO_3$ to create gypsum. These acids are also effective in soil pH reduction (if no free lime is present), dissolution of calcite, and prevention of caliche formation.

Irrigation water reactions of sulfuric acid and SAGs to remove carbonate and bicarbonate ions are as follows:

Sulfuric acid:

$$H_2SO_4 + 2HCO_3^- \rightarrow 2CO_2\uparrow + 2H_2O + SO_4^{-2}$$

$$H_2SO_4 + CO_3^{-2} \rightarrow CO_2\uparrow + H_2O + SO_4^{-2}$$

Sulfurous gas generators (SAGs): These generate sulfuric acid and then the reactions as given above would proceed. The process for SAGs is

$$S + O_2 \rightarrow SO_2$$

$$2\ SO_2 + H_2O \rightarrow 2\ H_2SO_3 \text{ (sulfurous acid)}$$

The sulfurous acid can by oxidized (oxygen added) from oxygen in water and become sulfuric acid, or it can remain in the sulfurous acid form and remove the CO_3^{-2} and HCO_3^{-1} ions in the irrigation water.

As noted in the section on elemental S and its oxidation to sulfuric acid, once sulfuric acid or sulfurous acid is added via the irrigation system to the soil, they react in the following manner:

$$H_2SO_4 + CaCO_3 \text{ (added lime or native calcite)} \rightarrow CaSO_4 \text{ (gypsum)} + H_2O + CO_2$$

12.3.3 Other Sulfur-Based Amendments

12.3.3.1 Ferric Sulfate, Ferrous Sulfate, Aluminum Sulfate, and Iron Pyrite

Soil application of several sulfate-based amendments (ferric sulfate, ferrous sulfate, aluminum sulfate, and iron pyrite) is possible as preestablishment treatments with lime added at the same time if calcite is not present in the soil (Table 12.2). Similar to the use of sulfuric acid on soils with a strong calcitic layer that impedes future salinity leaching, these materials applied with mechanical tillage can greatly fracture and weaken the calcitic layer. Abrol et al. (1988) give the chemical reaction of these materials in the soil. Each reacts to produce sulfuric acid by chemical processes and will react as sulfuric acid once it is formed; however, they are slow acting due to initial solubility and chemical reactions required to become sulfuric acid. These are not widely used, but may be available at reasonable cost in some locations. Iron pyrite is naturally present in acid sulfate soils, and when exposed to oxygen, it starts to form sulfuric acid. Since high rates of lime are necessary to alleviate the extreme acidity and with many acid-sulfate sites being sodic or saline-sodic, the lime will eventually convert to gypsum and aid in improving sodic conditions. Iron pyrite may also have a biochemical process as well as chemical means for conversion to sulfuric acid (Abrol et al., 1988).

12.3.3.2 Lime Sulfur and Calcium Thiosulfate ($Ca[S_2O_3]$)

Lime sulfur (CaS_x, or calcium polysulfide) is Ca-based polysulfide and is a liquid containing unoxidized sulfur in the sulfide form. It is also referred to as *lime sulfur* (Table 12.2), and the Ca in it is beneficial as a Ca source (see Section 12.2.4 concerning Ca as a sodic soil amendment). It normally is applied as a liquid through the irrigation system, but it could be soil applied at high rates for sodic soil reclamation prior to turfgrass establishment. Calcium thiosulfate is not generally used to acidify alkaline conditions, but can be used to supply soluble Ca in the irrigation water, and when applied to the soil the S will aid in the creation of gypsum when combined with lime. When these materials are applied via the irrigation system, they do not act as water acidification agents for CO_3^{-2} and HCO_3^{-1} removal from the water. Lime sulfur is highly alkaline and can be corrosive. Since the S is in the sulfide form in either product, it must be oxidized by *Thiobacillus* to transform the S into sulfuric acid, and thus climatic (temperature) conditions influence the rate of transformation as discussed in the section on elemental S (Section 12.3.1).

12.3.3.3 N-Based and K-Based Polysulfides

An N-based polysulfide is *ammonium polysulfide* ($[NH_4]_2S_x$, also called Nitro-Sul) that contains N (Table 12.2). The NH_4^+ in the Nitro-Sul also is an acidifying agent once it is in the soil and provides an N fertilizer source. *Ammonium thiosulfate*, [$(NH_4)_2S_2O_3$, or Thio-sul), is also a liquid S source and contains N. Similar to lime sulfur, the S in these materials is in the sulfide form and must be oxidized by *Thiobacillus* to transform the S into sulfuric acid, and S effectiveness is controlled by climatic (temperature) conditions to influence the rate of transformation as discussed in the section on elemental S (Section 12.3.1). However, the N in these products is immediately available. While these liquids can be injected into an irrigation system, the N content limits the total concentrations that should be applied for the two ammonium-based materials in order to not exceed specific turfgrass N requirements. Due to the NH_4^+ content, they would not be considered for high application rates prior to turfgrass establishment—in contrast to lime sulfur with Ca instead of NH_4^+. Since the NH_4^+ limits total quantity of these materials that can be applied on either turfgrass or bare-soil situations, these materials should be considered more as N fertilizer sources and not sodic soil amendments—they will have some effects once in the soil by the S oxidation to sulfuric acid plus the NH_4^+ acidification, but it will be minimal compared to other products that can be applied at higher rates. Similar to lime sulfur, these do not act as water acidification agents for CO_3^{-2} and HCO_3^{-1} removal from the water.

Potassium thiosulfate ($K_2S_2O_3$) is 17% S and 25% K_2O, and is a neutral-to-basic (pH 7.4–8.0) liquid formulation that is used more for the K addition (each gal. contains 3 lb. K_2O and 2.1 lb. S) and generally not as a supplement to minor acidification processes in irrigation water.

12.3.3.4 N-Based Sulfate Liquid Acids (N-Phuric Acid, pHairway, and N-Control)

These liquids are widely used in the turfgrass industry as combination N-fertilizers and for acidification of irrigation water for the purpose of reducing CO_3^{-2} and HCO_3^{-1} content in irrigation water. As with direct sulfuric acid or SAGs (see Section 12.3.2, "Sulfuric Acid and Sulfurous Acid Generators [SAGs]"), these are acidic and contain S in the SO_4 form. Thus, as discussed in Section 12.3.2, the SO_4 can be used to create gypsum if a lime source is available—calculating the pounds of SO_4 applied per 12 inches of irrigation water, lime can be surface applied at the same number of pounds as $CaCO_3$ to create gypsum (1 lb. SO_4 + 1 lb. lime $\rightarrow$ 1.36 lb. gypsum). The N contribution from these materials must be factored into the overall N-fertilization program to avoid applying excess N, especially on grasses very efficient in N-uptake and with low N requirements, such as seashore paspalum. Quantities of these products to achieve adequate reductions in CO_3^{-2} and HCO_3^{-1} content in irrigation water may well differ from the N application levels appropriate for the grass, especially when continuously used.

Characteristics are as follows:

- *pHairway* is a combination of urea and sulfuric acid (monocarbamide dihydrogen sulfate [MCDS]) formulated for turfgrass irrigation systems. Depending on the formulation, various N and S rates are applied.
- *N-phuric* is a combination of urea and sulfuric acid containing 10%–28% N and 9%–18% S; but for water treatment, the concentrations are usually 15% N and 16% S. It is also an MCDS. Forms urea sulfate when injected into the irrigation water, similar to pHairway.
- *N control*: 28-0-0-9S, urea sulfate (dicarbamide dihydrogen sulfate [DCDS]), fertigation with alkaline water, 3.4 lb. N and 1.06 lb. S/gal.

12.3.4 Acid Substitutes and Organic Acids

12.3.4.1 Acid Substitutes

In recent years, products called *acid substitutes* or *synthetic acids* have been marketed in the turfgrass industry for injection in the irrigation system as replacement for sulfuric acid or SAGs for removal of CO_3^{-2} and HCO_3^{-1} ions. Whitlark (2010) presented an excellent discussion on these products, and turfgrass managers are encouraged to read his article containing comparisons with some traditional acidification materials. Claims of being more cost effective are often made, but this claim is not necessarily true, especially if a science-based rate cannot be determined. Some problems with these materials are as follows:

- Unknown ingredients. To date, the practice has been to provide no information on the label of the specific product chemical components and quantity of specific components. As will be discussed in Chapter 13, "Amendment Application Options and Guidelines," the common means to determine how much acid is needed to remove residual carbonates from irrigation water is by using the RSC method. This depends on knowledge of the product's specific Ca, S, or HCO_3 equivalent (Tables 12.2 and 13.2), which allows product rate to be accurately adjusted to the specific percent of CO_3^{-2} and HCO_3^{-1} ion removal desired. Without knowledge of the strength of the acid, normal calculations used to determine acid needs, such as those illustrated by Kidder and Hanlon (2009), cannot be done.
- Without the RSC method to use, the other means to determine proper rate of product is by titration with the specific irrigation water to determine relationships between the quantity of product to remove the desired level of carbonate ions and final pH of the water.

Knowledge of irrigation water pH is not sufficient to estimate acid use even if the components and their concentrations in the product are known since acid requirement depends on the bases present in the water.

- For the traditional sulfate-containing amendments that could be applied through the irrigation system, we noted how soil lime applications could be made in order to create gypsum and to remove high SO_4 from the soil. This is a cost-effective means of achieving several benefits from these amendments. However, for products without label information, it is not possible to do this with any confidence of success since the chemistry is not known. Also, without full label disclosure, there may be present additional chemicals that should be known, such as any N, Ca, Fe, or other nutrients that should be accounted for in the fertilization program—or that may produce a "growth response" that could be attributed to the acid, but is really a nutrient reaction in the plant. If a product is not sold as a fertilizer, it can still contain N or Fe, which could provide a greening response, but that plant response would not result from the acid.
- Sometimes these materials have been promoted to reduce *soil* bicarbonates and carbonates, as have some of the weak organic acids discussed in the next section. The most soluble soil carbonates are Na-carbonate and Na-bicarbonate, with both being relatively soluble and therefore leachable; and these are the primary "pool" for soil solution CO_3^{-2} and HCO_3^{-1} ions since these solution ions are in equilibrium with Na-carbonate and bicarbonate precipitates. As was noted in the section on Ca amendments to the soil, the Ca rate is based on Na on the CEC sites—the Na that adversely affects soil structure. Sodic soils normally contain Na-carbonate forms, but the best means to remove them is by leaching so that carbonate, bicarbonate, and Na are leached. While a weak or diluted acid addition dissolves some of the Na-carbonates and Na-bicarbonates, it is a very small quantity—just as it requires very high rates of elemental S to dissolve calcium carbonate from soil. Also, for the Na carbonates and bicarbonates, it must be remembered that water alone dissolves these products. Some labels may claim increased dissolution of soil lime, but this actual dissolution would be very small in magnitude to the total concentration in the soil at current recommended label rates on these products, and also would only be equal to that obtained by strong acids (such as sulfuric acid) if the product rate was applied at equivalent acid concentrations to the S in the strong acids (see Tables 12.2 and 13.2).

As with any other type of product marketed for salt-affected turfgrass sites that do not contain full label disclosure, we do not recommend their use (Carrow, 1993). Claims are difficult to evaluate as to the accuracy with such products without specific product information. Rates listed on the label are questionable and not specific to complex alkaline site conditions.

12.3.4.2 Organic Acids

Various weak organic acids have been promoted for irrigation water acidification materials, especially acetic acid (CH_3COOH and vinegar) and citric acid ($C_6H_8O_7$), with others being lactic acid, formic acid, oxalic acid, and others. The most common organic acids are carboxylic acids with their acidity associated with their carboxyl group –COOH. These are classified as weak acids since, when in solution, only a small quantity of H^+ (active acidity for acidification of the water) protons are present in the water at one time. In contrast, strong acids such as sulfuric acid are fully dissociated with all their protons (H^+) in active form. In order to determine a certain quantity of acid to use to achieve adequate CO_3^{-2} and HCO_3^{-1} ion removal from irrigation water with these materials, titration with the onsite irrigation water is required; or if acid strength is known along with irrigation water bases, there are formula methods to provide the information to determine the rates needed (Bailey and Bilderback, 1998; Kidder and Hanlon, 2009)—that is, quantity of acid versus water pH versus degree of CO_3^{-2} and HCO_3^{-1} ion removal (note: water ECw would decrease as these ions evolve off as CO_2 gas). The rates listed on the label of many of these products are questionable and not specific to alkaline site conditions.

Since these products are organic acids, there may be claims related to reducing soil solution CO_3^{-2} and HCO_3^{-1} ions—as with the acid substitutes and synthetic acid claims discussed in the previous section; the quantity would be miniscule compared to the total ion concentrations in the soil, and the major pool of these ions is in the form of Na carbonate and Na bicarbonate that an effective leaching program can dissolve and remove from the rootzone without the use of these products. Also, the effect on increasing dissolution of lime would be very minor compared with plant root–excreted organic acids present in the rhizosphere at much higher rates and these "natural" acids continually being "applied." Yang et al. (2010) noted the importance of organic acid by root secretion in sodic stress on *Chloris vigata*, a halophyte, especially for pH adjustments and for protecting the root from pH injury during high sodic conditions. In this case, the organic acids are being excreted on a daily basis directly adjacent to root tissues.

As to other claims that may be made for organic acid application that are not directly related to sodic soil reclamation, but may be claimed as additional benefits for alkaline turfgrass ecosystems, it is important to put application of these at "their recommended label rate" in perspective to normal plant excretions of organic acids by roots (see Chapter 17, Section 17.4.5, "Organic Amendments"), where the key points within this section are as follows:

- Plant roots secrete organic acids to solubilize insoluble soil nutrients.
- Plant roots respire CO_2 that in the presence of water may form carbonic acid (H_2CO_3).
- Plant roots may also exude H_2CO_3 directly to solubilize soil nutrients.
- Organic acids release H^+ protons and organic acid anions, lowering rhizosphere pH in calcareous soils, breaking up nutrient complexes, and alleviating anaerobic stress.
- Root exudates include lower molecular weight compounds (sugars, amino acids, phenolics, and organic acids: citric, malic, oxalic, malonic, acetic, fumaric, succinic, lactic, tartaric, and piscidic).
- Root exudates include high molecular weight gelatinous materials (mucigel and mucilage) containing polysaccharides and polygalacturonic acids that are secreted by root cap and epidermal cells.
- Root exudates include sloughed-off cells, tissues, and their lysates that decompose and supply carbon for microbial activity.
- Organic acid anions (mainly oxalate^{-2}, citrate^{-3}, and malate^{-3}) form organometallic complexes with metal ions such as Fe, Zn, Mn, Cu, and others to modify their mobility in a detoxification process.
- High concentrations of citrate are more effective than oxalate and malate in mobilizing Ca, Fe, Mn, and Zn in the rhizosphere.

12.3.5 Combination of Amendments

When the irrigation source has high adj SAR or the existing soil is highly sodic, a combination of amendments and application procedures may be beneficial. For example, sulfuric acid injected into the irrigation water could remove high carbonates and bicarbonates and generate gypsum at the soil surface if $CaCO_3$ is present or added. Additional gypsum could be broadcast over the surface, injected to 4 to 12 inches (10 to 30 cm) at 4- to 12-inch spacing in slits with or without sand, and applied in vertical holes (via deep drill operations). Also, the turfgrass manager may include calcium nitrate as a routine fertilizer source. Other alternatives are possible, such as using gypsum or $CaCl_2$ applied through an irrigation system (but not with acid injection). Some turfgrass managers mix very fine gypsum into an irrigation pond just prior to irrigation as a means of applying larger quantities of very fine particle-sized gypsum in suspension. As the seriousness of salt-affected problems increases, more options become cost-effective.

12.4 ORGANIC AMENDMENTS FOR SODIC SOIL RECLAMATION

In this section, *the focus is on the role of organic amendments for the reclamation of sodic and saline-sodic soils, especially to improve soil physical conditions*. Use of organic amendments in salt-affected turfgrass sites is discussed elsewhere in this book (a) with respect to site assessment issues in Chapter 8, Section 8.1.1, "Organic Amendments"; and (b) in routine management on salt-affected sites in Chapter 17, Section 17.4.5, "Organic Amendments."

Application of organic matter on sodic and saline-sodic soils is considered as a secondary amendment, since they are best used in conjunction with a Ca source and are much less effective if used as the sole amendment. However, organic amendments often enhance the effectiveness of direct application of a Ca amendment or creation of gypsum via an S-material + lime source. Additionally, organic amendments can enhance cultivation effectiveness that is often used with application of a Ca source. Common organic amendments that may be applied prior to turfgrass establishment can be composted or noncomposted crop residues, peat, farm manures, organic manures, and just about any composted "aged" organic matter.

Placement of organic amendments in sodic soils depends on the specific purposes and available equipment, but includes (a) bare soil–surface mulch, mixed into the surface, or deep mixed; and (b) established turfgrass—since a healthy turfgrass provides an ongoing source of green manure, it would suggest that organic amendments are not needed. However, recent research in agronomic situations implies considering deep injection of organic matter as pellets on deep sodic soils. This is discussed later in this section. In all of the above situations, a synergistic effect of organic matter with gypsum and/or cultivation has been consistently shown in agricultural research.

For sodic and saline-sodic soil reclamation, organic matter amendments can provide several direct or indirect benefits, such as the following (Abrol et al., 1988; Adcock et al., 2007; Clark et al., 2007, 2009; Kaur et al., 2008; Gill et al., 2008, 2009):

- Increased macroporosity and total porosity by the mixing action of the amendment into sodic soils with poor structure. The initial improvement in physical properties from the organic matter may be difficult to separate from those attributed to the cultivation operation; however, organic additions maintain these conditions longer than cultivation alone. Macroporosity is very important on sodic soils for increased aeration, infiltration and percolation of water, and root channels.
- One reason for longer term positive benefits of organic matter on physical properties is because it enhances stabilization of soil aggregates by stimulation of the soil microbial population, with the soil microorganisms decomposing the organic matter and organic matter stabilizing the aggregates, making them stronger.
- Organic matter complexing Na.
- As surface mulches, reducing evaporative losses that concentrate salts and increasing infiltration for better leaching.
- Decomposing organic matter increases organic C, organic matter content, release of nutrients, and retention of nutrients by chelation and increased CEC. The nutrient content of organic amendments can be substantial depending on the source, especially for N and micronutrients.

An example that illustrates the influence of organic matter in sodic soil situations is the research by Tejada et al. (2006). Application of poultry mature, which has variable concentrations of fulvic acid depending on the product source, as an organic amendment to remediate saline soils resulted in 70% higher water-soluble carbohydrates and enhanced soil biological properties (18%–40% greater urease, protease, B-glucosidase, phosphatase, arylsulfatase, and dehydrogenase activities) compared to a cotton gin–crushed composted organic addition (Tejada et al., 2006).

Soil pH was increased 0.6 pH units by the addition of poultry manure in a sodic clay subsoil (Clark et al., 2007). In a subsequent experiment in which organic amendments and gypsum were compared for stabilization of macroaggregates (>2 mm) after deep incorporation into a high-clay sodic soil, the largest growth in aggregates was found in canola and chickpea stubble plus chicken manure–amended soils (Clark et al., 2009). The gypsum amendment did not form large, slaking-resistant aggregates (>2 mm) deep in the soil profile, and this was attributed to the inability of gypsum to stimulate soil biological processes down to 0.90 meter deep.

Application of sheep and poultry manure to arid and semiarid region soils irrigated with sodic water up to 40 SAR resulted in an increase in CEC as well as greater adsorption of Ca, Mg, and K cations and less adsorption of sodium (Na^+) (Jalali and Ranjbar, 2009). The addition of these organic amendments led to increased Na leaching and a lower exchangeable sodium percentage. The rate of soil sodification in treated soils declined in the following order: control soil (highest sodification) > sheep manure–treated soil > poultry manure–treated soil > gypsum-treated soil (lowest sodification). Leaching of cations and anions in poultry- and sheep-treated soils was greater than in gypsum-treated and control soils (reflecting improvements in porosity that can also promote salt leaching).

Gill et al. (2008, 2009) investigated using pelletized organic matter applied manually through a pipe attached to a deep ripper to achieve application to 12- to 16-inch (30- to 40-cm) depths in highly sodic, heavy clay subsoils. They achieved impressive increases in wheat production, which was attributed to a combination of N additions, increased soil aggregation, increased macroporosity, better deep rooting, and increased deep water extraction by roots. The improved soil physical conditions were the result of deep ripping and enhanced microbial activity (influencing soil structure formation and stabilization), and results for the combination were much better than deep ripping alone. This suggests that for turfgrass situations on high-clay content, sodic soil profiles that extend at least to the potential turfgrass rooting depth that pelletized organic amendments alone or in combination with gypsum or sand via a Water Wick deep sand injector can reach would be beneficial since this device can inject to at least 12 inches (30 cm). Pelletized organic material would allow easier mixing and deep injection.

12.5 PHYTOREMEDIATION

12.5.1 Phytoremediation by Higher Plants

Phytoremediation (vegetative bioremediation, phytoamelioration, and biological reclamation) by the use of *live, higher plants* for reclamation of sodic and saline-sodic soils is really not a choice on perennial turfgrass areas—the grass is part of the ecosystem and makes the same contributions to the "system" as it does in a nonsaline soil. In contrast, agricultural soils and other nonturfgrass reclamation situations (Chapter 20, "Reclamation, Drainage Water Reuse Schemes, and Halophytic Forage Sites") may have specific, salt-tolerant plants established for 1 to 2 years to aid sodic soil remediation, followed by the incorporation of selected plants in ongoing annual cropping sequences. Qadir et al. (2007) present an excellent review of phytoremediation in these situations. Shukla et al. (2010) also provide an updated review of bioremediation.

Qadir et al. (2007) reported the benefits of phytoremediation by higher plants in sodic and saline-sodic soils within nonturfgrass situations, but these would also be present in perennial turfgrass systems. Benefits were as follows:

- CO_2 produced in the soil by root respiration or from microbial respiration as soil microorganisms respond to plant root exudates and debris as food sources can increase calcite (lime) dissolution by the following reaction: $CaCO_3 + CO_2 + H_2O \leftrightarrow Ca^{2+} + 2HCO_3^{-1}$.
- Production of organic acids as exudates from root and soil organisms, which aid in dissolving calcite. Whether by CO_2 or organic dissolution of calcite, soluble Ca becomes available to displace Na.

- Organic matter added to the soil by plants will improve soil structure and aid in the stabilization of structural aggregates.
- Physical effects of the plant roots in creating biopores or structural cracks for improved water movement and aeration.
- Organic matter additions to the soil also contribute to nutrient recycling, organic matter–based CEC, and organic matter chelation of micronutrients.
- Na removal by plant uptake of Na when the shoots are harvested and removed from the site. This is considered as a very minor contribution to reclamation. Qadir et al. (2007, 223) found that "Na removal by shoot harvest of crops such as alfalfa would contribute to only 1% to 2% of the total Na removed during phytoremediation of sodic soils." If the irrigation water contained Na, this removal amount would be lower.

It is not unusual for products to be promoted for salt-affected turfgrass areas (or nonsaline areas) to provide some of these benefits—for example, organic acids to increase calcite dissolution and organic compounds to provide greater CEC or chelation of micronutrients. Also, the authors have observed some individuals promoting halophytic turfgrasses for their phytoremediation benefits or for extracting Na from the soil. As Qadir and colleagues (Qadir, Oster, et al., 2006; Qadir et al., 2007) demonstrated for agricultural systems, perennial plants on the site already provide these attributes. Preliminary estimates of actual salt *in situ* concentrations in seashore paspalum, the most salt-tolerant halophytic turfgrass, have ranged from >8% to <18% total salts on a dry-weight basis, and if all biomass (roots, shoots, and clippings) are removed, the total salt removal would range from >0.25 to <2.0 tons per acre per year, which would be miniscule compared to total concentrations accumulated in the total soil profile when using saline irrigation water (Duncan and Carrow, 2000).

12.5.2 Phytoremediation with Soil Microorganisms

In the previous sections, it is apparent that the presence of a healthy, perennial turfgrass on a sodic or saline-sodic site will positively influence the soil microbial population; in fact, the organic matter debris and plant root exudates are the primary reason for vigorous, sustained microbial populations. For information related to direct or indirect influence of microbial activity and the organic matter fraction, see the following:

- Section 12.3.4, "Acid Substitutes and Organic Acids," deals with application of organic acids in sodic soil reclamation.
- Section 12.4, "Organic Amendments for Sodic Soil Reclamation," discusses the addition of organic matter amendments during the reclamation process.
- Section 17.4.5, "Organic Amendments," deals with use of organic amendments in routine culture of sodic and saline-sodic soils with an established turfgrass on the site.
- Section 17.4.6, "Microbial Amendments and Bionutritional Products," discusses the use of microbial amendments or applications on sodic and saline-sodic soils during routine maintenance with an established turfgrass present on the site.

Soil microorganisms are an essential part of the balanced plant ecosystems as recently noted by Hayat, Ali, et al. (2010) in their review of beneficial soil bacteria and their role in plant growth enhancement.

The question of concern is whether soil microorganisms applied in salt-affected soil situations may cause unintended problems as noted in Chapter 17, Section 17.4.6, where problems were noted that can be associated with the addition of certain soil microorganisms, especially those that may be favored by a saline environment and close-cut turfgrass conditions. These same problems are possible if either known or unknown (i.e., not on the label) microorganisms are applied to sodic soils for reclamation of turfgrass soils.

Microbial-based remediation for salt-affected soils has been a very minor component within the "phytoremediation" approach, but some have recently encouraged greater use of this method for agricultural reclamation situations, such as by *Cyanobacteria*, which have traditionally been problem organisms in turfgrass ecosystems, as noted in Chapter 17, Section 17.4.6 (Singh and Dhar, 2010). Research by Nadeem et al. (2010) and Egamberdieva (2009) also provides recent examples of applying bacteria in salt-affected agricultural soils and obtaining positive results on wheat (an annual crop). In these studies, the particular bacteria species were reported. Turfgrass managers are cautioned that application of organisms beneficial for phytoremediation in agricultural soils is not the same as applying to perennial turfgrass sites where the moisture surface, ample fresh organic matter, fertility, and persistent light exposure may foster rapid proliferation, especially since turfgrasses already support active organic matter and microbial populations.

One additional marketing ploy is the addition of "salt-eating microbes." The claims on these products indicate that salt loads in the soils can be reduced or the salinity chemistry can be altered. The problems with these organisms are that they are not removed from the soil and that when they die (assuming they have ingested large quantities of salts), the salts remain in the soil. Therefore, it is questionable how effective these particular products are for managing excess salts in turfgrass ecosystems.

13 Amendment Application Options and Guidelines

CONTENTS

13.1 AMENDMENT APPLICATION: OVERVIEW

13.1.1 Overview

Sodic and saline-sodic soil properties are diverse and dynamic, and they can be positively changed by natural processes and by planned reclamation practices. However, soil properties can be rapidly degraded with continuing use of saline water, especially if that water is sodic. Rengasamy (2010a) has illustrated by a flow sheet the dynamic changes that can occur between the broad categories of salt-affected soils and then within each category under acidic, neutral, and alkaline pH situations. Initial reclamation of sodic and saline-sodic soils as well as ongoing maintenance of these sites after the grass is established, especially when the irrigation water is saline, entail a comprehensive and integrated amendment program so that changes are positive and turfgrass performance is sustainable. It was noted in Chapter 12 that more than one "amendment program" is required in most situations, for example, irrigation water acidification, use of lime to transform the SO_4 in the irrigation water to gypsum, gypsum applied to the soil surface, and deep injection of gypsum. Additionally, other chemicals, as discussed in Chapter 16, "Nutritional Practices on Saline and Sodic Sites," and Chapter 17, "Additional Cultural Practices," will be required. Thus, a "whole-systems" strategy-oriented approach must be taken with the purpose for application of each amendment clearly understood so that all cultural inputs are carefully integrated together to create the desired objectives. Practical amendment-related questions or concerns are as follows:

- What amendment(s) should I use?
- How should it or they be applied?
- What application equipment is needed?
- What are appropriate rates to apply?
- When are the best times of year for treatment?
- How frequently should those amendments be applied when using saline water?
- What is the solubility of each product?

The first question was the primary focus of Chapter 12, "Remediation Approaches and Amendments" (i.e., what amendments to use for different situations). In Chapter 13, the remaining questions are addressed. The answers to these questions are strongly influenced by (a) the quality of the irrigation water and whether it requires treatment; (b) the particular site-specific sodic or saline-sodic soil conditions as well as type of soil, especially clay content and clay type; and (c) the particular amendments chosen for use.

13.1.2 Getting Started

For reclamation both prior to turfgrass establishment and on an existing mature turfgrass stand, a stepwise approach to determine soil and water amendments is developed from the site assessment information discussed in Section II, "Site Assessment BMPs for Saline and Sodic Soil Sites" (Chapters 4 through 8). Assessment of the irrigation water indicates what future problems may develop from use of the water, whether treatment of the water is warranted, and how much of the water may be needed to leach salts (i.e., the salinity level of the irrigation water is a major factor influencing the leaching requirement). Assessment of the soil chemical status is just as important as the water quality evaluation because it indicates what problems currently exist and their spatial location. In terms of the soil, important factors to consider are as follows:

- Type of salt-affected soil: saline, saline-sodic, and sodic.
- Soil texture: sand and particle size composition, silt + clay fractions, organic components, whether nonbiodegradable inorganic amendments should be or have been added to the soil profile, and whether sands are calcareous or quartz or silica types.
- The magnitude of the problem: soils with very high total salts or Na levels will require more intensive reclamation efforts than soils with lower levels.
- Location of the problem: a saline or sodic condition only at the surface is much easier to correct than (a) high-salinity or -sodicity conditions throughout the rootzone, or (b) when a distinct sodic B or C horizon exists.
- The need for cultivation practices to sustain pore continuity, and to ensure water movement (leaching) and appropriate depths for cultivation.
- The need for drainage to enhance salt removal and for water table control.

In terms of the irrigation water quality, the essential aspects to take into consideration along with these soil factors are as follows:

- Quality aspects of total salinity (ECw), sodium permeability hazard (adj SAR and RSC), presence of toxic ions (especially Na, Cl, sulfates, and B), and any other factors (such as

bicarbonates, carbonates, and nutrient deficiencies) that may influence turfgrass management and soil conditions. It is the irrigation water that determines the type and magnitude of future problems.
- Expected duration of the salt problem: a soil with a serious salinity or sodic problem can be reclaimed over a 1- to 3-year period, if high-quality (lower salinity) irrigation water is present. But when the irrigation water quality is poor, a continuous, long-term salt management program is necessary. Poor irrigation water quality coupled with a highly sodic soil will require a very intensive initial remediation program of 2 or 3 years followed by a long-term program to maintain adequate soil conditions for sustained salt management.
- Opportunities for improving quality through chemical amendments or blending with lower salinity water (such as desalinized water).
- Quantity of water: is the quantity sufficient to provide plant ET needs and acceptable for leaching of salts? If not, priorities must be established for which areas to irrigate and leach.

It is best to start with the irrigation water quality test since this has long-term influence on what the soil conditions will equilibrate to, and it also determines whether any irrigation water treatments are necessary—which is the topic of Section 13.2. A review of Chapter 6, "Irrigation Water Quality Tests, Interpretation, and Selection," will be helpful in understanding the various water quality issues.

When the quality of the irrigation water indicates that a problem may occur, certain amendment options arise as best management choices. These will be illustrated as we discuss the water quality problems normally encountered. It is essential, however, to realize that treatment of irrigation water or the soil with amendments will be ineffective for alleviating salt problems unless a good leaching program is followed to remove excess soluble salts and, in the case of sodic soils, to remove the excess Na.

13.2 IRRIGATION WATER ACIDIFICATION

When the irrigation water is the primary source of Na added to the soil on established turfgrass sites, it is best to start by determining whether any irrigation water amendment is required, especially acidification to remove excess $HCO_3^{-1} + CO_3^{-2}$. Acidification for neutralization of excess $HCO_3^{-1} + CO_3^{-2}$ is the topic of Sections 13.2.1 (high Na) and 13.2.2 (low Na). In Section 13.2.3, a brief discussion of irrigation water acidification for the purpose of soil acidification is presented—this is not a "sodic" issue, but is included to avoid confusion with this purpose for acidification (which can in some instances reduce the problems caused by some pathogens) versus acidification related to removal of carbonates from water.

13.2.1 Acidification of Water with Moderate or High Na and $HCO_3^{-1} + CO_3^{-2}$

13.2.1.1 Problems and Indicators

The most serious irrigation water problem is when Na concentration is moderate to high coupled with moderate to high $HCO_3^{-1} + CO_3^{-2}$ content. A Na > 125 ppm (5.4 meq/L) in conjunction with a residual sodium carbonate (RSC) greater than 1.25 meq/L is an indicator of potential problems, where RSC is defined as follows (Table 6.7):

$$RSC = (HCO_3 + CO_3) - (Ca + Mg)$$

Ion concentrations are expressed in meq/L (Tables 6.6 and 6.7).

Another set of indicators for this set of conditions is (a) an adjSAR > 10 meq/L (Table 6.6), plus (b) HCO_3 > 120 ppm or 2.0 meq/L or CO_3 > 15 ppm or 0.5 meq/L. Under these irrigation water conditions, a high percentage of the Ca and Mg in the irrigation water will react with the HCO_3^{-1} + CO_3^{-2} to form lime, which leaves the Na in solution to displace Ca and other cations on the cation exchange capacity (CEC) sites, thereby resulting in a sodic soil with poor soil physical conditions. Acidification is necessary to reduce the $HCO_3^{-1} + CO_3^{-2}$, which then allows more of the Ca and Mg to act as counterions against Na.

Usually when the $HCO_3^{-1} + CO_3^{-2}$ concentrations are high, the irrigation water will contain significant quantities of Ca and Mg. But, even when the irrigation water is high in Na and HCO_3^{-1} + CO_3^{-2}, but the Ca and Mg content are relatively low, acidification is beneficial since high inputs of $HCO_3^{-1} + CO_3^{-2}$ into the soil would allow these compounds to react with excess Na (the Na concentration that does not go onto the CEC sites) and form Na-carbonate and Na-bicarbonate, which act as "storehouses" for relatively soluble Na and can contribute to soil sealing. Additionally, the $HCO_3^{-1} + CO_3^{-2}$ can react with any Ca released from Ca amendment applications and convert it into much less soluble lime forms.

13.2.1.2 Acidification Amendments

Several water acidification amendments were discussed in Chapter 12. When selecting an acidification amendment for neutralization of $HCO_3 + CO_3$ ions in the water, important considerations are as follows (Whitlark, 2010):

- What rate to use—amendment rates are primarily a function of water pH and $HCO_3 + CO_3$ concentrations, but other water constituents may influence acid injection rates. Rates are obtained by use of the RSC value; laboratory titration with the irrigation water; estimated rates from published tables; or, for some products, use of their label rate. Estimated application rates based on results as a water amendment or soil amendment are presented in Table 12.2 for the amendments for which data can be determined. The remaining application options are discussed under each type of acid amendment.
- Will the rate be effective? This is best determined by a laboratory titration of the acid with the irrigation water or by use of the RSC method. As Whitlark (2010) reported, some products termed "acid substitutes" or "synthetic acids" that do not identify the chemistry of the acid may have label rates that are not sufficient to achieve any appreciable results.
- Cost based on effective rates for decreasing carbonate and bicarbonate concentrations to acceptable levels.
- Field comparisons based on visual observation are not reliable because some acids contain N or K, which influence turfgrass visual quality response after application. The acid effectiveness is based on what it does in terms of water and soil chemistry in alleviating sodic and saline-sodic conditions, and not the immediate visual response on the turfgrass.

The acid-forming amendments most often used to add to the irrigation water to remove excess $HCO_3 + CO_3$ ions are *H_2SO_4 acid and SAGs injectors*, where SAGs burn elemental S to create sulfurous acid that then reacts the same as sulfuric acid (Table 12.2). For these two water acidification approaches, they do not contain any N or other nutrients; therefore, nutrient application rates are not of concern as to the overall quantity of acid applied. Removal of excess $HCO_3 + CO_3$ concentrations by chemical breakdown to CO_2 gas and water is the primary factor influencing rate. The *RSC method* for the irrigation water quality test can be used to determine the quantity of acid in water acidification to reduce the residual bicarbonate and carbonate for these strong acids. The formula for 100% H_2SO_4 is

$$\text{RSC} \times 133 = \text{pounds of 100\% } H_2SO_4 \text{ per acre-foot irrigation water}$$

If a 93% H_2SO_4 was used instead of 100% purity, the correction for purity is

$$\text{Equivalent amount} = \frac{100\%}{\%\text{Purity}} \times \text{Pounds equivalent}$$

$$\text{Equivalent amount} = \frac{100\%}{93\%} \times 133 = 143 \text{ lb. of } 93\% \text{ acid per acre-foot water}$$

In practice, sufficient acid is added to achieve 75% to 80% depletion of the residual carbonates, but to maintain water at pH > 6.0. In general, water pH adjustment to pH 6.5 will neutralize about 50% to 65% of the $HCO_3 + CO_3$ ions; while at pH 6.0, about 75% to 85% depletion is achieved. These are general estimates regardless of the acids. Acidification is normally between pH 6.0 and 6.5 and not below pH 6.0 to protect components of the irrigation system from extreme acidity and to leave some bicarbonates for water buffering against rapid pH changes. Sometimes in subsurface irrigation injection systems with high alkalinity water, lime deposits may occur in the lines and water pH is adjusted to more acidity levels for a period of time to dissolve the lime deposition and clean out the lines.

Another method for determining the quantity of a specific acid needed to adjust irrigation water pH to a level sufficient to remove most of the $HCO_3 + CO_3$ ions as evolved CO_2 gas can be by *laboratory titration*. In this case, a sample of the irrigation water is submitted to a laboratory, which then uses an acid titration to obtain a curve depicting water pH versus percent $HCO_3 + CO_3$ ions remaining in solution versus quantity of acid required for each pH level adjustment. Whether the RSC or the titration method is used, it is important to recognize that if irrigation water quality varies within a year, including the $HCO_3 + CO_3$ concentrations, then water acidification rates should be adjusted accordingly.

With the S being in the oxidized H_2SO_4 form, no microbial transformation is required to oxidize the acid from sulfuric acid and SAG injection systems. If the soil does not contain any lime, then it would result in soil acidification (see Section 13.2.3, "Acidification of Water for Soil pH Control"). When a soil is not buffered by the presence of free $CaCO_3$, the soil pH would be pH < 7.0. When $CaCO_3$ is added to a soil and it starts to accumulate, pH will gradually rise until it reaches about pH 8.3, and then it stabilizes unless high Na accumulation is present.

Some N-based acids used for irrigation water acidification plus N fertilization claim a two-phase acidification benefit—first in the water, then in the soil by acidification processes from the S and acid-forming N forms in the products. Essentially, H_2SO_4 sources could also make the same claims of water acidification and then soil acidification. When there are even relatively small quantities of free lime in the soil, it requires a very high quantity of acid to dissolve it—which is necessary before pH will decrease on alkaline soils. Sanden, Fulton, and Ferguson (2010) noted that on an elemental S basis, it would require about 3.5 tons of S to neutralize a 1% lime content in an acre 6-inch profile of soil—that is, 160 lb. S per 1000 sq. ft., where for established turfgrass, the normal recommendation for elemental S application is no more than 5 lb. per 1000 sq. ft. twice per year.

However, with the presence of a lime source either applied or native in the soil, gypsum can be formed. As was discussed in Chapter 12, Section 12.3.2, "Sulfuric Acid and Sulfurous Acid Generators (SAGs)," once the H_2SO_4 in the irrigation water from sulfuric acid injection or SAG is applied to the soil, it can react with lime to form gypsum. With the S being in the H_2SO_4 form, no microbial transformation is required to oxidize the acid from sulfuric acid and SAGs injection systems. The authors recommend that for every pound of H_2SO_4 applied per 12 inches of irrigation water, 1 lb. of lime be applied to the soil surface, which will result in the formation of 1.36 lb. of gypsum per 1000 sq. ft. For example, if a system added 15 lb. of H_2SO_4 per 1000 sq. ft. over the time period that 12 inches of acid-treated irrigation water was applied and the turfgrass manager applied 15 lb. of lime at the beginning of the period, this would create about 20 lb. gypsum per 1000 sq. ft.

The lime should be applied periodically based on about when 12 inches of irrigation water are applied since it is important to have available free lime at the surface to prevent sodic conditions in this critical zone.

This procedure is very effective on sites with high-Na irrigation water since it has a threefold benefit: (a) it acidifies the irrigation water to remove CO_3^{-2} and HCO_3^{-1} ions so that Ca and Mg in the water are not complexed into low-solubility lime forms and, therefore, can counter excess Na from the irrigation water; (b) it allows the SO_4 to be used to create gypsum, which contributes to the total gypsum need of the site; and (c) it "scrubs" out excess SO_4 that could contribute to black layer if anaerobic conditions occur. If the irrigation water already has appreciable SO_4^{-2} present before acidification (as a nutrient), such as a saline-reclaimed water source often does, the lime addition should be based on total pounds of H_2SO_4 plus SO_4 applied per 12 inches of irrigation water—the SO_4 does not need to be calculated as H_2SO_4 since the formula weight is very similar. This quantity of gypsum may not be enough to meet the total gypsum requirement in order to displace the Na from the CEC clay sites, but it can be a very significant part of the total volume needed for effective Na salinity management on the specific site—see Section 13.3.4 for a case study.

Other acidic liquids that contain N and are commonly used to acidify irrigation water include *N-Phuric, pHairway, and N-Control* (Lubin, 1995; Whitlark, 2010) (Table 12.2). These materials are less harsh than direct sulfuric acid injection or SAGs, but they contain significant quantities of N; this N addition must be considered in their use for injection into the irrigation water for CO_3 and HCO_3 neutralization. The controlling factor on total quantity of acid and whether they are injected with every irrigation event is the amount of N being applied to avoid excessive N application. For the irrigation events where one of these acids is injected, however, they would reduce HCO_3 and CO_3 levels depending on the water pH achieved (Whitlark, 2010). The best means to determine the appropriate rate for HCO_3 and CO_3 neutralization is by the water titration method as discussed previously in this section. This requires a balance between what to use for turfgrass N needs and what to apply for water acidification requirements.

These products will have (a) an immediate water acidification impact; (b) a soil acidification effect from the SO_4 content, which will be rapid assuming that a lime source is present at the surface; and (c) a somewhat longer acidification effect due to H^+ ions generated by the urea ammonification to NH_4^+ and the NH_4^+ nitrification to NO_3^- (the process generally occurs in 2–4 days). The ammonification process requires urease, a naturally occurring soil enzyme, and the nitrification reaction requires *Nitrosomonas* and *Nitrobacter* bacteria. The H^+ ions can displace Na from the CEC sites, but are not good soil structure flocculation agents. A possible stimulation of lime dissolution by this acidity source would be more beneficial. Each of these N-based acids inhibits N volatilization in alkaline soils by their surface acid reactions.

While other liquid sulfate or sulfide chemicals can be injected into irrigation water, this would be primarily for their fertilizer content with examples being *potassium thiosulfate, calcium thiosulfate, ammonium thiosulfate, and ammonium polysulfide* (Table 12.2). Each of these would exhibit some solution acidity that could neutralize HCO_3 and CO_3 during the specific irrigation events when injection occurs. Also, they would result in formation of gypsum from the S within each compound as it reacted with lime in the soil. For the ammonium-based materials, additional soil acidification would occur from the NH_4^+ ion nitrification to NO_3^-. Calcium thiosulfate would provide Ca, but the water rates used in irrigation would result in this soluble form of Ca being washed into the soil where it could be taken up by plant roots or complexed in the soil. Application rates for these products are normally dictated by the ammonium or potassium fertilization needs and not by the quantity to neutralize HCO_3 and CO_3 ions. Titration could be used with the irrigation water source to determine neutralization effects at the rates used for fertilization.

Another acidification material is *phosphoric acid solution*, but this acid does not add S and can result in formation of insoluble Ca and Mg phosphates at $pH_w > 6.0$. Also, there would be concern over excessive soil P application as well as potential for water resource contamination over time if the P moves. Thus, for salt-affected sites, this acidification method is not recommended.

Whitlark (2010) compared acid substitutes and synthetic acids to traditional acidification acids (sulfuric acid and N-pHuric) in terms of effectiveness for HCO_3 and CO_3 ion neutralization and costs. The terms *acid substitutes* and *synthetic acids* are used in the product information material and do not have a defined science-based meaning. Information on specific acid chemistry or strength is not given on most of those product labels, but these alternative acids are claimed as "proprietary"—however, patent protection allows "proprietary" chemicals to be specifically identified (as is apparent for new pesticides). The only guidance on rate to use during acidification application for these products is by the label rate, which is much lower than the rate used with traditional acids. Whitlark (2010) found that the published rates were too low to be effective for HCO_3 and CO_3 acidification purposes, and when application rate was based on titration data, the product costs greatly exceeded those of the traditional acids. Relative to these products and for other similar ones that may arise in the future, turfgrass managers should recognize that the acid standard is sulfuric acid since it is a strong acid that can be used at high concentration, and the H^+ ions are fully dissociated into solution and thus are immediately active. Due to these characteristics, for acid products that claim much lower application rates but equal effectiveness whether in water or in soil, it is a chemical impossibility.

Weak organic acids have been sometimes promoted for turfgrass irrigation water acidification materials, especially acetic acid (CH_3COOH; vinegar), citric acid ($C_6H_8O_7$), or unspecified "proprietary" organic acids (see Chapter 12, Section 12.3.4, "Acid Substitutes and Organic Acids"). To determine quantity of acid to use to achieve adequate CO_3^{-2} and HCO_3^{-1} ion neutralization from irrigation water with these materials, titration with the irrigation water is required, or if acid strength is known along with irrigation water characteristics, there are formula methods to provide the information needed that are used by the greenhouse industry (Bailey and Bilderback, 1998; Kidder and Hanlon, 2009) (i.e., quantity of acid vs. water pH vs. degree of CO_3^{-2} and HCO_3^{-1} ion removal). The rates listed on the labels of many of these products are questionable and not specific to site conditions. There may be claims related to reducing soil solution CO_3^{-2} and HCO_3^{-1} ions by these organic acids—as with the acid substitute and synthetic acid claims discussed in the previous section—but the effectiveness of these weak organic acid products would be minuscule.

Sometimes the question arises about acidification of calcareous sands, especially on golf greens. Calcareous sands can arise by several means: (a) sand-sized particles that developed from limestone parent material, in which case the lime content is likely to be rather high; (b) quartz or silica sands that have some coating or depositions of lime due to high bicarbonate and Ca irrigation water content; (c) a silica sand with coral pieces such as from a seashore; or (d) a silica sand, quartz being the most common type, that has been heavily limed and still contains lime particles. Each of these would exhibit effervescence if an acid such as acetic acid was applied to it. Using water acidification or applying acidification chemicals to any of these sands except the calcareous sand derived from limestone would not be a problem. However, calcareous sands derived from limestone tend to be softer sands (as opposed to quartz or silica sands with lime from the other sources) and are more subject to acidic action altering particle sizes and creating layers from small particle migration, especially in greens complexes. Use of silica sand topdressing would help maintain the desired particle size range on golf greens if acidification was absolutely necessary. For other higher cut turfgrass areas, this issue would not be a concern.

13.2.2 Acidification of Water with Low Na but High HCO_3 and CO_3

The discussion in this section does not apply to sodic or saline-sodic soils, but is an issue of excessive alkalinity. It is included because irrigation water acidification is a possible control measure. On sandy soils, irrigation water high in HCO_3, CO_3, Ca, and Mg can cause calcite precipitation to the degree that water infiltration may decline (Carrow et al., 1999). This can occur even when the water is low in Na; thereby, in this situation, Ca and Mg precipitation would not contribute to a sodic problem. Acid injection would be an option to help decrease calcite formation and even dissolve it over time. As stated, this example of acid treatment of irrigation water is not a Na problem, but is one of salt-induced (i.e., calcite) sealing of soil pores, normally at the depth of irrigation water penetration in arid climates.

Depending on the balance of HCO_3, CO_3, Ca, and Mg ions, it is possible to have various RSC values, but if it is at RSC > 1.25 meq/L, this does not mean there is a Na problem since we are discussing a situation where Na is low. Concentrations of these constituents in irrigation water typically are 100–400 mg/L HCO_3, 0–5 mg/L CO_3, 25–200 mg/L Ca, and 20–40 mg/L Mg. Assuming that an irrigation water is very high in each of these aspects at 811 mg/L HCO_3, 200 mg/L Ca, and 40 mg/L Mg (13.3 meq/L, 10 meq/L, and 3.28 meq/L, respectively), there is sufficient Ca and Mg to react with all of the 13.3 meq/L of HCO_3. This water would result in a combined total of 2104 lb. of $CaCO_3$ + $MgCO_3$ per acre-foot of applied irrigation water or 48 lb. $CaCO_3$ + $MgCO_3$ per 1000 sq. ft. per 12 inches of irrigation water. As a comparison, a soil with 1% free lime in the surface 4-inch zone would contain about 230 lb. per 1000 sq. ft. of lime. In an arid climate where rainfall would not assist in dissolving the precipitated calcite and Na carbonates and the irrigation program was consistent, calcite accumulation could occur, especially at the bottom depth of routine irrigation water front penetration. However, with use of acidifying fertilizers, periodic deeper leaching by irrigation or rains, and normal cultivation practices, all of these management options would assist in preventing a distinct calcite and Na bicarbonate zone that could inhibit infiltration or percolation. If a layer does form, an indirect result is reduced oxygen flux (and loss of air porosity) and several of the root-borne diseases, such as take-all or decline and Curvularia fading out, can become problems as the water infiltration and percolation problems escalate above this zone (refer to Section 17.6.2, “Predisposition to Diseases”).

13.2.3 Acidification of Water for Soil pH Control

Another use of acid injection is to reduce soil pH by adding an S source to the soil using the irrigation system. As with the previous example of water acidification, this is not a sodic or saline-sodic issue, but is mentioned since it relates to irrigation water acidification. For sodic and saline-sodic reclamation, it was stressed that lime was required to form gypsum and the net result would be no pH change except for highly sodic soils that may decrease from within the pH 9.5 to 10.4 range to about pH 8.3 with gypsum applications, but with no further pH reduction.

All soils have a buffer system against rapid pH change via the CEC sites of clay and organic matter; however, alkaline soils have an additional buffering system due to lime when present due to added lime or native calcite. Earlier in the chapter, it was stated that on an elemental S basis, it would require about 3.5 tons of S to neutralize 1% lime content in an acre 6-inch profile of soil (i.e., 160 lb. S per 1000 sq. ft.) (Sanden, Fulton, and Ferguson, 2010).

In the absence of lime to buffer for soils at pH < 7.0, acidifying agents can decrease pH rather easily, especially on low-CEC soils, and with substantial application of less volume of the S-acidifying agent. The implications for using an irrigation water injection system to add an acidifying agent are as follows:

- If the soil has any significant quantity of lime, pH will not change unless very high rates of acid are added where the surface few centimeters could become depleted of the lime.
- If the soil has a pH > 7.0, indicating that lime is present, but it is not at the surface, soil pH in that zone can decline with acid addition.
- If the soil does not contain lime, the pH will be at <7.0, and there is no reason to reduce pH further.

Carrow, Waddington, et al. (2001) noted that pH adjustment was very important for excessively acid soils, but for alkaline soils, the agronomic reasons for reducing pH are much less important. This is true especially when the cost is considered.

13.3 GYPSUM REQUIREMENT: SODIC SOILS AND IRRIGATION WATER INJECTION

When speaking of the *total gypsum requirement* for sodic or saline-sodic soils, this relates to the total quantity of gypsum to apply to displace the Na on the soil CEC sites in order to improve soil structure to an acceptable level within the soil profile depth that is chosen for amendment. The "gypsum requirement" does not necessarily mean gypsum is actually used as the amendment, since a number of Ca sources and S-acidifying agents + lime can be used to generate the gypsum or to provide the quantity of available Ca needed for soil reclamation (Table 12.2). Often, more than one amendment option is used, and these combinations achieve the desired total Ca level for Na displacement. The total gypsum requirement, amendment choices, and application procedures vary between bare-soil applications (Section 13.3.1) on a sodic soil versus those on an established turfgrass site (Section 13.3.2). In Section 13.3.3 are discussed various reasons for gypsum injection or injection of other Ca, Mg, or K salts into the irrigation system—some of these reasons are not related to sodic soils, but are presented briefly for a full understanding of "gypsum injection."

13.3.1 Total Gypsum Requirement for Bare Soil: Prior to Establishment

In this section, total gypsum requirement for application prior to turfgrass establishment is discussed for three conditions, namely, (a) soils containing <95% sand (i.e., most soils), (b) high sand content soils (>95% sand), and (c) sand-capping interface with fine-textured subsoils. Procedures for determining total gypsum requirement for the initial soil condition are presented.

13.3.1.1 Soils Containing <95% Sand

When a sodic soil is to be reclaimed prior to turfgrass establishment, the total gypsum requirement can be applied, such as large quantities of gypsum, elemental S + lime, sulfuric acid + lime, or other soil amendments, at high rates and mixed into the soil (Table 12.2). Two different approaches for determining the total gypsum requirement are (a) estimated from published tables, and (b) estimated by calculation procedures. Both methods consider the major factors that influence the gypsum requirement, which are as follows:

- Initial exchangeable sodium percentage (ESP) or Na base saturation level on the soil CEC
- The total CEC, with more fine-textured soils exhibiting higher CEC
- The soil depth to be reclaimed

For the published table estimates of gypsum requirements, the information in Table 13.1 can be used as a general guideline. The rate varies with soil texture and percent clay and initial ESP. The maximum rate per application should be 4 to 6 tons per acre (9000 to 13,500 kg/ha) followed by leaching to remove the Na. Another option is to use the data in Table 12.2 that lists the pounds per acre 6 inches of soil to replace 1 meq/kg of Na on the CEC for each potential amendment. If the soil ESP is known from a soil test, this can be used to estimate the specific gypsum requirement based on the particular amendment used.

The best means to estimate the total gypsum requirement is by calculation based on laboratory soil test data. The U.S. Salinity Laboratory (USSL; 1954) and others have developed a formula for estimating the gypsum requirement. The procedure recommended by Hanson et al. (1999) calculates

TABLE 13.1
Estimated Gypsum Requirements per 12 Inches (30 cm) of Soil Depth to Reduce ESP to Below 10% Based on Soil Texture and Initial ESP

		Initial Exchangeable Sodium %				
Soil Texture	**Clay Content**	**15**	**20**	**30**	**40**	**50**
	%			Tons per Acre[a]		
Sand, loamy sand	0–15	0.5–2.0	0.7–3.0	1–4	2–5	3–8
Loams	15–55	2–3	3–4	4–6	5–8	7–10
Clays	>55	3–4	4–6	6–8	8–11	10–14 (+)

Source: After Carrow, R. N., and R. R. Duncan, *Salt-Affected Turfgrass Sites: Assessment and Management*, John Wiley, Hoboken, NJ, 1998.

Note: These values can vary considerably based on clay type with clays containing high CEC requiring more gypsum.

[a] 1 ton = 2000 lb.; 2000 lb. per acre = 2240 kg ha^{-1}; and 1 ton per acre = 46 lb. per 1000 sq. ft.

the gypsum requirement for a soil that would be more accurate than the table methods presented in the previous paragraph, as illustrated with an example where a soil has an ESP 22% Na and a total CEC is 18 meq/100 g soil. The desired ESP is 4% Na.

- Submit a soil sample for laboratory analysis to determine the ESP on the soil CEC as well as total CEC from the normal soil fertility test. It is important to obtain accurate CEC and ESP results, as discussed in Chapter 5, Section 5.3.2, "Determining Total CEC and Exchangeable Cations."
- Calculate the initial exchangeable Na in meq/100g as
  Initial exchangeable Na (meq/100 g) = (initial ESP × CEC)/100
  Initial exchangeable Na = (22 × 18 meq/100 g)/100 = 3.96 meq Na/100 g
- Calculate the exchangeable Na (in meq/100 g soil units) needing replacement to attain the desired ESP of 4% Na.
  Final exchangeable sodium = (final ESP × CEC)/100
  Final exchangeable Na = (4 × 18 meq/100 g soil)/100 = 0.72 meq Na/100 g
- Calculate the Ca requirement (in meq/100 g) needed to displace the Na down to the desired level of ESP = 4% Na. The calculations are in meq/100 g units since 1 meq of Ca equals 1 meq Na in terms of chemical equivalency. This is the difference between the initial exchangeable sodium and the final desired exchangeable sodium where
  Ca requirement = initial exchangeable Na − final exchangeable Na
  Ca requirement = 3.96 − 0.72 = 3.24 meq Ca/100 g needed to displace the 3.24 meq Na/100 g soil from the soil CEC
- Convert the calcium requirement from meq/100 g to tons per acre-foot using Table 13.2. If we wish to use gypsum as the amendment, then it would require about 5.6 tons of 100% pure gypsum per acre-foot of soil since the values in this table are in acre-feet. This would be 257 lb. of gypsum per 1000 sq. ft. mixed to a depth of 12 inches.

Once the "gypsum" rate is determined, other possible amendments can be selected as shown in Table 12.2. If it was decided to add all the gypsum during this preestablishment period, then 75% may be spread over the site and mixed to 12 inches, followed by 25% mixed in the surface 4 inches. If elemental S was used, this would require a relatively long period of favorable weather (temperature > 55°F or > 13°C) in order to allow microbial transformation of the S, plus there would need to be enough lime or calcite in the soil to allow gypsum formation. On sites with hard calcite layers in the surface 12-inch zone, the choice of amendment may be sulfuric acid with the calcite providing

TABLE 13.2
Converting from meq Ca/L to Pounds Amendment/Acre-Foot of Applied Water

	Pounds Amendment per Acre-Foot Water				
meq Ca/L	Gypsum (100% pure)	Sulfuric Acid (100% pure)	Lime Sulfur (23% S)	Nitro[a] Sul (20% N, 40% S)	Urea–Sulfuric Acid[a] (10% N, 55% acid)
1.0	234	133	192	50	107
2.0	468	266	383	100	214
3.0	702	399	576	150	321
4.0	936	532	768	200	428
5.0	1170	665	959	250	535
6.0	1404	798	1151	300	642

Source: After U.S. Salinity Laboratory, *Diagnosis and Improvement of Saline and Alkali Soils, Handbook 60*, U.S. Government Printing Office, Washington, DC, 1954; Hanson, B., S. R. Grattan, and A. Fulton, *Agricultural Salinity and Drainage* (Division of Agriculture and Natural Resources Publication No. 3375), University of California, Davis, 1999.

[a] One mole of ammonium is assumed to replace two moles of sodium.

the Ca source. Deep ripping to at least 12 inches should be done followed by acid addition. Irrigation should be applied to leach the Na as it is displaced from the CEC sites.

If the soil is of intermediate texture but sodic, gypsum addition to 6 inches may be adequate. Whether a full 12 inches of amendment depth is required depends on the site conditions. Conditions that favor a deep amendment depth are the presence of any impervious layer such as a high soil strength B horizon (hard pan) or calcite layer; when the soil is high in clay content; the clay is a 2:1 type; and the future irrigation water contains significant Na and the soil is already highly sodic. Likewise, a deeper treatment is recommended when physical impediments to water movement exist such as a hard calcite layer. These same conditions, with the exception of the hard calcite layer, would also suggest consideration of $CaCl_2$ amendment if it is cost-effective for at least some of the Ca requirement. The Ca in this amendment is immediately available and, when mixed in the surface, gives a more rapid enhancement of infiltration rate due to increasing ECe, which suppresses the adverse effects of Na on infiltration (Figure 3.1). For example, increasing the soil ECe to 5.0 dS/m can result in very little effect of Na on infiltration even at SAR = 24.

The gypsum amendment strategy rather than using $CaCl_2$ is appropriate on saline-sodic soils, where the soluble salts are already present; therefore, $CaCl_2$ would not provide the same benefit as in a sodic soil. Leaching should not be initiated on saline-sodic soils until gypsum and any other amendments have been applied and mixed, and adequate gypsum remains on the surface. On these soils, the initial infiltration rate may be rather good due to the high soil ECe, but if leaching is initiated before amendment application, the soil can convert to a sodic soil that can exhibit much less favorable soil physical properties.

Due to cost for such large quantities of gypsum or Ca amendment, lower rates may be used before turfgrass establishment. However, the surface 4 inches of the soil profile should receive adequate amounts of Ca on a regularly scheduled basis to insure improved and sustainable infiltration in this zone. Even when the total gypsum requirement cannot be added, there should be improvement in surface zone physical conditions. When the future irrigation water that will be applied on the established turfgrass does not contain sufficient Na to recreate a sodic soil condition, then only part of the total gypsum requirement needs to be applied prior to establishment, and the remainder can be applied in 2–4 applications over the next couple of years.

On bare soils, gypsum or an S-source + lime as the sole amendment will be assisted by any tillage operations, which promote better conditions for water movement by creating temporary structure and pore continuity improvements. However, the addition of organic matter throughout the depth of gypsum treatment will be of great benefit and provide more pronounced positive results (Jayawardane and Chan, 1994; Levy, 2000; Adcock et al., 2007; Gill et al., 2008, 2009). The organic matter will help with structure formation and stabilization. If economics do not allow sufficient composted organic matter to be mixed into a 12-inch zone, then it should be strongly considered for the surface 4-inch zone. Organic matter should be applied at approximately 5%–15% by volume or 1%–3% by dry weight to the surface 4-inch zone and incorporated within this range for the deeper zone of treatment.

13.3.1.2 High Sand Content Soils (>95% Sand)

The initial (Na content before establishment) and ongoing (due to irrigation water Na) gypsum requirements for very high sand content soil (>95% sand that has good infiltration and percolation rates) may be less than that presented in Table 13.1 or by the calculation method of Hanson et al. (1999). In these soils where soil structure by aggregate formation is functioning due to the single-grained nature of sand, Na does not degrade the sand "structure." When coupled with good inherent infiltration and percolation, Na initially present as Na carbonate can be leached and the Na on CEC sites does not represent a high total Na quantity, since sands generally have low total inherent CEC. Thus, the gypsum required to displace the Na on the CEC will be relatively low. However, some gypsum is necessary to (a) displace the Na on the CEC sites, (b) counter Na displacement of Ca in the root tissues of the turfgrass, and (c) maintain adequate Ca as a nutrient in quantity and balance with other cations.

13.3.1.3 Sand-Capping Situations

When deep sand capping is placed over a high clay content subsoil, the interface of the sand cap and underlying soil is important (see Chapter 15, "Drainage and Sand Capping"). The subsoil should be tilled with high quantities of gypsum worked into the surface 2 to 4 inches of the subsoil before applying the sand cap. This strategy will help insure better water drainage into the underlying subsoil to aid in water and salt removal plus provide a zone for chemical interaction for potential upward movement of Na from the underlying clay subsoil. Gypsum rates should be in the range of 100 to 150 lb. per 1000 sq. ft.

13.3.2 Gypsum Application on Established Turfgrass Sites

For established turfgrass areas, soil properties and characteristics of the irrigation water influence reclamation decisions. On some sites, the soil is sodic or saline-sodic, but the irrigation water is not high in Na; therefore, the reclamation process is relatively short term (i.e., from weeks to a year or more, depending on the soil). However, where irrigation water contains significant Na to cause sodic soil conditions, vigorous measures must be adopted, both initially and on a continuous basis. Most challenging (from the management standpoint) will be when the irrigation water has high Na and this Na addition is in conjunction with one or more of these situations: the presence of any impervious layer such as a high soil strength B horizon or calcite layer, when the soil is high in clay content throughout the profile, or when the clay is a 2:1 type.

13.3.2.1 Dealing with the Sodic Soil in the Short Term

On an established turfgrass site that already has a sodic soil and the irrigation water continues to add additional Na, the soil will never really become "clean" of Na. Rather, the goal is to reduce Na on the soil CEC sites to a level that is acceptable—the goal is to achieve soil physical conditions that allow adequate infiltration, percolation, and drainage rates of water. When water moves downward properly, several benefits are achieved: irrigation is much easier; leaching of total soluble salts and Na ions is enhanced; during rain periods, waterlogging is reduced; soil aeration is improved; soil-borne disease problems are reduced; and turfgrass physiological health and rooting are enhanced.

Probably the single most recognized problem with sodic turfgrass soils is at the very surface with *low infiltration rates*. When the irrigation water contains appreciable Na, it is a constant force fostering soil structure degradation, and the turfgrass manager must look to this surface zone as an area of first management priority. Various practices can be used to improve infiltration (Chapter 14, "Cultivation, Topdressing, and Soil Modification," and Chapter 15, "Drainage and Sand Capping"), but soil and water amendment choices and application methods have a major effect. Amendments can improve infiltration on sodic soils by a combination of (a) creating better soil structure and stabilization of structure, and (b) suppressing the adverse effects of Na on soil structure by creating higher surface salinity (ECe) (see Figure 3.1). Some amendment options to achieve better water infiltration are as follows:

- When irrigation water acidification is necessary, application of lime at the surface to react with SO_4 in the irrigation water will produce Ca to displace Na as well as increase surface ECe to some extent in the top inch or 2–3 cm. Earlier, we noted that for every 1 lb. H_2SO_4 or SO_4 applied per 12 inches of irrigation water, there should be added to the soil surface 1 lb. of lime per unit area to create about 1.36 lb. gypsum.
- Periodic granular broadcast application of gypsum at the surface will also provide similar benefits. More frequent application of gypsum using smaller particle sizes will be better than less frequent application with a large particle size range; but if application can only be made 2–3 times per year, a source with a wider range of particle sizes is best. Smaller particles dissolve more rapidly. The key point here is to apply Ca products frequently enough, based on solubility, in order to provide a relatively constant supply of available Ca to displace Na and maintain soil structure.
- Injection of soluble $CaCl_2$ or gypsum in suspension will also increase Ca content and ECw and soil ECe which will foster better infiltration of water.

Enhanced water infiltration rate is important, but then the water must be able to percolate through the rootzone and drain past the rootzone in order to remove total soluble salts and excess Na. To deal with sodic conditions in the rootzone or part of the rootzone, a starting point is to determine the gypsum requirement for the zone of amelioration, and then to apply this in a manner that will consistently affect as much of the soil rootzone profile as possible.

The total gypsum requirement on established turfgrass sites where irrigation water with applicable Na plus an existing sodic condition is present is different than for a bare soil, as discussed in the preceding section. To estimate the gypsum requirement for the soil alone (not considering any irrigation water issues) of established sites, the calculation procedure based on soil test ESP and CEC data for bare soil can be used, but then adjusted from a 12-inch (30-cm) soil depth rate to a 4-inch (10-cm) soil depth (i.e., one third the rate). This gypsum requirement is to ameliorate the soil for currently existing sodic conditions and does not consider future problems from irrigation water additions of Na.

This total gypsum requirement could be applied to the soil surface over time in several applications as gypsum or with part being with other Ca amendment options, especially when the soil is a sandy loam or higher in sand content. Gypsum can be safely used at high rates on a mature turfgrass stand—50 to 100 lb. per 1000 sq. ft. per application (2440 to 4880 kg/ha) for mowing heights over 0.5 inches, and 10 to 50 lb. per 1000 sq. ft. for close-cut turf. However, surface applications alone will have limited impact on the deeper soil profile of highly sodic, fine-textured soils except over a long period of time. To maintain macropores for continuous water infiltration, percolation, and drainage on these soils will necessitate the use of surface cultivation and deep cultivation in conjunction with amendment additions in the cultivation channels—these are the future "macropores." Options include broadcast applications made in conjunction with core aerification that will allow some of the amendment to be worked somewhat deeper into the soil, which is especially important on high clay content sodic soils.

Another option is deep injection of gypsum into slots that can be accomplished with sand-slitting (sand-banding and sand-grooving) devices that result in a combination of cultivation, drainage, and sand injection. A couple of examples are (a) the Blec Sandmaster, with injection depths up to 10 inches × a 0.8-inch width and 10-inch spacing; and (b) the WaterWick Machine, with injection up to a 12-inch depth × 0.62 inches × 12-inch spacing. For sodic soils, we would suggest that either 100% gypsum or 40%–60% gypsum with the remainder sand be injected. In similar agricultural situations, a number of researchers have reported improved, long-term results using *gypsum slotting* (Rhoades and Loveday, 1990; Jayawardane and Chan, 1994; Levy, 2000; Prathapar et al., 2005). Jayawardane and Chan (1994) did pioneering research in Australia on what they termed "gypsum slotting," which was gypsum applied into slots or bands, similar to sand slitting except with gypsum.

Gill et al. (2008, 2009) reported that pelletized organic matter in combination with deep ripping and gypsum was very effective on fine-textured sodic soils with highly sodic B horizons. They did not inject within slots, but their work suggests that the incorporation of organic matter pellets plus gypsum with or without sand should be considered. One benefit of gypsum slotting compared to tilling it into the soil in agricultural cases has been that the macropores created in the slots are protected by the soil between the slots and the pore spaces remain open longer.

The devices previously noted are really drainage more than cultivation devices—and that is how they have often been used in turfgrass systems, as discussed in Chapter 15, "Drainage and Sand Capping." The slots intercept water and allow infiltration and percolation to deep depths. Thus, when conducting a gypsum-slotting operation, it should be done with consideration to intercepting runoff and subsequent effects on water flow. In the previous discussion, the emphasis has been on remediation of the existing soil, especially at the beginning of using irrigation water with moderate to high Na. In reality, in the worst-case situations, gypsum slotting with or without sand or organic matter may be required on a periodic basis, at least at some of the most difficult areas on a site. One caution is not to use any inorganic, nonbiodegradable sand substitute that has a high percentage of internal microporosity within the particles for inclusion in the gypsum-slotting media in order to avoid excess retention of Na and soluble salts in that material.

13.3.2.2 Dealing with the Irrigation Water and Soil in the Long Term

In addition to the total gypsum requirement for the existing (i.e., initial) sodic soil condition, Na that is being added via the irrigation system (where the water contains appreciable Na) must also be dealt with by applying additional Ca amendments, since the irrigation Na influences the long-term nature of the soil. In this case, the meq/L of Na added per 12 inches of irrigation water per unit area (per 1000 sq. ft. or per acre) can be countered by a three-step process:

- First is using the Ca and Mg in the irrigation water that is not precipitated by HCO_3 and CO_3 as part of the Ca requirement. When acidification is required to neutralize 75% to 85% of the HCO_3 and CO_3, the final RSC value of the treated irrigation water will be the available Ca and Mg in meq/L. This quantity of Ca and Mg will counteract some of the Na in the irrigation water—in our example, where we are assuming high Na content in the water.
- Second, additional Na in the water can be countered by Ca created by generating gypsum using the SO_4 in the irrigation water plus surface-applied lime (i.e., the sulfur-scrubbing technique).
- Third, the remaining Na that is not dealt with by the two previous means will then need to be countered with applied Ca from gypsum or another Ca source, normally a granular broadcast of Ca amendment to the surface. This process and calculations are presented in Section 13.3.4, "Case Study: Irrigation Water Calculations."

13.3.2.3 Additional Application Guidelines

In the previous sections, various soil or irrigation water amendment guidelines were presented. Many of these were based on use of 100% gypsum. When other amendments are selected, the equivalent quantities are given in Table 12.2. However, a few precautions must be observed for some of these amendments, such as the following:

- Elemental S, sulfuric acid, and other S-based amendments can be applied on a bare soil at high rates as long as they are mixed into the profile and a few weeks are allowed for chemical reaction and equilibration. During these weeks, the soil should be kept moist, and soil temperatures should be above 55°F (13°C) to allow conversion of elemental S into H_2SO_4 when the S is in the sulfide (reduced) form in the amendment. In the case of S in the SO_4 or H_2SO_4 forms, these start to react immediately.
- Sulfuric acid cannot be applied on a mature turfgrass unless diluted in irrigation water so that pH_w is at or above 6.0.
- Elemental S and other granular S sources should not be applied to a mature turfgrass at above 5 lb. S per 1000 sq. ft. (245 kg S/ha) to avoid potential foliar burn and, more importantly, excessive acidity in the surface 1 inch—even on alkaline soils with lime, very high elemental S applied could cause a zone of excessive acidity at the turfgrass crown level or an acidic thatch zone for a period of time that may cause injury—crown damage and root pruning.
- If S-based amendments are applied to the turfgrass, it is important that lime or native calcite be present in the surface zone to avoid excessive acidity (pH < 4.8) (Carrow, Waddington, et al., 2001).
- All amendments in Table 12.2, except gypsum, phosphogypsum, and FGDG, have high foliar burn potential when applied to a turfgrass sod as granular or suspensions unless immediately watered into the soil or applied with sufficient water. This limits the quantity that can be safely applied per application, and often cultivation should be considered to integrate these products into the upper soil profile to minimize layering or burn potential problems. Turfgrass managers may wish to try out rates in a trial area, starting at 2 to 5 pounds of product per 1000 sq. ft. (98 to 245 kg ha^{-1}), which is the upper limit for each application of salt-type fertilizers before foliar burn potential increases. All materials, including gypsum, PG, and FGDG, should be watered immediately into the soil with sufficient irrigation to wash the materials off of leaves.
- Amendments containing N are limited per application and on an annual basis by the turfgrass N requirement (i.e., N controls rates and timing, and not the Ca availability or water acidification needs).
- S-based products that contain N will result in additional acidity in the soil from N reactions that product active H^+ ions, and this contributes to dissolution of lime.

Limitations on rates of many of the gypsum alternatives in Table 12.2 mean that many of these alternatives are best used to supplement Ca from gypsum, PG, or FGDG, which are the Ca forms that can be applied at high rates and are moderately slow-release forms of Ca. Also, where sulfuric acid or SAGs are used for continuous irrigation water treatment, the Ca, Mg, and gypsum that can be obtained via these operations would be considered as "base" Ca addition programs. For example, calcium nitrate used as a routine N fertilizer could add soluble Ca to supplement gypsum. Also, the material may be applied through the irrigation system. Or another example could be a grower who finds that elemental S is more cost-effective than gypsum for his or her location. A couple of applications of elemental S at <5 lb. per 1000 sq. ft. per application as a broadcast could be made each year.

13.3.3 Gypsum, Other Ca, or Salt Amendment Injection Options

Injection of saturated gypsum solution, gypsum suspension, or other dissolved soluble salts into the irrigation water can be used for two primary reasons: (a) to add Ca as part of the overall

Ca amendment program for sodic soils, and (b) to add salts to ultrapure low-salinity irrigation water to increase ECw and water infiltration. This latter reason is not a sodic soil issue, but is briefly presented in order to understand the uses of "gypsum injector" systems. Along with either of these reasons, additional soluble fertilizer can be added as needed as a component of the fertilization program.

Injection application methods that can add higher quantities of amendments via the irrigation system than traditional fertigation systems include (a) devices that create a saturated gypsum solution by percolation of water through fine-grade gypsum; (b) devices that use agitation in a tank to suspend and dissolve Ca from fine-grade gypsum or any other water-soluble Ca source or salt source, such as calcium chloride; and (c) a system that uses a cone-shaped tank with a wobbler spray head at the bottom with a stainless steel cone on top, which allows water-soluble dry material to be turned into a suspension with further dissolution into the irrigation lines. This latter system is simple in design, operation, and cost.

13.3.3.1 Ca Injection for Sodic Soils

Sometimes, the RSC is moderate and some addition of a soluble Ca source would decrease it into an acceptable range (Table 6.7). In this instance, soluble Ca could be injected to lower RSC and provide Ca ions. However, if RSC is high from excessively high HCO_3 and CO_3 content, addition of a soluble Ca source may cause excessive precipitation of lime in the irrigation lines. The use of the pHc provides guidance relative to potential for lime precipitation in lines, especially low-flow and low-pressure drip irrigation lines, where pHc is a theoretical, calculated pH of the irrigation water in contact with lime and at equilibrium with soil CO_2. The pHc value is provided in some irrigation water quality reports or can be calculated using the tables in Ayers and Wescot (1976) or Carrow and Duncan (1998). When the pHc is <8.4, this suggests a tendency for lime precipitation.

The RSC can be used to determine the quantity of gypsum to add per acre-foot of irrigation water to supply Ca to react with the remaining bicarbonates and carbonates to achieve an RSC = 0. The formulas used are as follows:

- *RSC × 234 = pounds of 100% pure gypsum per acre-foot irrigation water*
- RSC × 86 = kg of 100% pure gypsum per 1000 m^3 irrigation water
- Gypsum injection could be used to achieve the required gypsum and subsequent soluble Ca. Alternatively, the water could be acidified to remove excess carbonates.

If the irrigation water has high Na, but is low in HCO_3 and CO_3 content, irrigation system injection could be a means to deliver a Ca amendment on a routine basis. Since the HCO_3 and CO_3 levels are low in this situation, water acidification would not be needed to neutralize these ions, nor would any SO_4 within the water from product additions be available to create gypsum from a lime source—only the base level of SO_4 in the water. However, S-based acidification plus soil-applied lime or native lime could be used to generate some of the gypsum requirement as explained below.

In this case, a soluble Ca source is added as a way of providing Ca ions to replace Na, as well as to increase EC_w to suppress Na permeability hazard, or acidification with a lime source is used as an alternative to gypsum injections. However, in this section, we will deal with gypsum injections. For every 1.0 meq Ca/L added to irrigation water, the ECw should increase by 0.075 dS/m (Oster and Grattan, 2002). Tables 13.2 and 12.2 list equivalent rates of calcium salts, acids, and acid-forming materials for irrigation water treatment. The objective would be to balance the meq/L of Na in the water with the meq/L of Ca where the Ca comes from what exists in the irrigation water along with what is added as a water treatment. As a general guide, the amendment rate to increase total Ca to 2.5 meq/L is considered a moderate rate, and rates to increase to 5.0 meq/L of Ca are high. From Table 13.2, 234 lb. (106 kg) of 100% gypsum dissolved in an acre-foot of irrigation water would raise Ca in the water by 1 meq Ca/L. Thus, if the irrigation water contains 4.5 meq Na/L and 1.5 meq/L Ca and 0.5 meq/L Mg, it would require 2.5 meq/L additional Ca to balance the Na. From Tables 13.2 and 12.2, 2.5 × 234 lb. = 585 lb. 100% gypsum per acre-foot of irrigation water would be required. If $CaCl_2 \cdot 2H_2O$ (27% Ca) (see Table 12.2) was used, it would require 2.5 × 200 lb. = 500 lb. per acre-foot. If the Na content in the irrigation water is high enough that over 5.0 meq Ca/L would be required, the additional quantity applied should be by granular broadcast on the soil. Table 12.2 also contains the quantity of S-based amendments required—assuming lime is available in the soil surface.

The Ca added via gypsum injection devices would not only supply Ca directly to the surface of the sodic soil to displace Na but also increase irrigation water ECw. In Figure 3.1, the goal would be the portion of the soil SAR versus ECw figure where there is "no reduction in rate of infiltration." For the example above, where 2.5 meq/L Ca was added, this would increase irrigation water ECw by 2 × (0.075 dS/m) = 0.15 dS/m; or for 5.0 meq/L of Ca per acre-foot, this would result in a 2 × 0.075 dS/m = 0.30 dS/m. This does not seem like a very significant increase in irrigation water ECw; for example, if added to an irrigation water that already had an ECw = 2.0 dS/m, the total after addition would be ECw = 2.3. In Figure 3.1, the soil SAR tolerance to maintain adequate water infiltration would increase from about SAR 10 to 12.2. However, this degree of ECw increase is very significant when the irrigation water is ultrapure, at ECw < 0.50 dS/m (Figure 3.1).

13.3.3.2 Ca or Salt Injection for Ultrapure Water

Injection of Ca or salt injection for ultrapure water is generally not considered a "sodic or saline-sodic" remediation practice. However, there are locations such as southern California where sodic soils are present, and one of the irrigation water sources used for some of the irrigation requirements during periods of the year may be snow melt with very low ECw. In these situations, consideration of the quantity of Ca addition in irrigation water to offset low EC_w by considering effects on soil SAR, as illustrated in Figure 3.1, is appropriate. If the initial irrigation water has ECw = 0.20 dS/m and 5.0 meq/L Ca was added, the resulting ECw would be 0.20 dS/m + 5.0 (0.075 dS/m) = 0.575 dS/m, which is in the range where soil infiltration would go from severe reduction to no reduction. The primary reason for poor infiltration when the irrigation water is too pure or after a high rainfall volume monsoon season is that many of the salts are leached from the surface few centimeters, which fosters soil crust formation, and where addition of salts restores surface structure.

Duncan et al. (2009) provide a good discussion of ultrapure water issues. An alternative may be irrigation water blending with a water source containing higher ECw (see Section 13.4, "Irrigation Water Blending and Salt Loads"). Also, Hanson et al. (1999) provide excellent information on determining gypsum needs in ultrapure water, including example calculations.

13.3.4 Case Study: Irrigation Water Calculation

This example is for determining the Ca and gypsum needed to counteract and balance Na in irrigation water. Additional Ca from gypsum or other Ca sources may be needed to deal with Na already accumulated in the soil on CEC sites. The information below is only for dealing with gypsum quantity needed, based on the irrigation water applied per 12 inches acre-foot of applied water.

13.3.4.1 Situation

Existing irrigation water has the following properties based on an irrigation water quality test:

- HCO_3 700 ppm = 800/61 = 13.1 meq/L
- CO_3 0 ppm = none
- Na 230 ppm = 230/23 = 10 meq/L
- Ca 110 ppm = 110/20 = 5.5 meq/L

- Mg 23 ppm = 23/12.2 = 1.9 meq/L
- RSC = (HCO_3 + CO_3) – (Ca + Mg) = (13.1 + 0) – (5.5 + 1.9) = 5.7 meq/L

13.3.4.2 For Water Acidification

RSC × 133 = 758 lb. 100% of sulfuric acid per acre-foot water; but we will use about 75% of full-rate and lower water pH to about 6.5. Some HCO_3 is good to buffer rapid pH changes of water. Assume 75% efficiency to remove HCO_3 = 0.75 × 13.1 meq/L = 9.8 meq/L removed and 3.3 meq/L remaining.

The 75% sulfuric acid rate is 568 lb. of H_2SO_4 per acre-foot of water (i.e., for every 12 inches of irrigation water applied per acre) = 556 lb. of SO_4 per acre-foot of water = 12.7 lb. SO_4 per 1000 sq. ft. = 4.2 lb. S per 1000 sq. ft.

The 75% efficiency of HCO_3 removal would result in a 25% HCO_3 level of 0.25 × 13.1 = 3.3 meq/L remaining, which would react with 3.3 meq/L of the Ca + Mg. This would leave (5.5 + 1.9) – 3.3 = 4.1 meq/L of Ca + Mg that can counteract 4.1 meq/L of the total 10 meq/L of Na. This leaves 10 – 4.1 meq/L = 5.9 meq/L of Na to counteract with applied gypsum. Thus, we need to obtain another 5.9 meq/L of Ca to counteract the remaining 5.9 meq/L of excess Na.

13.3.4.3 Total Gypsum Requirement

The total gypsum requirement is the quantity needed of gypsum or gypsum equivalent that will be enough to provide 5.9 meq/L of Ca.

- Since ppm or mg/L of Ca = meq/L × Equivalent Weight.
- 5.9 meq/L × 20 = 118 ppm of Ca. See Tables 6.9 and 6.10 for conversions of ppm nutrients to lb. nutrient per 1000 sq. ft.
- 118 ppm Ca = 7.4 lb. Ca per 1000 sq. ft. per 12 inches of irrigation water.
- Gypsum with 23% Ca content is common. For gypsum with 23% Ca content, it would take about 100/23 × 7.4 = 32 lb. of actual 100% gypsum per 1000 sq. ft. for every 12 inches of irrigation water of total gypsum to displace the Na.
- Part of this 5.9 meq/L of Ca (or 32 lb. needed) will come from the irrigation water. The goal is to use the free SO_4 in the irrigation water from the acidification treatment (and any SO_4 in the original or untreated irrigation water). The 12.7 lb. SO_4 per 1000 sq. ft. will react on almost a 1:1 ratio with lime ($CaCO_3$) to create 1.36 lb. gypsum for each lb. of SO_4.
- Thus, 12.7 lb. SO_4 + 12.7 lb. fine particle size lime (added to the surface of the soil) = 17.2 lb. gypsum per 12 inches of irrigation water per 1000 sq. ft. from the irrigation water SO_4 reacting with the applied lime.
- Total remaining gypsum needed per 1000 sq. ft. per 12 inches irrigation water per acre-foot = 32.0 – 17.2 = 14.8 lb. per 1000 sq. ft. per 12 inches irrigation water applied.

This gypsum or Ca equivalent would often be added as a granular broadcast application with timing based on how often 12 inches of irrigation water are applied.

13.4 IRRIGATION WATER BLENDING AND SALT LOADS

Sometimes, a poor-quality irrigation source can be blended with a better quality water to achieve acceptable irrigation water on turfgrass ecosystems. Blending is usually achieved by pumping both sources into a common lake or pond at the appropriate ratios and allowing mixing to occur. Quality of the resulting blend can be calculated using this relationship:

(Water Quality Factor of Source 1) (Proportion of Source 1 Used) + (Water Quality Factor of Source 2) (Proportion of Source 2 Used) = Resulting Level of Blended Water

TABLE 13.3
Influence of Irrigation Blending Using Ocean Water on Total Dissolved Salts (TDS) and Na Concentration in Specific Blends with an Effluent Water Source

Total Dissolved Salts						
% Blend = Ocean Water (34,500 ppm):Effluent (1000 ppm) TDS						
100:0	90:10	70:30	50:50	20:80	10:90	0:100
PPM						
34,500	31,150	24,450	17,750	7700	4350	1000
Na Concentration						
% Blend = Ocean Water (460 meq/L Na or 10,580 ppm):Effluent (9 meq/L or 207 ppm) Sodium Concentration						
100:0	90:10	70:30	50:50	20:80	10:90	0:100
meq/L						
460	414.9	324.7	234.5	99.2	54.1	9
PPM						
10,580	9543	7468	5394	2282	1244	207

Source: After Duncan, R. R., R. N. Carrow, and M. Huck, *Turfgrass and Landscape Irrigation Water Quality: Assessment and Management*, Taylor & Francis, Boca Raton, FL, 2009.

An example is if Source 1 has an ECw of 3.8 dS/m (very high) and Source 2 has an ECw of 0.70 dS/m (low), then the resulting blend ECw for various ratios would be as follows (Table 6.5; Australian Water Authority [AWA], 2000):

Source 1:Source 2 (%)	**EC_w of Blended Water (dS/m)**
75:25	(3.8)(0.75) + (0.70)(0.25) = 3.03 (high)
50:50	(3.8)(0.50) + (0.70)(0.50) = 2.25 (medium)
25:75	(3.8)(0.25) + (0.70)(0.75) = 1.48 (medium)

This procedure can be used to determine the value of other factors (adj SAR, RSC, Na content, etc.) as long as the values for each source are in the same units. Based on the ECw criteria of AWA (2000) in Table 6.5, a 25:75% blend would result in an ECw just within the medium salinity classification.

Table 13.3 illustrates an extreme case of water blending, which is ocean water (34,500 ppm salts; 46.6 dS/m using a conversion factor of 740) blended with a reclaimed water source of 1000 ppm or 1.35 dS/m. At the blending ratio of 10% ocean water and 90% reclaimed water, the final mix would still contain 4351 ppm salt and an ECw = 5.88 dS/m. If the ocean water contained 460 meq/L Na (10,500 ppm Na) and the effluent 9 meq/L Na (207 ppm Na), the final 10:90% blend would contain 54.1 meq/L Na and 1244 ppm Na (Duncan et al., 2009).

To put this in perspective, the blended water with 1244 ppm Na would mean that 3386 lb. of Na would be applied to the grassed areas with each acre-foot of water. Since 20 lb. of Ca are needed to counter 23 lb. of Na, you would need 2944 lb. of Ca per acre-foot of water to counter that blended sodium load in the irrigation water. The calculations depicted in Section 13.3.4 can then be used to determine the approximate additional calcium to add beyond the concentrations that are already present in the ecosystem.

14 Cultivation, Topdressing, and Soil Modification

CONTENTS

14.1 SOIL PHYSICAL PROBLEMS: OVERVIEW

On all turfgrass soils, especially on recreational sites that receive intense traffic, good soil physical properties are essential (Carrow, 1989, 1991, 1992b). But, when a high salt load and excess Na have accumulated in the soil, good physical conditions are absolutely critical for long-term turfgrass sustainability. Sodic and saline-sodic soils are the most challenging stresses because Na becomes an active agent that degrades soil structure, as noted in Chapter 3, "Sodic, Saline-Sodic, and Alkaline Soils"; while for saline soils that have high salt loads, there must be physical conditions (pore–space continuity) that allow rapid leaching of these salts to avoid physiological drought stress.

In later sections of this chapter, the discussion will be on cultivation, topdressing, and soil modification. It is beyond the scope of the book to cover these practices in detail, but the general principles will be presented along with comments related to how these practices may require adjustment in saline or sodic soil situations. However, before any of these practices are implemented, the *turfgrass manager should start with identifying what specific soil physical problems exist on a site*. Then, the best management options can be made. Thus, Section 14.1 deals with soil physical problem identification—part of the salinity best management practice (BMP) comprehensive site assessment.

Turfgrass managers often think of a soil physical problem as a specific field problem like "soil compaction." But, a more basic approach that is essential on salt-affected soils is to think in terms of the soil pore–space distribution as related to potential and actual salt accumulation or leaching from the soil surface to the drain lines or region of salt disposal—that is, water infiltration into the soil surface, percolation through the rootzone, and drainage of applied water to where the salts can no longer affect the plant rootzone. The presence or absence of macropores throughout the soil profile is the most important physical attribute.

14.1.1 Macropores: Essential for Salinity Leaching

The primary soil physical goal on salt-affected soil, whether sandy or fine textured in nature, is to maintain continuous macropores throughout the infiltration, percolation, and drainage spectrum. How the *continuous macropore spectrum* is achieved will require a combination of the various soil physical approaches (repeated cultivation, soil modification, functionally effective drainage, etc.) available to each turfgrass manager based on his or her resources. Within the above approaches to enhancing soil physical properties, there are a wide variety of options available to deal with any specific problem depending on budget, labor, and other considerations.

Mesopores and micropores are essential for soil moisture retention, but macropores are necessary for rapid water movement after rain or excess water application, gas exchange in the soil with the atmosphere (i.e., aeration), and root channels (Table 8.1). The soil profile on each site on a golf course or other recreational turfgrass area should be assessed in terms of any barriers to water movement (i.e., layers in the soil without adequate macropores) throughout the whole water movement spectrum. Not only should macropores be present, but also they should be able to drain in a timely fashion—not be waterlogged in order to properly function. An impervious horizon or even a thin layer with few internal macropores will result in water ponding, saturation of macropores, and salts accumulating above the layer.

Managing macropores will require minimizing factors that degrade soil structure while fostering conditions that create and stabilize structure. Situations that foster loss of macropores or soil physical degradation are (a) soil compaction from vehicle and foot traffic; (b) the weight of the overlying soil creating compacting forces that can compress unstable soil aggregates deeper in the soil profile; (c) a lack of aggregate stabilization, especially by microbial activity and organic matter content; (d) the presence of excess Na that weakens aggregation (see Chapter 3, Section 3.1.2, "Process of Physical Degradation of Soil Structure by Na"); and (e) the specific type of clay, with 2:1 expanding and contracting clays being more subject to rapid loss of macropores than 1:1 nonexpanding clays. Aggregation is fostered by the action of wetting and drying, freezing and thawing, root penetration pressures, high sand content soil modification that creates and stabilizes macropores (especially at >85% sand content), and cultivation operations. Aggregates are stabilized or held together by various mechanisms involving organic matter polysaccharides and gels, divalent (Ca) or trivalent (Al) ions, Fe/Al/Mn oxides that act as cementing agents, and removal of Na.

Successful management to prevent, alleviate, or minimize soil physical problems requires a systematic approach (Carrow, 1989). First is the correct identification of the primary field problems on the site that are limiting salt movement and turfgrass performance. Usually more than one primary problem is present. Secondary problems or symptoms arising from the basic or primary problem are often mistaken for the primary problem, especially for soil physical problems. For example, surface soil compaction of a fine-textured soil (i.e., one with appreciable silt and clay) results in low water infiltration, waterlogged or low soil O_2 conditions, poor rooting, and, when dry, a hard surface. The primary field problem is surface compaction; but, more specifically, the destruction of the larger soil pores (macropores) has occurred, while all other visual expressions are secondary problems. Secondary problems must be dealt with as they occur, but prevention of these limitations on turfgrass sustainability in the future depends on correcting the underlying soil infrastructure problem. Only after the primary soil physical problems are identified, as well as their location in the profile and their specific causes, can wise decisions be made in terms of what salinity BMP approaches are best—cultivation, soil modification, drainage, amendment applications, and so on.

Identification of soil physical problems requires "good information." This arises from onsite inspection with emphasis on (a) *visual symptoms* of the grass and soil surface (many times, a basic problem is lack of surface water drainage either from the whole surface or in "pockets" or low-topography spots); (b) *observation within the soil profile* for any layers or horizons that would impede water, air, or root movement, along with evidence of anaerobic conditions such as a black layer or high water table, and location and health status of roots; (c) for salt-affected sites, the *chemical soil tests* are essential to determine magnitude and location of excessive salts and high Na accumulation; (d) *soil physical analyses* may need to be performed in a laboratory; and (e) *evaluation of the "surrounds"* may be needed to determine if conditions off the site are contributing to problems on a sports field, which, for example, could be in the form of surface drainage coming onto a field, subsoil water movement into the field soil by lateral movement or capillary rise from a water table, shade, or impediments to air drainage across the field surface (Carrow, 1989; Carrow and Duncan, 1998; Neylan, 1994). It is useful to ask the question "What really is the cause of this problem that is being expressed negatively on the turfgrass ecosystem?" in order to help determine the root or basic issue (i.e., the site-specific primary problem).

14.1.2 Common Soil Physical Problems on Fine-Textured Soils

Fine-textured soils are those containing >10% silt and/or clay. The most important soil physical field problems on these soils when they are salt-affected are discussed below (Carrow, 1989, 1991; Adams and Gibbs, 1994; Carrow and Duncan, 1998; Handreck and Black, 2010).

14.1.2.1 Salt-Related Problems

Salt-affected turfgrass soils will have soil physical problems that nonsaline soils do not have. These have been addressed in Chapter 2, "Saline Soils," and Chapter 3, "Sodic, Saline-Sodic, and Alkaline Soils." Basically, high salinity will require a degree of drainage effectiveness that goes beyond that of a similar soil that is not salt affected. On nonsaline soils, removal of excess soil moisture from the rootzone is normally sufficient as long as the drained soil has adequate aeration in the rootzone. However, with excess soluble salts, the focus is to move the salts in a continuous manner well past the bottom of the rootzone where any short- or long-term impediment to water movement will cause salts to accumulate above this barrier. On sodic soils, the Na ion becomes active, degrading soil structure and working to destroy the essential macropores necessary for leaching, aeration, and rooting. Thus, soils, layers, or horizons that do not present soil physical problems in a nonsodic soil can become major challenges with excess accumulated Na present.

14.1.2.2 Excessive Quantities of Silt and Clay

This problem is a common theme across all fine-textured soils. High silt or clay content results in too few macropores (>0.075 mm diameter) unless they are highly structured. In addition to *lack of macropores* (desirable to have at least 15% by soil volume of macropores), these soils also often exhibit *high soil strength*, particularly as the soil dries. This means that roots are exposed to a "hard" medium to grow into unless cracks (i.e., these are macropores) occur. Two important *clay types* react differently:

- 2:1 clays (montmorillonite, illite, and vermiculite) are shrink-swell clays or cracking clays. They "self-cultivate" when dry, but when exposed to excess moisture, they swell and seal the cracks as well as many other macropores.
- 1:1 clays (kaolinitic, allophane, and Fe/Al oxides) are nonexpanding and become very hard as they dry. However, when macropores are created (by cultivation, for example), they remain open for longer periods even when exposed to excess moisture.

Excessive silt and clay quantities may be positioned throughout the whole soil profile or may be accentuated in the B horizon. Many times, the B horizon is a zone of clay accumulation, and when this horizon becomes sodic, physical conditions deteriorate even more. Soils where structure development occurs will exhibit better properties because structure formation results in "macropore formation"; but Na will hinder this positive aggregate-forming process. Fine-textured soils often have additional primary problems that further increase their inherent problems of lack of sufficient macropores and high soil strength, namely, they are more susceptible to soil compaction, sodic conditions, or poor natural structure development (resulting in "massive" structure).

14.1.2.3 Soil Compaction

On recreational turfgrass sites, the pressure applied to the soil by vehicles and foot traffic when it is at field capacity or greater in moisture content causes breakdown of natural structure, and compaction of the soil into a more dense mass occurs, whether the soil is salt affected or not (Carrow, 1992a). Soils most susceptible to compaction have appreciable silt and clay content and have 2:1 clays, but when these soils are exposed to high Na application, they will be even more prone to compaction. This is a problem because of further loss of macropores and creation of an even more "hard soil" by enhancement of soil strength. Soil compaction can occur in different soil zones depending on the history of traffic events, namely:

- Surface 0 to 3 inches (0 to 8 cm) (i.e., thin surface compaction zone) is almost always present on fine-textured recreational turfgrass soils.
- Deeper surface zone of compaction from 0 to >3 inches (0 to >8 cm) due to heavy equipment operating on a site.
- A buried zone of compaction. One type is where the surface 0 to 12 inches (0 to 30 cm) becomes compacted by heavy equipment during construction and the surface is tilled prior to establishment, but not to the depth of compaction. This leaves a compacted subsurface zone.
- Another example is a "cultivation pan" created by routine core aeration to the same depth, such as to the 3-inch (8-cm) depth, and the pan forms at the 3- to 5-inch (8.0- to 12.7-cm) depth.

14.1.2.4 Presence of Layers

Recreational turfgrasses are often grown on soils with layers that differ substantially in texture, organic matter content, wettability or compactness, or any combination of these three (Carrow, 1992a). These layers may be relatively thin (<0.25 inch or 0.63 cm) or sometimes much thicker. Most often, they are continuous over the sites, but sometimes a discontinuous layer is formed. Discontinuous layers are less of a problem than those that are uniform throughout the site. Also, a layer may be at the surface, or it could be a subsurface layer. Often deeper layers, such as at 12 inches (30 cm), on a non-salt-affected soil do not present much management challenge except during unusually wet periods, but with salinity, this layer can become a major impediment to salt movement and a subsequent zone that is prone to persistent salt accumulation with those soluble salts being capable of rising back into the rootzone through capillary pores, especially during prolonged high-evapotranspiration (ET) environmental conditions.

Procedures that help identify the presence of a layer problem are (a) observation of any distinct textures, distinct soil profile color differences, organic matter localization, or compacted zone; (b) observation of impeded water movement, rooting, or both; (c) use of a penetrometer to distinguish areas of high resistance or hardness; (d) a layer that has an odor indicating a lack of oxygen—an example would be "black layer" and a rotten egg or hydrogen sulfide odor; and (e) any evidence of salt accumulation above a suspected layer.

Some of the primary problems previously discussed were layer problems—surface compacted layer, subsurface compacted zone, and a heavy B horizon due to high clay content or high Na (sodic) accumulation. Other potential layer problems in turfgrass soils are as follows:

- Any distinct texture or organic matter content change from the native soil has the potential of being a layer problem.
- Sand over a "heavier" soil leaving a distinct interface. When sand topdressing is applied to a fine-textured sports field to alter the surface texture and help prevent surface soil compaction, core aeration is also recommended to prevent a distinct interface from developing and to open macropore channels into the heavier soil zone.
- Sand buried 1–4 inches (2.5–10.0 cm) below the soil surface that leaves a distinct interface.
- Wind, water, topdressing, or sod deposition of finer textured material than the original soil (such as sand-based profiles) at the soil surface.
- Algae scum layer formation at the surface or dominant surface layer formed by any photosynthetic microorganism.
- Black layer formation at a 1- to 10-cm depth.
- Thatch or mat over a soil surface or within the soil profile.
- Soil profile distinct color differences, such as a darker subsurface zone with a glassy or slick appearance, compared to soil above and below the zone.

Each of these layer situations can prevent water movement across the layer and sometimes impede root penetration. If water ponds above a layer after every irrigation or rainfall event, a temporary perched water table causes an anaerobic zone above the layer, prevents O_2 diffusion to roots below the layer, and will be a zone of salt accumulation. This is especially serious when temperatures are high on cool-season grasses or on overseeded warm-season grasses.

14.1.3 Common Soil Physical Problems on High-Sand Soils

High sand content soils are those containing >90% sand. The most important soil physical field problems on sandy soils when they are salt affected are discussed below (Carrow, 1989, 1992b; Snow, 2004; Adams and Gibbs, 1994; Neylan, 1994; Carrow and Duncan, 1998; Baker, 2006; Handreck and Black, 2010).

14.1.3.1 Salt-Related Problems

Accumulation of high total salt levels within the rootzone, especially the surface 4-inch (10-cm) zone, will limit water uptake and induce drought stress. Since most sands already have low water-holding capacity, salts can rapidly concentrate in the soil solution upon drying, and the limited water availability will be further restricted by the salt. The accumulation of salts in the soil profile inhibits plant root water uptake due to the osmotic pressure of saline soil solutions lowering its potential energy (since water always moves from a higher to a lower energy potential in the soil–water–plant continuum) and the pressure in the roots subsequently being higher than the surrounding saline soil solution.

Salts can accumulate at the surface by the addition of irrigation water containing even modest salt loads under conditions where leaching does not occur. Hot, dry weather greatly enhances salt accumulation. Also, when a net downward movement of water does not occur, capillary rise of salt-laden water from below the rootzone may rise into the rootzone with high-ET conditions. If this is allowed, the salt within the salt-rich subsurface zone is often very concentrated and may even contribute to direct root injury as well as salt-induced physiological drought stress due to inhibition of plant root water uptake. In the case of sands, high Na accumulation is detrimental not only because it adds to the total salinity problem but also because, while few structural aggregates are present in sands, high sodium will also cause dispersion of any silt, clay, or organic colloids, which may then migrate to the depth of routine irrigation water penetration to reform a minilayer with subsequent salt accumulation above this zone.

14.1.3.2 Low Water-Holding Capacity

Many sands have low water-holding capacity due to few mesopores and micropores. Contributing factors include too coarse of a sand particle size (high volume > 1.0 mm), low organic matter content (generally <1%), depth of rootzone mix (inappropriate for the physical characteristics of the sand; i.e., exceeding 12 inches or 30 cm depth in USGA specification green), and susceptibility to hydrophobicity. When saline irrigation water is used, the salts reduce water activity with the result of less plant-available water.

14.1.3.3 Excessive Organic Matter in the Surface

Accumulation of excessive organic matter as a thatch or as a mat (organic matter plus sand), which does not contain sufficient sand (>85%), can lead to reduced infiltration and aeration (Carrow, 2004a, 2004b). The reduced aeration is most evident within the zone, but low oxygen can occur below the layer if no macropores are continuous through this layer to allow gas diffusion (i.e., a primary reason for "venting" cultivation operations). Even when constructed properly, the rootzone profile changes over time (Neylan, 1994). The most consistent and dramatic change that occurs on high-sand greens or athletic fields after establishment is accumulation (with favorable growth, a 1%–2% increase per year in the establishment period) of organic matter (OM) within the surface 1.25 inches (32 mm), to increase above the 1%–3% by weight of OM (often in the form of good-quality peat moss) used in initial rootzone mixes. This "layer" no longer has the same soil physical properties as the original rootzone mix. Thus, it is not a problem of original specifications, but one of management changes as well as organic composition dynamics that have occurred after establishment. Research has consistently demonstrated that as organic matter content in a sand mix increases to about 4.0% to 4.5% by weight, the percent of larger soil pores (macropores, and aeration pores) >0.075 mm in diameter between sand particles decreases due to plugging by organic matter, which includes live roots and rhizome tissues.

Also, the nature of the surface organic matter can sometimes adversely change in a short period of time. For bentgrass and annual bluegrass greens during hot, humid periods, a substantial portion of the roots can die in a relatively short period of time. The dead roots create a temporary period of sealing the macropores by the decaying, gel-like dead tissues. The combination of low soil oxygen, limited water uptake, and high temperatures can cause summer bentgrass decline (Carrow, 2004b). The higher the organic matter content in the surface zone, the more likely this will occur in hot, humid periods because the excessive organic matter can retain too much moisture, resulting in low oxygen (air porosity) within the zone if there is frequent rain or irrigation.

When salts are present in association with the surface organic matter layer, any shift to light, frequent irrigation will (a) allow salts to accumulate in the surface, and (b) inhibit leaching or flushing of the salts into and through the profile, including salts retained above the perched water table of a USGA green (Snow, 2004). Thus, salt osmotic stress becomes an additional stress during this time and can increase the degree of injury (root dieback, reduced water uptake, and decreased plant respiration and cooling).

14.1.3.4 Layers in Coarse-Texture Soils

The presence of fine-textured layers within the sand rootzone (i.e., surface or subsurface) is another common soil physical problem. Even small quantities of silt or clay can seal a zone within a sand matrix if the particles migrate and accumulate at a microsite. Layers arise from many sources, but common ones are from sod, by the use of topdressing media with fines, from incomplete onsite mixing of soil and amendments, by the use of inorganic sand substitutes that sequester water and soluble salts, from continuous wetting front-induced migration of fines and salts down to a certain depth, and from wind or water fines deposition. Sometimes clay lens (areas of silt and clay deposition) are present deeper in sand soil profiles due to water-deposited fines from previous periods. In arid climates, irrigating so that water moves to the same depth of penetration on a routine basis may move fines or salts to a zone where a layer may start to form. As water comes into contact with a layer containing few macropores, movement is decreased and water starts to pond or perch above the layer. A subsurface layer can easily transform into a black layer situation as anaerobic substances accumulate to further seal soil pores (refer to Chapter 17, Section 17.5.1, "Black Layer"). Not all layers in a sand profile are from fines or precipitated minerals. Sometimes a *distinct textural change* (such as a very coarse sand layer within a fine sand rootzone mix) or *composition change* (a buried organic layer such as from a buried thatch zone, buried native flora and roots during initial course construction, or organic layer from organic sod that becomes buried; a zone of precipitated salt accumulation) can disrupt water movement to create a perched water table. As with fine-textured soils, any layer that impedes downward water movement will enhance salt accumulation and make leaching more difficult.

14.1.4 BMP Tools to Address Soil Physical Problems

The reason for the overview of the most common soil physical problems on fine-textured and high sand content soils is to emphasize how important soil physical properties are in managing salt-affected soils. Not only do the salts contribute to physical problems by the accumulation of total soluble salts and excess Na, but also they often make the existing turfgrass ecosystem conditions worse—for example, problems on a nonsaline soil from a compaction or an excessive organic matter surface layer on a sand profile will be greater in magnitude with the presence of salts. Accurate identification of soil physical problems and their location in the profile is the first essential step in order to make an informed decision on which approach or combination of best management approaches should be used.

To achieve adequate sustainable soil physical conditions often requires the full set of *soil physical management tools* from the turfgrass manager's toolbox for addressing soil physical issues:

- Proper selection and application of water and soil amendments (i.e., specific calcium products) on sodic and saline-sodic soils to aid in counteracting the adverse effects of Na on soil physical conditions, as addressed in Chapters 12 and 13.
- Cultivation programs: surface and subsurface, and flexibility and frequency in implementation.
- Soil modification: partial modification of the existing soil by topdressing, sand capping, sand, and/or organic matter additions; complete modification where the rootzone is totally constructed without the native soil.
- Drainage: surface and subsurface. Drainage and salt disposal options should also be considered as part of an overall water management plan (Carrow and Duncan, 1998).
- Proper irrigation system design and water distribution uniformity, and efficient application plus scheduling, as addressed in Chapters 10 and 11.
- Additional practices: wetting agents, nonbiodegradable inorganic sand substitutes when used properly for salt-affected sites (otherwise, adverse effects can occur), polymers, altering soil gases by suction or positive air pressure systems, altering soil temperature by blowing air or by heating systems in cold climates, and physical stabilization materials for sports fields.

As noted earlier, the first part of the chapter was on the identification of soil physical problems, and the last part will give attention to some of the practices to alleviate or prevent those problems—cultivation, topdressing, and soil modification. It is beyond the scope of the book to cover these practices in detail, but the general principles will be presented along with comments related to how these practices may require adjustment in saline or sodic situations.

14.2 CULTIVATION OF SALINE AND SODIC SOILS: GUIDELINES

Cultivation on salt-affected sites must be performed at greater intensity and frequency compared to nonsaline sites. The same general principles of soil cultivation that apply to any non-salt-affected soil would also apply to salt-affected sites. But, there are also differences that should be understood. A good cultivation program is a necessity on sites with saline irrigation water in order to maximize infiltration, percolation, and drainage. All cultivation devices do at least one task and that is to create temporary macropores—some devices create small macropores, while others develop very large pores depending on the equipment used (Nektarios et al., 2004). Adequate macropores are especially important at the surface in order to allow adequate infiltration of water. Even with pulse irrigation, it is important to allow as large of a "pulse" water quantity as possible to infiltrate before the surface starts to saturate. Also, adequate macropores throughout the rootzone soil profile allow more efficient capture and movement of natural precipitation. Essentially, cultivation is used to allow for better irrigation programs both from the water-use efficiency and from the salt management aspects. On sodic soils, cultivation not only is for the enhancement of leaching but also is necessary to maintain soil aeration, which is dependent on macropores for gas diffusion.

A cultivation program should be developed for each specific primary soil physical problem identified on the site based on the location of the problem, the frequency of reoccurrence, and how the problem influences salt management as well as degree of salt stress. The turfgrass industry has available a number of different types of cultivation equipment (Table 14.1). Some guidelines for cultivation on salt-affected turfgrass soils may assist in developing effective programs. *Guidelines or considerations are as follows:*

1. As a rule of thumb, any cultivation operation on a sodic soil will last in effectiveness only about one third to one half of the time compared to the same soil without Na accumulation due to the Na actively degrading structure. Thus, cultivation must be performed more often on sodic sites. Macropores developed by cultivation at the soil surface are the first to collapse and seal, and this sometimes happens very quickly depending on traffic, soil type, and soil moisture.
2. The negative action of Na would also suggest that any practices to keep the cultivation holes open (maintaining the macropores) should be considered, especially in the critical surface zone. For either hollow- or solid-tine core aeration on sodic soils, filling the holes with sand would be beneficial; but on sodic soils with appreciable 2:1 clay content, the sand should be mixed with gypsum and/or organic material pellets for greater effectiveness (i.e., in order to obtain both physical and chemical benefits). This was noted in Chapter 13, Section 13.3.2, "Gypsum Application on Established Turfgrass Sites," where deep injection of gypsum into slots or holes using cultivation or drainage devices capable of injecting granular materials was discussed. These practices are called by different terms: *gypsum slotting*, *sand banding*, *sand slitting*, *sand grooving*, and *sand injection*; but more than sand injection alone is what gives greater and sustainable salt management effectiveness in the soil profile. Careful consideration should be given to not inject some inorganic amendments that are designed to retain water and soluble salts, since leaching of accumulated salts from these amendments is extremely difficult, especially in native soils. Sand topdressing, especially when the result is a 1- to 3-inch (1.5- to 7.6-cm) sand layer over fine-textured soils, can help maintain the integrity of the surface macropores created by cultivation operations.
3. On sodic soils, deep cultivation devices should be incorporated into the equipment inventory along with any sand, gypsum, and organic matter injection devices that may serve dual purposes for cultivation or drainage. Sodic problems arising from the irrigation water are ongoing and not "fix it and go on" salt management issues; thus, amendment injection via cultivation equipment is an ongoing matter.

TABLE 14.1
Types of Turfgrass Cultivation Devices That Offer Options for Depth of Cultivation, Spacing, Soil Removal, Degree of Surface Injury, Soil-Shattering Types, and Injection of Liquids, Granular, or Air

Type of Cultivation Device | **Comments and Examples**

Coring with Hollow-Tine, Spoon, or Screw Device

All of these units remove soil from the hole.

- Tractor-drawn units with spoons or tines that enter the soil at an angle
- Drum-type
- Verti-Drain or Soil Reliever with hollow tines
- Core aeration vertically operated tine (VOT) units
- Deep-Drill units: Floyd McKay or Green Care

Coring with Solid-Tine Device

This offers coring with round solid tines, not flat tines or blades.

- Verti-Drain or Soil Reliever with solid tines
- Vertically operated solid tines, surface venting, or shatter core tine units
- Aera-vator: solid tines on an articulated shaft
- Planetair curved solid tine

Slicing

Slicing units have solid tines or blades that penetrate into the soil, but they are not power-driven tines or blades.

- Straight-line tines that push into the soil; most common
- Straight-line blades pulled through the soil
- Offset tines (Aerway Slicer, fairway and greens models)
- Hand forking: the "original" cultivation device

Spiking

Spiking blades (pin spiking) are not power driven (i.e., do not cut through the soil) but penetrate by the weight of the device. Units may have power drive to move across any area but not to power the spikes or small blades.

- Pull-behind units with small spikes, pins, or knife-like blades; generally do not penetrate very deep (<2 inches, often <1 inch)
- Green mower spiker attachments

Grooving and Subaerification

Grooving and subaerification units use vibrating blades or rotary blades to loosen the soil. *Groovers* use power-driven blades to cut into the soil. If a unit just cuts into the thatch, it is called a *dethatching unit*; and if it is adjusted to cut only into the surface of the turf, it is called a *verticutting unit*. Other types of cultivation devices also have "subaerification" effects such as the Verti-Drain, Soil Reliever, and Aerway Slicer, since all can result in a soil-uplifting action that has a shattering effect.

- Decompactors: Blec Groundbreaker, Blec SandMaster (sand injection option), Tremor Earthquake, Redexim Charterhouse Verti-Quake, and others
- ShatterMaster: blades vibrate in direction the unit is going and "shatter" the soil.
- Graden: rotary blade cuts groove for organic matter removal, sand injection option.
- WaterWick: more of a drainage device but has deep sand or amendment injection capability.

High-Pressure Water Injection

Devices use high-pressure water to create cultivation macropores of 1/8-inch to 1/4-inch diameter.

- Hydro-Ject walk-behind and truckster units
- Dry-Ject: has sand injection option

continued

TABLE 14.1 (Continued)
Types of Turfgrass Cultivation Devices That Offer Options for Depth of Cultivation, Spacing, Soil Removal, Degree of Surface Injury, Soil-Shattering Types, and Injection of Liquids, Granular, or Air

Type of Cultivation Device	Comments and Examples
High-Pressure Air Injection	
High-pressure air is injected to create cultivation. Sometimes Styrofoam beads or sand are used to fill voids created by the air. These are modifications of equipment used to remediate the soils of septic field systems. • Terra Lift • Aerragreen	

4. On sodic soils, surface core aeration operations where gypsum can be integrated into the soil as a topdressing amendment either alone or mixed with sand are very beneficial in allowing higher rates of gypsum to be applied; getting the gypsum deeper into the soil, since Na will be moving downward and the subsurface soil also requires gypsum for remediation; and creating macropores for more effective leaching and capturing rainfall events.
5. For saline soils that are naturally fine textured and exhibit inherently low infiltration, core aeration coupled with the cores being filled with appropriate sand sizes will keep holes open longer. Periodic deep cultivation will benefit these soils by penetrating through any salt accumulation and/or layered zones and thereby enhancing salinity movement.
6. Consider that macropores must be present throughout the whole infiltration, percolation, and drainage spectrum for maximum salt leaching. Thus, select cultivation methods and timing to achieve this on an ongoing basis. Cultivation targeted in this manner will allow more deep and infrequent irrigation that contains a leaching fraction. Certain deeper cultivation methods allow for sand, gypsum, or pellets of organic matter to be injected during the operation—for example, sand slotting or drill-and-fill devices. In some areas, cultivation during winter dormancy periods may be needed when irrigating with saline water.
7. Related to guideline 6, above, any layer or horizon that impedes water movement for leaching should receive cultivation on a schedule to maintain macroporosity, which means that the cultivation devices should penetrate completely through the layer or horizon to be effective.
8. On high-sand-content greens, the surface 1- to 2-inch zone is where water movement rate is generally the least. If a good maintenance leaching requirement (LR) program is followed to where salts have not been allowed to accumulate in the surface, periodic surface cultivation to maintain vertical "macropores" or holes across this zone is beneficial to allow rapid water infiltration during heavier rain events. Green et al. (2001) demonstrated the effectiveness of cultivation to enhance infiltration rates on golf greens even in very arid climates in terms of salinity leaching. But if salts are allowed to accumulate at the surface to the point where a reclamation leaching is necessary, and leaching is attempted by heavy irrigation applications, cultivation may result in poor salinity leaching between the cultivation holes at the surface. Areas immediately surrounding open aeration holes become damaged and are subject to brown discoloration from irrigation water salt desiccation. Cultivation will not hinder salt leaching if a pulse irrigation regime is followed since the whole surface area is subjected to slow water infiltration and percolation, not just in the aeration holes (macropore size).
9. On fine-textured soils, when salts accumulate in the surface zone and/or deeper in the rootzone to a point where reclamation leaching is required, (a) a good surface cultivation

program is necessary to allow rapid infiltration; (b) deep cultivation is needed to allow water percolation—also, this strategy will allow water penetration during heavy rains; and (c) additional drainage, such as through tile lines, may be needed to keep the salts moving away from the rootzone. Additionally, this cultivation program strategy will help promote some evaporation from the soil and turfgrass canopy and hopefully minimize problems with surface pathogens and algae and moss buildup.

10. Most cultivation equipment is confined to the surface 16 inches (40 cm), but sometimes soil layers or horizons impede drainage below the depth of cultivation tine penetration and may need to be broken through for salinity drainage. Options are deep drilling with a small hand drill system, drilling with water pressure in a pipe with the nozzle at the end, or air injection devices used for septic field soil fracturing such as the Terralift, which penetrates down to 6 feet (1.9 m) and may be needed for maximum salt management. For sodic conditions, deep drilling with the holes backfilled with sand and gypsum would be preferred. Another option could be using a water pressure device, but with applying $CaCl_2$ solution in the final injection as the device is pulled out. The key point here is that on highly saline sites, one or two initial cultivation methods may shrink the salt-affected site down to smaller seepage or high salt accumulation micropockets over one or more years, but additional more aggressive methods (such as drill and fill, the WaterWick, or similar devices mentioned above) may be required to successfully micromanage and eventually remediate these smaller salt-affected areas.
11. Cultivation must be a priority. Implementing scheduled cultivation events can be especially challenging under certain conditions. But, if cultivation is essential for salts to move in a timely fashion and not accumulate, then this maintenance salinity BMP must be understood by players and management. Cultivation and irrigation require the mind-set of "keep the salts moving." Challenges include the following:
    - *During normal site use.* Cultivation is disliked by golfers and most likely any site user. However, when salts accumulate, there is only one means to remove them from the soil and that is by leaching, where there are no surface macropores = no leaching = increased grass stress and persistently wet surfaces that negatively affect golf play or recreational use.
    - *Before and during golf tournament or scheduled events.* One aspect to consider is whether an overseeded grass is present, since these grass species and cultivars most often have inherently lower salt tolerance compared to the basic foundational or understory grass, and these overseeded plants are juvenile with minimal or no actual salt tolerance mechanism activation (refer to Chapter 9, "Selection of Turfgrass and Landscape Plants"). With frequent, light irrigation cycles involving saline irrigation water containing even modest concentrations of salts and Na, the soils can rapidly accumulate salts at the surface while the Na seals the surface.
    - *During persistent rainfall cycles.* Rainfall is excellent leaching water. If there are consistent rainy periods, it is important to cultivate prior to the rain. One dilemma on fine-textured sodic soils is that deep cultivation can open the soil for better water intake, but this can lead to waterlogging if it rains too much and the subsoil has a low drainage rate. This is a case where some deep drilling on selected low areas or the presence of drainage techniques discussed in Chapter 15, "Drainage and Sand Capping," may be necessary prior to the rainy period.
    - *During winter slow-growth cycles.* During the winter, whether the turfgrass is dormant or actively growing at a slow rate, salts may be moving, especially upward by capillary rise if it is a dry winter with only limited or light irrigation and high-ET conditions. Cultivation prior to going into the cold period along with extra leaching at that time may be needed if there is limited rain up to this period. Another cultivation issue during cold periods, especially on actively growing turfgrass that is being used, is concern

over cultivation injury to the dormant or semidormant plant that remains obvious for a long duration. If this is of concern, then venting operations such as solid quad tines (0.25-inch diameter), needle tines, or use of Hydro-Ject or Dry-ject operations allows macropores to be created with minimal injury. These cultivations may be required on a relatively frequent basis depending on the site, such as a 1- to 3-week cycle.

14.3 TOPDRESSING

Topdressing salt-affected soils in terms of sand types, rates, and application practices is similar to topdressing nonsaline soils. O'Brien and Hartwiger (2003) provide an excellent discussion of golf green topdressing with much of the material applicable to sand-based sports fields. Sand selection on close-cut, high-sand turfgrass areas is similar to that used in new construction (Adams and Gibbs, 1994; Snow, 2004; Baker, 2006). Gilhuly (1999) and Skorulski et al. (2010) have developed very good articles related to fairway topdressing, which also applies to sports fields built with native soils. Baker and Canaway (1990) and Spring et al. (2007) report on topdressing in sports fields to develop a sand layer at the surface on nonsaline sites. McCoy and Danneberger (2008) and Crum et al. (2003) discuss the importance of using a sand quality that will have good surface-bearing capacity and stability to traffic (i.e., avoiding narrow particle size ranges or rounded sand particle shapes).

While topdressing practices are similar on normal and salt-affected soils, the reasons differ substantially. On soils receiving irrigation water high in soluble salts or Na, topdressing is especially important in order to maintain infiltration for salt leaching and to counter structure degradation by Na; more specifically, reasons for topdressing in saline soils are as follows:

- Altering the soil surface 1 to 3 inches (2.5 to 7.6 cm) of fine-textured soils to be less prone to compaction by changing the surface to a high-sand content in order to maintain infiltration. Salt accumulation at the soil surface can easily occur when the soil has low infiltration rates.
- Creating a surface 1–3-inch high-sand content zone will allow better irrigation scheduling, especially pulse irrigation, since this surface zone will retain water until it can infiltrate during pulse irrigation cycles.
- Improving the surface soil condition by dilution of organic matter or maintaining sand as the dominant medium in the surface of high-sand sites so that infiltration will be maintained while minimizing high-moisture upper soil problems.

For sodic soils or sites using irrigation water high in Na, the above benefits and reasons for sand topdressing on high-sand rootzone mixes as well as fine-textured soils apply; but they are even more critical due to the Na fostering degradation of soil structure. The higher the clay content, especially with 2:1 clay types, the more difficult it becomes to maintain basic water infiltration and percolation; thereby, the more important sand topdressing is to create a surface layer or application of a layer during establishment. The presence of a 1- to 3-inch (1.5- to 7.6-cm) surface sand zone is very useful. In Chapter 15, "Drainage and Sand Capping," the use of a deeper sand cap (i.e., >3.0-inch, especially 10- to 12-inch caps) will be discussed for many of the same reasons.

Since soil compaction is greatest in the surface 0- to 3-inch zone, creation of a 3-inch sand layer by topdressing is the goal in order to achieve significant improvement in the surface physical conditions. Frequent topdressing is the usual means to develop this layer over time. It is important to avoid a distinct interface of the soil and sand, but mixing of the soil into the sand should be avoided—as happens with hollow-tine core aeration where the cores are deposited on the surface. Even with core removal, there will be some soil intermixing with the sand. By using cultivation with solid tines, followed by topdressing, this can achieve the goal of breaking though the interface, while allowing sand penetration deeper than the 3-inch zone and avoiding soil intermixing with the sand. In areas with high earthworm activity, intermixing may occur; but topdressing often suppresses earthworm activity (Backman, 2002). Sand topdressing can be combined with sand injection, such as with a Blec SandMaster, which allows a good transition with the underlying soil. Drainage is often essential on sodic sites, and sand topdressing can maintain macropore continuity with appropriate underlying drainage features.

Sometimes "manufactured" soils are created by crushing existing hard surface material such as *decomposed granite* and *pulverized volcanic pumice*. When crushed, these materials tend to exhibit a wide particle size distribution. Initially, they may demonstrate acceptable infiltration and percolation rates, but over time as fines migrate and the materials settle, these rates decline and aeration porosity declines—with very negative effects on salt-affected soils. Sand topdressing of these materials coupled with sand injection, such as by the Blec SandMaster, is very beneficial in increasing infiltration rate and creating macropore channels for percolation and aeration. Addition of gypsum as well as zeolite may be beneficial when the irrigation water is high in Na.

14.4 SOIL MODIFICATION

14.4.1 Complete and Partial Soil Modification

Complete soil modification is where the rootzone medium is constructed without using the existing soil, and usually with an installed drain system. *Partial soil modification* normally involves (a) mixing of sand, organic matter, or inorganic sand substitutes into the existing soil; (b) using a sand layer over the existing soil with or without sand slits; or (c) amending a sand soil for greater moisture-holding capacity. A subsurface drainage system may be incorporated depending on the specific partial soil modification method. Topdressing was discussed in the previous section, especially for creation of a surface sand layer of <3.0 inches, and this would be an example of partial soil modification—this example was treated separately because it is the most prevalent type of partial soil modification.

In partial soil modification, deep-soil cultivation methods with granular amendment injection capabilities are often used in order to provide better drainage, such as noted in Chapter 13, Section 13.3.2, "Gypsum Application on Established Turfgrass Sites," describing deep injection of gypsum, sand, or organic matter pellets into slots or holes using cultivation or drainage devices. As mentioned above, different terms are used for this operation such as *gypsum slotting*, *sand banding*, *sand slitting*, *sand grooving*, and *sand injection*. Then a sand layer may be applied at construction or via sand topdressing to create a 1- to 3-inch (1.5- to 7.6-cm) sand layer over fine-textured soils to maintain continuity with the macropores in the sand, materials in the slots or holes, and the underlying drainage. Adams and Gibbs (1994), McIntyre and Jakobsen (2000), and Baker (2004, 2006) illustrate and describe various partial soil modification options.

Turfgrass managers using saline or high-Na irrigation water applied to fine-textured soils that are susceptible to becoming sodic are encouraged to be familiar with the athletic field partial and complete modification procedures discussed and illustrated by Adams and Gibbs (1994) and Baker (2004, 2006). These same procedures are often what must be used on the extreme salt-affected sites such as saline and sodic marine-clay soils and high clay content sodic soils in arid regions. Many of these systems use sand layers of various depths at the surface, sand slitting, and underlying drainage since the specifications usually were developed for climates with high rainfall. Such systems have performed well, primarily because a drainage system was included.

The USGA specifications for golf greens are considered the "gold standard" for complete soil modification specification for high-sand rootzones. Included are specifications for individual amendment components, the final rootzone medium, and the drainage system, and additional guidelines to insure the final product is installed according to plan (Snow, 2004). Other publications that contain various complete soil modification specifications are Davis (1981), Adams and Gibbs (1994), and Baker (2006), while Murphy (2007) and Adams (2008) discuss rootzone mix components.

In most instances, complete soil modification and partial soil modification methods are similar for salt-affected soils and nonsaline soils, but there are some differences to consider. One issue, as has been pointed out, is a good drainage system. Sometimes drainage systems are omitted for nonsaline soils, but this should not be the case for saline and sodic sites where removal of soluble salts and Na must be spatially separated from the root system without potential for resalinization of the rootzone. Surface and subsurface drainage methods are discussed in Chapter 15, "Drainage and Sand Capping."

Previously, the potential for salt retention in some of the inorganic amendments was noted in Chapter 8, Section 8.1.2, "Inorganic Amendments." The inclusion of sand substitutes that have the potential for salt retention should be avoided, especially at high-volume use, or certainly at 10% or greater volume when the irrigation water is highly saline. The <10% by volume refers especially to any surface area, and care must be taken not to allow a higher percentage in a layer or specific spot. Leaching of salts within the internal micropores is very challenging in terms of volume of water required and the time required for effective leaching. Some zeolites (Chapter 17, Section 17.4.4) and Lassenite (Section 17.4.4) appear to have an internal pore size distribution with mesopores in the plant-available range; thereby, these pores are depleted of water and replenished with normal irrigation and rain events, which avoid salt accumulation. Zeolites increase cation exchange capacity (CEC) significantly and preferentially retain K^+ and NH_4^+ ions against leaching, but these nutrients are still available to plants. The quantity required to enhance CEC is within the 1%–2% by volume range on sands in the surface 0 to 4 inches (0 to 10 cm); so the physical influence would be minor compared to that of sand substitutes used at 10% to 20% by volume. Also, zeolite or Lassenite materials can be selected with particle size ranges compatible with the rootzone mix without adding significant additional fines or microporosity.

Regardless of whether straight sand, organic amendments, or inorganic sand substitutes are used, the final mix should be evaluated by soil physical analyses (Table 14.2). Sometimes a sand profile that appears to have good potential sand qualities for a mix or a 100% sand rootzone can exhibit very different results in a laboratory test. For example, Sand 1 in Table 14.2 has a 97.44% sand content, but exhibits low saturated hydraulic conductivity and low air-filled porosity. These negative qualities are in response to the high content of very fine sand sizes (0.053- to 0.150-mm diameter) that, when combined with the silt and clay composition, give a total of 20.9% fines, which is well above the USGA specifications guideline of <10%. When irrigation water is saline or sodic, it is best to use a sand that is above the USGA specification lower end for aeration porosity (macroporosity) at 15% (the USGA specifications recommended range is from 15% to 30%), such as using 18% to 20% aeration porosity as the lower limit for salt-affected sites. Aeration porosity of 10% is sufficient to stop the root growth of grasses, but the rate of root growth starts to decline at about 15%. Infiltration rates normally decline rapidly as aeration porosity declines. By being at the lower limit for aeration porosity, it does not take much accumulation or migration to plug some of the pores with organic matter (including viable plant roots), inorganic fines, surface algae activity, or root dieback. However, *using a lower limit for air-filled pore space of about 18% to 20%, rather than 15%*, would provide greater protection from a rootzone mix easily sealing. Sealing will decrease salt leaching, and the resulting low-oxygen condition also reduces the grass tolerance to salinity.

The authors have observed substantial improvement of rootzone mixes by surface stripping that were started at the 15% aeration porosity in the laboratory measurements but, when used in the field, soon started to exhibit unacceptable aeration and infiltration as the grass established when irrigated with saline water. While the same degree of decline in aeration and infiltration may be manageable with good irrigation water, it is not sustainable with saline water. Improvement was achieved by stripping the surface 3-inch zone and replacing with a coarser sand. For example, sand no. 2 has an air-filled porosity of 16.8%, which is marginal for salt-affected sites. If an acceptable sand that is more coarse (for example, with sand sizes >0.25 mm) in nature cannot be found, then sieving out some of the fines of the sand being used can achieve the same response. A turfgrass manager, for

TABLE 14.2
Soil Physical Analyses of Two Sands to Illustrate the Importance of Obtaining Laboratory Data on the Rootzone Mix

Analysis			Sand 1	Sand 2	USGA Specifications
Particle Size Analysis (%)					
Clay < 0.002 mm			1.05	0.59	<3.0%
Silt 0.002–0.05 mm			1.49	0.32	<5.0%
Sand 0.05–2.0 mm			**97.44**	**99.04**	
Gravel > 2.0 mm			0.02	0.04	
Organic Matter (% wt. ash)			0.62	0.10	
Sand Fractions (%)					
Sieve No.	**Size mm**		**% Retained**		
10	2.0	Fine Gravel	0.02	0.04	0% best; <3.0%
18	1.0	Very Coarse Sand	0.26	1.88	0% best; <10% at >1.0 mm
35	.500	Coarse Sand	3.21	24.01	At least 60% must fall in the 1.00 to 0.25 mm range
60	.250	Medium Sand	24.17	23.92	
100	.150	Fine Sand	49.34	23.92	Not more than 20%
270	.053	Very Fine Sand	**18.37**	2.93	<5.0%; <10% in very fine sand + silt + clay
—	.002 Silt		1.49	0.32	
—	Pan Clay		1.05	0.59	
Soil Moisture Measurement					
Saturated Conductivity (in/hr)			**1.7**	25.1	Greater than 6.0 inch/hr
30 cm Moisture Retention (%)			22.4	13.8	
Soil Pore Space					
Air-Filled Pore Space (%)[a]			**4.4**	16.8	15% to 30%
Capillary Pore Space (%)[b]			35.6	22.4	15% to 25%
Total Pore Space (%)			40.0	39.2	35% to 55%

Note: In this case, only sand was to be used in the mix.

[a] Air-filled pore space = aeration porosity = macropores.

[b] Capillary pore space = moisture-holding capacity = mesopores and micropores, where the mesopores are those that hold plant-available water.

example, was able to have the sand supplier sieve out the very fine sand fraction of sand no. 2, and aeration porosity increased to 20%. This approach is especially useful in locations where sands tend to contain appreciable quantities of very fine sand particle sizes (<0.15 mm). The same practice of stripping the surface 3 inches and replacing with better quality sand can be used when other problems exist in the surface 0- to 3-inch zone, such as excessive fines or organic matter, whether the site is salt affected or not. The somewhat coarser sand layer will not cause a layering problem because it will still be similar enough in particle size distribution to prevent a perched water layer situation as in a USGA green.

On older greens where the irrigation water source is changed to a more saline and Na-ladened water, any soil physical issue that would be adversely affected by total soluble salts or high Na will soon be exhibited. If the older greens were originally push-up greens, they may demonstrate good surface properties due to long-term topdressing, but the subsurface may be susceptible to reduction in percolation and drainage if appreciable 2:1 clay or silt components are present. In other cases, the surface conditions may not be suitable, perhaps due to excessive organic matter accumulation. Thus, for older greens where the irrigation water is now more saline, the greens profile will need to be carefully evaluated as to any hindrances to downward water movement and for any layers that may be affected by Na or that would become a potential zone of salt accumulation. Lower quality sand (especially the fines components) will increase the necessity for more frequent aerification.

These same principles of site assessment would apply to any other turfgrass area exposed to higher soluble salts or Na, such as fairways, athletic fields, and so on (see Chapter 8, "Assessment for Salt Movement, Additions, and Retention"). But on locations where an initial site assessment is conducted, a long-term sand-topdressing program will likely not have been practiced. Based on site assessment evaluation (visual observation for specific problems, and field and lab soil and water quality tests), partial or complete soil modification may be selected as the best salinity BMP option for alleviation of the problems along with consideration of cultivation and drainage options.

Sometimes a site with saline or sodic irrigation water deteriorates to a point where grass replacement is under consideration, such as perhaps selecting a more salt-tolerant grass. Preplanting options may include no-till, shallow upper soil profile excavation, as discussed previously; no change in the existing soil; or complete soil profile modification. The critical site-specific assessment is to determine the reason or reasons that the grass deteriorated. If there are adverse soil physical problems that contributed to the grass decline, these must be dealt with for long-term success and sustainability prior to planting. For example, a more salt-tolerant grass may be desired as the "salinity solution," but unless the underlying soil infrastructure problems are addressed, those same soil salinity problems are apt to arise later. Do not expect the salt-tolerant turfgrass to fix the soil salt accumulation and infrastructure problems.

Hartwiger (2007) provides a good overview to no-till planting of turfgrasses and the factors to consider when replacing one grass with another. Also, Anonymous (2010) noted that physical, chemical, and biological soil conditions can all affect golf course greens. If you find problems in these areas on older greens, unless you take corrective action, those problems are going to persist and probably get worse especially with saline irrigation water use. Minor organic matter buildup can usually be dealt with after you no-till, but no-till is by no means a magic bullet for most problems and especially not for saline ecosystems. Be wary of excessive organic matter buildup and application of organic amendments (such as sewage biosolids) especially during high biomass-producing periods. Look for contouring problems, poor drainage, waterlogging, standing water, dry spots, soft spots, layering, diseases, pests (especially nematodes), poor air movement, excessive wear, shade issues, persistent weed problems, seepage micropockets, and high salt accumulations (Anonymous, 2010).

14.4.2 Biochar as a Potential Amendment

Biochar is a soil organic amendment that is being used in agricultural systems and is being touted for its biosequestration or atmospheric carbon capture and storage plus oxygen release within the climate change arena (Winsley, 2007; Woolf, 2008). It has been suggested as an alternative for peat or other organic amendments in soils, including turfgrass ecosystems. This black carbon (it is a modified form of charcoal) is formed by the pyrolysis of biomass through heating in an oxygen-free or low-oxygen environment to the point that no combustion occurs (Woolf, 2008). Biochar soil amendments can be produced from any biomass feedstock, including plant residue, woody residue and biomass, composted materials, municipal wastes including sewage sludge, and manures. Comprehensive reviews on this soil amendment can be found in Woolf (2008), Sohi et al. (2010), and Lehmann and Joseph (2009).

The bioavailable soil organic carbon will support soil biota (such as arbuscular mycorrhiza fungi and others), maintain soil quality (especially on badly eroded soil or soil undergoing remediation), and combat erosion (Woolf, 2008). The biochar-mediated enhancement of mycorrhizal abundance and functionality is discussed in Warnock et al. (2007, 2010), with four possible mechanisms operating: (a) alteration of soil physicochemical properties, (b) indirect effects on mycorrhizae through effects on other soil microbes, (c) plant–fungus signaling interference and detoxification of allelochemicals on biochar, and (d) provision of refugia from fungal grazers.

Amelioration of agricultural soils with poor soil fertility characteristics (including sandy textures, acidic pH values, kaolinitic or 1:1 clays, low CEC, and diminutive soil organic carbon contents) have utilized applications of mulches, composts, and manures to increase soil fertility (Novak et al., 2009). However, those amendments are short term in tropical environments due to rapid oxidation of the organic amendments plus tendency for leaching of added fertilizer nutrients. Biochar provides an alternative organic amendment that has slower biodegradation characteristics in the soil, yet the amendment maintains high surface area per unit mass and a high charge density for cation sorption with the objective of enhancing soil fertility. These biochar products can range from neutral to alkaline pH depending on the biomass source (Singh et al., 2010).

Addition of pecan shell–based biochar to a sandy, low-fertility, acidic Norfolk soil increased soil pH, soil organic carbon, plus selective sorption of Ca, K, Mn, and P while decreasing exchangeable acidity, S, and Zn (Novak et al., 2009). Soil CEC was not significantly increased. Leachates were found to have increased EC, K, and Na concentrations but decreased Ca, P, Zn, and Mn concentrations (which supports the use of this product on saline sites). Manure-based biochars have been found to result in higher water-soluble leachate salts compared to leaf- and papermill sludge–based (highest calcite concentration) and wood-based biochars (Singh et al., 2010).

Use of a biochar amendment as a substitute for peat moss in sand-based turfgrass sites has been investigated for nutrient movement and moisture retention (Slavens et al., 2009; Brockhoff, 2010). As biochar concentrations increased from 0% to 25%, water retention was 73% higher than the pure-sand control and 63% higher than 5% biochar composition at field capacity; however, bentgrass rooting depth decreased 46% with biochar concentrations >10% (Brockhoff, 2010). Saturated hydraulic conductivity of sand media containing 25% biochar had a K_{sat} of 6.6 cm/hr, whereas the 5% biochar concentration resulted in a K_{sat} of 55.9 cm/hr and pure sand was 84.8 cm/hr. Leachate results as biochar additions increased from 0% to 25% where ECe increased from 1.5 to 3.4 dS/m; dissolved total organic carbon increased from 20 ppm to 340 ppm; nitrate and ammonium concentrations decreased from 5 ppm to 0 ppm and 0.8 to 0.2 ppm, respectively; and soil P and K increased from 0 ppm to 118 ppm and 21 ppm to 892 ppm, respectively.

Additional research will be needed to resolve the salinity impact issues with this organic amendment addition to saline turfgrass ecosystems, but the preliminary results indicate that the product does sequester critical nutrients (Ca, P, K, Mn, and Zn) that are essential for salinity-exposed turfgrasses, reduces N leaching, allows leaching of total soluble salts and Na (including some K, which is highly mobile and will move with most leaching events), has the potential to stabilize some of the moisture dynamics in sandy soil profiles, and should benefit microbial functionality while not biodegrading as rapidly as many other organic amendments. The key issues will be determining (a) the best biochar product (such as poultry manure–based biochar) for organic amendment substitution for peat moss in sand-based recreational turfgrass sites in saline ecosystems; (b) the rate to use, especially related to any adverse effects on saturated conductivity and aeration porosity; and (c) the long-term response when using high-saline and Na irrigation waters.

15 Drainage and Sand Capping

CONTENTS

15.1 DRAINAGE AND SALINITY

15.1.1 Drainage Goals in Salt-Affected Soils

Drainage is critical on salt-affected soils, whether in humid, semiarid, or arid climates, or in climates with wet and dry seasons—drainage becomes more than control of excess water, but also includes control of excess salts and Na. An intensive coverage of drainage is beyond the scope of this book, but our discussion will reveal the most important best management practice (BMP) considerations related to saline and sodic soil sites. On saline and sodic soils, the primary goals for surface and subsurface drainage differ from those for non-salt-affected soils. Primary goals on saline and sodic soils for drainage entail the following:

1. Control surface and subsurface water and salt flow to avoid soil salt accumulation areas on the turfgrass site. This would include interception of lateral salt movement on the surface or subsurface.
2. Facilitate removal of salts from the root system and the area below the rootzone.
3. Allow water drainage from soil macropores so that they can function properly within the rootzone for both water and salt leaching. Saturated macropores do not allow gas diffusion, nor do they receive water (i.e., infiltration) when saturated due to impeded flow from below.
4. Controlling a rising water table on some sites that may result in resalinization of the rootzone soil profile.
5. Develop an acceptable means to dispose of subsurface drainage water, and particularly to avoid contamination of an aquifer of high (drinking) water quality.

Drainage on non-salt-affected soils would involve a combination of the goals delineated in Table 15.1 and are different from the reasons listed for saline and sodic soils. However, the same benefits of good drainage on normal nonsaline soils are also applicable to saline and sodic soils, but the previous five goals are the primary drainage goals.

Which of the primary goals for a salt-affected site that will be of most importance varies based on site and climate conditions: to achieve adequate water and salinity control will require a well-designed, site-specific, and integrated drainage system—a system with multiple types of surface and subsurface drainage features. This will necessitate use of a drainage specialist with knowledge of drainage for saline and sodic soils—one who can integrate an understanding of lateral, horizontal, and upward salt flow; fluctuations of water tables on a seasonal time frame and over a long period; subsurface tidal influences along coastal venues; effective salt disposal; the interrelationship between irrigation and drainage in terms of maintaining acceptable salinity levels in the rootzone while minimizing excessive soil moisture movement into the underlying soil or water table; and consideration of the long-term turfgrass ecosystem sustainability of the site.

Even with the best drainage design developed and implemented prior to establishment, there will be drainage issues that become apparent over time that will require additions and adjustments. This is especially true for salt-affected sites since both water and salt movements are involved.

TABLE 15.1
Primary Reasons or Goals for Good Surface and Subsurface Drainage on Non-Salt-Affected Turfgrass Sites

Drainage Goals:

Avoid	Comments
Compaction and rutting	Moist surfaces are more prone to soil compaction and rutting.
Poor-quality playing surface	Moist surface easily track or rut and are soft.
Interference with maintenance	Surface stability of soil is reduced.
Poor wear tolerance	Succulent tissues are more prone to physical injury.
Poor abiotic stress tolerance	Enhances succulence in turf tissues that reduces high temperature, low temperature, and drought resistance.
Excess thatch	Reduced aerobic soil microbial activity in moist surfaces reduces thatch decomposition; thatch accumulates.
Poor rooting	This is in response to low soil oxygen.
Black layer	Anaerobic soil and thatch conditions foster black layer.
Weed encroachment	Certain weeds such as clovers and annual bluegrass thrive in moist soils.
Algae and moss	Moist surfaces favor the development of algae and moss.
Excess growth	Grass grows faster, especially with high N.
Wet wilt	Waterlogged roots do not take up water when they are under low soil oxygen stress.
Scald	Waterlogged roots that do not take up water during high-temperature periods exposed the turf to high-temperature stress and injury.
Enhanced disease activity	Diseases increase that are favored by moist surfaces or anaerobic conditions, especially root rot types.
Runoff	Greater runoff loss of water since saturated macropores have low infiltration when the macropores are saturated several inches into the soil.
Denitrification loss of N	Anaerobic conditions foster N loss by denitrification.
Cold soils	Saturated soils are slow to heat up in spring periods.

Note: These same benefits would also apply to salt-affected sites.

15.1.2 Resources

Basic reference material on drainage for non-salt-affected turfgrass soils includes the following: (a) O'Brien (2005) provides a good overview of drainage considerations for a whole golf course, while Baird (2005) and Snow (1994) concentrate on golf greens; and (b) Pira (1997) provides guidelines to drainage on golf courses, as do McIntyre and Jakobsen (2000) and Adams and Gibbs (1994) for practical drainage on different turfgrass sites. All of these publications illustrate different drainage techniques and integration of these drainage options into specific types of management situations.

As noted in Chapter 14, Section 14.4.1, "Complete and Partial Soil Modification," turfgrass managers using saline or high-Na irrigation water applied to fine-textured soils that are susceptible to becoming sodic are encouraged to be familiar with the athletic field partial and complete soil modification procedures discussed and illustrated by Adams and Gibbs (1994) and Baker (2004, 2006). These modification procedures are often what must be used on the extreme salt-affected sites such as saline and sodic marine clay soils or high clay content sodic soils in arid regions. These are "systems" that integrate subsurface drainage with the use of sand layers of various depths at the surface, sand slitting, and surface contouring with the specifications adapted to climates with high rainfall as well as arid regions.

Agricultural Drainage by Skaggs and van Schilfaarde (1999) is a good source for agricultural drainage in general. Publications that consider the interaction of drainage and salinity include Hanson et al. (1999), Hoffman and Durnford (1999), Gafni and Zohar (2001), Christen and Skehan (2001), and Christen and Ayars (2001). Understanding the diverse means for saline seep development is useful for placement of drainage features, and this topic is presented well by Brown et al. (1983), Seelig (2000), and Wentz (2000). Chapter 8, "Assessment for Salt Movement, Additions, and Retention," also contains information related to drainage that is important at the site assessment phase; this includes identifying saline seeps in Section 8.1, "Assessing Soil Physical Properties," and Figures 8.1 and 8.2.

15.2 SURFACE DRAINAGE

In this section, our objective is to briefly present the most common types of *surface drainage features* that are used on turfgrass sites with comments related to salinity control. Surface drainage techniques involve catching and removing surface water, whether flowing on the surface or ponding in a low area. Common examples are given throughout this section.

15.2.1 General Contouring

Proper contouring of the surface landscape is the beginning strategy for developing good drainage systems. When the surface water contains salts, either from irrigation water or by dissolving soluble soil salts, these salts will move with the gravity-induced water flow and accumulate where the water accumulates. Normally, "proper contouring" is considered the creation of diversion channels and channels of conveyance to direct and carry excess water flow, which are discussed in the next two sections. However, proper contouring involves more than the creation of a specific drainage feature. It includes contouring (a) to avoid low-topography areas without surface drainage such as bowl-greens or bowl-like depressions within a turfgrass area; (b) to insure adequate surface drainage from an area, such as three surface drainage patterns off a green, a crown on a football field, a back or side slope on tees and bunker surrounds, and the like; (c) to avoid features where large volumes of surface water rapidly flow without adequate removal so that standing water or erosion easily results; (d) to spread water uniformly over slopes in order to minimize erosion problems; and (e) to provide adequate fall or drop so that gravity-induced water can move to the target collection site.

One contouring issue that is often not considered for potential sodic or saline-sodic sites is to avoid flat expanses with no surface drainage slope, even if minimal, especially on fine-textured soils. Such areas are prone to develop sodic scald areas where there are very small depressions (initially often not visible) that water can start to collect within. Initially these depressions collect more saline or Na irrigation water, and eventually surface sodic conditions cause structural deterioration, which causes a deeper depression and escalates into a full scald area. A slight slope will help prevent such areas, and when they start to occur, the slope configuration enhances the potential for sand slitting through the depressed site into surrounding downslope areas for surface drainage.

The basic contouring principles previously noted, coupled with specific surface drainage techniques discussed below, are the most important management approaches to minimizing many soil physical problems. Ignoring these aspects will greatly reduce the effectiveness of subsurface drainage, soil modification, turfgrass cultivation procedures, and ultimately salt management. Basic contouring principles plus surface drainage features are best installed before turfgrass establishment, but many features can be added as needed. The authors have observed design features on golf courses in which steep slopes on bunkers or surrounding mounds and the subsequent accumulation of soil salts from saline irrigation water have resulted in subsurface migration of excess salts onto greens cavities and fairway areas. Interceptor (French) drains must be installed to capture and rechannel those salts to primary drainage outlets in order to reduce the salinization of these high-traffic areas.

15.2.2 Diversion Channels

Grassed waterways to catch water runoff and divert it into an outlet are called *diversion channels*. Many times on golf courses, mounds can function as diversion features with a grassed waterway channel at the base. One example of a common problem is saline irrigation water applied to a golf course rough on a sloped area that runs onto the fairway and becomes a salt accumulation area. If the irrigation water was not saline, the water would disperse over the fairway area and not be a problem, but the salts left behind from saline water are a different matter. For saline irrigation water applied on a site adjacent to a high-quality turfgrass area, this water should be prevented from flowing onto the quality turfgrass area. Diversion channels are often the best means to achieve this goal with careful attention to where the diversion channel outlet is positioned in order to avoid salt accumulation problems at this location. Ditches also can be diversion features to intercept and divert water flow from an area.

15.2.3 Conveyance Channels (Outlet Channels)

Conveyance channels are waterways that carry water from a collection point or structure (such as several diversion channels feeding into a channel of conveyance) to an outlet. These could be larger grassed waterways, cement-lined channels, rock-lined channels, waste bunkers, buried solid pipe, or tree- or grass-lined ditches. The outlet for conveyance channels must be environmentally acceptable, especially for saline irrigation water.

15.2.4 Catch Basins

These collect water in a low spot or from diversion channels. A catch basin may feed directly into an underground drain pipe, a sump pump pit, or a dry well. A *dry well catch basin* allows water to move into the subsoil, usually by removing soil several feet deep and backfilling with porous rock, gravel, or sand. Such a basin can only drain a small area and will work only when there is not a water table near the surface. A catch basin emptying into an underground drain line can drain a much larger site. Dry well catch basins are very useful in arid regions to handle limited size areas of water and salt accumulation. For example, in an arid climate on a site with a high clay content soil, subsurface moisture (laden with salts) had moved laterally to an irrigation trench line and then along the line to the lowest area where salts and water had accumulated. A dry well was installed to allow drainage at the low spot, which drained both excess water and soluble salts over time away from the critical turfgrass rootzone. This is an example of an unusual saline seep situation, but mini–dry wells are often an option for other types of saline seeps either directly or as the outlet for a diversion channel, French or interceptor drain, or slit trench.

15.2.5 French Drains (Blind Inlet Drains)

French drains allow water to enter from the surface since the drain extends to the surface. Typical French drains may be 3 to 8 inches wide (7.6–20.0 cm); 8 to 16 inches or deeper (20–40 cm), filled to the surface with porous material such as pea gravel or very coarse sand, with a slotted drainage pipe for carrying water over 100 feet (30 m) or for transporting larger volumes of water; or, when small quantities of continuous water are anticipated, such as a French drain to intercept seepage at the base of a slope, no tile drainage may be needed. French drains are inexpensive interception features for saline seeps. The surface of French drains must be kept open for water to flow into the drain line.

15.2.6 Slit Trenches

Slit trenches are similar to French drains, but are much smaller. These drains were noted in Chapter 14, Section 14.2, "Cultivation of Saline and Sodic Soils: Guidelines," under the various terms of *gypsum slotting, sand banding, sand slitting, sand grooving*, and *sand injection*, where more than sand alone provides greater salt management effectiveness in sodic soils. They are most often utilized on sports fields, but can be used on other turfgrass sites. Slit trenches generally are 1/2 to 1 inch wide (12 to 25 mm), 4 to 12 inches deep (10 to 30 cm), and filled with sand to the surface; the base of the slit trench normally extends into a gravel layer or crosses a gravel layer every few feet. The gravel layer or lines would contain drain pipe. Slit trenches can be part of the original complete soil modification design or installed later on fields with poor surface drainage. Various types of sand-slitting arrangements are given by Adams and Gibbs (1994), McIntyre and Jakobsen (2000), and Baker (2006).

15.3 SUBSURFACE DRAINAGE

Surface drainage is often focused on handling large volumes of water in relatively short periods of time or surface interception of slowly seeping water and salts. Normally (in agriculture), subsurface drainage is targeted to remove excess soil moisture from the rootzone, and this can be a relatively slow process (several days). However, in high-use turfgrass situations, especially on sand rootzone mixes, subsurface drainage is also expected to be relatively rapid and able to move large volumes of water in short time periods.

Subsurface drainage is designed to remove excess water (surface water, gravitational water, and water from a high water table) from the rootzone and, sometimes, for water table control. As noted in previous chapters, but repeated here for emphasis in the drainage section, for excess water to reach the subsurface tile lines, the water must be able to do the following:

- Infiltrate the soil surface.
- Percolate through the rootzone.
- Drain below the rootzone.

In each of these cases, *saturated flow* (via macropores) is especially important for rapid water removal during precipitation periods, when irrigation water is applied at a high rate, or in low areas where water accumulates. *Unsaturated flow* of water downward to the drain lines or up from any saturated or moist zone above tiles, a perched water table, or a natural water table (i.e., capillary rise of water) can also occur after saturated drainage. Any practices to increase saturated flow will enhance drainage rate (i.e., such as by the creation of macropores by cultivation, soil modification, applying a vacuum, or the use of wetting agents in some cases).

15.3.1 Types of Subsurface Drainage

15.3.1.1 Tile Drainage

When most people think of "drainage," it is subsurface tile drainage, and this is the most important form of subsurface drainage. In contrast to agriculture, turfgrass sites generally utilize (a) shallower placement of tile drains (1.5 to 3 feet vs. 3 to 5 feet; 1 foot = 0.30 m), and (b) closer spacing of tile lines (10 to 20 feet vs. >20 feet). Shallower and closer spaced drain lines facilitate more rapid removal of standing and gravitational water as long as water permeability through the soil to the drain is sufficient.

However, tile spacing and depth when using saline irrigation water require greater attention to the water table and irrigation system capabilities. Salinity leaching requires additional water beyond evapotranspiration (ET) replacement and compensation for the uniformity of the irrigation system. Thus, more water may be moving past the rootzone into the underlying soil, which may influence water table location. In arid regions, an aquifer may be rather deep into the soil, but increased water input may cause the natural water table to rise over time; or, if an impediment to drainage exists, a new perched water table may start to form. Christen and Ayars (2001) report on the substantial areas in Australia where rising water tables have brought salinity into the rootzone—in these cases, usually from changes in flora-cropping systems that reduced water extraction and allowed more drainage water into the subsoil that eventually reached the water table.

Hanson et al. (1999) summarized four approaches for determining optimum water table depths (i.e., for decisions on depth and spacing of tile drains) in agricultural situations, but these also apply to turfgrass areas. The situations are noted as follows, where the *depth* refers to the critical water table depth and, thereby, the depth and spacing of tile systems to maintain water levels at or below the critical depth of the water table for each situation. They provide suggested tile depths based on each situation:

- Depth (i.e., water table depth) necessary to lessen waterlogging of the rootzone, usually involving very shallow water tables. In cases where the water table is very shallow, drainage systems must be deep enough to keep the water table below a critical depth that does not allow capillary rise of water (and soluble salts) into the rootzone. Even moderate seasonal fluctuations of a very shallow water table may induce waterlogging. Salts in the irrigation water will impact (resalinize) the aquifer water rapidly, and this water plus dissolved salts may rise back into the rootzone during a wet period.
- Critical depth of water table for salinity control. In this situation, the water table is rather shallow, but not as much as in the previous case where potential for waterlogging was of concern. Rather, the aquifer water is saline and shallow enough to allow saline groundwater to come into the rootzone by increased capillary rise of water as the plant extracts water in response to ET. A net upward flux of saline aquifer water into the rootzone would require a sufficiently deep water table to be below the normal capillary rise limits of the soil.
- A "downward flux of salt" situation: this case is most appropriate to many turfgrass sites. On irrigated turfgrass sites with a pulse irrigation regime involving an irrigation system with high distribution uniformity and careful water conservation irrigation programming both spatially and for quantity applied, the infiltrated water allows a small net downward

flux of water and salts (i.e., the leaching requirement). This minimizes salt accumulation in the rootzone, minimizes the water drainage fraction, and reduces upward flow of groundwater. This situation allows for use of shallow drainage and closer tile spacing than typical of nonsaline irrigation water sites. Minimizing water drainage flow while controlling salts by this irrigation approach also has long-term benefits in not causing the water table to rise. Sometimes, when water tables are considered relatively deep (e.g., 50 feet, or 15.6 m), long-term drainage of a water quantity beyond that volume required to achieve good leaching can cause a water table to rise to within the capillary fringe of a grass. In Australia, Christen and Ayars (2001) reported that after pre-land clearing, water tables depths of 32 to 96 feet (10 to 30 m) were common in many areas. With poor irrigation and water losses from water distribution and drainage channels, water tables rose by 0.32 to 3.2 feet per year (0.1 to 1.0 m), with the result of many water tables rising to within 9.6 feet (3 m) of the surface. The rising water table also brought accumulated salts. The capillary fringe above the water table for many soils is from 1 to 8 feet, so capillary rise of water and salts into the rootzone occurred. Thus, due diligence attention to long-term water and drainage management sustainability is essential. The type of irrigation described above (pulse, highly uniform, spatially adjusted, and adjusted properly for minimal water application) is a critical BMP on salt-affected sites.

- Maximize water use from the shallow groundwater. This is a drainage water reuse method practiced on agricultural crops in some areas. Duncan et al. (2009) discussed drainage water reuse potential in turfgrass situations. This particular approach, noted by Hanson et al. (1999), is not appropriate for perennial turfgrass sites.

These approaches for deciding on tile depth and drainage based on water table and irrigation method with saline irrigation water use illustrate the importance of considering salinity issues in designing tile drainage. For complex turfgrass sites, there may be areas of very shallow, moderately shallow, and deep water tables that would require flexibility in drainage design across the landscape. For sod production fields using saline irrigation water, these four approaches should be assessed.

Perforated plastic pipe is the most prevalent pipe used, especially corrugated pipe with slots for water entry. The most common mistake in tile drainage is to use too coarse a material to surround the tile (i.e., gravel rather than coarse sand), thereby allowing finer particles from the upper rootzone or adjacent subsoil to migrate through the coarse fill material and eventually get into the tile. McIntyre and Jakobsen (2000) recommended use of coarse sand and provided practical specifications. Coarse sand with a saturated hydraulic conductivity of 80 to 160 inches hr^{-1} (2000 to 4000 mm hr^{-1}) is very suitable, and such sand is an excellent filter against the migration of fines (very fine sand, silt, and clay) to the tile line compared to gravel-sized fill.

An exception to these guidelines would be on high-sand rootzone mixes similar to the USGA green construction (Snow, 2004) where a gravel layer is suitable because (a) the gravel is carefully chosen to prevent rootzone particle migration and to create a perched water table where the intermediate layer (the choker layer, >90% particles between 1 and 4 mm diameter) is used; and (b) where the choker layer is not used, an even more stringent guideline is used to prevent rootzone particle migration. In the latter case without the intermediate layer, a perched water table may form if the underlying gravel layer has predominant particles five to seven times the diameter of the rootzone sand.

Another common mistake that can limit the effectiveness of tile drainage is to use a geotextile fabric over the top of the tile line or trench. The flat surface of a fabric can easily be clogged by very fine sand, silt, and clay (in extreme cases, roots can clog the fabric). Movement of colloidal fines (clay and organic colloids) is more common in sodic soils than comparable non-salt-affected soils. Even on well-selected rootzone mixes, some fines occur, and these often migrate into the drainage layer and eventually to the tile line. This is normal, and these fines will flush from the tile lines with water movement. It is only when the rootzone mix collapses into the drainage layer and allows large quantities of fine particles to seal the drain lines (as well as move into the lines) that problems occur, or when fines move from the subgrade into the drainage layer and tile lines.

With any tile drainage system, the rate of water infiltration and percolation controls the drainage rate. Thus, on fine-textured soils, compacted soils, and salt-affected sites, a good, continuous, deep cultivation program will substantially improve drainage.

When the irrigation water or soil contains appreciable Na, the Na will eventually move downward with an effective leaching program. When the Na comes from the irrigation water and there is surface-applied Ca to displace it from the soil cation exchange capacity (CEC), the Na moves downward and creates a deeper sodic condition until sufficient Ca is available throughout the rootzone to move the Na to the drainage zone where it will also create sodic conditions until Ca reaches this zone. This sequence from surface-applied Na and Ca is why we recommend vigorous surface and subsurface Ca amendment applications (e.g., gypsum slotting, and high rates of gypsum mixed prior to establishment) to facilitate more rapid displacement of Na into the drain system. When installing tile drainage, *it is important to add coarse gypsum at a high rate over the top of the coarse sand covering the tile and within the backfill soil of the trench.* This will help insure soil structure stabilization as Na drains near the tile.

15.3.1.2 Mole Drains

Mole drains are temporary (i.e., lasting from a few months to several years) drainage channels formed by pulling a mole plow through the soil. The mole plow has a pointed bullet at the end of a blade that creates a round channel of 2 to 4 inches diameter (5 to 10 cm). These channels are most stable when in a B horizon with high clay content. One version of this drainage system is to form slots 4 to 6 inches wide (10 to 15 cm), spaced at 3 to 7 feet (1 to 2 m) apart to a depth of 15 to 20 inches (40 to 60 cm) with a mole drain below each slot. On sodic soils, the soil within each slot would be gypsum enriched. This procedure is obviously not well suited to established turfgrass sites due to surface disruption, but may be of benefit on some sod farms. If the mole drain can be fitted to inject gypsum into the channel, this additional amendment would be beneficial, especially on sodic soils where mole channels do not remain open as long as other sites.

15.3.1.3 Other Subsurface Drainage Approaches

In addition to tile drainage and mole drains, three other practices can enhance excess water movement from the rootzone as well as aid in salt removal. These are as follows.

Deep cultivation techniques, as noted in Chapter 14 in the cultivation section and previously in this chapter. Even when no tile drains are present, deep cultivation can increase water drainage if the soil has excessive fines, is sodic where structure has been destroyed, has layers, or is compacted in any zone. These same conditions may occur on a soil with installed tile, and therefore deep cultivation would improve the tile drainage performance. Filling the core holes with topdressing coarse sand enhances water movement.

Water table control was discussed in the tile drainage section, but some additional comments are warranted. The presence of tile drains will drop a high water table to the depth of the tile line outlet as long as the outlet is free to discharge. Sometimes the height of the outlet from a localized area is controlled to raise or lower the water table. One example is where a plastic lining prevents water drainage into the surrounding soil or subsoil, and all water must drain through laterals joined to a main drain line that is tied into a sump pit. The outlet height of the main drain line can be raised or lowered. The Prescription Athletic Turf™ (PAT) and Purr-Wick (which includes compacted sand, and the top 2 inches [50 mm] are generally modified with organic peat moss and inorganic calcined clay aggregates) systems use this approach (Adams and Gibbs, 1994). Another example is the creation of a "perched water table," where drainage is inhibited until water pressure is sufficient to initiate rapid drainage. The USGA green construction method allows this feature if the profile contains an abrupt change of five to seven particle-sized diameters from the rootzone to the intermediate sand layer (normal case), from the intermediate sand to gravel layer, or from the rootzone to gravel layer when a choker layer is omitted. A perched water table can also form above a layer containing few macropores such as a high clay or silt layer. However, this type of barrier is an undesirable situation since drainage cannot be controlled, as in the USGA green where a water pressure head will cause rapid saturated flow when excess water occurs. A third example is on level sites, where water can be backed into tile lines for control of water table depth such as in some organic soil sod farms, or in other permeable soils.

Some additional comments on the perched water table of the USGA (Snow, 2004) related to salt-affected sites are as follows: (a) when applications of saline irrigation water are applied during prolonged dry periods that are not sufficient to flush the greens by breaking the perched water table, then salts accumulate and can cause salinity stress, especially on cool-season grasses in the summer; and (b) during prolonged wet weather, it is possible for the perched water table to become anaerobic enough to start black layer formation from the perched water table upward—we have seen this in a few cases. However, if this is expected, we suggest that the drain line be inspected to make sure it is functioning and not backlogging water into the green cavity, since this is a more typical cause of deep black layer in a USGA green. Obviously, during the wet period, irrigation would not be applied, but if saline irrigation water is applied as the weather changes to hot and dry (persistently high-ET conditions), salinity leaching will be impeded until the black layer dissipates.

Subsurface moisture removal devices capable of applying a vacuum to the soil rootzone mix via the drainage lines can increase the water drainage rate under saturated or near-saturated conditions. There are units that are mobile or can be used in place. Also, the PAT sports field construction method has this feature as a part of the drainage system. When systems allow reverse flow of the fans, air can be blown back through the drainage lines for subsurface aeration.

15.3.2 Green Drainage

Golf greens are the most prone to high-soluble salt accumulation and subsequent stress due to their limited CEC-buffering capacity, frequent irrigation, high traffic, and close mowing height. Some issues related to greens were discussed in the previous section, but there are some additional considerations. For older greens in arid climates that started out as pushup greens using native soil and have been amended with topdressing sand over many years, if the irrigation source is changed from a high-quality (low-salinity) water to a saline water, it is necessary to make sure that they actually do drain (i.e., when higher rates of water are applied, does water come out of the drain lines?). With saline irrigation water, it is necessary to apply sufficient water volume to obtain a leaching fraction that must ultimately reach the drain lines. With the same soil situation, but when moderate- to high-Na irrigation water is now used rather than high-quality water, adequate internal drainage is even more problematic due to the soil structure degradation effects of Na on the original soil base. Continuous functional drainage is an essential BMP salt management requirement.

For high-sand green profiles that were constructed to foster high infiltration, percolation, and drainage, it is important to remember that salts will accumulate if water flow is stopped. The first potential barrier is any perched water table built into the soil profile to retain extra water, as was noted in the prior section. Another potential barrier is at the green cavity and collar interface involving one of the surface-contoured drainage patterns. If the collar develops a lip over time (often from sand-topdressing deposition) that slows or impedes surface drainage, then more water will drain downward, along with the soluble salts. The drainage underneath these areas at the bottom of the rootzone must be sufficient to remove the excess deposited water and salts. If a barrier was installed to separate the green rootzone and drainage from the surrounds, then the water in the green drain must be able to functionally drain. This area that can become a problem is often a low-topography point, and drains may not have been placed directly against the separation barrier, thereby leaving an area of slow drainage. This drainage problem not only can foster black layer development from deep in the rootzone but also is an area of salt accumulation that can allow salts to rise by capillary action during high-ET periods. Flat drainage tile can be vertically installed immediately next to the separation barrier, but it will require an outlet (connection to the conventional drainage system or to the French drain emerging off the greens cavity) for rapid drainage. If a layer develops along the edge of the greens cavity between the soil surface and the drain lines, then the edge of that cavity will act like a dam and salts will accumulate rapidly, often decreasing turfgrass canopy density and eventually killing the grass in lower topography areas. The vertically positioned flat drainage tile at the edge of the greens cavity can provide a more permanent conduit to move those gravity-induced salts down to the appropriate drain lines below the grass root system.

Another potential barrier to drainage is clogging or collapse of the drain lines, either underneath the green or in the outlet line outside the green perimeter. These conditions will block removal of the salts and force salt-laden water back into the greens cavity. Occasionally, a drain line outlet is sited such that water comes back through the line during wet periods. This could bring salts contained in the line or not yet drained from the lower part of the greens rootzone back into the upper rootzone. This problem can also occur along coastal venues, where high tidal influences can potentially bring excess salts back through these drain lines.

It is very beneficial on salt-affected sites to have an access to the main drain line from a green that is located just off the greens perimeter. This allows flow conditions to be monitored after leaching events or during normal irrigation with a leaching fraction. Water samples from the drain line can reveal the salt concentration and constituents in the sample such as Cl, SO_4, Na, Ca, and Mg and provide quantitative details on leaching program effectiveness. Even in sand-based profiles, soil salt accumulation can often range two to three times higher than the incoming irrigation water quality depending on irrigation scheduling, site-specific leaching requirements, microclimate conditions, and the whole turfgrass ecosystem's salt management program.

"Chronically wet soil" problems, especially in low-topography areas, occur from layering in putting greens containing a sandy rootzone overlying a finer textured native soil (generally from topdressing over years) or a USGA specification green containing the sandy rootzone over a gravel layer that creates a perched water zone above the layer interface (McCoy, 2005). When distinct soil particle texture differences occur across a defined boundary (layer interface is near the soil surface or the sand contains excess fines), a sharp discontinuity in pore sizes occurs throughout the soil profile. Even though the soil may be saturated, drainage is a problem since the perched water is held at a slight suction and often is unable to enter the conventional drainage systems. The FloWick passive capillary drainage (PCD) system employs a treated and woven fiberglass rope (pores within the weave are sized similar to the rootzone sand) as the drainage conduit with connections to the underground conventional drainage system, and this system is only installed on the lower wet zones in a greens cavity (McCoy, 2005; McCormick, 2006). The PCD fiberglass rope has a permeability of approximately 1000 inches per hour with a *hanging water column* suction increase of about 70% of the actual gravity-induced elevation drop in a sand-over-gravel soil profile with a capability of reducing soil water content in these problem wet zones by 5%–10% in a 24-hour period while reoxygenating the subsoil (McCoy, 2005).

15.4 SAND CAPPING SALT-AFFECTED SITES

15.4.1 Soil Conditions Favoring Sand Capping

Sometimes, turfgrass sites irrigated with saline or sodic water (or in which the soil is already sodic) exhibit substantial barriers to deep salinity leaching or maintenance of the acceptable surface physical conditions, such as *marine clays*, *vertisol soils*, hard and deep *caliche layers*, *acid-sulfate soils*, and many *fine-textured soils in regions with a monsoon season*. In these cases, the most cost-effective and sustainable approach over time is often deep sand capping, usually by a 10- to 12-inch (25- to 30-cm) sand cap with an underlying drainage system necessary in most situations, which is discussed later. Sand topdressing or addition of a sand layer up to 3.0 inches (as noted in Chapter 14, Section 14.3, "Topdressing," and Section 14.4, "Soil Modification") and deep cultivation methods can help, but with low-permeability subsoil and lack of subsurface drainage, these salinity management situations may not be sufficient on the most challenging salt accumulation sites.

In Chapter 14, it was noted that when a sand layer of <3.0 inches is used on turfgrass sites receiving saline or high-Na irrigation water, this infrastructure installation must be done in conjunction with (a) sand slotting or drill and fill with sand to create adequate percolation rates for salt leaching when applying saline irrigation water, and (b) irrigation water containing appreciable Na content. There must be an aggressive surface and subsurface Ca amendment program, with the subsurface Ca amendment required to counter the adverse effects of excess Na on soil structure degradation. Usually, the Ca is applied with sand-slotting or deep drill operations (with gypsum or sand plus gypsum) to fill the slots or holes in order to provide subsurface macropores with longevity.

When deep sand capping (defined at >3.0 inches) is used on the extreme sites listed above, the same attention to subsurface drainage is required as for thinner sand layers, but drainage is not by sand slotting as may be used for a shallow sand layer; rather, it is by tile systems. McAuliffe (2009) noted that tile drainage is usually necessary for successful deep sand capping on non-salt-affected sites, particularly when this practice is used on fine-textured soils in high-rainfall regions. When the underlying drainage is omitted from sand-capped areas, water will move downward to the interface and then laterally via gravity to low areas (unless the subsoil has high drainage capability), which then become waterlogged. However, with saline irrigation water, appreciable salts move with the gravity-induced water flow and accumulate in these areas. Thus, tile drainage is essential in most situations to rapidly capture water and soluble salts for movement off the area in planned fashion.

The specific site conditions determine whether a tile drain system is required in conjunction with deep sand capping, assuming that saline and high-Na content irrigation is being used. Examples are as follows:

- *Deep, high clay content, 2:1 clay types.* These are soils with inherently low infiltration, percolation, and drainage, and they are the most prone to soil structure degradation by Na. Vertisols and marine clays would be in this category. Marine clays are usually saline-sodic due to their coastal locations, while some vertisols in arid regions may also be sodic or saline even before turfgrass establishment. These soils definitely require a drainage system. In an arid climate, there may be economic temptation to omit the drainage system, but it is essential to insure downward movement of salts; and if Na is present in the irrigation water, these soils will seal in the subsoil and drainage will be even more limited. Long-term turfgrass sustainability requires effective and functional drainage for salt management.
- *Fine-textured soils receiving high-Na and saline irrigation water in climates with pronounced wet seasons (monsoons).* In this case, the fine-textured soils may not be as compositionally extreme as vertisols in clay content, but they still have sufficient clay and silt, and are prone to Na-induced structure degradation. Under moderate total rainfall, a sand layer of <3.0 inches along with sand–gypsum slotting may be sufficient, but with prolonged rainy weather, site conditions deteriorate due to the Na effects deeper in the profile and salts only being leached to just below the root system. Thus, deep sand capping may be considered on these soils, much like it is on many Asian golf courses with fine-textured soils and high rainfall periods even without the poor irrigation water quality (McAuliffe, 2009). As with vertisol soils, a drainage system is essential for water removal during the rainy season.
- *Acid-sulfate soils.* These have the same characteristics as marine clays with initial saline-sodic properties, but at times may have lower clay content. The major difference is that when exposed to oxygen, they become very acidic (see Chapter 3, Section 3.3, "Acid-Sulfate Soils"). Tile drainage is necessary not just for salt and Na removal, but also to

provide net downward movement of any acidic water and acid water constituents (high Mn and/or Al) during the very acidic phase (pH < 4.0) of these soils. High-gypsum and -lime applications should be mixed into the soil before sand capping of these soils to ameliorate the acidity problems.

- *Caliche.* This is cemented calcite (lime) that can occur at or near the surface of arid region soils. These calcite zones can be several feet thick and are impervious to water drainage. Deep sand capping may be necessary to provide a suitable rootzone medium that is deep enough to prevent salts from accumulating at the upper soil surface. Due to the high cost of installing drain lines in hard caliche soils, the caliche may be broken or fractured by deep ripping and pulverization, then sand capped without tile drainage. The deep-ripped channels generally provide good drainage.
- *Gypsiferous soils* are soils very high in gypsum; some of these may be hard at the surface when dry, similar to caliche, and they are often saline. However, due to the higher solubility of gypsum, if irrigated turfgrasses are established on these soils, their response is much different. The FAO (1990) publication on gypsiferous soils should be consulted prior to considering any construction on these sites. They are mentioned here only as a caution to avoid these sites if possible, and since they may be confused with caliche soils, sand capping these sites is not recommended.
- *Decomposed granite and pulverized volcanic pumice* are "manufactured soils" created by crushing the base minerals to form a soil in arid locations. These are discussed in Chapter 14 as situations where a sand layer of <3.0 inches (often integrated over years, such as by using the Blec SandMaster equipment) coupled with sand and gypsum slotting or injection are used to improve infiltration, percolation, and aeration if these materials start to seal over time. Deep sand capping is not necessary on these materials.

15.4.2 Procedures for Sand Capping

On large turfgrass areas such as a golf course fairway, the site is graded, usually with removal of the existing topsoil to be used on other course locations, and then a well-designed and extensive drainage system should be installed in the remaining subsoil. Drains are usually within the 25- to 50-feet required spacing. Tile drains must have a suitable outlet since these drains are functional for both water and salt disposal. Depth of the sand layer depends on the physical properties of the sand, and it is essential to obtain quality soil physical analyses, which should be the same as those used for golf greens. Sands used on salt-affected sites should not contain silt, clay, or organic fines in the colloidal range; and they should be within the aeration porosity range of 15% to 30% (preferably in the 18% to 30% range), be within the capillary porosity range of 15% to 25%, and have a saturated conductivity of at least 6 inches per hour prior to planting turfgrasses. The laboratory will determine aeration porosity at different water tensions (depth of sand), and by plotting the results, a proper depth can be determined for the particular sand used for sand capping. If sand depths less than about 10 inches (25 cm) are used, these should be somewhat coarser in particle size than those normally used in USGA specifications to insure good surface aeration. Narrow particle size ranges and sand that is highly spherical should be avoided to make sure the sand provides a stable surface. Sand amendments that have a high percentage of their internal porosity within the micropore range should be avoided, since these sands can retain soluble salts that are difficult to leach. Lassenite (refer to Section 8.1.2, "Inorganic Amendments," and Section 17.4.4, "Lassenite") may be considered if greater plant-available moisture is required, but this inorganic amendment can always be incorporated in the upper 3 to 4 inches (75–100 mm) of the soil afterward for drought-prone areas. Fertilization will need to be adjusted to compensate for lower CEC. Fairway grasses may develop roots that penetrate into the underlying soil, which aids in moisture and nutrition stabilization. Zeolite (refer to Section 17.4.3) can be incorporated on areas exhibiting nutritional challenges, and for highly saline irrigation water, it may be worth the cost to incorporate 225 to 450 lb. per

1000 sq. ft. to raise the CEC by about 1 to 2 meq/100 g soil, respectively, to achieve a total CEC of about 3 meq/100 g. The zeolite should be incorporated into the surface 2 to 4 inches of the rootzone mix, where most of the root absorption of nutrients occurs. Since this quantity of zeolite is minimal relative to the total sand in the surface 4-inch zone, it will not appreciably affect physical conditions as long as the zeolite particles are primarily in the 0.25 mm or greater range.

The interface of the sand cap and underlying soil is important. Except for the caliche subsoil, this area should be tilled with high quantities of gypsum worked into the surface 2 to 4 inches of the subsoil before applying the sand cap. This strategy will help insure better water drainage into the underlying subsoil to aid in water and salt removal. Gypsum rates should be in the range of 100 to 150 lb. per 1000 sq. ft. As noted for caliche, it should be ripped or fractured before applying a sand cap.

Deep sand capping is usually done prior to turfgrass establishment. However, on established turfgrass sites with the soil conditions noted above that favored sand capping, it may be necessary to consider excavation and sand capping at least in the worst localized salt accumulation areas. Procedures would be the same as described above, but with due attention to a drain system installed with proper slope to promote salt movement and outlets to rapidly remove water.

A sand-capped area receiving high Na content irrigation water will still require gypsum or other granular Ca sources, but not at the rate necessary for the original, more finely textured salt-affected soil. Since sand has stable structure, Na will not cause structural deterioration unless clay or silt comes onto the site (i.e., be careful of what sod is used for initial grass establishment or renovation to ensure that it does not bring "fines" onto the site that can create a salt accumulation layer). Otherwise, Na will disperse any clay and silt, and they will move downward to start to plug the sand pores, negating the sand cap effectiveness. Soil-applied granular Ca amendment will be required to maintain adequate available Ca for plant roots and nutritional needs. Cultivation requirements will also be much less on sand-capped areas, and fewer closed days will be needed for excessively wet soils. It is these factors that often result in a proper sand cap being economical over time.

16 Nutritional Practices on Saline and Sodic Sites

CONTENTS

16.1 NUTRIENT- OR ION-RICH IRRIGATION WATER: CHALLENGES

Irrigation water constituents are part of the overall fertility program on a turfgrass or landscape site (Duncan et al., 2009). When irrigation water quality is of normal quality (low salinity generally <750 ppm TDS) and consistent in ion components, this salinity challenge issue is often overlooked because soil fertility and plant nutrition are affected primarily by the combined concentrations in specific products that are applied by fertilization, liming, and any nutrients contained in management products applied to the turfgrass, as well as soil chemical properties. Changes in fertility status (fine-tuning on a site-specific basis) are rather slow and predictable. However, certain irrigation water quality situations complicate nutritional programs because of excessive additions of one or more nutrients or elements, imbalances between nutrients or elements, or lack of available nutrients or elements (Duncan et al., 2009). This is why one of the primary types of salinity stresses is considered to be nutritional imbalances.

Two types of irrigation water that may be high in nutrients and/or ions and, therefore, can profoundly affect turfgrass fertilization are (a) reclaimed irrigation water, and (b) saline irrigation water. In arid regions, it is common for reclaimed water to become more concentrated in salts (often as a result of the treatment processes) over time, so reclaimed water issues and salinity issues are often combined together. Therefore, even though the primary focus of this chapter is on saline irrigation water sources, we will discuss both irrigation water sources together.

16.1.1 Factors Contributing to Nutritional Challenges: Reclaimed Water

Excluding any excessive soluble salts in reclaimed water, which are covered in the next section (Section 16.1.2), the usual nutritional issues associated with reclaimed water are (a) N, P, or S concentrations if they are high; and (b) possible micronutrient low or high levels and imbalances. Often, the levels of these constituents do not change appreciably over the year in most reclaimed water sources. The levels of these nutrients per unit area per 12 inches (30 cm) of irrigation water applied must be included in the overall fertilization best management practice (BMP) program. Table 6.9 indicates the contribution of reclaimed water to the turfgrass ecosystems based on nutrient concentration and quantity of irrigation water applied.

Use of reclaimed water for irrigation, including nutrition aspects, has been reviewed by several scientists for agronomic and horticultural crops (Pettygrove and Asano, 1985; Pescod, 1992; Ayers and Westcot, 1994; Bond, 1998; Grattan and Grieve, 1999; DEC/NSW, 2003; Scott et al., 2004; Stevens, 2006). For turfgrass and landscape systems, reclaimed water use and nutritional issues have been reviewed by Harivandi (1991), Pepper and Mancino (1994), Snow (1994), Wu et al. (1995), and Marcum (2006). Turfgrass and landscape irrigation water nutritional problems (as well as other irrigation water issues) and management options have been noted by Carrow (2011) and Duncan et al. (2009), with the latter covering the continuum from source to delivery to storage to application on plant to soil to surface and subsurface waters.

The primary ions of interest in reclaimed water relative to nutritional challenges are N, P, SO_4, and total soluble salts. However, any nutrient that is unusually high in concentration may require management attention (Tables 6.9 and 6.12). While in arid regions, it is common for reclaimed water to contain appreciable total soluble salts; in many other locations, reclaimed irrigation water has only slightly more nutrients than other water sources such as groundwater or stormwater runoff into lakes. In this section, we focus on reclaimed water with higher than normal concentrations of N, P, and SO_4, while issues related to high-soluble salts in some reclaimed water sources are addressed in the remainder of this chapter.

Normally, a wastewater treatment facility must treat the water in a manner to reduce N and P to acceptable levels for release of effluent into public water features (i.e., receiving waters) based on federal or state guidelines. However, when golf courses or other large landscape areas are primary customers for disposal of reclaimed water, treatment plants may be able to legally transfer more N- and P-rich effluent directly to the turfgrass facility for irrigation than would be possible if the effluent was disposed of in a public storage facility for receiving water (Gross, 2008; Duncan et al., 2009). While treatment facilities would save costs associated with not needing to reduce N and P concentrations to environmentally acceptable levels for disposal in public waters, costs and problems related to excessive N and P are transferred to the turfgrass irrigation facility—from a public cost to a private facility cost.

16.1.1.1 Nitrogen

The quantity of N added in the irrigation source will directly contribute to the nutritional needs of turfgrass and other landscape plants receiving irrigation (Table 6.9). Thus, seasonal and annual N fertilization must be adjusted accordingly, and turfgrasses should be used that can tolerate the N level applied. Some turfgrasses deteriorate rapidly when overfertilized with N, especially those with low N requirements (such as centepedegrass, creeping bentgrass, and seashore paspalum) or under either abiotic or biotic stresses (Carrow, Waddington, et al., 2001). On golf greens, high N results in more growth than desired, expressed as excess clippings, scalping, slower putting speeds, thatch accumulation, greater succulence, enhanced disease susceptibility, decreased wear and traffic tolerance, and reduced hardiness. Creeping bentgrass (*Agrostis stolonifera* L.), a cool-season species used on golf greens, receiving excess N during hot summer periods is especially prone to performance deterioration from overfertilization.

Eutrophication in the reclaimed water storage pond is more likely when N is high. Water containing even 1.1 ppm N can result in algae and aquatic plant growth flourishing. Barley straw is a management option to tie up excess NO_3 in these water features and to reduce algae growth (Gaussoin, 1999; Lembi, 2002; Duncan et al., 2009).

Another irrigation lake storage issue related to high N in reclaimed water is potential formation of ammonia, a highly toxic N form if applied directly to grasses (Duncan et al., 2009). Ammonia toxicity has been reported in marine sediments in eutrophic settings (Burgess et al., 2003). Reclaimed water containing relatively high total N stored in a lake situation that allows anaerobic conditions can result in transformation to the nonionized ammonia ion (NH_3) form, and application through sprinkler systems can result in bleaching of grass areas where ammonia-ladened water droplets hit the grass surface.

The author has observed a couple of situations where this problem was expected based on site conditions that favored ammonia presence, reported ammonia odor, and observed rapid turfgrass injury. Conditions that would enhance the potential for ammonia accumulation are high pH > 9.5, low oxygen, high temperatures, lack of aeration in storage ponds or lakes, pump intakes placed on the bottom of the lake or pond, and relatively high N in the reclaimed water that could convert to ammonia. These conditions can occur within strata of a lake, especially if the lake bottom is rich in organic deposits (Burgess et al., 2003; Arauzo and Valladolid, 2003). Lake aeration with bottom diffusers as well as any means to control N and P levels to reduce eutrophication potential would also inhibit ammonia formation. Irrigation pump intakes should not be placed within these stratified zones in lakes or ponds. Also, with direct treatment facility connections of reclaimed water, where the water goes directly into irrigation lines or the wet well, care should be taken to insure that anaerobic, high-pH, and high-temperature conditions do not result in ammonia toxicity during periods that the system is not operating.

16.1.1.2 Phosphorus

Limits on P concentrations in irrigation water are lower than those of other macronutrients because P is a primary promotion factor for algae and aquatic plant growth. If reclaimed water containing high P is stored in an irrigation lake, the P could cause eutrophication with algae blooms, proliferation of aquatic plants, low-oxygen conditions, and odor problems for the end user. The combination of high N plus P would be especially conducive to causing eutrophication. Costs associated with P control in the stored irrigation water are passed from the treatment facility to the end user. Additionally, it is not unusual for overflow of reclaimed water to surface waters to be considered an unpermitted discharge, even when the pond is specifically constructed to allow only reclaimed water and direct harvested water falling on the surrounding surface to enter the pond. Potential options for reclaimed water users are to (a) negotiate with the treatment facility to reduce P to a level that is less of a problem or to provide a cost adjustment contingency for the additional costs that the end user must bear related to the higher P level (Stowell and Gelernter, 2001); (b) negotiate with the water authorities to allow overflow of water during unusual rain storms from the irrigation lake to not be considered an unpermitted discharge; and (c) consider barley straw or other products for algae control measures (Lembi, 2002; Duncan et al., 2009).

16.1.1.3 Sulfate

Sometimes irrigation water is naturally high in SO_4 ions, but reclaimed water often exhibits high levels. The normal range of SO_4 is 30–90 ppm, where 90 ppm = 1.87 lb. S per 1000 sq. ft. per 12 inches of applied irrigation water (0.92 kg S per $100 m^2$ for every 30 cm irrigation water)

(Table 6.9) Also, irrigation water acidified with a SO_4 acid or a SO_3 generator may be high in SO_4 ions. One concern is that under anaerobic soil conditions, high levels of SO_4 can be reduced to H_2S, FeS, or MnS forms, which contribute to black layer development in soil profiles. The FeS and MnS precipitates are gel-like and seal the soil, which ultimately interferes with water infiltration and percolation through pore spaces and contributes to further anaerobic conditions (Carrow, Waddington, et al., 2001). However, if the acidification is not conducted on sites with moderate to high Na and sufficiently high HCO_3/CO_3 levels in the irrigation water to react with Ca and Mg, then the soils will become increasingly sodic with poor soil physical conditions—resulting in low water infiltration and low aeration problems. Thus, removal of excessive levels of SO_4 is important to minimize black layer problems while allowing acidification to reduce Ca/Mg bicarbonate and carbonate complexation problems. An effective leaching program is the best means to control SO_4 since it is one of the most easily leached soluble salts.

16.1.2 Factors Contributing to Nutritional Challenges: Saline Irrigation Water

Saline irrigation water (including saline reclaimed water) has a number of additional issues of concern beyond what is typical of reclaimed low-saline water. A short description of fertility programs under saline irrigation water is *dynamic* and *challenging*. Discussion of plant nutrition across all plant types as affected by salinity is given by Naidu and Rengasamy (1993), Ayers and Westcot (1994), Marschner (1995), Grattan and Grieve (1999), Alam (1999), and Barker and Pilbeam (2007). Kelly et al. (2006) discuss crop nutrition when using reclaimed water for irrigation. Carrow, Waddington, et al. (2001), Carrow and Duncan (1998), and Duncan et al. (2009) focus on turfgrass nutritional issues under salinity conditions.

Each irrigation water source is unique as are the soil and other site conditions, but soil fertility and plant nutrition with saline irrigation water are characterized by (a) high additions of certain nutrients or elements; (b) imbalances, where one nutrient or element may suppress uptake or alter the availability of another; (c) root-toxic or shoot-toxic ions (Na, Cl, and B); and (d) the availability of micronutrients as well as macronutrients being affected by water quality. The net result is that soil fertility and plant nutrition become very changeable, unpredictable on a site-specific basis, and much more complex when making adjustments (Duncan et al., 2009; Carrow, 2011). Proactive monitoring of soil, water, and plant tissue are critical BMPs for making these fertility adjustments.

Factors contributing to a much more dynamic and challenging fertility program are as follows:

- *Saline irrigation water* can easily become the one input that adds the most chemical constituents to the soil–plant ecosystem—higher in total quantity over a season than all other additions.
- *Irrigation water treatments.* In addition to the natural constituents in the irrigation water, any added treatment chemicals contribute to the total, such as SO_4 from acidification.
- *Soil amendments.* When Na is present, it is common for gypsum to be added to the soil directly as a granular product or applied in suspension via the irrigation system.
- *Leaching programs* are implemented that leach not just undesirable soluble salts but also essential soil nutrients.
- *Changes in soil nutrient and element status* become more dynamic (a) spatially within the soil profile, since nutrients and elements are applied to the surface; (b) spatially across the landscape, since irrigation system design and scheduling result in differing quantities of water and constituents in the water applied over an area; and (c) temporally across a season as water quality may change with time, such as from rainy to dry seasons.
- Irrigation water quality often exhibits *spatial and temporal variability.* When irrigation water quality is poor, there can be more substantial differences across irrigation lakes from the influent side to the outlet as well as by depth. Seasonal variability in water quality may

be due to weather patterns, such as a dry and rainy season, or a single source may vary over time in quality when delivered from treatment facilities or when additional water is delivered via upstream sources.

The specific reasons or goals for fertilizing turfgrasses are more complex compared to sites with good irrigation water quality. When dealing with salinity, there are several primary reasons for developing a flexible nutritional program with fine-tuning of the fertilizer compositions and associated amendment applications based on proactive soil, water, and tissue sampling. For example, fertilization must be targeted to achieving the following goals:

- *To meet basic plant nutrient needs* for the specific glycophytic (with variable salt tolerance levels) or halophytic turfgrass species and cultivar(s) planted onsite. When salinity additions are high, the grass species may need to be changed to a more salt-tolerant one, which requires turfgrass managers to adjust in terms of nutritional needs to new species and to specific cultivars in some cases.
- *To correct nutrient deficiencies, toxicities, and imbalances* induced by irrigation water, irrigation water treatments (e.g., acidification and salt additions), soil treatments (such as gypsum and lime), and other amendments (wetting agents, zeolite, and cytokinins) required to correct irrigation water problems. With variable and poor water quality, deficiencies, toxicities, and imbalances all become more frequent occurrences and may be ongoing problems that must be continuously addressed because of regularly scheduled irrigation applications.
- *To correct for leaching* of soluble nutrient salts and the interactive solubility and mobility of specific nutrients.
- *To maximize plant stress tolerance responses*, especially salinity, drought, wear, heat and cold, and pests. Many plant stress tolerance mechanisms are affected by particular nutrients, such as K, Mn, Zn, and Ca, and their roles in actual inherent genetically controlled drought or salinity stress tolerance mechanisms.
- *To alleviate soil chemical problems* involving low or high pH, sodic soil conditions, acid-sulfate soils, chemical complexation problems involving bicarbonates and/or carbonates, excess sulfates, toxicity concentrations (boron, Na, and Cl), and so on.
- *To regulate or control plant growth rates* for adequate wear tolerance and to minimize excessive organic matter accumulation that can develop with excessive N and irrigation applications.
- *To sustain turfgrass root volume and enhance root system redevelopment* under salinity-challenged conditions, especially related to Na root toxicities, Al/Mn toxicities (acid-sulfate soils), and Na deterioration of the soil structure.
- *To compensate for inherently low cation exchange capacity (CEC) soil profiles* and soils with either very low or very high organic matter concentrations that may occur in sandy soils and that are designed and managed to allow leaching of total soluble salts.

When considering all the factors that influence plant nutrition and soil fertility when using saline irrigation water, it is important to view each nutrient and its specific concentration, but also to look at the whole ecosystem (i.e., a holistic or whole-systems approach). Fertilizers also contribute to the overall salt load and total soluble salt additions since many products contribute additional salts (Table 16.1). Adjustments must be made in several areas on an ongoing basis. A whole-systems approach must be implemented when dealing with saline irrigation water that encompasses (a) water quality and specific nutrient load and imbalance; (b) buildup of salinity in the soil; (c) dominance and imbalance of toxic ions on the CEC sites; and (d) actual uptake of nutrients by the turfgrass plant, availability of each nutrient, and the fertilizer's form (liquid or granular), method of application, and timing of application.

TABLE 16.1
Salt Indices of Selected Fertilizers

	Salt Index	
Fertilizer Source	**Based on Equal Amounts of Material**	**Based on Equal Amounts of Plant Nutrients**
Sodium chloride	153	
Potassium chloride	116	1.94 K_2O
Ammonium nitrate	105	2.99 N
Sodium nitrate	100	6.06 N
Calcium chloride	82	
Urea	75	1.62 N
Potassium nitrate	74	5.34 N/1.58 K_2O
Ammonium sulfate	69	3.25 N
Calcium nitrate	65	
Ammonia	47	0.57 N
Potassium sulfate	46	0.85 K_2O
Magnesium sulfate (Epsom salts)	44	
Sulfate of potash-magnesia (potassium magnesium sulfate)	43	1.97 K_2O
Diammonium phosphate (DAP)	34	0.64 P_2O_5/1.61 N
Monoammonium phosphate (MAP)	30	0.49 P_2O_5/2.45 N
Triple (concentrated) superphosphate	10	0.22 P_2O_5
Slow-release carriers	<10	
Gypsum	8	
Normal (ordinary) superphosphate	8	0.39 P_2O_5
Potassium monophosphate	8	0.16 P_2O_5/ 0.24 K_2O
Limestone	5	
Natural organic (5% N)	3.5	0.70 N
Dolomitic lime	1	

Source: From Carrow, R. N., D. V. Waddington, and P. E. Rieke, *Turfgrass Soil Fertility and Chemical Problems: Assessment and Management*, John Wiley, Hoboken, NJ, 2001. With permission.

16.2 MONITORING NUTRITIONAL STATUS: PROACTIVE OR REACTIVE

As the concentrations and diversity of elemental and soluble salt constituents in irrigation water increase, so does the need for more frequent monitoring of soil, water, and plant status. Monitoring must be proactive (i.e., before a soil or plant problem becomes evident). Reactive monitoring after a problem starts to become evident in the soil and expressed in the form of plant symptoms is not uncommon due to the diversity of nutritional challenges, but as much as possible, a good nutritional-monitoring program must be based on a proactive approach. Since soil, plant, and water testing are covered in other chapters in detail, only brief comments are made here.

Irrigation water quality testing is essential and must be conducted more frequently when a new irrigation water source is used until the levels of constituents are determined and the consistency of values has been documented over time (see Chapter 6, "Irrigation Water Quality Tests, Interpretation, and Selection"). It is very common for irrigation water quality to be variable over time. Irrigation water tests have a major influence on whether water treatments are necessary, which soil amendments should be used, and subsequent soil chemical status. Additionally, nutrient additions from the irrigation water source must be accounted for within the overall plant nutrition and soil fertility program. This is especially important when using reclaimed water. Data in Table 6.9 contain normal levels of nutrients in reclaimed irrigation water and conversions that can be used to determine nutrients added per 1000 sq. ft. per 12 inches of irrigation water applications. Remember that these concentrations are applied with each irrigation cycle.

Soil testing is also required on a more frequent basis, including field monitoring of soil salinity as discussed in Chapters 4 and 5. Soil tests are of two general categories: (a) routine soil testing (Chapter 5, "Routine Soil Test Methods"), and (b) specially requested analyses (such as the saturated paste extract [SPE] salinity test) that relate to salt-affected soils (Chapter 4, "Salinity Soil Tests and Interpretation"). Some salt-challenged sites may require a complete physical analysis be conducted on selected soils (such as on older greens or sports fields) to help in determining whether the old soil profile can be retained or a new soil should be constructed to improve salt management when saline irrigation water is being used.

Tissue testing is also commonly used to identify plant response problems (Chapter 7, "Plant Analysis for Turfgrass"). With saline irrigation water applied over the shoots and going into the soil, there are ample opportunities for nutrient deficiencies to occur, especially micronutrients, Ca, and K. The key point here is whether critical nutrients are actually available for uptake and whether the plant has sufficient root volume for adsorption. Was the plant able to absorb the nutrients, despite the accumulation of soluble salts, at concentrations that meet sufficiency requirements for that particular cultivar? How much total and specific concentrations of salt ions (such as Na or Cl) are actually foliarly adsorbed into shoots and stolons in the canopy? A good resource for understanding tissue testing on turfgrasses and their interpretation is provided by Plank and Carrow (2010).

16.3 SALINE IRRIGATION WATER NUTRITIONAL CONSIDERATIONS

In Chapter 6, "Irrigation Water Quality Tests, Interpretation, and Selection," individual items in irrigation water analyses were discussed. Readers are encouraged to review these critical points. In this section, additional aspects are presented that are directly related to nutritional concerns in salt-challenged sites. While it is good to start with the irrigation water quality test results, soil tests and plant analyses also must be integrated into the salinity BMP turfgrass decisions.

16.3.1 Water pH

The water pH can alter soil surface pH and thatch pH over time. Soil nutrients are most readily available in the soil pH range from 6.0 to 7.5. However, the chemical constituents that cause irrigation water to exhibit a pH outside of this range are more important than pH by itself. Another secondary effect of low-pH irrigation water used in conjunction with acid-forming fertilizers that can be found on turfgrass sites is acidic thatch or a thin, highly acidic microzone at the surface that affects microbial population dynamics and subsequent transformations of fertilizers. Acidification of irrigation water can also cause similar problems. One additional concern when applying acidified water on calcareous sands and highly alkaline soils is the potential for deterioration of soil physical characteristics and subsequent changes in soil structural components that affect infiltration, percolation, and leaching capabilities due to the enhanced dissolution or re-precipitation of carbonates by acids.

16.3.2 High Chloride

Excessive chloride does not cause direct turfgrass root tissue injury except at very high levels (generally >355 mg/L) that are well above the guidelines in Table 6.8 for more sensitive landscape plants. Instead, Cl inhibits water uptake and, thereby, nutrient uptake. Additionally, high Cl may reduce NO_3 uptake by altering *Nitrosomonas* conversions of ammonium-N and urea-N granular products to nitrates. If the irrigation source has consistently high Cl content (such as found in water sources or blends involving seawater, brackish water, or saltwater inundation as well as some recycled or effluent sources) (Duncan et al., 2000, 2009), then N rates may need to be increased by 10% to 25% using primarily NO_3 forms applied foliarly using a "spoon-feeding" strategy.

16.3.3 High Total Salinity and Sodium Permeability Hazard

The presence in the irrigation water of excess total salts or high Na concentrations that may induce a sodic soil condition will necessitate extra water to be applied for leaching. This will result in *leaching of all nutrients* to a greater degree and require somewhat higher supplemental nutrient levels, especially on sandy soils. Fertilizers are not applied at higher rates than normal per application. But, fertilization should be more frequent using a spoon-feeding approach (modest amounts of slow-release fertilizers applied less frequently, or smaller amounts of fast-release fertilizers applied more frequently) so that annual rates are 10% to 50% higher. Slow-release nutrient forms can be incorporated using a prescription philosophy to aid in maintaining adequate nutrient availability levels. Fertigation through the irrigation system is another excellent prescription fertilization strategy that allows easy prescription adjustments based on nutrient contents in the water.

When high Na content in irrigation water requires appreciable Ca to be supplied to dislodge Na from the CEC sites, extra Mg and K will be needed to maintain adequate soil test levels and nutrient balances for these nutrients. Light, more frequent applications are better than heavier and less frequent treatments. Potassium is exceptionally mobile, readily being displaced from the CEC sites by Na, and rapidly moving in solution down through the soil profile with irrigation water. Weekly Mg and K applications may be warranted in extreme cases (often with a granular Mg and K source one week and a liquid Mg plus K application the next week, then alternate source applications until sufficiency levels are attained internally in the plant as designated by tissue testing) where saline water is used for irrigation and with sandy soil profiles.

16.3.4 Potassium

Since recreational turfgrass sites require ample K, any K in irrigation water is often viewed as beneficial. If K is high in reclaimed water, there is normally adequate Ca and Mg to prevent any nutrient imbalances, but K will contribute to total salinity. The key to fertility adjustments is to supply sufficient levels on a continuing basis for root uptake, taking into consideration water percolation rates (since K is highly mobile) and Na levels in the irrigation water, which can dislodge K from CEC sites and displace it into soil solution. Potassium is then quite susceptible to leaching because of its mobility and can rapidly become deficient. Key ratios within the irrigation water to consider include K:Na (2–4:1 on meq/L basis), Ca:K (10–30:1); $N:P_2O_5:K_2O$ basis (2–3:1:4–8), and Mg:K (2–10:1). Potassium needs to be 3%–8% base saturation on the CEC. As salinity increases, 1.5–3.0× rates may be required to maintain K sufficiency in turfgrass plants due to the combination of Na-suppressing K uptake; Na-enhancing K displacement from the CEC sites; Na-enhancing potential for K leaching; Na often requiring high Ca additions that also compete with K at the CEC and uptake levels; greater leaching on saline sites, where K is one of the easiest to leach; and high K being needed to maximize salinity tolerance mechanisms. Potassium is an essential nutrient for root system maintenance plus all other abiotic (drought, heat, cold, and traffic and wear) stress tolerances. For salinity, K is a very important ion governing osmotic adjustment, and for the halophyte turfgrass seashore paspalum, it is essential for full osmotic adjustment and cannot be substituted by another inorganic or organic osmolyte (Lee et al., 2007, 2008). Potassium works synergistically with calcium internally in the plant to maximize stress tolerance responses.

16.3.5 Calcium

When saline irrigation waters are used, Ca is one of the most important nutrients and the one causing the most confusion. Turfgrass managers should be aware of the total Ca added by the water source since groundwater, surface water, reclaimed water, and even rainwater (1 to 8 ppm Ca) all provide Ca. As noted in Table 6.9, 60 ppm Ca would add 3.75 lb. Ca per 1000 sq. ft. per 12 inches of irrigation water (equivalent to 16 lb. of $CaCO_3$). Thus, rainwater at 8 ppm Ca would add 0.50 lb. Ca per 1000 sq. ft. (2.2 lb. Ca CO_3 equivalent) per 12 inches of rain. Key considerations include Ca:Mg (3:1 meq/L basis), Ca:K (10–30:1), and Ca:Na+Mg (2–3:1) ratios as *indicators* of potential nutrient imbalances with increasing salinity.

Calcium is a critical nutrient to keep balanced in salt-challenged systems because of its soil stabilization function (dislodging excess Na from CEC sites and flocculation of colloids), importance for root cell membrane integrity, synergism with other nutrients (such as K), and actual turfgrass nutritional requirements. The key consideration is the actual availability of calcium for uptake by the turfgrass with fluctuating soil chemistry and salinity interactions and the need for a multipronged application approach. Granular sources such as gypsum or lime (the latter product with an S source to create gypsum) should be applied to the soil to counter excess Na on the soil CEC, reduce Na root toxicity, and provide available Ca for root uptake.

From the authors' experience, it appears that frequent applications of a high Na content irrigation water over the shoot tissues may actually strip Ca from leaf tissues and reduce internal Ca tissue content. Thus, foliar applications of Ca with actual adsorption are often needed under these conditions for nutritional balance in the turfgrass shoots even when soil Ca levels seem to appear to be adequate. Also, Ca amendments are commonly applied to soils at high rates to alleviate sodic conditions and to prevent Na toxicities to root tissues (Na displacement of Ca in root cell membranes with subsequent deterioration of the root tissues). Thus, as the use of saline irrigation water has increased, so has the number of Ca products. Unfortunately, the formulations and rates recommended by some of the manufacturers of the products are not agronomically sound. One example is *foliar fertilization*, which is an excellent spoon-feeding approach to enhance nutrient use efficiency of Ca and other nutrients under saline irrigation; but a distinction should be considered between foliar feeding, fertigation, and foliar application, and this adsorption issue is particularly critical for Ca (Table 16.2) (Carrow, Waddington, et al., 2001).

- *Fertigation.* Applies nutrient via the irrigation system, and almost all is washed off leaves and enters the plant through the root system.

TABLE 16.2
Calcium Fertilizer Materials, Relative Solubility, and Suitability for Foliar Uptake[a]

Fast release and solubility (foliar application for foliar uptake): Liquid products
Calcium nitrate
Calcium chloride
Calcium gluconate/glucoheptonate
Calcium complexed with sugar alcohols or amino acids
Calcium acetate

Intermediate release and solubility (suspension or granular products; root uptake or clipping removal)
Calcium sulfate (gypsum) (regardless of sieve particle size)
Calcium thiosulfate

Slow release and low solubility (suspension or granular products; root uptake or clipping removal)
Calcium hydroxide/oxide
Calcium carbonate (lime) or powdered coral
Dolomite (calcium/magnesium carbonate)
Calcium silicate

[a] Foliar application can be by any Ca liquid or suspension material, but foliar uptake is only by dissolved Ca (i.e., liquid forms).

- *Foliar application.* Uses 1–2 gal. of water per 1000 sq. ft. to apply a nutrient, and a high percent (>90%) of the nutrient usually stays on the leaf tissues if the grass has a reasonable shoot density.
- *Foliar applied versus foliar uptake.* A suspension can be foliarly applied, but the suspended particles will not be foliarly taken up since they are not soluble. A liquid material can be foliarly taken up through the leaf system. Thus, not all products that are foliarly applied are taken up through the foliage.

Several nutrients are internally mobile in the plant (N, P, K, Mg, Cl, and Na) and can be translocated downward to the root system. Other nutrients are somewhat mobile (S, Cu, Mo, Zn, and B), while a few (Ca, Fe, Mn, and Si) are relatively immobile. The immobility of Ca and its slow internal movement within the turfgrass plant can be attributed to its primary functions: cell membrane stabilization, constitution of cell walls, carbohydrate translocation, protein synthesis, activation of enzyme systems, and enhancement of nutrient uptake in roots and movement of those nutrients into cells (especially its synergism with potassium).

It is not unusual for finely ground gypsum, lime, or other insoluble Ca forms to be put into a suspended formulation and then sold at a high price for "foliar feeding." Normally the product literature points out how Ca foliar feeding can prevent Ca deficiency of tissues and how Ca can displace Na in the soil and alleviate sodic conditions. The question is whether such a product can really perform either of these claims. The list in Table 16.2 summarizes the solubility and suitability for foliar applications of various Ca fertilizers. Effective Ca foliar feeding under salt-challenged conditions involves at least 10% Ca in the product that is applied to the foliage or leaves at a rate of 0.10 to 0.25 lb. Ca per 1000 sq. ft. using 1 to 2 gal. water per 1000 sq. ft., and the nutrients are water soluble and not in suspensions. Uptake is rapid through the ectodesmata pores, cuticle cracks, and stomata pores. Once inside the leaves, the nutrients pass directly into cells through the cell wall and plasma membrane or enter the apoplasm (space between cells) and then may be transported in the xylem (upward). More mobile nutrients can enter the phloem (upward or downward movement) and be transported to the root tissue, but not immobile Ca.

Not only the Ca product form is important, but also the actual Ca concentration and recommended rate for application. If the target is to apply foliar Ca to alleviate Ca deficiency in shoot tissues, the rates given above are appropriate as long as they are foliarly taken up. However, does the Ca that is not taken up help to alleviate sodic soil conditions by displacing Na from soil colloids? For remediation of sodic conditions, the application rates are normally at 10 to 20 or more pounds Ca per 1000 sq. ft. Thus, the claim that Ca at 0.10 to 0.25 lb. Ca per 1000 sq. ft. will assist in remediating sodic soil conditions is misleading.

Application of granular sources such as gypsum or lime that involve root uptake will entail a lag period of 3–4 weeks before actual stabilization in the turfgrass shoots, while liquid sources that involve actual foliar uptake can take 1 to 4 days with visual results within a week. Hot, dry conditions will limit uptake. Mowing of leaves or excessive irrigation or rainfall before uptake will wash the nutrients into the soil for possible uptake. If clippings are returned to the soil, the nutrient can become available for uptake after microbial breakdown; if clippings are removed after uptake, the nutrient will not be recycled. When collecting clippings for tissue analysis, allow 1 to 2 weeks after liquid product applications to the leaves before sampling and submitting for tissue nutrient analysis—or use a more vigorous tissue-washing protocol, as discussed in Chapter 7.

In salt-challenged ecosystems, a dual-product application strategy is often required to meet Ca requirements in managing excess accumulated salts in the soil (granular products applied at least monthly depending on rainfall and irrigation scheduling) and soluble liquid products not in suspension that are applied for foliar uptake into shoots to meet sufficiency levels. To counter soil accumulated Na, good-quality granular products should slowly solubilize and release Ca in pulses with the irrigation cycles and rainfall events into the soil ecosystem. Some granular Ca products are very finely ground with higher solubility and can result in excessive flushing of Ca with high rainfall events, often resulting in loss of potentially available Ca with leaching through the soil profile.

16.3.6 Magnesium

Most often, Mg is present in irrigation water at lower levels than Ca. Sometimes Mg content will be relatively high (infusion from ocean water, brackish water, or saltwater intrusion into wells), which can reduce Ca on CEC sites and restrict K availability (Duncan et al., 2000). Exceptionally high levels of Mg will mimic excess Na effects on soil structure and internally in turfgrass plants, affecting nutritional balance. In these cases, supplemental Ca may be needed to maintain adequate Ca for maintenance of good soil physical conditions and to counter Na toxicities. Seawater has a high Mg content, so saltwater intrusion sites may exhibit this excess concentration problem. Also, supplemental K will be necessary to maintain ample K nutrition. Normally, a 3 to 8 Ca:1 Mg ratio (meq/L) is suitable.

A more typical problem than water sources containing excess Mg is low Mg content in irrigation water. Low Mg caused by the addition of high-Ca amendments for irrigation water that contains too much Na is another common situation. Another problem of increasing frequency is Mg deficiency induced by application of unneeded Ca on sandy sites. As with Ca, knowledge about Mg content and rates applied in the irrigation water are very useful in avoiding deficiencies or excessive Mg problems. Some turfgrasses such as seashore paspalum actually require high levels of magnesium and will thrive in environments where higher than normal concentrations are found in irrigation water. But balancing these high levels with Ca supplements is still required in the fertility maintenance program, especially where turfgrass color expression (Mg as the core molecule in chlorophyll) is important.

16.3.7 Sulfur

It is not unusual for SO_4 content in reclaimed water to be 100 to 200 mg/L, and groundwater influenced by seawater intrusion may be even higher (seawater contains about 2600 mg/L SO_4). Sulfur deficiencies may occur in high-rainfall climates on exposed, sandy soils that do not receive SO_4 from rainfall due to their location relative to industrial activity. Normally 2 or 3 lb. S per 1000 sq. ft. per year are sufficient for turfgrass nutritional needs, and this is often provided by SO_4 containing N, K, or Ca fertilizers. Irrigation water at 200 mg/L SO_4 would supply 4.2 lb. S per 1000 sq. ft. per 12 inches of water.

16.3.8 Iron (Fe)

In addition to macronutrients in irrigation water, *micronutrients* (Fe, Mn, Cu, Zn, Mo, Ni, and B) can affect turfgrass fertilization. The list in Table 6.12 has recommended maximum concentrations of trace elements in irrigation water for long-term values (LTV) and short-term values (STV) based on the Australian Water Authority (AWA; 2000) and Westcot and Ayers (1985). The 5.0 mg/L guideline in Table 6.12 for Fe in irrigation water is not related to any potential "toxic level," but continued use could cause (a) precipitation of P and Mo and deficiency problems for turfgrasses (P) or landscape plants (P or Mo); (b) staining on plants, sidewalks, buildings, and equipment; (c) potential plugging of irrigation pipes by anaerobic Fe bacteria sludge deposits or potential buildup in lakes and ponds, which can be a problem at >1.5 mg/L Fe; and (d) high, continuous rates of Fe, which may induce Mn deficiency or (much less likely) Zn and Cu deficiencies. On heavily leached sands, where Mn is often low, Mn deficiency may become a problem. At 5.0 mg/L Fe, 12 inches of irrigation water would add 0.31 lb. Fe per 1000 sq. ft., while a typical foliar application to turfgrass is 0.025 lb.

Fe per 1000 sq. ft., but in only 3 to 4 gal. water per 1000 sq. ft. In most instances, Fe concentrations are low and turfgrasses will respond to foliar Fe. When total salinity is high, Fe plus a cytokinin (such as from seaweed extracts with >30% cytokinin concentration) as a foliar treatment are often beneficial, since salt-stressed plants exhibit low cytokinin activity. Critical indicator ratios include Fe:Mn:Mg (1:1:1 with pH < 8.0 and 3:1:1 when pH > 8.0).

16.3.9 Manganese (Mn)

Manganese can become toxic to roots of many plants, so use of water high in Mn (0.20 mg/L) can contribute to this problem, especially on poorly drained, acidic soils. Acidic, anaerobic conditions transform soil Mn into more soluble (i.e., toxic) forms. If water is high in Mn, liming to pH 6.0 to 7.5 and good drainage greatly reduce the potential for Mn toxicities since the Mn forms less soluble compounds under that pH range. At >1.5 mg/L Mn in irrigation water, Mn can contribute to sludge formation within irrigation lines. Also, high Mn may inhibit Fe uptake and promote Fe deficiency. Supplemental foliar Fe applications would prevent this problem. With most turfgrass situations, Mn is very low in irrigation water, and supplementation over and above a regularly scheduled micronutrient application may be needed as salinity increases. Many micronutrient packages have Mn concentrations near or less than 1%, and additional supplementation is often required in highly saline turfgrass systems. Mn:Zn ratios should be 1:1, since both are essential nutrients for activating salinity tolerance in turfgrasses, and both have other critical functions (disease suppression, enzyme activation for growth, and color enhancement).

16.3.10 Copper (Cu), Zinc (Zn), and Nickel (Ni)

The irrigation water levels in Table 6.12 are based on the potential to develop toxicities on sensitive landscape plants over time. Turfgrasses can tolerate relatively high rates due to the mowing of leaf tips. Unusually high Cu and Zn could inhibit Fe or Mn uptake and, thereby, induce deficiencies of these nutrients, even on grasses. In this case, these nutrients would need to be supplemented for maximum turfgrass performance.

16.3.11 Molybdenum (Mo)

Molybdenum toxicity would be very unlikely in plants, but livestock feeding on grasses high in Mo can be detrimental. In turfgrasses, as salinity increases, Mo has direct competition with divalent oxyanions (sulfates and phosphates) for exchange sites. Excess applications of gypsum or lime, single superphosphates, and sulfur acidification can negatively affect Mo availability. Mo acquires hydrogen ions and becomes less ionic as soil acidity increases, that is, Mo is less readily absorbed by turfgrass roots or forms Mo polyanions that are completely unavailable for uptake.

16.3.12 Boron (B)

Boron is often associated with saline hydrogeological conditions and is another element that can be a toxicity problem if too high in irrigation water. Toxicity can occur from irrigation water, wastewater, composted sewage sludge, or native arid soils. Leaching is easiest in acidic sodic soils that are sandy. As soil pH increases from 6.3 to 7.0, B is more tightly adsorbed on clays and Fe/Al oxides. Thus, at pH < 7.0, leaching may prevent B accumulation, while at pH > 7.0, light lime applications to maintain high available Ca levels can help fix the B in less available forms. Leaching is more effective on coarser textured soils than on fine-textured ones.

16.3.13 Other Trace Elements

Reclaimed water may contain excessive levels of some elements such as heavy metals. These are reported by Westcot and Ayers (1985) and Snow (1994), as given in Table 6.12. These elements would not directly influence turfgrass nutrition, but would be of concern for toxicities on some landscape plants. Some halophytic species, such as seashore paspalum, are phytoaccumulators of heavy metals (refer to Chapter 20) (Duncan and Carrow, 1999).

16.3.14 Bicarbonates and Carbonates

High bicarbonates are relatively common in reclaimed water and some groundwater sources (Eaton, 1950). While $HCO_3 > 500$ mg/L can cause unsightly, but not harmful, deposits on foliage of plants, there are no specific HCO_3 or CO_3 levels that result in nutritional problems. Instead, it is the imbalance of HCO_3 and CO_3 in conjunction with Na, Ca, and Mg that is most important. When HCO_3 + CO_3 levels exceed Ca + Mg levels (in meq/L), both Ca and Mg can be precipitated as insoluble lime deposits in the soil and/or as scale in irrigation lines. Problems that may arise from bicarbonate and carbonate precipitation are:

- If Na is moderately high (from >100 to 150 mg/L), removal of soluble (and available) Ca and Mg by precipitation into the relatively insoluble carbonate forms will leave Na to dominate the soil CEC sites and create a potential sodic soil condition. High Na concentrations on the CEC sites will depress availability of Mg, K, and Ca. Acidification of irrigation water is the normal management strategy for this situation, and the acid aids in dissolving the insoluble lime precipitate as carbon dioxide and water, releasing Ca for counterion activity against Na. The extra calcium is then available to compete with Na for positioning on the CEC sites.
- On sandy soils, the calcite (lime) may start to seal some of the pores and reduce infiltration.
- Ca-Fe and Ca-P bicarbonate complexes can also form in highly alkaline conditions and affect nutrient availability.
- In dry periods or climates, high Na levels can react with bicarbonate and carbonate to form Na-carbonates that act to retain excess Na in the rhizosphere.

16.3.15 Root Toxicities from Na, Cl, and B

See Table 6.8. Specific ion toxicity (Na, Cl, and B) and miscellaneous chemical constituent problems in sprinkler irrigation water can affect sensitive plants (after Ayers and Westcot, 1994; Hanson et al., 1999; AWA, 2000). While the guidelines for root toxicities or soil accumulation of these ions in Table 6.8 are most appropriate for sensitive trees, shrubs, and other landscape plants, excessive levels of Na^+ can cause turfgrass root deterioration at higher levels than are indicated in the table, especially for glycophytic grasses.

16.4 PRODUCTS, LABELS, AND RECOMMENDATIONS

In the previous section regarding the Ca discussion, it was noted that some products actually would not work to remediate the problems described on the product label due to (a) a chemical form (low solubility suspension) that could not be taken up foliarly, even though it is sold as a foliarly applied product; and (b) the rate used is actually between 1000th and 10,000th of what is really required for the stated problem (in this case, the alleviation of sodic soils). Another commonly observed and increasing problem that seems to be stimulated by saline irrigation water sources is the "proprietary product" where the manufacturer does not completely list the active ingredients—patents were actually developed to allow protection plus disclosure, but this full disclosure protection seems to be insufficient for some manufacturers to provide on the label. Fertilizer and pesticide manufacturers have operated under patent laws with full disclosure of product materials in terms of chemical nature and quantity and have been able to compete in the marketplace. So the "proprietary product" nonlabel that is used for so many products is not scientifically or agronomically valid.

In the case of poor irrigation water quality, the turfgrass manager must make many management adjustments and use a multitude of products; but each product should be applied for a specific reason, in the correct formulation, and at the correct rate. *The authors strongly recommend, based on experience, not to apply products when the specific chemical ingredient (chemical and quantity) is not listed on the label* since developing good fertilization and salinity management programs is already complex and introducing unknowns into the equation is not a good maintenance practice when managing salt-challenged or other turfgrass ecosystems. One reason some products do not wish to list the active ingredient is because it is a common material available from much less expensive sources. The use of products that do not fully disclose all constituents on the label is not a BMP management strategy.

Another version of not listing an active ingredient is to include an ingredient that will give a fertility response, such as soluble N or foliar Fe, but those elements are not listed on the label. If a product is not sold as a fertilizer, it does not require a product label for disclosure on quantity and composition. The only reason for this practice is to insure a turfgrass response, which is attributed by the turfgrass manager to result from the "inactive active ingredient."

Confusion sometimes occurs between a true nutrient and materials that may result in a plant growth response similar to a "fertilizer" response in some situations such as cytokinins, biostimulants, humates, fulvic acids, vitamins, and minerals (Carrow, Waddington, et al., 2001). More recently, the term *bionutritionals* has been used by some in the turfgrass industry, where they are defined as follows: "bionutritional fertilizers is the name given to the wide range of living organisms, including microbes, bacteria, mycorrhizae, seaplant extracts and hormones, that can be added to a fertilizer prill or delivered as stand-alone products to nourish plants" (Aylward, 2010, 36). This implies that these materials function to provide nutrients or are fertilizers. In Chapter 17, "Additional Cultural Practices," these materials are discussed in more detail.

16.5 SUMMARY

In summary, when the irrigation water is nutrient rich and/or salt laden, it becomes the greatest source of desirable and undesirable nutrients and elements of all management inputs. The irrigation water cannot be ignored in terms of fertility and plant nutrition programs since the nutrients and the soluble salts are applied with each irrigation cycle. Irrigation water sources with high concentrations of chemicals will cause the most problems, and fertility programs must be adjusted accordingly.

It is instructive to note that for water quality testing, two of the four "salinity problem" areas are directly related to nutrients, namely, the nutritional status of the water and the status of toxic or problem interactive ions. While total soluble salts are the number one salinity problem that accounts for turfgrass manager success or failure on a salt-affected site, just behind this issue in importance is their ability to maintain soil fertility and plant nutrition availability and stability in an ever-changing environment. Most turfgrass managers learned about developing fertility programs under much more stable and tranquil conditions when using a good low-salinity irrigation water supply. For maintenance of a sustainable turfgrass ecosystem soil and plant nutritional program, the most important aspects are as follows:

- Soil chemical properties are primary (CEC level, nutrient balance, nutrient concentrations, pH, and salt control), realizing that these properties are changing over time.
- Soil physical problems are also important (especially excessive organic matter or any factor that limits water movement), and the addition of any inorganic amendments that sequester soluble salts can alter physical characteristics.

- Soil biological activity or biostimulants are least important since good turfgrass canopy density usually equates to good conditions for microbial activity.
- The ecosystem is dynamic and constantly changing (salinity magnifies the challenges).
- The diversity of products should be considered: whether granular or liquid and their solubility, whether they are actually foliarly absorbed, and whether they are biostimulants with fully disclosed composition on the labels.
- Staying with a commonsense basic fertility program (there are no magic bullets or miracle products), and the fertility program needs to be "dialed in" to your site-specific salinity challenges.
- Salts from saline irrigation water will accumulate in the soil over time unless leached, and constant proactive monitoring must be scheduled to minimize the overall interactive salinity impact on turfgrass performance, especially nutritionally.

17 Additional Cultural Practices

CONTENTS

17.1 SALINITY AND ASSOCIATED STRESSES ON A SITE

Salinity stress on turfgrasses does not occur in a vacuum—other stresses are usually present on the site, and turfgrasses are generally exposed to more than one abiotic and biotic stress at the same time. If salinity is the primary limiting abiotic environmental stress affecting grass performance, any additional stress, whether it is abiotic (drought, heat or cold, shade and low light intensity and light quality, or wear and traffic) or biotic (insects, diseases, and weed competition), will adversely impact the overall sustainability of that grass on a site-specific basis. These second or third environmental stresses coupled with the primary salinity stress will often kill the grass in random areas on the site. Management requires the turfgrass manager to address all stresses on a site.

Thus, in salt-affected ecosystems, recognition of additional primary stresses (i.e., beyond salinity stresses) as well as secondary stresses is critical. A *primary stress* is a basic stress on the site, but primary stresses almost always cause *secondary stresses* (i.e., stresses that would not be present if the primary stress was removed or alleviated). *Visual symptoms* of injury may be from direct injury of a primary stress or they may be symptoms of a secondary stress. For example, high Na levels can result in soil structure deterioration, which is exhibited by poor drainage and aeration problems. *Visual symptoms* are what many consider the primary problems since they "see" them—such as standing water, poor roots, anaerobic conditions causing take-all and decline disease problems, difficulty in irrigation due to low infiltration, and so on. But, on sodic soils, these visual symptoms are really secondary problems caused by excessive Na levels—this is the primary problem that must be dealt with in order to avoid the secondary problems and associated visual symptoms. As the Na displaces Ca on the soil cation exchange capacity (CEC) sites, the soil structure degrades and what is lost are the macropores. Thus, a management program targeted to the primary problem would involve Na displacement from the CEC sites by available Ca amendment applications to the soil, leaching, and restoring macropores (by cultivation).

On salt-affected sites, one or more of the four types of salinity stresses (total soluble salts, Na permeability hazard, ion toxicity and problem ions, and nutritional imbalances and deficiencies) will be primary stresses, but there can be (a) other primary problems in addition to salinity, and (b) secondary problems that arise out of any of the primary problems on the site. In summary, an important component of a best management practice (BMP) program on salt-affected sites is to recognize and address any (often multiple) stresses in addition to salinity that may enhance salinity injury.

Many of the "additional" cultural practices addressed in this chapter are targeted to abiotic and biotic stresses that may exhibit a significant interaction with salinity. Stresses that most often interact with salinity stress are (a) environmental, such as drought and heat; (b) traffic, such as wear and soil compaction; and (c) biotic, such as diseases, insects, or nematodes that are able to be more competitive under saline conditions due to salinity favoring their growth or reducing competition from other organisms normally competing with them, including weeds.

Visual plant and soil symptoms may provide insight into various primary or secondary problems present. Recognizing that the total ecosystem being managed has a salinity-induced problem is critical in deciding on the primary and secondary cultural management options for the specific site. Plant and soil symptoms that are typical on salt-affected sites are useful in recognizing salt-stressed areas, and these were noted in Chapter 2 ("Saline Soils") and Chapter 3 ("Sodic, Saline-Sodic, and Alkaline Soils"), but are summarized here and can include any combination of the following:

- Turfgrass goes off color despite your best efforts in fertility management.
- Canopy density decreases in random micropockets or areas.
- Increased surface compaction.
- Increased footprinting or equipment tire marks.
- Loss of wear and traffic tolerance.
- White surface crust from surface salt accumulation.

- Turfgrass has a bluish-green, drought-stressed appearance even after irrigation.
- Turfgrass exhibits heat stress symptoms more quickly than normal.
- Turfgrass is slower than normal in emerging from winter dormancy, and winter kill is more prominent.
- Black organic matter deposits on the surface are colonized by moss or algae.
- Standing water remains after irrigation or rainfall, with very slow infiltration and percolation.
- The incidence of localized dry spot and hydrophobic soil areas increases.
- The incidence of disease and/or insect problems increases.
- Pesticide (fungicides, insecticides, and herbicides) and other chemical expenses increase.
- Weed problems increase, due to less aggressive and less competitive turfgrass.
- Turfgrass manager is constantly reacting to challenges with limited progress in turfgrass performance or quality. Turfgrass manager is unable to implement proactive management protocols.
- Well-defined dark-colored layers appear in cup cutter plugs and other soil profiles.
- Main drain line intakes on fairways and roughs are overly wet or grass density and color are excellent, poor, or both.
- Halo appears around sprinklers with excellent grass density and color, but in peripheral areas with less water distribution uniformity are stressed turfgrasses.
- In overseeding situations, there is gradual deterioration of both the warm-season grass understory and problems with the establishment or persistence of the cool-season grasses from one year to the next.

A review of these plant and soil symptoms that can arise from salt-induced stresses reveals that many are in common with drought, high temperature, wear, soil compaction, certain diseases, and other possible site stresses. Proactive soil, water, and tissue testing when combined with the experience of the site manager will often be necessary to determine the combination of stresses present and to adjust the management program to deal with the multiple challenges.

Whenever total soluble salts or Na is present in sufficient quantity to create a salt-affected situation, the salinity stresses must be considered as primary—if these are not dealt with in an appropriate manner, the other stresses will be enhanced. When salinity is the dominant or primary stress, management programs must be focused on salt management in concert with managing the grass. There are no magic bullets, miracle products, single pieces of equipment, or simple strategies when managing salts. The turfgrass ecosystem is constantly exposed to potential salt ion accumulation in the soil profile plus potential foliar absorption directly into shoots and other plant parts from repeated saline irrigation water applications. The salinity stress BMP strategy is a holistic or whole-ecosystem approach.

Even though salinity is the most complex environmental stress that can be imposed on turfgrass ecosystems, the management program should be kept basic in strategy and scientifically oriented with regularly scheduled proactive monitoring of impending salt accumulation in the soil and overall salinity impact issues. One key question to ask is "Are the products that you are applying to the turfgrass providing the performance expectations that are required for the site?" A second question is "Is the cultivation program providing sufficient salt movement away from the turfgrass root system?" And a third question is "Are all drainage lines properly functioning?"

17.2 ENVIRONMENTAL CHALLENGES AND MANAGEMENT

Certain environmental stresses commonly interact with one or more of the salinity stresses on a relatively frequent basis on salt-affected sites. In this section, we review the most common interactions.

17.2.1 Drought Stress

Less salt-tolerant turfgrasses will adversely respond to the combined water-induced drought and salt-induced drought stresses rapidly with escalated reduction of growth and plant tissue injury. More salt-tolerant grasses will exhibit less dramatic response and may have time to adapt for survival. However, the addition of water-induced drought stress over the top of salinity stress causes greater potential for permanent damage than either stress alone. The following discussion of the sequence of plant responses to these dual stresses is focused on more salt-tolerant grasses.

First, how plants respond to salinity alone, which is a two-phase growth response to salinity, is reviewed (Munns, 2002). The first phase of growth reduction is due to salt accumulation outside the roots and the associated potential for desiccation of those roots leading to reduced water uptake. This is the *water stress* or *osmotic phase*, with growth reduction regulated by root hormonal signals, and the resulting cascade of chemical internal plant changes producing effects that are identical to those of water stress caused by drought. The second phase results from internal cellular injury and is due to excess salt accumulation in transpiring leaves, and lack of compartmentalization of salt ions in the vacuole (a key genetically controlled strategy that separates the salts from the photosynthetic machinery). This second phase inhibits growth of younger leaves since the carbohydrate supply to the growing cells is reduced, and the grass is subsequently using additional resources in a defensive mode to combat the increasing salinity exposure.

Comparing water stress effects and salt stress effects on plants is based on differential time exposure (Munns, 2002). When drought stressed, several minutes of drought stress exposure on salt-tolerant plants result in an instant reduction in leaf and root elongation rate, followed by rapid partial recovery. With multiple hours of drought stress exposure, a steady but reduced rate of leaf and root elongation results.

Multiple days of exposure to drought stress eventually result in a reduced rate of juvenile leaf emergence and eventually in leaf growth being more affected than root growth (this is a defensive compensation mechanism by the plant to reduce water loss in the leaves; leaf senescence and leaf drop can often result, especially in older leaves). Multiple weeks of exposure to drought stress on salt-tolerant plants result in substantially reduced leaf size (decreased leaf area from death of older leaves) and number of lateral shoots. Multiple months of exposure to drought stress lead to eventual death of the entire plant (Munns, 2002). Even short-term drought exposure (2 weeks) can significantly impact the survivability of halophytes with increasing salinity (Brown and Reza Pezeshki, 2007). The cyclic wet and dry cycles will gradually lead to reduced overall turfgrass performance and loss of grass density as the grass struggles to survive.

Proper irrigation scheduling is a critical component of salt management. If the turfgrass is allowed to dry down to the level at which actual drought stress impacts the grass, the grass will often die. Keeping the lower soil profile moist to allow a more favorable plant water status is an essential management strategy.

Short-duration, frequent saline irrigation water scheduling can increase the localization of excess salts from the irrigation water in the upper rhizosphere, causing desiccation of root hairs and loss of potential water extraction capability by the turfgrass. Drought stress symptoms in the grass can often result. Water use efficiency in tall fescue has been shown to decrease with increasing salinity, but this decrease did not occur in Bermuda grass (Dean, 1996), a turfgrass species that is considered quite drought tolerant and only moderately tolerant to salinity (Carrow and Duncan, 1998).

Excess soluble ions—especially Na, which is a prolific water-loving salt ion—can reduce water activity (availability for uptake), and as salt ion concentrations in the soil profile increase, the turfgrass cannot extract enough water and often will exhibit drought stress as well as nutritional imbalance symptoms. Excess total dissolved salts that have localized in the upper soil profile can cause "physiological drought stress" in turfgrasses even after irrigation events when those salts have not been properly leached below the turfgrass root system and the moisture in the soil profile may actually be near field capacity. When the soil has high salinity, the plant can exhibit wilt symptoms, yet the soil appears moist due to water retained by the salt ions.

Careful timing of cultivation events is an important consideration in minimizing drought stress conditions when managing excess salts. Aeration can help to dry out a highly moist and salt-ladened upper soil profile, but if high-evaporation conditions prevail (multiple windy days and full sunshine), as moisture evaporates, excess salts can move rapidly to the surface through capillary pores to desiccate the shoot canopy. If this evaporation cycle is prolonged, the subsoil can also be dried down too much, and will need to be rehydrolyzed to stabilize moisture throughout the soil profile. Moisture and salinity meters can be used effectively to proactively monitor the salt movement and moisture dynamics in the soil profile, especially in the turfgrass rhizosphere.

Running an irrigation cycle in arid or semiarid regions prior to an anticipated or predicted rainfall event to bring the soil profile up to field capacity is one strategy to improve the potential for maximizing the leaching of any excess salts in the soil profile. A solid-tine or needle-tine aeration coupled with application of a "penetrant" wetting agent can provide additional potential enhancement for salt leaching in the soil profile with the rainfall event, especially between aeration holes.

Interactive responses of creeping bentgrass (*Agrostis stolonifera* L.), roughstalk bluegrass (*Poa trivialis* L), and perennial ryegrass (*Lolium perenne* L.) to salinity and drought showed that all three grasses were more severely affected by drought than by salinity (Pessarakli and Kopec, 2008). Roughstalk bluegrass was more detrimentally affected by either drought or salinity stress, while creeping bentgrass showed less negative effects in response to either drought or salinity individually among the three cool-season grasses.

Among four *Agrostis* species, *Agrostis gigantea* has been found to be the least sensitive to salt and drought stress, *Agrostis stolonifera* showed some resistance to salt stress, *Agrostis capillaries* showed some resistance to drought stress, and *Agrostis canina* was the most sensitive to both drought and salt treatments (Ahrens and Auer, 2009). These results demonstrate the different genetically controlled stress tolerance mechanisms (for salinity and for drought) that are present within the *Agrostis* genus.

17.2.2 Heat Stress

The interaction among salinity, drought, and heat stress in turfgrass cultivars can affect long-term sustainability and performance. With high levels of soluble salts in the soil reducing water uptake and sodic soil conditions often limiting rooting, it is not surprising that turfgrass plants grown under these conditions have reduced transpirational cooling capabilities and, therefore, are more prone to high-temperature stress.

Turfgrass plants with greater salt tolerance and high-temperature tolerance would be expected to best tolerate the additive effects of these stresses. This interaction is especially critical on cool-season turfgrasses, which exhibit inherently lower tolerance to high-temperature stress than warm-season species. With the trend toward use of more saline irrigation water, the development of grass cultivars that exhibit enhanced salinity tolerance plus improved drought and high-temperature tolerances will become increasingly important. For example, survival of salinity-tolerant 'Seaside' creeping bentgrass (*Agrostis stolonifera* L.) under drought stress was greater for clones that had higher heat tolerance, higher root-to-shoot ratio, less leaf area, and thinner stolons (Wu and Huff, 1983).

Application of saline sewage effluent on creeping bentgrass (*Agrostis palustris* Huds.) can result in increased organic matter accumulation at the soil surface, creating persistent water retention areas, and loss of oxygen flux can result. With anaerobic conditions, bentgrass susceptibility to salt and heat stresses increases as the grass morphologically concentrates adventitious roots near the soil surface (Seth-Carley et al., 2009). With extended periods of drought and high potential evapotranspiration, excess salts tend to concentrate near the soil surface, leading to possible toxic root and shoot desiccation conditions (Carrow and Duncan, 1998).

Brassinosteroids (28-homobrassinolide) counterstimulated antioxidative enzyme and proline concentration production plus membrane stability and leaf water potential when *Vigna radiata* was exposed to high-temperature and sodium chloride stress conditions (Hayat, Hasan, et al., 2010). Both high-temperature and sodium chloride exposure increased electrolyte leakage and lipid peroxidation that was not detoxified by brassinosteroid supplementation. Ashraf et al. (2010) recently reviewed the roles of brassinosteriods and salicylic acid in salinity tolerance.

17.3 TRAFFIC STRESSES AND MANAGEMENT

17.3.1 Traffic Stresses: Wear and Soil Compaction

Traffic stress, wear tolerance, and soil compaction are interrelated stresses that are magnified when salinity is imposed as another dominant stress (Carrow and Petrovic, 1992). The primary influence of high total soluble salts on turfgrasses is to reduce water uptake, which alters the dynamics of water and soluble salt management in the soil. The main effects of high sodium accumulation in the soil profile are to deteriorate soil physical conditions, leading to sodic soils, and to disrupt nutrient availability for turfgrass root uptake. The results of both problems are to impede turfgrass root growth and functionality, predispose the grasses to additional biotic and abiotic environmental stresses, cause the grasses to become more succulent, and alter soil nutrient availability and nutritional balances in the soil and plants.

Wear injury in any turfgrass is caused by shoot tissue compression and pressure bruising, abrasion, or tearing actions from foot or vehicular traffic. The reasons for increased susceptibility to wear injury are decreased turgor pressure of cells due to salt-induced drought; reduced growth and recovery rates; plus reduced carbohydrate reserves from photosynthesis, since the grass will reallocate those carbohydrates to repair the injury at the expense of root and shoot maintenance.

On golf courses as well as other high-traffic areas, commonsense management of foot and cart traffic and extending cart paths on grassed areas are key management options to minimize these additional stresses that lead to loss of grass canopy density. Golf cart traffic outside of cart paths can result in loss of turfgrass canopy density from compaction, bruising of shoot tissue, and predisposition to surface and root-borne diseases. As salts accumulate in the soil profile and cause growth rate reductions in the turfgrass, these high-traffic areas do not recover rapidly and they lose grass canopy density. Roping off areas to redirect cart traffic is an absolutely important component when managing turfgrass sites impacted by salinity. Even though total exclusion of cart traffic on fairways and roughs is difficult to accomplish, some courses have adopted selected one-to three-alternating-days-per-week requirements for cart-path-only golf play.

Narrow walkways from cart paths to tee or green complexes require the same traffic management requirements that are mentioned for carts. Redirecting foot traffic is essential to provide sufficient recovery time in these critical play areas approaching greens and tees. The same traffic management criteria are required for mowers, sprayers, and cultivation equipment, especially when narrow or limited approach venues are available around greens and tee complexes. Salinity additions and the potential for accumulating salts in the soil profile change these management protocols.

Soil compaction problems increase with salinity additions due to the need for frequent irrigation cycles to manage salts, which keeps the surface moist and more susceptible to compacting forces; sodium accumulating to excess levels that cause sodic soil conditions and make the soil more prone to compaction; changes (often delays) in cultivation programs due to altered moisture dynamics in the soil profile (yet normal grass maintenance equipment is run over these wet areas); and improper soil modification problems (too little or too much sand depth after sand capping, organic matter additions, and/or high levels of inorganic amendments that sequester moisture). The tendency for salt-affected sites to be more prone to soil compaction coupled with the need to leach salts result in these site requiring very aggressive cultivation programs.

17.3.2 Amendment Additions to Enhance Wear Tolerance

Potential for wear injury is greater on salinity-stressed turfgrass because of increased osmotic stress, reduced turgor pressure, and decreased plant vigor. Thus, maintenance of intracellular water turgor pressure, which is strongly related to internal potassium concentrations, is critical for minimizing wear and traffic problems (Trenholm, Carrow, and Duncan, 2001). Potassium is also directly involved in turfgrass root development and redevelopment in concert with phosphorus in order to maximize the potential for root uptake of moisture and nutrients in the soil profile. Since potassium is highly mobile in the soil and when leaching salts, you will be moving this nutrient through the soil profile, and prescription fertilization is often recommended on a weekly basis.

Silica or silicon dioxide (SiO_2) amendments may improve wear and traffic tolerance, since this element may impart somewhat greater shoot tissue strength (Trenholm, Duncan, et al., 2001), may enhance salt movement within tissues (Liang et al., 2006; Levent Tuna et al., 2008), alleviates biotic and abiotic stresses (Liang et al., 2003, 2008; Zhu et al., 2004; Richmond and Sussman 2003; Voleti et al., 2008), and may improve resistance to some pathogenic fungal diseases (Fauteux et al., 2005). Silicon can be applied in the form of silicic acid (H_4SiO_4) products such as potassium silicate (20.8% SiO_2, 8.3% K_2O), calcium silicate ($CaSiO_3$), or calcium silicate slag ($CaAl_2SiO_8$) (Carrow, Waddington, et al., 2001). Applications are often made as growing seasons progress into fall and winter periods, on emergence from winter conditions, and preceding tournament events to enhance wear and traffic tolerance in turfgrasses. Trenholm, Duncan, et al. (2001) noted that K had substantially more effect on wear tolerance than did Si.

Turfgrasses generally absorb $Si(OH)_4$, and this silicon form is transported in the xylem. Silicon is deposited in the xylem cell walls and outer epidermal cell walls on both sides of leaves. This deposition creates a fungal infection barrier, reduces cuticular transpiration, and mechanically strengthens tissues. Silicon also promotes lignin biosynthesis, increases cell wall elasticity during elongation, and eventually strengthens wall tissues (Carrow, Waddington, et al., 2001). The frequent mowing of turfgrasses likely reduces the effects that Si has on cereals in accumulating and reducing lodging.

17.4 ADDITIONAL AMENDMENTS FOR SALINITY MANAGEMENT

17.4.1 Wetting Agents

Wetting agents can assist in salinity management on sandy soils (Nektarios et al., 2002; Moore et al., 2010; Soldat et al., 2010) (a) by alleviation of localized dry spots that are hydrophobic in nature and that increase uniformity of soil water content and associated salts, and (b) by reducing "fingered flow" that allows a more uniform wetting front that aids in salinity leaching. On all turfgrass soils, there may be hydrophobic microsites due to water-repellant organic matter associated with the thatch or mat, and wetting agents can aid in faster and more uniform rehydration of these areas, which fosters better infiltration and percolation. Wetting agents can be used on hydrophobic sandy soils to assist in moving salts between cultivation aeration holes during pulse-based leaching events.

Wetting agents are surfactants (*surf*ace *act*ive *agents*) that reduce tension of water up to 50%–60% and enhance the wettability of soil particles and organic matter (Karnok et al., 2004). The nonionic types of wetting agents are composed of various esters, ethers, and alcohols; have persistence in the soil; and are safest to use in golf turfgrass management. For salinity management, penetrants are preferred rather than retention aids.

Organic matter can become hydrophobic, causing localized dry spots due to the water-repelling nature. The organic matter may be in the form of thatch, mat, or organic coatings on sand grains, which are thought to arise from fungi mycelia. The typical wetting-agent molecule has a polar head that is hydrophilic, attracts water molecules, and forms bonds between water and the surface of soil particles, enhancing the soil-to-water interface (Karnok et al., 2004). The nonpolar tail of the wetting-agent molecule is hydrophobic and is generally water repelling, forming bonds with organic material that causes localized dry spots, promotes wetting of hydrophobic organic matter, and may aid in microbial degradation of organic matter.

When wetting agents reduce the leaf surface tension, moisture (guttation fluid or dew) rolls off leaf tips rather than forming a droplet. Application of the wetting agent therefore improves mowing and playing conditions early in the day, when high-humidity conditions may be prevalent. Reduced droplet formation and surface retention result in less frost on turfgrass shoots, providing the opportunity for golf play and maintenance practices to begin earlier in the day during low-temperature climatic conditions.

17.4.2 Cytokinins

The cytokinin class of plant hormones has key roles in regulating various physiological and developmental processes (Mok and Mok, 1994; Carrow et al., 2001). These small adenine-derived molecules are essential for plant growth and cell division (Skoog and Miller, 1957). Cytokinins regulate root growth (Werner et al., 2001, 2003) as well as root growth rate (Dello Ioio et al., 2007), and are essential for meristem maintenance and establishment (Doerner, 2007) as well as for controlling cell differentiation and root meristem size (Dello Ioio et al., 2008).

Cytokinins regulate responses to light conditions in shoots, to availability of nutrients and water in the root, and to abiotic and biotic stresses (Ferreira and Kieber, 2005; Werner and Schmulling, 2009). Cytokinins preserve chloroplast integrity by preventing degradation when plants are exposed to reduced light quality conditions by protecting cell membranes and photosynthetic machinery from oxidative damage (reducing reactive oxygen species [ROS]) and by enhancing antioxidant enzymes such as catalase and ascorbate peroxidase (Zaveleta-Mancera et al., 2007; Causin et al., 2009). Cytokinin-mediated intercellular signaling is related to control of development, protein synthesis, and acquisition of macronutrients (Sakakibara et al., 2006; To and Kieber, 2008), including nitrogen remobilization (Criado et al., 2009).

A strong tendency is for cytokinin levels to decrease with adverse environmental stresses, especially drought, high-temperature, and salinity stresses (Hare et al., 1997). One reason for this response is that roots are the primary tissue for cytokinin biosynthesis, and each of these abiotic stresses reduces root viability and quantity. When turfgrasses are irrigated with saline irrigation water, salinity is a continuous background stress on the roots, and, therefore, it is not unexpected that exogenous application of cytokinins may result in positive grass responses.

Nabati et al. (1994) reported alleviation of salinity stress on Kentucky bluegrass by cytokinin application. Recently, Ervin and Zhang (2008) reviewed hormonal application on turfgrasses and noted that cytokinin applications often provide positive responses when roots are stressed by salinity or other abiotic stresses. It has been our observation that when roots are absent or highly stressed due to any factor, including salinity, cytokinin applications often provide a stimulatory growth response of shoots and roots, especially with high cytokinin content (>30% concentration) materials. Khan et al. (2009) reviewed seaweed extracts as biostimulants on plants, including turfgrasses, and stated that cytokinins appeared to be involved in positive responses from these materials. Use of cytokinins for seed enhancement (seed priming) in salt-stressed environments has often shown positive responses, such as reported by Iqbal et al. (2006). Thus, application of seaweed products with >30% extracts (multiple natural cytokinins) is a key management strategy to sustain and redevelop root systems in all turfgrasses when exposed to abiotic stresses. In all these situations where cytokinins provide a positive plant growth or development response, this should be viewed as an additional management tool and not as a replacement for fertilization or other salinity BMPs. Observationally and in preliminary trials, the authors have observed that seaweed products that contain multiple cytokinins are more effective than individual synthetic cytokinins (such as the so-called rooting hormones like kinetin or benzyladenine) when managing high-salinity conditions in multiple turfgrass species and cultivars.

17.4.3 Zeolite

Natural zeolites (clinoptotilite) are hydrated alumino-silicates that encompass symmetrically stacked alumina and silica tetrahedra (1:5 ratio; silicon atom in the middle, and oxygen atoms at the corners), which results in a stable honeycomb structure with a negative charge. Zeolites are especially important to consider on high-sand and low-CEC soils that receive saline irrigation water primarily due to their positive effects on nutrient retention and balances. Zeolites can vary in properties, but from the authors' experience, the internal porosity (internal pore size distribution) of zeolites does not appear to retain soluble salts to nearly the extent of calcined clays (porous ceramics) or diatomaceous earth materials (see Chapter 8, Section 8.1.2, "Inorganic Amendments").

Zeolite has been found to effectively ameliorate salinity stress and improve nutrient balance in sandy saline soils (Al-Busaidi et al., 2008; Yamada et al., 2002). As an inorganic sand substitute and soil amendment, zeolite provides an increased capability for holding nutrients in soil profiles by permanently raising the CEC. The higher quality zeolites (CEC > 100 cmol/kg) have a preference for potassium and a nonpreference for sodium; but micronutrient retention is also improved (Nus and Brauen, 1991; Andrews et al., 1999; Robinson and Neylan, 2001; Ok et al., 2003). The negative charge within the pores is neutralized by cations such as potassium (K^+) and calcium (Ca^{2+}). Due to the pore size diameters, K^+ and NH_4^+ ions can fit into these spaces and are protected from leaching. This is especially important on low-CEC sands that receive saline and high-Na irrigation water and are regularly leached.

For establishment, zeolites can be loaded with specific nutrients to assist grow-in beyond just that attributed to greater CEC (Andrews et al., 1999; Al-Busaidi et al., 2008). Important nutrients for nutrient loading at establishment in salt-affected soils would be K, Ca, NH_4^+, and micronutrients. Al-Busaidi et al. (2008) also noted that greater water-holding capacity was attributed to the zeolite when applied at 5% volume, which can reduce salinity stress.

Other researchers have demonstrated positive results from inorganic zeolite amendment applications to sand-based turfgrass ecosystems in addition to those already mentioned (Ferguson et al., 1986; Nus and Brauen, 1991; Petrovic, 1993; Ok et al., 2003; Bigelow et al., 2004; Murphy et al., 2005). Zeolites can reduce nitrate leaching and improve nitrogen use efficiency in sand-based greens (Ferguson and Pepper, 1987; Huang and Petrovic, 1994). This inorganic amendment also has utility in soil remediation (Ming and Allen, 2001), including native soils (Wehtje et al., 2003). Water use can be decreased, clipping yields increased, and water use efficiency (WUE) improved with the addition of clinoptilolite zeolite to sand (Huang and Petrovic, 1991). The addition of zeolite to saline-sodic soils (kaolinitic, smectic, and allophonic) increased saturated hydraulic conductivity due to a decrease in exchangeable sodium percentage, improved wet aggregate stability, reduced soil aggregate dispersion, and controlled soil erosion (Moritani et al., 2010).

The zeolite rate recommendation for greens mixes is often 5%–10% by volume and is based on results from a complete physical analysis to determine if the sand-based mix data meet USGA specifications or other designated requirements for the site in concert with irrigation water quality. This rate is usually based on enhancing water-holding capacity and may or may not provide a significant increase in moisture retention depending on the source. Commonsense additions of zeolite are recommended when amending sands that normally have low CEC (usually < 1.0 cmol/kg), especially when salt deposition is an issue (Qian et al., 2001) and sand quality sizes (Yang et al., 1998; Huang and Petrovic, 1995) are considered. For CEC enhancement, zeolite with a CEC of 150 cmol/kg added at 225 lb. per 1000 sq. ft. (10,985 kg/ha) and mixed into the surface 4 inches (10 cm) of a sand would increase CEC by approximately 1 cmol/kg averaged over the 4-inch zone (Carrow, Waddington, et al., 2001). Thus, much lower rates can be used to increase total CEC (existing CEC plus that added by zeolite) to 2.5 to 3.0 cmol/kg within the surface 4 inches (10 cm), which is sufficient to stabilize nutrients in a sand profile. Applied at 225 to 500 lb. zeolite per 1000 sq. ft., this quantity zeolite can be tilled into the upper surface and will have little if any effects on physical conditions as long as most sand particles are in the 0.2- to 1.0-mm diameter range. A 4-inch (10-cm) zone of sand weighs about 31,000 lb. per 1000 sq. ft., so 500 lb. of zeolite per 1000 sq. ft. is only 1.6% by weight in this zone. On established turfgrass, lower rates can be metered out at 10 to 100 lb./1000 sq. ft. per application in conjunction with cultivation (solid- or hollow-tine, spiking, or slicing) events to integrate the zeolite into the top 2–3 inches (50–75 mm) of the upper soil profile in order to hold critical nutrients around the crown region and rhizome layers of grass plants. If sand topdressing is conducted in conjunction with the cultivation, then the zeolite can be integrated with the sand prior to topdressing at higher rates, depending on the quantity of topdressing sand to be applied—as a rule of thumb, use about a 30:70 (zeolite:sand, by vol.) ratio to allow the sand to be the predominate matrix. After application, the mixture can be drag matted into the cultivation holes. This can be repeated until the total CEC is sufficient.

17.4.4 Lassenite

Lassenite is a calcined (heated to 1500–1800°F) diatomaceous earth product composed of silica dioxide and is primarily added to soil profiles to increase water retention (hydraulic conductivity of 10.9 inches or 27.7 cm per hour) via microporosity. Lassenite (occasionally referred to as *pozzolan*) has 50% capillary porosity (micropore space), 18% noncapillary porosity (air-filled pore space), and 25.9 cmol/kg CEC (Stewart, 2009, 2010; http://www.westernpozzolan.com).

The critical question when managing soluble salts and saline irrigation water is the capability of inorganic amendments, which do not biodegrade, either to sequester the soluble salt ions within internal micropores inside the particles while retaining water or to readily release this salt-ladened water to the soil ecosystem. Compared with other water retention porous ceramic inorganic sand substitutes, Lassenite appears to release the solubilized salts in the water more readily for plant use, which facilitates leaching with careful amendment additions to the soil profile (Stewart, 2009, 2010; Weeaks, 2009; Weeaks et al., 2007, 2008). A majority of the moisture held by Lassenite is plant available and released prior to 2 bars tension, and physical analysis reveals that the hydraulic conductivity is 10.9 inches per hour (http://www.westernpozzolan.com).

Discussions of the importance of pore size distribution within sand substitutes and precautions on sites receiving saline irrigation water are presented in (a) Chapter 8, Section 8.1.2, "Inorganic Amendments," related to identifying salt retention sites during site assessment; and (b) Chapter 11, Section 11.2.3, "Soil (Edaphic) and Hydrological Factors," related to salinity-leaching programs and effectiveness. Essentially, sand substitutes used on salt-affected sites that exhibit most of their pore sizes in the smaller portion of the micropore range should be used with caution. Lassenite, as noted, has a more favorable internal pore size distribution to provide plant-available water and for effective pore leaching of soluble salts.

17.4.5 Organic Amendments

In the current section, the focus is on organic amendments in routine culture of salt-affected soils. The major routine management issues related to soil organic matter in salt-affected turfgrass sites are due to how the organic matter affects salt retention and leaching. Use of organic amendments in salt-affected turfgrass sites has been discussed (a) with respect to site assessment issues in Chapter 8, Section 8.1.1, "Organic Amendments"; and (b) for remediation of sodic and saline-sodic soils in Chapter 12, Section 12.4, "Organic Amendments for Sodic Soil Reclamation."

Soil organic matter (the organic fraction of the soil exclusive of living plant residues) is regarded as a sink for carbon dioxide, a filter for water, and a buffer to contaminants such as soluble salts (Li, 2004). Just as in normal soils, salt-affected soils benefit from soil organic matter stabilization of soil structure, enhancing physical properties by balancing water and air porosity, increasing CEC, and sequestering toxic ions such as sodium (Li, 2004). Sodic and saline-sodic soils especially benefit from these positive contributions of soil organic matter. Some problems can arise on salt-affected sites; for example, excess dispersion of organic colloidal substances >3%–4% (by weight) in greens sand-based soil profiles and >5% on all other soil profile areas will sequester and possibly layer in salt accumulation zones that can affect turfgrass performance—organic colloids can be dispersed just as clay colloidal particles can. Thatch management is essential to minimizing localization of excess salts, especially sodium, since organic matter has the propensity to attract that monovalent cation in the upper soil profile. Frequent, light irrigation with saline water on a turfgrass with excess thatch or mat can result in salts being retained in the higher CEC sites and with the high water content. Upon evaporation drying, the salts concentrate. Also, by retaining more moisture after rain or irrigation, excess thatch or mat can reduce the fraction of water moving past this zone, thereby limiting leaching effectiveness.

Many saline and sodic soils are alkaline in pH, but not in all cases. The alkaline pH can influence soil organic matter dynamics, but this is regardless of whether salt issues are present or not. For example, microbial decomposition and resynthesis processing of soil organic matter has been studied between alkaline and acidic turfgrass ecosystems using ^{13}C and ^{15}N (Yao and Shi, 2010). Soil carbon storage was ~12% greater in alkaline soils than acidic soils. Soil organic C and N accumulations were about threefold higher in alkaline soils compared to acidic soils. Soil organic matter was found to be more stable (mineralized more slowly) on the alkaline site than on the acidic site. Wong et al. (2010) has reviewed the soil carbon dynamics in saline and sodic soils, including soil microbial biomass and microbial activity.

Soil organic matter is a very heterogeneous substance with complex chemistry. Various organic-based chemicals (fertilizers, composts, humic and nonhumic soil or foliar applied materials, microbial inoculants, etc.) are promoted in the turfgrass industry for improvement of various soil physical, chemical, and/or biological conditions—for both nonsaline sites as well as salt-affected areas. However, the use of these organic-based chemicals in routine maintenance is similar on both types of sites in most cases.

Carrow, Waddington, et al. (2001) summarized the soil organic matter fractions, characteristics, applications, and typical levels in turfgrass situations. Soil organic matter consists of the following: (a) *nonhumic substances*, usually 20% to 30% of the total, are less complex and more easily decomposed and include fresh organic matter containing amino acids, carbohydrates, and fats; and (b) *humic substances* or *humates* (usually 60% to 80% of the total) consist of *humin*, *humic acid*, and *fulvic acid fractions* that are higher molecular weight structures usually dark brown in color and more resistant to decomposition than nonhumic materials. Of the humic fractions, fulvic acids are the least complex and most soluble at all pHs, while humic acids are more complex and soluble only at very acid pH (pH < 2). Humic molecules are bound together in supramolecular conformations by weak hydrophobic bonds at neutral and alkaline pH and by hydrogen bonds at acid pH. Humic substances have considerable CEC sites and chelate multivalent cations such as Ca^{2+}, Mg^{2+}, Fe^{2+}, Fe^{3+}, and Al^{3+} plus trace elements. In a soil with 2% total organic matter by weight based on a 4-inch soil depth, this would represent approximately 620 lb. of organic matter per 1000 sq. ft. On that basis, the quantity of each of the various nonhumic and humic substances already present in a normal turfgrass situation can be considerable, especially when contrasted to organic fertilizers typically applied at 10 to 16 lb. product per 1000 sq. ft. or other organic chemicals applied at rates of <1.0 lb. product per 1000 sq. ft. The point is that in a "healthy turfgrass system" when normal decomposition is occurring, the diverse organic fractions are present and active. Thus, on salt-affected sites as well as with nonsaline soils, the first management option to insure adequate nonhumic and humic substances (humin, humic acid, and fulvic acid fractions) is to maintain a healthy turfgrass. Then, site-specific use of an organic-based product can be used as needed when it benefits overall ecosystem management.

Continuing the discussion on various soil organic matter fractions, humin, as the most complex and the most resistant to decomposition, is composed of long, cross-linked carbon chains that are relatively inert, and reactivity in the soil is based on surface area exposure. The humic acid fraction of humin has shorter carbon chains and has high ion exchange capacity, but is not normally soluble in water unless saturated with monovalent ions such as hydrogen, potassium, and sodium. Humic acids provide increased adhesive and cohesive water retention characteristics in the soil profile (DePew, 1998). In fragile unstable soils, humic acids form a hydrophobic coating around aggregates, thereby reducing slaking when in contact with water (Mbagwu and Piccolo, 1989). This would be beneficial in sodic or saline-sodic soils as well as for soil structure stabilization in all soils. However, humic acid coating on a sand particle is different than on a soil structural aggregate. In a greenhouse sand-based green study with creeping bentgrass (*Agrostis stolonifera*), pure humic acid caused a decrease in water-holding capacity in the soil and contributed to lower moisture retention than pure water, thereby reducing the amount of water available to turfgrass roots (Van Dyke and Johnson, 2007). This study also revealed that high sodium levels were concentrated in bentgrass plant tissue treated with pure humic acid.

Fulvic acids are the acid radicals found in humic matter and, as noted, are soluble regardless of pH or ionic strength. Fulvic acid, acting as a natural organic electrolyte, actively chelates metallic minerals and functions as effective micronutrient carriers, especially on low soil organic content situations, such as during establishment. Since soluble fulvic acids may be extracted from various organic materials, other soluble organics may be present at low levels, such as vitamins, coenzymes, auxins, hormones, and natural antibiotics. As with the nutrient responses, when applied to a low organic content soil system, hormonal responses from fulvic acids (which have some hormone-like activity) or other extracted hormones may give a growth response. Normally, after soil organic matter accumulates to normal levels, such responses are not observed—most likely because high levels of these substances already exist in the ecosystem compared to typical application rates.

17.4.6 Microbial Amendments and Bionutritional Products

17.4.6.1 Soil Microbial Activity

Microbial amendments have been present in the turfgrass and agricultural industries for many years. Research has demonstrated that in healthy sand-based rootzones, microbial colonization is rapid during establishment, soil microbial populations are at levels found in many native soils, and microbial populations are stable (Bigelow et al., 2000). Also, soil organic matter under mature turfgrass ecosystems gradually increases over time, but the rate is affected by site conditions such as pH and management regimes (Qian et al., 2003; Yao and Shi, 2010).

These studies were conducted on nonsaline soils, so the question may arise about salinity and soil microbial interactions. Kaur et al. (2008) reviewed the research status of salinity on soil microbial aspects and reported on a study of microbial biomass and C as influenced by remediation options—gypsum and organic amendments—under agricultural situations. They found the following:

- On sites with existing saline and sodic conditions sufficient to cause substantial stress on the plant, microbial biomass and soil organic matter decreased.
- When appropriate management practices, including gypsum with or without organic amendments (farm manure or green manure), were adopted, soil microbial biomass and organic matter increased.

To apply this to turfgrass ecosystems, if stresses, including salt stresses, are sufficient to reduce turfgrass growth and development (i.e., biomass production of shoots and roots), then soil microbial biomass and soil organic matter may decline, regardless of whether the site is nonsaline or salt affected. Thus, to "enhance" soil microbial biomass, with all its positive benefits, and soil organic matter, with all its benefits, the maintenance of the turfgrass ecosystem by appropriate salinity BMPs is the most important approach. This has been our observation on turfgrass sites, even with those using highly saline irrigation water sufficient to require halophytic grasses. Aerobic soil microorganisms require certain conditions in the microsite in order to exist within a turfgrass soil. Maintaining these conditions will insure a well-balanced and vigorous soil microbial population, where factors that influence microbial dynamics are:

- For nutrients, energy supplies, and sources, irrigated turfgrass sites normally supply ample fresh organic matter residues from plants and soil microbial turnover as well as adequate nutrient supplies. Bacteria, heterotrophic: organic matter is the energy and C source. Bacteria, autotrophic: the energy source comes from oxidation of inorganic substances (NH_4, S, and Fe) and most C from CO_2. Actinomyces obtain energy and C from organic matter. Soil microorganisms required the same nutrients as the turfgrass. The C:N ratio in most turf soils is <15:1, and C:N > 25:1 can hinder decomposition rates.
- Regarding temperature, each microbial organism (MO) type has a minimum, optimum, and maximum temperature range for activity. Total soil microbial activity is greatest between 75°F and 95°F. Climates with day and night temperatures >75°F often exhibit little organic matter accumulation—unless some other factor limits activity, such as a dry surface, low oxygen, or low pH. Bacteria and many other MOs exhibit greatly reduced activity below 55°F. Organic-matter-accumulating climates have long periods between 32°F and 55°F. As temperatures increase every 18°F between 32°F and 95°F, decomposition rate increases 1.5- to threefold.
- Moist conditions at field capacity (FC) or just below FC favor most organisms. Dry thatch, mat, or soil surfaces caused by surface drying can limit MO activity—such as in hot, dry periods with high evapotranspiration. With moisture consistently above FC, algae and anaerobic bacteria are favored, while fungi and actinomycete populations greatly decline. Waterlogged turfgrass soils or excessively wet soils often exhibit organic matter accumulation (thatch buildup) due to limited microbial activity of fungi, actinomycetes, and aerobic bacteria.
- Regarding aerobic (ample oxygen) versus anaerobic (oxygen is limited) conditions, aerobic conditions result in the highest organic matter decomposition rates, and total soil microbial populations start to decline as air-fill porosity decreases below 20% (USGA Greens, 15%–30% aeration porosity). At air-filled porosity (macroporosity) <10%, there is a shift from aerobic to anaerobic microbial populations.
- With media pH (soil, thatch, and mat), actinomyces are functional at pH 7.0 to 7.5 (preferred range); fungi operate best at pH 4.0 to 8.0; and bacteria, between pH 6.0 and 8.0. Each specific organism has a preferred range. At thatch, mat, or soil pH < 5.5, the bacteria and actinomyces populations greatly decline. Both organism populations are important for decomposition of more resistant forms of organic matter.
- As microclimatic conditions favor development of a specific MO population, some other organism populations may decline due to competition for nutrients and C source. Fungicides specific to a particular target organism(s) can reduce microbe populations even when microclimatic conditions favor the MO. Altering microclimatic conditions (pH, oxygen, and moisture) will influence the type of microorganisms and total microbial activity. Adding a specific MO population does not mean that they will remain present and competitive—this depends on consistent microclimate conditions suitable to the MO and interaction with multiple abiotic stresses that may be present.

17.4.6.2 Photosynthetic Microorganism Amendments

A major concern on salt-affected sites (and nonsaline sites as well) is the intended or unintended proliferation or application of photosynthetic microorganisms to the surface of an established perennial turfgrass. Periodically, photosynthetic algae or bacteria have been promoted for use on turfgrass areas for various reasons such as promoting soil structure, adding sugars, or adding organic matter via microbial growth (Petrovic, 1975; Baldwin and Whitton, 1992; Carrow, 1993), and for suppression of nematodes and pathogenic organisms. Photosynthetic microorganisms normally require adequate light and moisture for growth, but some organisms may have the capability of being active with or without light. Application of a photosynthetic microorganism product onto the surface of a turfgrass site where there may be ample light for the organism to proliferate should cause serious reconsideration of that decision by turfgrass managers with respect to potential adverse responses, regardless of whether on a saline or nonsaline site; but on salt-affected sites, there are real long-term ecosystem sustainability concerns. If such a photosynthetic microorganism is able to obtain its growth needs, it may actually proliferate and produce microbial biomass in the surface zone. If that photosynthetic organism is adapted to saline conditions, the control of this organism becomes even more challenging when using saline irrigation water. This can have adverse effects on water infiltration, surface aeration, salt accumulation, and salinity leaching. Since it has been relatively common for some companies to promote products without listing the active ingredient (in this case, specific microorganisms), the potential for adverse and unexpected ecosystem response is very real in the perennial turfgrass market.

These same products may provide benefits to agricultural soils, especially ones that are low in organic matter. But, application on a bare soil with the ability to till and integrate the photosynthetic organism product into the soil profile is different than on a low height of cut turfgrass site. Some examples will illustrate.

Years ago, an aggressive algae product (*Chlamydomonas mexicana*) was promoted with the claims that the organism was photosynthetic, produced considerable sugars (carbohydrates), and greatly increased in microbial biomass, and this green algae product was marketed to improve soil structure on agricultural sand turfgrass soils (Petrovic, 1975; Carrow, 1993; Nus, 1994). Benefits were demonstrated on agricultural soils, particularly in the southwestern United States under irrigation. The algae in the product are found in desert algae blooms—algae that can grow very rapidly when exposed to moisture and sunlight during desert or arid-region rains. What we found was that when the product was applied to turfgrass soils where the surface was moist from irrigation and the turfgrass was cut at normal fairway heights, this particular vigorous algae did exactly as was claimed in terms of growth—they grew rapidly and produced a slippery algae layer on the surface, similar to what Baldwin and Whitton (1992) noted. If the turfgrass canopy density was thin, algae growth was even more pronounced and persistent, with an algae scum forming since moisture and sunlight conditions were consistent.

Most "normal" photosynthetic algae, which are less vigorous than the algae Carrow (1993) and Petrovic (1975) worked on, will do exactly the same thing when exposed to ample sunlight, especially on thinning turfgrass canopies and under close mowing height conditions (such as greens where sunlight can penetrate to the soil surface)—these organisms rapidly produce microbial biomass consisting of manufactured carbohydrates plus the algae cells as they rapidly increase in number. The net result for photosynthetic algae is algae scum on the surface, and the subsurface gel-like materials produced from the carbohydrates and dying cells can also contribute to black layer development. The algae layer becomes anaerobic very rapidly and then fosters an increase in anaerobic-facilitated root-borne organisms (such as pathogens that cause decline or take-all patch) while suppressing beneficial and competitive aerobic microorganisms.

Another example of a photosynthetic microorganism is the blue-green algae (really a *Cyanobacteria*) found by Dr. Clint Hodges (1987, 1989a, 1989b, 1992a, 1992b) to contribute to black layer in sandy soils due to the high biomass production in initially small "microsites," which then grew into a larger anaerobic surface layer or upper soil profile zone. Hodges found that this *Cyanobacteria* also was very vigorous under sunlight and high-moisture conditions in turfgrass situations. Normally, good turfgrass canopy density is the major factor suppressing photosynthetic microorganism development due to shading and competition for nutrients. But if ample sunlight is present and penetrates into the soil surface, photosynthetic organisms will do their job and respond with vigorous growth and persistence. Baldwin and Whitton (1992) and Gelernter and Stowell (2000) noted the adverse effects of *Cyanobacteria* and algae to turfgrasses.

The authors do not know of any product containing *Cyanobacteria* for turfgrass use, such as how the algae product previously noted was sold, but this has been done in agriculture—again, illustrating that what can be used in one ecosystem may have very different effects in another. Zahran (1997) reported on the diversity and difference of bacterial flora in saline environments and suggested isolation of bacteria from saline environments to use for salt remediation. Recently, a review of *Cyanobacteria* for reclamation of salt-affected agricultural soils was written by Singh and Dhar (2010). In agricultural soils, with sufficient surface moisture and light, the *Cyanobacteria* can produce significant microbial biomass that provides the benefits of an "organic amendment"; but the same biomass production in the surface of a mature turfgrass is what turfgrass managers call *algae scum*. Some mention of the ability of *Cyanobacteria* to sequester or accumulate salts was noted, but this is short-term sequestration that lasts until the algae cells die and then release the salts to accumulate back into the soil profile. Effective salinity remediation requires salts to be leached; see also Chapter 12, "Remediation Approaches and Amendments."

Recently, products have been promoted for turfgrass situations containing a species of purple nonsulfur photosynthetic bacteria where it was possible that the products contained *Rhodospirillium* spp.—since this was an example of limited label information regarding the specific organisms. This class of bacteria, such as the *Rhodospirillium* spps., is found in its most active form in the surface ocean marine mudflats—a saline and anaerobic environment where there is ample sunlight at the surface for aggressive photosynthetic activity. We will use the *Rhodospirillim* spp. as an example of where this particular organism can subsequently grow: in anaerobic (low- or reduced-oxygen) conditions with light (its most active form), anaerobic conditions without light, and aerobic conditions, all of which make it very broadly adaptable and persistent. For maximum growth and persistence, the most important requirements are as follows:

- Long-duration exposure to sunlight and even the ultraviolet spectrum coming through cloud cover, especially in thin grass canopies
- Anaerobic condition—this only has to be microsite anaerobic conditions such as in a surface microbial mucilage layer in a small zone of limited oxygen (such as between aeration holes).
- High moisture
- Food source
- Sulfur in reduced forms as the electron source

When these conditions are present, this organism seems to do as claimed and grow rapidly (the claim for this organism is that it has the capability for reproduction every 5 minutes). In fact, in shallow saline ocean areas, this organism and similar ones contribute to microbial "mats" high in microbial biomucilage. The best way to think of this organism in the anaerobic form is to compare it to algae that cause algae scum or to blue-green algae that can contribute to black layer. To our knowledge, the turfgrass industry has not really had a bacterial product of this nature, where microenvironmental conditions (with enhancement from high-salinity conditions) that could foster maximum growth of the organism are coupled with an organism that really appears to be what is claimed in terms of vigor and with virtually no effective chemical control if it actually gets out of control (we see certain organisms increase out of control all the time—we call them *pathogenic diseases*).

Whether applying any product containing aggressive photosynthetic algae, blue-green algae, or photosynthetic bacteria, questions for a turfgrass manager are as follows:

- "What if site-specific or even microsite conditions occur for maximum growth of these organisms?" Such conditions include a sufficiently thin turfgrass canopy that allows more sunlight penetration for these photosynthetic-oriented organisms. When does algae scum occur? It develops after thinning of the turfgrass canopy, colonizes the soil surface or shallow zone in the upper soil profile, and, with consistent sunlight and moisture and with minimal shading, can become a major dominating and detrimental problem for recovery of the turfgrass in those thin canopy areas. One difference with the purple nonsulfur photosynthetic bacteria is that this species does not seem to cause an "above the soil surface" layer similar to algae scum, but instead, the colonized layer seems to be in the upper soil profile layer(s) of the soil at the crown or rhizome level, where it creates the anaerobic zone that can plug the soil pores and help promote decline, take-all, or pythium root dysfunction root-borne disease problems. One critical question is "Does this organism dominate the microbial ecosystem, which means reduced competition from more favorable microorganisms?" The answer in this case was "yes."

- "If ideal growth conditions for the organism occur and the photosynthetic organism really does what is claimed (growing rapidly, producing sugars, and producing microbial biomass) and if a problem occurs, what chemical controls exist?" Without chemical control for suppression, control measures include sand topdressing (for shading), mechanical operations (cultivation and vertical mowing) to reduce high-moisture conditions, and controlling secondary diseases such as take-all or decline that are fostered by the anaerobic upper soil profile zonal environment created by these organisms. But will mechanical operations help control the organisms or cause more problems during high-use periods? Any organism that rapidly adapts between aerobic and anaerobic conditions in the upper soil profile will usually have dominant control over that ecosystem and negatively impact turfgrass performance while suppressing competitive favorable microorganisms.
- "Since this organism is located at the soil surface and responds to sunlight, if you perform mechanical operations that cause any further openings, will the organism just respond to the higher light exposure and become even more of a problem?" Another way to put this is as follows: without a chemical means to suppress a very rapidly adaptive (to both aerobic and anaerobic conditions) organism that truly is aggressive and creates its own dominating preferential anaerobic microsites and zones near the soil surface (1/8 inch plus), the traditional mechanical operations (except heavy sand topdressing, which covers the zone, and venting with small-diameter tines or the Hydroject or dryject) will actually increase the problem and not lead to grass density recovery. Most grass density recovery success has been from controlling the secondary decline or take-all and other diseases to prevent further turfgrass thinning; but once any thinning does occur, the turfgrass canopy density reverts back to ground zero and a subsequent long turfgrass recovery process.
- "Does research on appropriate turfgrass ecosystems support the use of the specific microbial amendment for the intended purpose?" Turfgrass managers do not want to be the "trial" site, especially on salt-affected sites. This additional biotic stress in concert with multiple salinity challenges is very difficult to manage once it dominates the microbial ecosystem. When the organism tolerates salinity conditions, the problem is even more of a challenge to eliminate.

The bottom line is that we do not recommend *any photosynthetic microorganism product* to be applied to close-cut turfgrass (especially greens) or any turfgrass ecosystem where canopy thinning may occur, because these organisms may well do exactly what is claimed, which is to grow very rapidly at the surface where the sunlight and moisture are predominant. If the product does not have documented labeled chemical controls, our recommendation is a double "Do not apply!" Photosynthetic microorganisms are not good products in our estimate since they negatively affect the surface soil physical conditions involving oxygen and moisture flux through the upper profile of close-cut turfgrasses, where good oxygen and moisture flux conditions are critical. On salt-affected sites, any adverse turfgrass responses would be expected to be greater in magnitude, even to the point of contributing to the death of the turfgrass.

17.4.6.3 Bionutritionals

Recently the term *bionutritionals* has been used by some in the turfgrass industry, where this term is defined as follows: "bionutritional fertilizers is the name given to the wide range of living organisms, including microbes, bacteria, mycorrhizae, seaplant extracts and hormones, that can be added to a fertilizer prill or delivered as stand-alone products to nourish plants" (Aylward, 2010, 36). This classification of bionutritionals contains many materials that have been lumped under the broad *biostimulant* category over the past 25 years (Carrow, Waddington, et al., 2001). Experience with biostimulants illustrates that one cannot make broad acceptance or rejection decisions regarding these amendments based on such a broad range of diverse products. For example, in previous paragraphs of the current section (Section 17.4), some products such as cytokinins provide a biostimulant response similar to a fertilizer response, but due to biostimulation of plant growth by hormonal activity rather than as a nutrient. However, other materials such as microbial additives may be added at minuscule rates compared to what the healthy plant is already producing; or, in the worse case, these microbially amended products may actually cause unintended detrimental effects. Thus, a range of responses from positive, to no response, to detrimental can occur when speaking of broad categories, such as biostimulants or bionutritionals. Critical points related to salt-affected sites for these materials as a group are as follows:

- Some of the products in the biostimulant and bionutritional groups (which essentially overlap each other and appear to be a "renaming" of the older biostimulant group) can be beneficial in specific situations, such as cytokinins. A specific observed field problem should be identified (i.e., lack of root production of natural plant-developed cytokinins due to salt stress on the roots), then a product is chosen that is targeted to be able to address this issue.
- Any bionutritional products without a very clear, detailed label of contents should not be used due to the potential for adverse effects and the inability of a turfgrass manager to determine what the expected response should be without such product information.
- In the quote above, the term *bionutritional fertilizers* is used. The reality of the situation is that if the product is to be claimed as a nutrient or fertilizer, it must contain a legal fertilizer label, which is not provided by most of these products. The actual label information is normally devoid of any nutrient claims to avoid these fertilizer laws. But, this type of terminology confuses the differences between actual fertilizers and nonfertilizer materials where "some" can give a positive growth response, usually via hormonal action or unreported nutrients (such as low N content) in the material. Whether this terminology has evolved to confuse or not, it is definitely confusing.
- It is important to remember that on salt-affected sites, one of the most critical "salinity stresses" is nutritional balances, levels, and imbalances. Thus, the fertilization program must be very specific and carefully monitored.
- Bottom line: turfgrass managers are encouraged to concentrate on developing a sound fertilization program using traditional fertilizers, lime, and gypsum. Then, if a specific issue arises that a particular "bionutritional"-labeled product can resolve, it is appropriate to use it—if the label actually describes the details of what is in the material. However, applying a broad range of these products without a science-based need is not economically or agronomically sound.
- Many of these products emerge from other agricultural crop or agronomic uses, but with no unbiased, university-based external research scrutiny; additionally, very seldom are actual research data provided involving turfgrass trials and long-term use. In salt-challenged turfgrass situations, the application of "unknown" products such as these is generally not recommended.

17.5 GREENS MANAGEMENT CONSIDERATIONS

Greens management does differ under saline irrigation water. With low height of cuts on greens surfaces, salinity can quickly become a dominant environmental stress, especially with reduced leaf area plus plant growth rate reductions coupled with decreased photosynthetic activity; turfgrasses can exhibit salt stress symptoms quickly depending on actual inherent genetic salinity tolerance and increasing topical and rhizosphere exposure to salts. In addition, sustaining root volume on low mowing height turfgrass cultivars (even in nonsaline soils) is a constant challenge, since most grasses are genetically programmed to maintain the shoots first (via carbohydrate allocations), and especially when accumulated salts in the upper soil profile often desiccate root hairs, which affects grass water and nutrient uptake.

The focus of this section is to highlight management issues that frequently arise on saline-irrigated golf greens or similar high-sand situations receiving saline irrigation. Some management considerations are discussed in other chapters, such as the following:

- Chapter 16, "Nutritional Practices on Saline and Sodic Sites"
- Chapter 14, "Cultivation, Topdressing, and Soil Modification"
- Chapter 11, "Irrigation Scheduling and Salinity Leaching"

17.5.1 Black Layer

Black layer can occur on high sand content rootzone mixes as well as fine-textured soils (Carrow, Waddington, et al., 2001). A continuous or discontinuous subsurface layer can occur and often exhibits a rotten egg (hydrogen sulfide) or similar odor. The black layer condition can be initiated when water percolation is impeded, creating a water-logged anaerobic zone in the soil profile. Causes of black layer include migration of colloidal particles (clay or organic material), salt deposition, naturally occurring sulfur-containing organic matter layers or irrigation water sources that are high in sulfur as a secondary contributor, and biological contributors (*Cyanobacteria* that produce gel-like substances and other bacteria that produce biofilms that plug pore spaces, creating anaerobic subsurface conditions) (Carrow, Waddington, et al., 2001). Black layer is more prevalent in salt-affected sites because many of the conditions noted above are more evident on these sites.

With anaerobic conditions in the black layer zone, H_2S (hydrogen sulfide), FeS (iron sulfide), FeS_2 (iron pyrite), and MnS or MnS_2 (manganese sulfides) compounds occur when SO_4^{-2} is reduced to S^{-2} by *Desulfovibrio desulfuricans*, creating sludge-like materials that further reduce water percolation and gas exchange (Carrow, Waddington, et al., 2001). Additional publications concerning black layer problems in turfgrass ecosystems are available (Berndt and Vargas, 1992, 1996, 2006, 2007, 2008, 2010; Berndt et al., 1987; Cullimore et al., 1990; Hodges and Campbell, 1997; Lindenbach and Cullimore, 1989; Borst et al., 2007; Baldwin and Whitton, 1992; Smith, 2001; Carrow, 1993; Hodges, 1992a, 1992b). Possible management strategies for reducing the potential for black layer formation include the following:

- Avoid anaerobic conditions by maintaining adequate macropores through regularly scheduled cultivation events that promote infiltration, percolation, drainage, as well as gas movement through the soil profile.
- Avoid applying "fines" (clay, silt, very fine sand, colloidal, or small particle size organic matter) to the rootzone media via topdressing or composts.
- Improve surface and subsurface drainage to remove excess water from the site, especially on green locations that tend to retain more soil moisture in the surface or in the subsoil.
- Irrigate at frequencies and at rates that avoid excess moisture accumulation in the soil profile that reduces air porosity.
- Apply wetting agents to enhance water infiltration and percolation.
- Avoid composts and organic fertilizers to reduce readily available carbon sources.
- Reduce use of sulfur-containing amendments. However, on sodic or saline-sodic sites gypsum is necessary, and most of the S is chemically tied up in the low-solubility gypsum or calcium sulfate molecule. It is the soluble SO_4^{-2} ion in water that is subject to being reduced and not the S within the $CaSO_4$.

- SO_4 or SO_3 ions present in the irrigation water from acidification treatment of bicarbonates where high Na in the water is a problem can be "scrubbed" by light additions of lime where the S and lime react to form gypsum (see Chapter 12, "Remediation Approaches and Amendments").
- For soils with existing high sulfate concentrations, aerate and apply lime to chemically form gypsum, thereby scrubbing excess sulfur out of the soil profile and reoxygenating any previous anaerobic zone (the SO_4^{-2} form is quite stable as long as aerobic conditions are maintained).
- Develop a good leaching program since SO_4^{-2} is one of the most easily leachable ions.
- Fertilize lightly (prescription program) and frequently, including nitrates (however, application of N products will depend on sufficiency requirements of specific turfgrass cultivars).
- Avoid application of any anaerobic-dominance microorganisms such as green bacteria (*Chlorobium*) or purple bacteria (*Chromatium* or *Rhodospirillum*) that use H_2S for photosynthetic C fixation as photoautotrophs in soil profiles exposed to high-moisture and reduced-oxygen conditions.
- Control *Cyanobacteria* (blue-green algae) accumulation in lakes or ponds that are used for turfgrass irrigation and can colonize soil surfaces. Some of these organisms produce hepatotoxic microcystin toxins in lakes and ponds (Healy, 2008; Wang and Chen, 2008; Okello et al., 2010). The *Cyanobacteria* and eukaryotic algae genera include *Nostoc*, *Phormidium*, *Coccomyxa*, *Cosmarium*, *Cylindrocystis*, *Klebsormidium*, *Lyngbya*, *Mesotaenium*, and *Zygogonium*, with *Coccomyxa*, *Cylindrocystis*, and *Mesotaenium* generally being more dominant under acidic conditions (Baldwin and Whitton, 1992).

17.5.2 Grooming and Vertical Mowing Practices

Aggressive grooming or verticutting is not recommended when excess salts accumulate in the soil profile, since the grass may not recover quickly due to reduced growth rates. The strategy is to not damage rhizomes and crown regions where carbohydrates are stored for use in recovery from injury. Light "tickle" grooming to trim "fat" leaves that may contribute to reduced ball speed (and light sand topdressing) is an acceptable option.

Any additional mechanical injury to the turfgrass can rapidly deplete the carbohydrate reserves needed for injury repair, and this depletion often occurs when plant growth rates have been reduced as a result of salt exposure in the roots and shoots. Any additional abiotic (drought, heat, cold, and wear and traffic) or biotic (insect, disease, and weed competition) stress often results in loss of canopy density and, in extreme cases, death of the turfgrass. Salt remediation and regrassing are then required to reclaim the turfgrass ecosystem.

Topical application of saline irrigation water through sprinkler irrigation systems can also potentially damage turfgrass shoots, especially after mowing. Exposed damaged cells from mower blades can often turn brown if irrigated with highly saline water after mowing. In addition, soluble salts (such as sodium and chloride) can foliar feed directly into the canopy and, over time, accumulate at concentrations that can disrupt internal nutritional balances.

17.5.3 Reel Mowers and Rollers

The issue of walk mowers versus triplex mowers on greens is another consideration when salt accumulation in the greens soil profile can reach performance-limiting levels. Depending on the salt tolerance level in the specific turfgrass cultivar, the general rule is to walk-mow as frequently as possible and to only use the triplex mowers (because of mower weight and compaction issues) once or twice weekly on low-height-of-cut greens. Since salinity stress can enhance wear injury, the perimeter mowing area is especially susceptible to injury.

Some turfgrass species, such as seashore paspalum, should only be mowed with smooth rollers (not Weihle rollers, which are highly ribbed rollers) on greens mowers since this grass has shoots that are softer in nature due to relatively high water content, reduced cell constituent content, and minimal lignin and cellulose in the shoots, which mean that bruising damage potential must be considered in the cultural management program. When Weihle rollers are used on low-height-of-cut seashore paspalum, the grass bruises easily and the grass is often predisposed to diseases because of compression damage from these rollers.

Rolling greens with high salt accumulation on green ecosystems should be done with caution. If the greens grass is growing and is not subject to temperature-induced growth reduction, then rolling to supposedly enhance greens speeds is acceptable. But when both salinity and cold temperatures have reduced growth rates on the turfgrass mowed at greens height, rolling should generally be done once or maybe twice weekly to minimize compression damage to the shoots. The grass will generally not recover from this type of damage until temperatures have increased and promoted regrowth.

Additionally, wet rainy cycles during the winter time frame can make the greens surface softer and excess rolling can imbed the turfgrass shoots into the soil surface, creating the potential for those leaves exposed to a cold soil to go off-color and negatively affect greens ball roll. Commonsense rolling of greens during these periods is recommended.

Potassium is a critical ion to maintain turgor pressure on salt-stressed grasses. If tissue K levels are too low, the grasses become more prone to scalping by the mower due to having a less firm turfgrass canopy surface. This is especially likely when high salinity is already imposing osmotic stress and if the soil surface (thatch or mat) layer is moist.

17.5.4 Plant Growth Regulators (PGRs) and Salinity

Salts are excellent plant growth regulators and actually act as gibberellin inhibitors, similar to some marketed PGRs. The higher the genetic salt tolerance level in a turfgrass cultivar, the less the impact from increasing exposure to salinity-induced growth reductions; these cultivars will require prescription applications of chemical PGRs to minimize growth spurts and to tighten the surface canopy density. Lower genetic salt-tolerant turfgrass cultivars will require very careful PGR applications and normally at lower rates than cultivars with higher salinity tolerance, since the salts will be providing additional growth regulation.

An additional consideration when applying PGRs to the more salt-tolerant grass cultivars is that when canopy density is lost due to injury, disease attack, or scalping problems, no additional chemical PGRs should be applied until full canopy density has been achieved and recovery is complete. Otherwise, recovery can take two times or longer to achieve, since fill-in is controlled mainly by gibberellin activity in the shoots and surface stolons. Fertility programs must be adjusted to promote recovery of the turfgrass, and PGR applications (both frequency and concentrations applied) must be altered to facilitate grass density recovery.

17.5.5 Salt Accumulation and Salt Monitoring

While salt accumulation and salt monitoring have been discussed in greater detail in Chapter 11, "Irrigation Scheduling and Salinity Leaching," we again emphasize the importance of salt leaching on greens. Salt accumulation in the greens soil is generally not a highly visible occurrence, but a slow methodical addition to the soil, with many variables contributing to horizontal and vertical dispersion of salts throughout the soil profile. Some visual grass and soil surface symptoms will gradually begin to appear (refer to the first section in this chapter), but proactive monitoring before those symptoms become apparent in the turfgrass ecosystem is a critical maintenance protocol to prevent a significant concentration of accumulated salts that would require a reclamation salt management program when the turfgrass starts to die. Reclaiming salt-ladened soil profiles is time-consuming plus expensive, and changes in soil chemistry plus soil physical conditions do not occur rapidly, especially when additional saline irrigation water is being applied regularly to the ecosystem.

The proactive monitoring program should include soil moisture monitoring both on a daily basis on greens complexes and scheduled regularly on tees and fairways as needed on a monthly to quarterly basis in order to properly schedule irrigation cycles for effective salt management. Salt movement monitoring in the upper soil profile can be accomplished with various handheld and in-place sensors, as noted in Chapter 4, Section 4.3, "Field Monitoring of Soil Salinity."

17.5.6 How Often to Leach and Flush the Greens Cavity

Every site is different when it comes to salt accumulation and leaching volume plus leaching frequency. Guidelines are given in Chapter 11, "Irrigation Scheduling and Salinity Leaching." The rule is to schedule routine irrigation with more water but less frequently with the water applied by pulse cycles plus with the anticipation of having sufficient infiltration and percolation rates pushing the wetting front containing soluble salts down below the turfgrass root system in the soil profile. This scheduling strategy is contrasted to light, more frequent applications. Frequency can be adjusted to allow as much time between irrigation events as possible without creating an excessively dry surface that favors hydrophobic conditions. Rooting viability and depth and climatic conditions will need to be factored in scheduling decisions. Check cup-cutter plugs for any potential soil profile moisture and salt zone layering that might develop.

The ideal salt management program would push those salts down to the drain lines in the greens cavity. Some courses with high-salinity irrigation water often flush one time on a monthly basis with less saline water, but if high evapotranspiration persists, more frequent flushing may be needed. Proactive monitoring of salts and moisture in the upper soil profile and rhizosphere will help determine the frequency for flushing greens on a site-by-site basis. The "ideal" strategy is pulse irrigation with a maintenance leaching fraction during every irrigation event when using highly saline irrigation water (refer to Chapter 11, Section 11.3, "Maintenance Leaching and the Leaching Requirement").

17.6 BIOTIC STRESS × SALINITY INTERACTIONS

On salt-affected turfgrass sites, biotic stresses may differ to some degree relative to nonsaline or sodic locations. This can be in response to microenvironment changes, different species or cultivars used, and different management practices.

17.6.1 Weed Competition

When turfgrass growth rates are reduced due to salinity exposure and the canopy thins out, the result is normally an increase in competition from weeds. (*Note*: This section adapted from McCarty et al., 2001.) A less aggressive turfgrass sward will also result in highly competitive weed infestations. On sites with a salt-tolerant grass, however, that is appropriate for the salinity levels, and with good salinity and general management, weeds found on salt-stressed grasses or with other site stresses may not be found. Also, salinity can suppress some weeds such as annual grasses on one site, but actually facilitate colonization of the more salt-tolerant weedy species on another site. Wiecko (2003) noted that ocean water could be used for weed control on salt-tolerant grasses. Weeds that are indicators of specific poor soil conditions are listed below:

- *Saline soils*: alexandergrass (signalgrass), kikuyugrass, torpedograss, annual bluegrass, nutsedges, spurges, kyllingas, some *Cynodon* spp., some *Zoysia* spp., *Sporobolus* spp., *Distichlis* spp., *Spartina* spp., some *Paspalum* spp., and *Juncus* spp. (refer to Table 9.6)
- *Low-pH soils*: red sorrel and broomsedge
- *Compacted soils*: goosegrass, prostrate knotweed, annual bluegrass, and slender rush
- *Poor sandy soils*: poorjoe, sandspur, quackgrass, spurges, black medic, prostrate knotweed, and yellow woodsorrel

- *Droughty soils*: bahiagrass and smutgrass
- *Poor drainage*: sedges, alligatorweed, annual bluegrass, barnyardgrass, pennywort, and rushes
- *High-pH (alkaline) soils*: plantains, sedges, crabgrass, and goosegrass
- *High-nematode populations*: spurges, Florida parsley, and prostrate knotweed

A listing of highly competitive weeds in turfgrass ecosystems that are subjected to abiotic stresses, especially multiple stresses such as salinity, compaction, and drought, is as follows:

Nutsedges and other sedges	Carpetweed
Crabgrass	Khakiweed
Goosegrass	Chickweeds
Smutgrass	Thistles
Kyllingas	Dogfennel
Carpetgrass	Cudweeds
Alexandergrass (creeping signalgrass) and other signalgrasses	Dandelions
Dichondra	
Sandbur	Spurges
Orchardgrass	Henbit
Crowfootgrass	Clovers
Sprangletop	Beggarweed
Panicums	Plantains
Torpedograss	Pusleys
Dallisgrass	Annual bluegrass
Several weedy paspalums	

17.6.2 Predisposition to Diseases

Increasing soluble salt accumulations both in the soil profile and internally in the turfgrass plant from topical sprinkler saline irrigation water applications can predispose the plant to any pathogen when environmental conditions are favorable for disease development, especially since the turfgrass will have reduced growth rates as a result of salt accumulation, is less aggressive in a competitive and diverse ecosystem, often has more succulent leaf blades, and frequently will have a problem with salt-induced root desiccation. (*Note*: This section adapted from Vargas, 1994; Tani and Beard, 1997; Smiley et al., 2005; Fech and Gaussoin, 2009; Kammerer and Harmon, 2009.) Additionally, irrigation scheduling (Chapter 11) is a key secondary contributor to disease development, especially when areas are overwatered or underwatered. Overfertilization and underfertilization based on turfgrass species and specific cultivar nutrient sufficiency requirements are also secondary contributors to increased disease problems when applying saline irrigation water. Salt-stressed grasses that receive excessive N and too frequent irrigation will generally exhibit more disease symptoms, just as on nonsaline areas. Additionally, adverse soil physical conditions caused by Na can favor disease activity. Inherent disease susceptibility or resistance at the species and cultivar levels is important on salt-affected sites since there are a number of factors, as noted above, that may favor disease development, especially if management is not at the maximum level. Several broad disease categories are listed below that outline the potential for predisposition to pathogen attack as salts accumulate in the turfgrass ecosystem:

- Soil-inhabiting fungi capable of causing seed rot and seedling diseases include *Pythium*, *Fusarium*, *Rhizoctonia*, *Microdochium nivale*, *Microdochium bolleyi*, *Alternaria*, *Bipolaris*, *Botrytis*, *Cladosporium*, *Colletotrichum*, *Curvularia*, *Drechslera*, *Nigropsora*, *Septoria*, *Stagonospora*, and *Trichoderma*.

- Root-infecting and root-decline fungi include *Phialophora graminocola*, *Ophiosphaerella herpotricha*, *O. agrostis*, *O. korrae*, *O. narmari*, *Gaeumannomyces graminis* var. *graminim* or var. *avenae*, *G. wongoonoo*, *Phialophora radicicola*, *G. incrustans*, and *Magnaporthe poae*.
- Primitive root-infecting fungus-like organisms include *Olpidium brassicae*, *Rhizophydium graminis*, *Lagena radicicola*, and *Polymyxa graminis*.

As salts increase in the soil profile, the secondary biotic stress responses on the turfgrass root system and canopy are variable and multidimensional. If soil salt accumulations and topical adsorption of salt ions reach internal concentrations that exceed the genetic salt tolerance levels of a specific turfgrass cultivar, and the climatic conditions are favorable for escalation of pathogen populations, then disease attack is a typical result, leading to increased biotic stress symptoms. *The potential disease problem categories are listed below with potential pathogenic organisms.*

17.6.2.1 Increased Soil-Borne Pathogen Problems

Take-all/decline and ETRI [or ectotrophic root-infecting fungi] (*Gaeumannomyces graminis* var. *avenae or* var. *graminis or G. incrustans*) and **DPEH (Dark Pigmented Ectotrophic Hyphae) fungal organism complex/root infecting fungi associated with patch and spring dead spot diseases** (*Magnaporthe poae; Ophiosphaerella korrae, O. narmari, O. herpotricha; Phialophora graminicola or radicicola; Leptosphaeria korrae or L. narmari)*]. These fungi are associated with patch diseases that cause root, crown, stem, and leaf rot in affected turfgrass plants. This disease lives in the alkaline soil surface or upper soil profile around the crown and upper rootzones of the turfgrass plant. Pathogen attacks are enhanced by prolonged wet conditions (poor internal soil drainage, elevated soluble salts, and soil compaction) and reduced oxygen influx (rootzone anaerobiosis) into the upper soil profile. Control strategies include a multipronged aggressive approach involving cultural as well as chemical programs.

- Fungicidal application (curative rate) that is flushed into the crown region since topical application often does not reach the critical crown and root initiation zone. Dual-chemistry fungicides (involving Heritage, Banner Maxx, Rubigan, Eagle, and Fungo—see lists for fairy ring, take-all, and Curvularia control) are more effective, or combine with biofungicide products like Rhapsody, Spot-Less, Actinovate, or EcoGuard.
- Core- or needle-tine venting aerification to reduce soil profile wetness and promote downward movement of both oxygen and fungicides.
- Irrigation scheduling that is less frequent and has a longer duration or cycle.
- Judicious fertility program: maintain a balanced fertilizer application program and monitor tissue concentrations (wet chemistry and spectrophotometric analysis) for appropriate adjustments on a cultivar basis. Excessive nitrogen applications and use of PGRs can increase disease incidence. Use sulfur-based products, and avoid urea fertilizers since they enhance the disease problem.
- Soil pH management: these take-all and DPEH organisms thrive in alkaline conditions. Careful application of acidification products should be implemented. Acidification may not be effective if soil Mn is inherently low, which is common on sandy soils, since it is not pH per se, but greater soil solubility of Mn under lower pH that suppresses these organisms (Gelernter and Stowell, 2004).
- Manganese concentration maintenance in the soil: the take-all, decline, DPEH, ETRI, and soil-borne fungal complex are difficult diseases to control in turfgrass ecosystems due to their positioning below the soil surface and around the crown and rhizome zones in the upper 1 to 3 inches of the soil profile, especially in anaerobic soil conditions. An

effective BMP strategy requires additional soil Mn applications under salinity and soil-borne biotic problem situations, since most soils with pHs > 5.5 are characteristically low in Mn concentration, most irrigation water sources also have low Mn concentrations, and most liquid micronutrient packages contain approximately 1% or less Mn in the formulation. These fungi are thought to convert low concentrations of Mn into an unavailable oxidized form in the soil (Heckman et al., 2003), and a higher concentration of granular Mn, such as manganese sulfate, should be applied directly to the soil to supply sufficient quantities (30 ppm Mn with Mehlich III extraction) to suppress these fungal pathogens (Anonymous, 2008; Gelernter and Stowell, 2004). The key BMP strategies with this critical plant nutrient under saline conditions include aeration and syringing of the soil-applied Mn into the 2-inch zone of the upper soil profile, soil applications of Fe and Zn to enhance the synergistic effects with these two key nutrients (a 3:1 ratio of Fe:Mn and 1:1 ratio of Mn:Zn), and proactive soil and tissue monitoring. The synergistic nutritional effects include genetic activation of multiple salt tolerance mechanisms in the root system (Mn and Zn), soil-borne disease suppression (Mn), and color enhancement (Mn and Fe). Since Mn is immobile internally in the plant and will not translocate from shoots to roots, liquid applications of Mn to the shoots have generally not been topically applied at sufficient concentrations even with irrigation applications immediately after spray applications to meet the disease suppression requirements in the upper soil profile. Granular Mn applications are needed to meet these suppression requirements. A 2.25 kg Mn/ha (2.0 lb. Mn/acre or 0.046 lb. Mn/1000 sq. ft.) rate of granular Mn fertilizer applied to the soil in April and/or October of each year has been found to be effective in take-all and decline suppression when the Mehlich-III Mn Availability Index is low (e.g., <21) (Heckman et al., 2003). Zinc should be applied at about half that rate to activate the genetically controlled salt tolerance mechanisms in conjunction with Mn in the plant. Potassium sufficiency levels should be maintained to offset salinity-induced turgor pressure alterations. One additional note: these take-all and decline organisms tend to escalate in population when oxygen flux through the soil profile is decreased (soils at or near field capacity with little to no air porosity, compacted soils, and/or heavy winter play leading to spring disease symptoms), and regular aeration is recommended to assist Mn in suppressing these organisms.

- It has been our observation that the *Helminthosporium–Curvularia* complex is often present after take-all infection and may inhibit recovery if appropriate control measures are not included as part of the take-all fungicide program. On salt-affected sites, the combination of salinity stress and the *Helminthosporium–Curvularia* complex can also delay recovery from other diseases such as *Rhizoctonia* brown patch.
- Effective thatch management is necessary to avoid an excessively moist surface during prolonged wet periods.
- A commonsense sand-topdressing program with aerification will aid in diluting excessive organic matter in the thatch or mat that may enhance anaerobic conditions.
- Tree shade reduction will allow better drying of sites.
- Ensure that above-ground and subsurface air movement are adequate on problem sites.
- Allow some upper soil surface dry-down if possible, but not to the point of stressing the turfgrass under drought conditions with minimal root volume.
- Do not scalp (potassium nutrition is critical to enhance plant turgor pressure) or expose areas to heavy compaction or cart traffic.
- Apply weekly to twice-weekly applications of a >30% concentration of seaweed extract containing cytokinins to enhance root system redevelopment if needed with less frequent applications when roots are healthy.
- If using Primo®, delay any further applications in order to stimulate GA-induced stolon growth that is needed for achieving any lost canopy density at the surface.

Pythium root dysfunction (*Pythium volutum* or *P. torulosum*; other organisms associated with turfgrass root diseases are *P. aristosporum*, *P. arrhenomanes*, *P. catenulatum*, *P. debaryanum*, *P. irregular*, *P. rostratum*, *P. tardicrescens*, *P. vexans*, and *P. volutum*)_

Anthracnose root and basal rot (*Colletotrichum cereal*, formerly called *C. graminicola*)

Curvularia fading out (*Curvularia geniculata*, *C. lunata*, *C. lunata* var. *aeria*, *C. verruculosa*, *C. eragrostidis*, *C. inaequalis*, *C. intermedia*, *C. emmoseto*, *C. protuberate*, *C. senegalensis*, and *C. trifolii*)

Leptosphaeria spring dead spot (*Leptosphaeria korrae*+, *Leptosphaeria narmari*, *Leptosphaerulina australis*, or *Leptosphaerulina trifolii*): +note that the causal organism has been renamed *Ophiosphaerella korrae*.

Brown ring patch or waitea reddish brown patch (*Waitea circinata* var. *circinata*)

Necrotic ring spot (*Ophiosphaerella korrae* and *Ophiobolus herpotricha*)

Summer patch (*Magnaporthe poae*)

17.6.2.2 Increased Problems from Salt-Tolerant Nematodes

The particular nematodes that may be predominant on a salt-affected site may differ with irrigation water quality to some extent. As with diseases, turfgrass species and cultivar susceptibility or resistances are important as a management BMP. Common salt-tolerant nematode populations include the following:

Lance (*Hoplolaimus galeatus*) endoparasitic
Sting (*Belonolaimus longicaudatus*) ectoparasitic
Spiral (*Helicotyenchus* spp.) ectoparasitic
Root-knot (*Meloigogyne* spp.) endoparasitic
Stunt (*Tylenchorhynchus* spp.) ectoparasitic
Needle (*Longidorus* spp.) ectoparasitic
Dagger (*Xiphinema* spp. and *Longidorus* spp.) ectoparasitic

17.6.2.3 Predisposition to Surface Drought and Desiccation Problems

Because high-soluble salts can induce physiological drought and sodic conditions foster soil drought problems, the same diseases that are favored by normal drought are also favored by these salt problems.

Helminthosporium melting out leaf spot disease complex (*Bipolaris* spp., *Drechslera* spp.) (*Bipolaris cynodontis* = leaf blotch; *B. sorokiniana* = leaf spot; *B. spicifer a* = stem or crown rot; *B. australiensis*; *B. buchloes*; *B. hawaiiensis*; *B. micropus*; *B.rostratum*; and *B. stenospila*) (*Drechslera siccans* = brown blight; *D. poae* = melting out; *D. dictyoides* = net blotch; *D. catenaria*; *D. dictyoides*; *D. erythrospila*; *D. gigantean*; *D. nobleae*; *D. phlei*; and *D. tritici-repentis*) and (*Exserohilum rostratum*); and associated organism: *Marielliottia triseptata*.

Secondary association: Curvularia fading out (*Curvularia geniculata*, *C. lunata*, *C. lunata* var. *aeria*, *C. verruculosa*, *C. eragrostidis*, *C. inaequalis*, *C. intermedia*, *C. emmoseto*, *C. protuberate*, *C. senegalensis*, and *C. trifolii*).

17.6.2.4 Soil Hydrophobicity and Localized Dry Spot Tendencies

These conditions can predispose turfgrass plants on both nonsaline and saline sites to certain diseases.

Fairy ring (*Lepiota sordida*, *Marasmius oreades*, *Lycoperdon perlatum*, *Agaricus campestris*, *Marasmius oreades*, *Tricholoma sordidum*, and *Vascellum pratense*)

17.6.2.5 Consistently High Upper Soil Profile Moisture Conditions

Consistently high soil moisture in the upper profile creates the ideal microenvironmental conditions for foliar pathogen populations to increase and subsequently attack on the turfgrass:

Pythium blight (*Pythium graminicola*, *P. vanterpoolii*, *P. ultimum*, *P. aristosporum*, *P. iwayamani*, *P. paddicum*, *P. periplocum*, *P. torulosum*, *P. aphanidermatum*, and *P. myriotylum*)
Fusarium blight diseases (*Fusarium acuminatum*, *F. avenaeceum*, *F. oxysporum*, *F. tricinctum*, *F. crookwellense*, *F. culmorum*, *F. equiseti*, *F. graminearum*, *F. heterosporum*, *F. poae*, *F. pseudograminearum*, and *F. semitectum*)
Microdochium patch (or Fusarium patch) (*Microdochium nivale*)
Patch diseases: brown (*Rhizoctonia solani* AG-2-2 III B and AG-1), **large** (*Rhizoctonia solani* AG-2-2 LP), **yellow** (*Rhizoctonia* cerealis AG-D [I]), and **pink** (*Limonomyces rosipellis*)
Other *Rhizoctonia* spp. causing diseases: *R. cerealis*, *R. oryzae*, and *R. zeae*
Dollar spot (*Sclerotinia homeocarpa*), **false dollar spot** (thought to be caused by *Poculum henningsianum*, an anamorph of *Sclerotinia homoeocarpa* [Bennett]), and **creamy blight** (*Limonomyces* spp. [biotype of *L. rosipellis*])
Algae (green algae [*Chlamydomonas*, *Cosmarium*, and *Cylindrocystis*] prefer acidic (pH < 6.0) conditions; blue-green algae [*Lyngbya*, *Nostoc*, *Oscillatoria*, and *Phormidium*] prefer pH > 7.0 alkaline conditions)
Mosses (*Bryum argenteum* [Hedw.] or silvery-thread moss, *Bryum lisae* De Not., *Amblystegium trichopodium* [Schultz] Hartm., *Brachythecium* spp. B.S.G., *Entodon seductrix*, *Amblystegium serpens*, *Selagimella* spp., *Ceratodon* spp., *Hypnum* spp., and *Polytrichum* spp.)
Wet wilt (shoots desiccate when rate of transpiration exceeds rate of water absorption by turfgrass roots even when soil moisture is adequate)
Bacterial wilt (*Xanthomonas campestris* and *Xanthomonas pv. Graminis*)
Slime molds (*Mucilago spongiosa* or *crustacea*, and *Physarum cinereum*; *Fuligo* spp.)
Powdery mildew (Blumeria graminis; formerly *Erysiphe graminis*)
Leptosphaerulina leaf blight (*Leptosphaeria korrae*+, *Leptosphaeria narmari*, *Leptosphaerulina australis*, or *Leptosphaerulina trifolii*): +note that the causal organism has been renamed *Ophiosphaerella korrae*.
Red thread (*Laetisaria fuciformis*)
Yellow tuft (downy mildew) (*Sclerophthora macrospora*)

17.6.2.6 Direct Salt Ion Concentration in the Irrigation Water Source

This section is adapted from Porter (1990), Muehlstein et al. (1991), Camberato et al. (2005, 2006), Olsen et al. (2003, 2004, 2006, 2008a, 2008b, 2008c), Peterson et al. (2006), Entwistle et al. (2005), Kopec et al. (2003, 2004), Entwistle and Olsen (2007), and Martin et al. (2009).

Rapid blight (*Labyrinthula terrestris* [D.W. Bigellow, M.W. Olsen, and Gilb.]). This organism causes high rates of disease with elevated sodium levels in water (Olsen et al., 2008a, 2008b, 2008c; Martin et al., 2009), especially in water that is low in calcium concentrations. Stress induced by water deprivation has not been a factor in rapid blight disease development (Olsen et al., 2006). Salinity tolerance of turfgrasses is correlated with rapid blight tolerance (Camberato et al., 2005).

Labyrinthula is in the marine slime mold family and prefers ECw > 2.0 dS/m and usually within the range of 2.0–10.5 dS/m (Olsen et al., 2004; Peterson et al., 2005, 2006). Warm-season grasses can be summer hosts, and cool-season grasses (*Poa annua*, *Poa trivalis*, *Lolium perenne*, *Agrostis tenuis*, and *Agrostis palustris*) are susceptible (Olsen et al., 2003). Some cultivars of hard fescue, intermediate ryegrass, and redtop are also susceptible (Kopec et al., 2004).

Turfgrass symptoms resemble *Microdochium* patch and take-all patch initially, and gypsum plus pyraclostrobin and mancozeb fungicide applications reduced the severity of the symptoms (Entwistle and Olsen, 2007). Leaf symptoms can range from water-soaked yellow-browning or bronzing discoloration to eventual red leaves. Preventative fungicide treatments (trifloxystrobin, mancozeb, and pyraclostrobin) coupled with salinity reduction and planting more salt-tolerant cultivars are suggested management strategies (Olsen, 2008). Additional information on the *Labyrinthula* organism can be found in Olsen (2007), Yadagiri and Kerrigan (2010), and Douhan et al. (2009).

17.6.2.7 Additional Comments

If the environmental conditions are favorable for pathogen population buildup, the turfgrass can suffer an attack from most disease organisms in both saline and nonsaline environments. Even so-called disease-resistant cultivars can be predisposed to pathogen attack if the primary stress is salinity and irrigation scheduling plus the fertility program are inadequate for the specific site when pathogen populations are elevated. Most of these pathogenic organisms need a weakened or less aggressive turfgrass (resulting from increasing salinity exposure) to cause significant damage since they are generally secondary- or tertiary-level pathogens in their mode of action. Salinity-induced growth rate reductions help to promote increased instances for plant disease damage.

Based on observations and global feedback concerning reoccurring disease problems when applying saline irrigation water, the most persistent disease problems include take-all and decline, fairy ring, nematodes, *Helminthosporium–Curvularia* complex, algae, and rapid blight. Except for rapid blight, which is favored by saline conditions, the other persistent diseases that seem to be more prevalent on salt-affected sites regardless of the grass species are likely due to salinity stresses predisposing the plants—with salinity from saline irrigation water a consistent, daily stress. Under the right conditions, any disease can become a limitation in salt-challenged turfgrass ecosystems just as in nonsaline sites.

17.6.3 Insect Interactions

Several categories below summarize the potential for predisposition to cyclic insect attack when turfgrasses are weakened by abiotic stresses such as salinity, and these same insects can attack nonsaline-grown turfgrass also. (*Note*: This section adapted from Potter, 1998.)

17.6.3.1 Root-Infesting Insects

Australian sod fly (*Inopus rubriceps* [Macquart])
Dichondra flea beetle (*Chaetocnema repens* McCrea)
European crane fly or leatherjacket (*Tipula paludosa* [Meigan])
Ground pearls (*Margarodes* spp.)
March flies (*Bibio* spp. and *Dilophus* spp.)
Mole crickets: tawny (*Scapteriscus vicinus* Scudder), southern (*Scapteriscus borellii* Giglio-Tos), short-winged (*Scapteriscus abbreviatus* Scudder), and northern (*Neocurtilla hexadactyla* (Perty])
White grubs: northern masked chafer (*Cyclocephala borealis* Arrow), southern masked chafer (*Cyclocephala lurida* [Bland]), black turfgrass ataenius (*Ataenius spretulus* Halderman), May beetle (*Phyllophaga* anxia, *P. fervid*, *P. fusca*, *P. hirticola*, *P. implicate*, *P. inversa*, *P. rugosa*, *P. crinite*, and *P. latifrons*), green June beetle (*Cotinis nitida* L.), southwestern masked chafer (*Cyclocephala pasadenae*), western masked chafer (*Cyclocephala hirta*), Japanese beetle (*Popillia japonica* Newman), European chafer (*Rhizotrogus majalis* [Razoumawsky]), oriental beetle (*Exomala orientalis* [Waterhouse]), Asiatic garden beetle (*Maladera castanea* [Arrow]), and Aphodius grubs (*Aphodius granarius [L.]* and *A. paradalis* Le Conte).

17.6.3.2 Insects That Damage Crowns or Burrow into Stems

Annual bluegrass weevil (*Listronotus maculicollis* [Dietz])

Billbugs: bluegrass (*Sphenophorus parvulus* Gyllenhal), hunting (*Spenophorus venatus vestitus* Chittenden), Phoenix (*Spenophorus phoeniciensis* Chittenden), and Denver (*S. cicatristriatus* Fahraeus)

Frit fly [*Oscinella frit* (L.])

17.6.3.3 Insects That Suck Plant Juices

Bermuda grass scale (*Odonaspis ruthae* Kotinsky)

Buffalograss mealybugs (*Tridiscus sporoboli* [Cockerell] and *Trionymus* spp.)

Rhodesgrass mealybug or Rhodesgrass scale (*Antonina graminis* [Maskell])

Chinch bugs: hairy (*Blissus leucopterus hirtus* Montandon) and southern (*B. insularis* Barber)

Greenbug (*Schizaphis graminum* [Rondani])

Leafhoppers: sharpshooters (*Dracuelacephala* spp.), and brown or grayish (*Agallia*, *Dikraneura*, *Endria*, *Exitianus*, and *Psammotettis*)

Two-lined spittlebug (*Prosapia bicinata* [Say])

Mites: banks grass (*Oligonychus pratensis* [Banks]), brown wheat (*Petrobia latens* [Muller]), clover (*Bryobia praetiosa* Koch), winter grain (*Penthaleus major* [Duges]), Bermuda grass (*Eriophyes cynodoniensis* Sayed), buffalograss (*Eriophyes slyhuisi* [(Hall]), and zoysiagrass (*Eriophyes zoysiae* Baker, Kono, and O'Neill)

17.6.3.4 Insects That Chew Leaves and Stems

Armyworm (*Pseudaletia unipuncta* [Haworth])

Fall armyworm (*Spodoptera frugiperda* [J.E. Smith])

Lawn armyworm (*Spodoptera mauritai* [Boisduval])

Cutworms: black (*Agrotis ipsilon* [Hufnagel]), bronzed (*Nephelodes minians* Guenee), variegated (*Peridroma saucia* [Hubner]), and granulate (*Felta subterranean* [Fabricius])

Sod webworms: bluegrass (*Parapediasis teterrella* [Zinchen]), larger (*Pediasia trisecta* [Walker]), silver-striped (*Crambus praefectellus* [Zincken]), striped (*Fissicrambus mutabilis* [Clemens]), western lawn moth (*Tehama bonifatella* [Hulst]), buffalograss (*Surrattha indentella* Kearfott), cranberry girdler (*Chrysoteuchia topiaria* [Zeller]), burrowing (*Acrolophus* spp.), *Crambus sperryellus*, tropical (*Herpetogramma phaeopteralis* [Guenee]), and fiery skipper (*Hylephila phyleus* [Drury])

Lucerne moth (*Nomophila noctuella* D. and S.)

Striped grassworm or grass looper (*Mocis latipes* [Guenee])

Vegetable weevil (*Listroderes difficilis* Germar)

Insects that have been noted by the authors on turfgrass sites that are consistently irrigating with saline irrigation water include the worm complex (fall armyworm, sod webworm, and cutworms), grubs (which generally come to the soil surface when salts accumulate in the soil profile, even during winter periods when they are normally hibernating), and billbugs. Mole crickets have been noted to be especially troublesome in bunkers and bunker surrounds as well as sand-based greens. One interesting fact is that seashore paspalum (*Paspalum vaginatum* Swartz) has a low to medium-low level of overall genetically controlled insect resistance, but exceptionally high beneficial insect activity that promotes integrated pest management (IPM) programs (Braman, 2009; Braman et al., 2003, 2004).

18 Proactive Monitoring of Progress

CONTENTS

18.1 ASSESSING PROGRESS IN SALINITY BEST MANAGEMENT PRACTICES (BMPs)

18.1.1 Initial versus Ongoing Site Monitoring

For turfgrass ecosystems receiving ongoing applications of soluble salts and Na by irrigation water, site evaluation does not stop with the initial site assessment; but it must be a continuous process. In Section II of this book, "Site Assessment BMPs for Saline and Sodic Soil Sites," the focus was on the critical role of the *initial site assessment* on a salt-affected landscape to determine the types of salinity stresses, causes, magnitude, dynamics, and spatial variability of salt-induced stresses. Such a site assessment is critical in order to develop a science-based, holistic (whole-ecosystem) salinity BMP program. Assessment "tools" included the following:

- "Salinity Soil Tests and Interpretation" (Chapter 4)
- "Routine Soil Test Methods" (Chapter 5)
- "Irrigation Water Quality Tests, Interpretation, and Selection" (Chapter 6)
- "Plant Analysis for Turfgrass" (Chapter 7)
- "Assessment for Salt Movement, Additions, and Retention" (Chapter 8)

In the current chapter, attention is directed to ongoing site evaluation (i.e., the proactive monitoring of progress). *Proactive* refers to a monitoring approach that allows adjustments in the salinity BMP program in a timely fashion before more serious salt issues arise. Thus, proactive monitoring is based on obtaining ongoing site information to determine turfgrass performance and soil stability changes before they become visual. The proactive approach is in contrast to a *"reactive"* approach, where timely monitoring is not done and management changes are made only at the point where visual salinity stresses are apparent. For sodic conditions, the latter approach is very costly since sodic conditions may not be readily apparent; but when they do become obvious, it can take years to remediate the permeability problems in many soil situations that are caused by excess accumulation of sodium ions. Many of the same site evaluation methods described in Section II are applicable for assessment of progress in salinity management. A comprehensive summary of soil, water, and infrastructure information to consider when characterizing or assessing a specific site for potential salinity challenges is found in Table 18.1.

18.1.2 Why Ongoing Monitoring Is Essential

With salt-induced stresses, proactive-monitoring progress of salinity management must be viewed as essential and not optional for several reasons. First, monitoring progress is costly, but unwise management decisions are substantially more costly. Wise decisions require accurate and timely science-based information that is specific to each salt-affected microsite. A management decision to omit a critical BMP for temporary cost savings does not stop salts from accumulating and degrading soil, plant, and water resources. Second, salinity movement, additions, and retention are dynamic over time; therefore, salinity BMPs must be dynamic to account for often subtle but persistent changes. The dynamic nature of salinity challenges at the microsite level within a property is due to the interactions among irrigation water quality, soil profile variability, turfgrass species and cultivar, and uncontrollable environmental extremes. While some of the changes can be predicted, the complexity of salinity stresses requires solid monitoring involving science-based data to accurately make BMP adjustments. Third, salinity is an environmental issue that impacts long-term sustainability of the site. Ignoring the implications of salinity inputs, movement, and retention does not merely set up the potential for ecosystem degradation; rather, it insures that degradation will happen that often goes beyond the site boundaries in terms of social, spatial, and economic impacts. Pillsbury (1981) noted the impact of salinization in ancient history and why the historical lessons must not be forgotten (see the quote in Chapter 1 from Pillsbury, 1981). In recent history, Christen and Ayars (2001) documented the substantial areas in Australia where rising water tables have brought salinity into the rootzone—in these cases, usually from changes in cropping systems and flora that reduced soil water extraction and allowed more drainage water into the subsoil. Environmental stewardship and sustainability on salt-affected sites require planning, proper infrastructure, flexibility in management, and continuous proactive monitoring.

18.2 PRACTICAL CONSIDERATIONS FOR ONGOING MONITORING PROGRAMS

When developing an ongoing monitoring program on a salt-affected turfgrass site, there are several considerations that will assist in this process. These considerations all relate to the ongoing and dynamic nature of salinity challenges, which do not always fit well with site management mentality such as that relating to the immediate year or season; a "fix it and go on to another issue" way of thinking; or a belief that "now that this issue is dealt with, we will not need to be concerned with it again."

18.2.1 Goal: Total Removal or Sustainable Levels of Saline and Sodic Conditions

What are realistic turfgrass ecosystem expectations in managing salts? Most turfgrass management problems are ones where corrective management is focused on elimination of specific issues, such as a disease outbreak or a specific weed contamination issue. While the turfgrass manager realizes that these problems may be "totally corrected" at that point of time, they are most likely to reappear again. However, others such as a golf club member or some management personnel may not view it in the same manner—they may think in terms of "we have corrected this problem and it is over; we can go to other issues." In reality, many turfgrass challenges, especially on high-use and high-quality recreational areas, are not a matter of total removal or correction, but of sustainable

TABLE 18.1
Types of Science-Based Data Typically Required for Environmental Site Characterization of Salt-Challenged Sites

Soil Parameters

- Type: sand, silt, clay, loam, pulverized volcanic pumice, pulverized caliche, pulverized decomposed granite (DG)
- Texture: complete physical particle size distribution, bulk density, degree of heterogeneity, nature of origin (alluvial, glacial, marine, etc.), organic amendments, inorganic sand substitute amendments, "fines" composition
- Clay structure: 2:1 or expanding and contracting (shrink on drying:swell on wetting), 1:1 or nonexpanding or nonshrink or -swell
- Distribution: thickness, topographic location, layering potential, native parent material
- Physical properties: air porosity or macropores, capillarity or microporosity, bulk density, cultivation and penetration capability, salt sequestration, particle migration
- Hydraulic properties: saturated and unsaturated hydraulic conductivity, permeability, porosity, moisture content and retention, infiltration rate, percolation rate, pore continunity
- Chemistry: cation exchange capacity (CEC), organic carbon content, pH, nutrient content, redox potential, salt ion retention, complexation capability (carbonates and bicarbonates), salt ion toxicity
- Microbiology: microbial population type and tolerance to salinity dynamics

Irrigation Water Parameters

- Groundwater depth
- Conditions of occurrence: confined, semiconfined, unconfined, perched, harvested, depth to water table, capillary fringe, water level fluctuations, coastal tidal influences, recharge or discharge fluxes, conservation regulations, interconnections between aquifers, saltwater influx, stratified salinity aquifers
- Conditions of movement: flow direction, horizontal and vertical gradients, flow velocity, seasonal or tidal variations, recharge frequency
- Physical properties: temperature, turbidity, total dissolved solids (TDS), debris, surface tension
- Chemistry: pH, major ions (nitrate, sulfate, iron, manganese, sodium, and chloride), dissolved oxygen, methane, carbon dioxide, hydrogen sulfide, ammonia, redox potential, specific conductance, total dissolved salts, salinity, miscellaneous contaminant levels, pharmaceuticals, point source and non–point source pollution
- Microbiology: microbes, nematodes, pathogenic and nonpathogenic spores, moss and algae potential, aquatic weed contamination
- Patterns of use: municipal, residential, commercial, industrial, agricultural, reclaimed (effluent), recycled, treated, blended, drainage reuse, subsurface drip, sprinkler distribution, flooding exposure

Surface Water Parameters

- Conditions of occurrence: static versus dynamic, drainage pattern and area, width and depth, elevation, obstructions to flow, stratification, relationship to groundwater, harvesting area, seepage zones
- Conditions of movement: flow direction, gradient, flow velocity, inflow and outflow volumes, sediment transport and deposition regime, flow frequency and duration, gravity issues, flooding potential, turnover rate
- Physical properties: temperature, turbidity, suspended solids (sediment), depth, volume, debris, surface tension, solubility, miscibility
- Chemistry: pH, major ions, dissolved oxygen, specific conductance, total dissolved salts, biochemical oxygen demand (BOD), chemical oxygen demand (COD), contaminants, upstream pollution sources, alkalinity (capacity of water to accept protons; ability to neutralize an acid; caused by bicarbonates, carbonates, hydrogen ions and compounds, and/or salts formed from organic acids such as humic acids; alkalinity <80 ppm leads to rapid fluctuations in pH plus is a corrosive water, while >200 ppm results in highly buffered water; and varying degrees of water hardness based on total concentrations of dissolved calcium and magnesium ions, reported as calcium carbonate) (Spellman, 2008)
- Microbiology: microbes, nematodes, spores, algae, moss, aquatic weeds
- Pattern of use: type, amount, pump intake positioning, blended, treated, reclaimed (effluent), recycled, drainage reuse, subsurface drip, sprinkler distribution, flooding exposure, chemical injection

continued

TABLE 18.1 (Continued)
Types of Science-Based Data Typically Required for Environmental Site Characterization of Salt-Challenged Sites

Contaminant Parameters

- Type: inorganic (acids, bases, oxides, salts, nitrogen, and phosphorus) or organic (fats, dyes, soaps, rubber products, plastics, wood, fuels, cotton, proteins, and carbohydrates) or biological or combinations, emulsions, gas composition
- Physical properties: solubility or miscibility, density or specific gravity, viscosity, surface tension, volatility (vapor pressure), adsorption coefficient (K_d for inorganic materials, and K_{oc} for organic compounds), dielectric constant, mobility, reactivity, ignitability, corrosivity, biodegradability, persistence, blending, stratification, catchment ponds, grassed filter strips
- Chemistry: nutrient composition, salt ion concentration, solubilization, suspension, metal speciation, degradation products or pathways, complexation potential, gases (dissolved oxygen, carbon dioxide, hydrogen sulfide, and ammonia-N)
- Distribution: media impacted, areal extent, vertical extent, phases present, dilution or blending, buffer zones
- Details of release: location; volume; release time; point source, non–point source, or diffuse source type; catastrophic, periodic, or long-term type of release; onsite treatment

Facility Parameters

- Type: landfill, surface impoundment, harvested from surrounding area, coastal dredged, acid-sulfate composition, debris screening, reclaimed water leaching fields, particle size screening, settling ponds, catchment facilities, grassed filter zones
- Location: above grade, below grade, property boundaries, accessibility, topography changes, exposure to tidal influences, position relative to water table, buffer zones
- Design and construction features: liners, leachate collection systems (dry wells, sump pumps), berms, mounds, dispensers, sloped native transitioning areas, waste bunkers, flood plain, blending options, size of storage facilities, water treatment options, reverse-osmosis size for the site, chemigation injectors, pure water influx (e.g., snow melt or high-volume rainfall)
- Operational details: waste product types (recycled or reclaimed effluent, and harvested water volume and frequency), onsite treatment, discharge points, reverse osmosis and throughput, blending capabilities, storage facilities, chemical injection
- History and period of use: seasonal, on demand

Other Important Parameters

- Area involved: grassed areas, landscape areas, out-of-play surrounding areas, topography discharge area surrounding the facility
- Geomorphology and topography: elevation changes, low collection and disposal areas
- Climatic conditions: water balance (irrigation schedule vs. precipitation vs. evapotranspiration); temperature; prevailing wind direction, speed, and persistence; cyclic extremes of cold, heat, drought, and/or flooding
- Vegetative cover: area grassed, diversity of landscape and native plants, seasonal changes in type of cover
- Surrounding land uses: residential, commercial, industrial, agricultural, history of use, activities, upstream discharge points, upstream water treatment facilities
- Presence and proximity of receptors: public and private water supply wells, utility corridors, wildlife (travel corridors, wetlands, sensitive ecological areas, and surface water barriers)
- Presence and proximity of anthropogenic influences: pumping wells, recharge basins, recharge zones for aquifers, injection wells, dewatering operations (quarries; sand, gravel, and/or stone operations; mines; and excavations), native historical protection areas, federal and state lands

Source: Adapted from Nielsen, D. M., G. L. Nielsen, and L. M. Preslo, Environmental site characterization, in D. M. Nielsen (Ed.), *Environmental Site Characterization and Ground-Water Monitoring*, 2nd ed., pp. 35–205, CRC Press, Boca Raton, FL, 2006.

management at an acceptable level. For example, wear stress on a site is an ongoing challenge in grass management just as are disease and weed pressures. Certainly, salts applied to a turfgrass area from any source (such as saline irrigation water or chemical amendments) must also be viewed as ongoing where total removal of the issue is not feasible. But, sustainable management is very possible—and *sustainability is the ultimate goal in a whole-ecosystem salinity BMP*. In Chapter 19, "Sustainable and Environmental Management Systems," the environmental, economic, and society or user components of "sustainable" management are discussed in greater detail related to the development of BMPs and overall sustainable programs that encompass all environmental issues on a facility.

Sustainable soluble salt management primarily means (a) maintaining the soluble salts in the rootzone at a level that does not cause undue stress on the plant (for example, root desiccation or alteration in nutrient uptake), and (b) assuring that salts removed from the rootzone are disposed of in an environmentally sustainable manner. *Sustainable Na management* is implemented to reduce soil Na concentrations to a level that soil structure deterioration is not a problem, especially with the surface zone being of most importance, but the Na permeability hazard must also be dealt with throughout the soil profile. *Sustainable nutrient management* is achieved by frequent adjustments in the fertility program in response to dynamic site conditions using supplemental prescription fertilization to insure that soil and plant tissue nutrient levels are adequate for each turfgrass species and cultivar.

Sustainable management is not easy, but is very achievable using a systematic salinity BMP approach based on the right site-specific information (see Chapter 19). Without ongoing site monitoring, sustainability is not possible. It is important to remember that sustainability is especially challenging with saline irrigation water, and primarily due to the salt loading (total salts, specific potentially toxic salt ions—sodium, chloride, sulfates, bicarbonates, and boron) that comes with each additional application of water on the turfgrass ecosystem. The subsequent problems in managing any excess salts in turfgrass ecosystems are the three exposure issues:

- Direct foliar application of total dissolved soluble salts and specific salt ions in the irrigation water directly on the turfgrass shoots
- The reoccurring accumulation of those excess salts in the soil profile
- Potential exposure of surface (ponds, lakes, and reservoirs) or subsurface (aquifers) waters to salinity leached from a site. Thus, a suitable environmentally sustainable salinity disposal program is essential.

18.2.2 Difficult Microsites

Three types of salt-affected situations are common: (a) a newly established turfgrass area, where saline irrigation water is to be used and salinity challenges were not present prior to use of the irrigation water; (b) a new site where saline irrigation water will be used and the soil already has salinity issues; and (c) an older location using saline irrigation water, where salinity has accumulated to the point that requires a more aggressive salt BMP program to be developed. Regardless of the situation, one common theme either will be present or will arise over time on all of these sites: *difficult microsite areas* that are more resistant to alleviation of the salt stresses.

Whether a new or an older salt-affected turfgrass landscape, the initial site assessment will allow a salinity BMP program to be initiated. This science-based BMP program will result in much of the area at a new site from exhibiting visual salt stresses; while on an older site, the salinity BMP program will result in acreage demonstrating salt stress to decrease. However, in each case, there will normally be microsites that do not respond to the initial BMPs. These difficult microsites must be evaluated individually in a more intensive manner over time to determine specifically the nature of salinity challenges and the underlying causes—it is not unusual for subsurface salt movement to be involved, as noted in Chapter 8, "Assessment for Salt Movement, Additions, and Retention," but there may be other causes.

Each microsite must be comprehensively and accurately characterized to determine the primary salinity limitations, the secondary biotic and abiotic constraints, and any other (such as infrastructure issues) challenges that are causing poor turfgrass performance. The approach to assessing these salt-induced limitations must be holistic (i.e., a whole-ecosystem assessment that can be used to make appropriate salt management and infrastructure improvement decisions geared to sustaining grass performance expectations on the site). When evaluating each microsite, it is necessary not only to observe conditions within the microsite, but also to step back and determine what above-ground and below-ground situations in the surrounding terrain may be affecting the site. Difficult microsites will require individual monitoring over time to make site-specific management adjustments to shrink the area affected and eventually allow that previously salt-challenged area to blend in with the surrounding landscape.

It is important for management to realize that all salt-affected areas and symptoms will not immediately disappear with the implementation of even the best initial salinity BMP program. Some areas will take longer. Often these microsites are areas of much higher accumulation than the surrounds and can involve unanticipated subsoil salt migration issues. When the whole site is exhibiting salinity stresses, these microsites are often not distinguishable from the surrounding turfgrass until initial BMPs correct the overall background salinity stresses affecting the broad areas. As the broad salinity-stressed areas are addressed, difficult microsites eventually stand out, and these persistent problem areas then can be assessed for causes and continuing remediation.

18.2.3 Indicator Area Monitoring

Monitoring of a salt-affected site soon after BMPs have been initiated usually remains rather comprehensive until positive results are evident. Once this point is achieved, monitoring can focus on (a) the most difficult microsite areas, as noted in the previous section; (b) periodic comprehensive monitoring, but at a reduced schedule, perhaps quarterly to annually; and (c) representative areas (i.e., *indicator areas*) with sampling scheduling appropriate for the turfgrass manager to determine any trends that are occurring. For example, on a golf course, this could be soil and tissue samples taken within the same locations on 1–3 greens, tees, and fairways every 1–3 months. These indicator areas are often selected because they are locations that are representative of other similar features or reoccurring turfgrass stress symptoms.

Another type of indicator area is potential salt accumulation (such as persistent seepage sites) locations. Recent developments in spatial mapping of soil salinity has made it possible to determine landscape microsite locations for installation of salinity sensors on areas that exhibit the most rapid salinity accumulation (i.e., where these are critical "indicator areas") (see Chapter 4, Section 4.3.2, "Approaches to Field Salinity Monitoring"; Figures 4.1 and 4.2; Krum et al., 2011). Obviously, soil salinity sensors that are real time and multiple depth in scope provide a continuous proactive-monitoring program for these potential hotspots.

18.2.4 Show-and-Tell Areas

It is not unusual for a site where saline irrigation water is used to become salt affected to the point that the turfgrass is deteriorating from combinations of total soluble salts, Na permeability hazards, nutrient challenges, and toxic levels of Na, Cl, or B. The point of unacceptable turfgrass performance is usually the result of several years of soil salt accumulation. A site assessment such as noted in Chapter 8, "Assessment for Salt Movement, Additions, and Retention," along with appropriate soil, water, and tissue analyses, may be conducted to determine a comprehensive salinity BMP. On seriously salt-challenged sites, the scope and costs can be considerable, especially for large facilities such as a golf course. In these cases, options that are less costly than a full implementation of a comprehensive BMP program can be initiated, and these options can become "show-and-tell" cases for future decisions. A couple of approaches are as follows:

- Initiate the least costly but potentially most effective BMP practices across the site. Examples could be an improved cultivation program, gypsum applications at a sufficient rate with the expectation to improve surface Na permeability issues, acidification of the irrigation water when it is essential to remove excessive bicarbonates and carbonates to allow Ca additions to become more effective, or fertilization adjustments to correct any nutrient imbalances or deficiencies. In this option of limited BMPs across the site, monitoring of soil, water, and tissue analyses should be sufficient to demonstrate any improvements as well as what may occur on a visual basis—pre and post pictures are good. The key point is that it is better to start with some BMP practices than to continue to wait for the "right time" to start.
- Another option is the "show-and-tell" site, where a specific *demonstration site* is chosen to implement a full BMP program in contrast to a limited BMP program noted in the previous paragraph. Depending on the site need, this may include sand capping, drainage, improvements on the irrigation system, intensive surface and subsurface Ca and other amendment applications, increased frequency in cultivation programs, regularly scheduled salinity-leaching programs (i.e., an essential strategy), and fertilization improvements. When there is widespread plant and soil deterioration across a site, the location for the demonstration area should be readily visible that will allow improvements to be easily observed in order to make future decisions based on expected grass performance results. A successful demonstration site management program is often the catalyst for additional budget allocations for more large-scale salinity BMP implementations on selected areas.

18.2.5 Be Dynamic

The dynamics that cause salinity problems on recreational turfgrass sites are constantly changing. The soil salinity chemistry and physical interactions in the ecosystem occur 24 hours per day, 7 days a week, and 365 days a year, regardless of temperatures or seasonal climatic changes. Thus, throughout this book, the dynamic nature of salinity stresses has been emphasized where salinity levels can change dramatically across the landscape, within the soil profile, and over time. Salinity stress is usually magnified with extreme environmental conditions:

- Prolonged windy conditions that significantly increase evapotranspiration (ET) rates and pull salts to the soil surface through capillary pores plus increase the demands for more frequent irrigation cycles with salt-ladened water
- Prolonged high temperatures that also increase ET rates and more frequent irrigation
- Severe drought conditions that challenge the capability for sustaining soil moisture requirements to meet turfgrass demands and irrigation scheduling for salt-leaching capabilities

The dynamics of salt movement either horizontally or vertically in the soil requires a sequence of monitoring events for subsequent implementation of turfgrass management criteria, namely, an initial baseline assessment (establishing a foundational data bank); validation monitoring on a rigid scheduled basis, thereafter based on salt load in the irrigation water and potential for salt accumulation in the soil profile; and a two-pronged long-term monitoring program that encompasses (a) turfgrass performance expectations in concert with budgets for salt management, and (b) compliance or contingency goals for environmental sustainability (International Audubon requirements; local, state, and federal environmental regulations; water conservation mandates; etc.).

Due to the dynamic fluxes of salts in the ecosystem, the turfgrass management philosophy must be equally as dynamic (i.e., proactive). As noted in the previous section, labor, equipment, and budget restraints may limit full implementation of a comprehensive salinity BMP approach. However, when salinity continues to be applied to a site at a level that can result in salt accumulation, limited action is better than no action (such as starting with the worst salt-challenged problem areas first). Delayed action will result in more costly and complex corrective management programs in the future. Proactive monitoring is a key part of responding to reoccurring salt challenges since it allows a better understanding of the nature of changing salinity stresses on specific areas of a site and allows wise science-based decisions to be made with the available resources and within budget constraints.

18.3 CRITERIA FOR PROACTIVE MONITORING

18.3.1 Summary of Proactive-Monitoring Criteria

In this section, the focus is on providing an overview and summary of proactive monitoring of a salt-affected site for the purposes of assessing salinity management progress and making BMP adjustments for long-term site sustainability. The various assessment criteria have been discussed in detail in other chapters, so only a summary is presented and key guideline tables are noted.

Any environmental characterization program should encompass site-specific physical conditions (soils and soil profile variability, geography, topography, hydrogeology, and microbiology) to assess the type, distribution, heterogeneity, and surface + subsurface salinity limitations in the turfgrass ecosystem. Basic questions to ask in characterizing the severity of the salinity problems include the following:

- Is the irrigation water the primary source for reoccurring salt additions?
- Are there inherent site infrastructure problems that are contributing to soil salt accumulation problems?
- Are all drain lines open and functioning properly?
- Are traditional amendment applications contributing to the salt problems?
- Are these amendments providing enhancement to actual turfgrass performance in concert with your current budget?
- What proactive science-based monitoring program will be needed in order to make effective and constructive salt management decisions for grass sustainability?

Soil fertility and salinity status is the center of a proactive-monitoring program. Routine soil fertility tests (e.g., Mehlich III or ammonium acetate extractions, depending on pH) plus base saturation data for assessing the nutritional status of salt-affected soils provide some important information for nutritional adjustments, but do not represent the full salinity impact on the turfgrass. The saturated paste extract (SPE: water extraction) is used as a soil chemical test to evaluate actual soil salinity status and its subsequent potential impact on turfgrass salt tolerance level, performance, and sustainability. Information provided by SPE tests includes the following:

- Total soluble salts in ppm or mg/L, measured by electrical conductivity (ECe in dS/m or mmhos/cm) of the SPE (generally ECe × 640 = total dissolved salts [TDS]; for soils containing appreciable gypsum, the conversion is ECe × 700 = TDS; for ocean water, the conversion is ECw × 744 = TDS)
- Sodium status (sodium permeability hazard), in terms of sodium adsorption ratio (SAR), which can be used to estimate exchangeable sodium percentage (ESP) (note: SAR and base saturation Na are generally similar values). The ESP is part of a routine soil test where the percent Na on cation exchange capacity (CEC) sites is the ESP value.

- Specific toxic ions (sodium, chloride, sulfates, and bicarbonates) in meq/L and ppm or mg/L. These water-extractable values are useful since these ions are primarily in water-soluble forms in the soil.
- Nutrients potentially in soil solution in meq/L and/or ppm or mg/L, but these are not reliable for determining soil nutrient status except for S and B. The sufficiency level of available nutrient (SLAN) values are solidly based in science, but water-extractable values should not be used for fertilization decisions (see Table 5.7).

A routine soil test should contain the information noted in Table 5.8, while a salinity soil test package should contain the information described in Table 5.9. Some general quick-reference guideline tables are presented below for determination of *soil salinity* challenges:

- Table 9.1, "Classification of Soil Saturated Paste Extract (SPE) for Total Salinity Problems"
- Table 3.1, "General Soil ESP and SAR Guidelines for Na-Induced Permeability Problems, Assuming Intermediate Irrigation Water Quality; and Approximate Conversions"
- Table 13.1, "Estimated Gypsum Requirements per 12 Inches (30 cm) of Soil Depth to Reduce ESP to Below 10% Based on Soil Texture and Initial ESP"
- Table 5.4, "Typical Soil Test SLAN 'Medium' Sufficiency Ranges for Macronutrients Using Common Extractants and with CEC < 15.0 cmol/kg"
- Table 5.5, "Micronutrient Extractants (Fe, Zn, Cu, and Mn) and 'Medium' Soil Ranges Used by Many Laboratories for Plant Micronutrient Availability"

Proactive-monitoring programs for salinity assessment must be a major part of the turfgrass management program. Grid sampling of soils to effectively identify salt distribution patterns in soil profiles is time consuming, but this sampling protocol is a critical salinity management requirement. On greens, 0–3-inch (0–7.5-cm) and 3–6-inch (7.5–15.0-cm) depth sampling in sand-based greens is a recommended sampling approach in concert with SPE testing to determine the magnitude of total salt and specific salt ion (sodium, bicarbonates, sulfates, and chloride) spatial distribution and possible upper soil profile layering that could be impacting turfgrass performance, especially around the crown region and root system. This dual soil depth sampling approach can also provide some indications regarding leaching effectiveness in those greens profiles. Since chlorides are highly mobile and sulfates move readily through the soil profile, if those two salt ions are generally higher in concentrations in the top 3 inches (7.5 cm) than the lower zone or in the irrigation water, your leaching program is probably not as efficient as it needs to be for effective salt management. The BMP decision then could involve possible aeration- or irrigation-scheduling changes or both.

Electromagnetic induction, surface grass performance mapping, and remote sensing have been used to identify possible soil profile salt challenges (Lesch et al., 1995a, 1995b; Dang et al., 2010). Electromagnetic induction has also been used to quantitatively evaluate soil salinity and its spatial distribution in field surveys (Yao and Yang, 2010). Recently, Krum et al. (2011) and Carson et al. (2010) have demonstrated electrical resistivity and capacitance and frequency domain sensors, respectively, as accurate means to determine spatial patterns of soil salinity with mobile sensor platforms designed especially for turfgrass situations.

Salt- and sodium-affected soil surfaces can be spatially discriminated using remote Landsat sensing technology (Metternicht and Zinck, 1997). Multispectral radiometry has been used to quantify stress responses in turfgrasses (Trenholm, Carrow, and Duncan, 1999; Trenholm, Duncan, and Carrow, 1999; Jiang et al., 2003). Small handheld TDS meters and electrical conductivity (EC) probes can be used to quickly and routinely monitor both soil and water when salt and drought stress symptoms occur in the turfgrass ecosystem. Soil penetrometer devices are available for monitoring compaction problems that can be associated with salt accumulation.

Table 6.13 lists the information that should be in irrigation water quality analyses. Some general quick-reference guideline tables are listed below for the determination of irrigation water salinity challenges:

- Table 6.5, "Irrigation Water Salinity Classification (i.e., Total Soluble Salts) Relative to Plant Salinity Tolerance, Infrastructure, and Management Requirements"
- Table 6.6, "SAR_w and adj SAR Guidelines to Determine Sodium Permeability Hazard and Classification as Affected by Clay Type and ECw"
- Table 6.8, "Specific Ion Toxicity (Na, Cl, and B) and Miscellaneous Chemical Constituent Problems in Sprinkler Irrigation Water for Sensitive Plants"
- Table 6.9, "Guidelines for Nutrient Content of Irrigation Water to Be Used for Turfgrass Situations; and Quantities of Nutrients Applied per Acre-Foot on Irrigation Water"

Suspended solids testing for "fines" can be determined with normal water quality testing, and with a complete particle physical analysis, such as with specialty testing labs such as Spectrex Technologies (http://www.spectrex.com).

Tissue testing is absolutely critical for making adjustments in your fertility program when growing turfgrass under salt exposure conditions. The data from water and soil samples are indicative of potential limitations and excesses in the specific grass ecosystem. Tissue data will indicate exactly what concentrations of nutrients that the turfgrass plant was able to absorb when salt ions are constantly interacting with the availability of those nutrients. However, a comprehensive data set—namely, irrigation water quality + normal soil fertility tests + tissue data over time—should be used together to make the fertility program adjustments in order to sustain and maximize turfgrass performance.

Clippings should be collected, dried, and shipped in paper bags to the testing laboratory on a regularly scheduled basis (i.e., monthly for greens, and quarterly for other stressed areas). When stress symptoms in the turfgrass areas are observed, a collection of clippings should be expedited to the laboratory to determine the primary nutritional insufficiency limitations to grass performance. The accepted and most accurate procedure for analyzing these tissue samples is wet chemistry and spectrophotometric procedure and not the near-infrared reflectance spectral (NIRS) procedure that some laboratories prefer to run because of time and cost issues. Wet chemistry techniques can include atomic absorption spectrometry, inductively coupled plasma emission (ICP) spectrometry, or direct-current (DC) plasma spectrometry (Happ, 1994).

Drainage is an essential infrastructure component that must be functioning continuously, since salts are constantly moving with gravity dispersal, subsurface and surface movement, or capillary movement upward in the soil profile. Drainage installation should extend beyond the normal herringbone and French drain designs to include interceptor drains and mini–dry wells that are positioned to collect migrating salts in the soil profiles, especially with the changing topography designs that are inherent on golf courses.

18.3.2 Cautions

Due to the serious nature of salinity stresses, there is no place for pseudo-science for unsubstantiated claims for products that are used. When such products become the basis for a salinity management program, monitoring of progress normally does not show the results anticipated—unless some other factor gives a positive response that is incorrectly attributed to the product in question. Science-based BMPs work, but they must be based on true science.

Water treatment that utilizes acid substitutes (not the traditional acidifying agents that include sulfuric acid, pHairway, and N-phuric acid, for example) for reducing bicarbonates and sodium, and for improving both water quality and the soil rhizosphere environment, has not proven beneficial either economically or chemically (Stowell et al., 2008; Whitlark, 2010). Many of these new synthetic acid replacement products have been used historically and extensively for industrial cleaning purposes. The products have not promoted Na^+ or HCO_3^- removal from sand-based greens when following the recommended and labeled application programs (Stowell et al., 2008; Whitlark, 2010).

Several nonchemical water treatment products are being marketed with claims regarding saline water conditioning. So-called inline water conditioners tout electromagnetic, electrolytic (alternating electrical fields), light-related or far-infrared, depressurization, oscillation and vibration, and catalytic redox treatments to "change the chemical and physical properties of water," but to this point, unbiased scientific data do not support the agronomic or economical use of these devices on turfgrass ecosystems (summarized in Duncan et al., 2009, 279).

Several critical points to remember for monitoring salinity challenges (also refer to Chapter 5 and Table 5.1):

1. The data will only be as good as the samples that are collected; your turfgrass management decisions will be developed and implemented based on science-based data from these samples.
2. Submit the samples to a reputable lab, preferably one that regularly analyzes samples for recreational turfgrass venues. Caution should be exercised when the only analytical testing protocol is performed by companies who sell products as a result of their sample-testing program.
3. Water samples should be analyzed for overall salinity and nutrient composition (see Table 6.9). In general, water distribution facilities and recycled (effluent) treatment facilities are testing the generated water based on local, state, and/or federal or country guidelines for exposure to humans and animals once the water leaves their facility. Chlorides, nitrogen, and phosphorus are generally included with oxygen composition data in those assessments because of regulatory guidelines; but major salt ions, such as sodium, bicarbonates, and sulfates plus other nutrients, are usually not tested. Send the water samples to a reputable agricultural laboratory for salinity testing.
4. Ask for testing procedures (e.g., Mehlich III or ammonium acetate) that are used for soil tests. Complete physical analyses of soil samples should only be sent to specialty laboratories that are sanctioned for performing those tests.
5. When you are concerned about the data, ask questions until you get a proper science-based answer to those concerns. Your management decisions are based on understanding the salinity constraints that are impacting your specific site.
6. There are no miracle products or devices that will solve salinity problems. The management program needs to be basic, be specific target oriented, and include a whole-system approach when making management decisions.

Section IV

Environmental Stewardship and Sustainability

19 Sustainable and Environmental Management Systems

CONTENTS

19.1 SUSTAINABLE ENVIRONMENTAL MANAGEMENT

Recreational turfgrass sites are diverse and dynamic ecosystems, being impacted by multiple environmental issues that in the 21st century require proactive, systematic, and comprehensive management strategies to deal with individual environmental challenges (requiring site-specific best management practices [BMPs]) on a site. The turfgrass industry, societal, regulatory, and political entities are concerned about how to manage environmental problems on a site-specific basis, and this is the driving force behind the growing interest in *sustainable development and management.* Therefore, turfgrass management should entail management of each individual environmental issue on a site—this requires development of an appropriate environmental management plan for each individual environmental issue (such as salinity issues) *and* a comprehensive environmental management plan or approach that encompasses and functionally integrates all environmental management issues on a site.

Salinity is not a minor environmental issue or challenge on facilities with the use of saline irrigation water or when managing salt-affected soils. *In this chapter, the concept addressed is how a BMP for salinity management for a specific site (i.e., an individual environmental issue) integrates into an overall environmental management plan for the site.* Increasingly, turfgrass managers (similar to managers of any business or manufacturing facility) are expected to develop individual BMPs for each potential environmental issue and to incorporate these BMPs into an overall sustainable management approach. These are living documents in the sense that (a) environmental management concerns are to become a part of daily management decisions; and (b) these environmental management strategies can evolve over time with new technology, knowledge, and resources. A starting place is to understand the nature and types of successful and sustainable environmental management plans.

19.1.1 UNDERSTANDING ENVIRONMENTAL MANAGEMENT PLANS

As previously noted, all enterprises (agriculture, turfgrass sites, manufacturing, businesses, etc.) are expected by society to effectively address any environmental issue that may arise. It is important to recognize that environmental issues can only be successfully addressed by site-specific, science-based management and not by "one-size-fits-all" bans or mandatory edicts. This was the reality that caused the U.S. Environmental Protection Agency (USEPA) in 1977 to evolve and adopt the BMP concept for protection of surface waters and groundwaters from pesticides, sediment, and nutrient pollutants (Carrow and Duncan, 2008). The BMP concept is the "gold standard" management approach for any single environmental issue since the characteristics remain the same, but the particular "strategies" are specific to each environmental issue. The BMP approach has been adopted by the turfgrass industry for both water quality and quantity challenges (Beard and Kenna, 2008; Carrow and Duncan, 2008), and specifically for water conservation in sports turfgrass ecosystems (Carrow, 2008). In this current book on BMPs for salt-affected sites, we have provided systematic details that outline the BMP approach for salinity issues.

Most sites have more than one environmental issue; therefore, the *environmental management systems* (EMS) concept started to become prevalent on a worldwide basis in the 1990s as a comprehensive environmental management approach for any facility (Carrow and Fletcher, 2007a, 2007b, 2007c). The EMS approach encompasses all environmental issues on a facility where (a) the site is assessed to determine specific environmental issues that are present; (b) for each environmental problem, BMPs are developed to manage it; (c) BMPs for all environmental issues are combined together to form the EMS plan and subsequent holistic document; and (d) management aspects in the BMPs are expected to be incorporated into daily management decisions (to be successful, implementation of each component is essential). The Australian Golf Course Superintendents Association in conjunction with Environmental Business Solutions (EBS; 2011) developed an EMS-based environmental management program for golf courses, called ePar, that allows golf courses to develop much of the documentation online. For individual golf courses or other turfgrass facilities, Audubon International (2011) and Audubon Lifestyles (2011) have similar comprehensive programs that foster environmental stewardship in a practical and science-based manner.

Sustainable management system terminology has increasingly replaced the EMS name in recent years as the best environmental management approach to address all environmental issues on a facility. Essentially, the EMS and sustainable management system concepts are the same, except that the sustainable management system approach implies and emphasizes a more balanced approach that considers not only environmental issues but also economic, social, and site use aspects.

Sustainability has become the key concept for many groups in recent years with the core concern being the environment. But, as the previous paragraph implies, the proper definition is essential, since sustainability is not always defined the same among the various enterprises. Some environmental activists groups define *sustainability* only in narrow terms, such as solely environmental considerations; or even on a single particular environmental goal, such as protecting the spotted owl or the California delta smelt; but this narrow definition leaves out critical adverse impacts on the economy and society, and even interactions with other environmental issues. *Sustainability* should be defined in the full dimension of potential impacts—sustainable resource management that is relative to all the environmental issues at a facility, not just one goal, and that includes economic effects and society impacts. The USEPA (2011) states,

> The 1970 National Environmental Policy Act (NEPA) formally established as a national goal the creation and maintenance of conditions under which humans and nature "can exist in productive harmony, and fulfill the social, economic and other requirements of present and future generations of Americans." Over the past 30 years, the concept of sustainability has evolved to reflect perspectives of both the public and private sectors. A public policy perspective would define sustainability as the satisfaction of basic economic, social, and security needs now and in the future without undermining the natural resource base and environmental quality on which life depends. From a business perspective, the goal of sustainability is to increase long-term shareholder and social value, while decreasing industry's use of materials and reducing negative impacts on the environment. Common to both the public policy and business perspectives is recognition of the need to support a growing economy while reducing the social and economic costs of economic growth. Sustainable development can foster policies that integrate environmental, economic, and social values in decision making. From a business perspective, sustainable development favors an approach based on capturing system dynamics, building resilient and adaptive systems, anticipating and managing variability and risk, and earning a profit. Sustainable development reflects not the trade-off between business and the environment but the synergy between them.

19.1.2 EMS or Sustainable Turfgrass Management Plans

The historical development of EMS is discussed in Carrow and Fletcher (2007a, 2007c) and USEPA (2007a, 2007b). The subsequent evolution of EMS into the golf course industry is documented in Carrow and Fletcher (2007b, 2007c). The 17 key steps for developing and implementing a salinity-oriented EMS or sustainable plan are listed in Table 19.1. The site assessment aspects (item 3 of Table 19.1) and development of individual BMPs for each environmental issue (item 5) are especially critical. In the EMS and sustainability concept, detailed individual environmental BMPs for each environmental issue identified on the facility must be comprehensively incorporated into daily standardized management decisions throughout the entire facility. This approach is used to facilitate overall environmental performance, prevent pollution, enhance environmental compliance standards, increase system management efficiency, and reduce or mitigate risks and liabilities.

From a turfgrass ecosystem standpoint, Carrow and Fletcher (2007b, 2007c) and Carrow et al. (2008) noted that many golf courses have at least 17 environmental issues requiring individual BMPs targeted toward overall sustainability of the facility (Table 19.2). A review of this list reveals that salinity is not often included; but for sites that are salt affected or receive saline irrigation water,

TABLE 19.1
Elements of an Overall EMS Plan

1. Document environmental principles and policy for the turfgrass facility.
2. Document legal and other requirements for the recreational turfgrass site.
3. Identify and assess significant environmental aspects and impacts (a comprehensive, site-specific, whole-ecosystem assessment)—this is the *site assessment phase*, and salinity site assessment is more complex than assessment for any other environmental issue. This step is critical in the development of realistic salinity BMP plans (see item 5).
4. Determine objectives and targets (including planning, designs, renovations, and daily operational guidelines) for the recreational turfgrass site.
5. Develop comprehensive environmental management programs (i.e., BMP-based ones) for each issue—include current management practices, necessary infrastructure improvements, and current and future monitoring practices culminating in an action plan for daily implementation. This is the *management phase*.
6. Establish project structure and responsibility (accountability).
7. Document issues dealing with training, awareness, and competence.
8. Communication and outreach.
9. EMS documentation (to demonstrate actual implementation of individual BMPs and their collective effectiveness on the whole environmental management system).
10. Document control.
11. Establish operational control.
12. Emergency preparedness and response.
13. Proactive monitoring and measurement.
14. Determine nonconformance guidelines plus corrective and preventative action (checks and balance, and flexibility in making necessary adjustments on a specific site) programs.
15. Environmental records.
16. EMS audit.
17. Management review and adjustment for improvement.

Source: From Carrow, R. N., and K. A. Fletcher, Environmental management systems: A new standard for environmental management is coming, *USGA Green Section Record,* 45 (4), 23–27, 2007a; Carrow, R. N., and K. A. Fletcher, *Environmental Management Systems (EMS) for Golf Courses* (an educational guidebook developed by the University of Georgia and Audubon International), 2007c, http://www.auduboninternational.org/e-Source/ and http://www.georgiaturf.com; USEPA, *Environmental Management Systems*, USEPA, Office of Water, Washington, DC, 2007a, http://www.epa.gov/ems/index.html; USEPA, *Key elements of an EMS*, USEPA, Office of Water, Washington, DC, 2007b, http://www.epa.gov/ema./info/elements.html.

Note: Key items are 3 and 5.

TABLE 19.2
Common Individual Environmental Challenges on Golf Courses That Would Be Addressed in a Comprehensive Environmental Management System (EMS) or Sustainable Turfgrass Management Plan

1. Environmental planning and design of golf courses, additions, and renovations (Carrow and Fletcher, 2007c)
2. Sustainable maintenance facility design and operation (Carrow and Fletcher, 2007c)
3. Turfgrass and landscape plant selection (refer to Chapter 9)
4. Water-use efficiency and conservation (Carrow et al., 2009c)
5. Irrigation water alternative sources and quality management (refer to Chapters 10 and 11; Duncan et al., 2009)
6. Pesticides: water quality management (Carrow and Fletcher 2007c)
7. Nutrients: water quality management (refer to Chapter 16; Carrow and Fletcher, 2007c)
8. Erosion and sediment control: water quality management (Carrow and Fletcher, 2007c)
9. Soil sustainability and quality (refer to Chapters 2 and 3)
10. Stormwater management (Duncan et al., 2009)
11. Wildlife habitat management (Carrow and Fletcher, 2007c)
12. Wetland and stream mitigation and management (Carrow and Fletcher, 2007c)
13. Aquatic biology and management of lakes and ponds (Duncan et al., 2009)
14. Waste management (Carrow and Fletcher, 2007c)
15. Energy conservation and management (Carrow and Fletcher, 2007c)
16. Clubhouse and building EMS concepts (Carrow and Fletcher, 2007c)
17. Climatic and energy management (Carrow and Fletcher, 2007c)

Source: Carrow, R. N., and K. A. Fletcher, *Environmental Management Systems (EMS) for Golf Courses* (an educational guidebook developed by the University of Georgia and Audubon International), 2007c, http://www.auduboninternational.org/e-Source/ and http://www.georgiaturf.com.

Note: Salinity BMP management can be included as a separate (no. 18) environmental challenge or can be incorporated into specific components (e.g., nos. 3, 4, 5, 7, and/or 9, to name a few challenges).

salinity must be included—in fact, it can be *the* most challenging of all the environmental issues to address due to the dominating complexity and scope of management required to address salinity challenges. On salt-challenged sites, the authors suggest developing the salinity BMP plan first before the other BMPs since it impacts many of these additional environmental issues in a significant manner.

For salt-affected sites, the development of a salinity BMP plan and incorporation of it into the overall EMS and sustainability plan can be addressed in one of two means. First, a comprehensive salinity BMP plan encompassing the areas discussed in this book and outlined in Table 1.4 can be developed and added as item 18 to the Table 19.2 list (i.e., salinity is highlighted as a specific environmental issue similar to the other 17 environmental concerns). For sites with significant salinity stresses that will be ongoing, the authors suggest that this approach be used since it highlights the importance of managing salinity in a science-based manner using carefully developed salinity BMPs as outlined in Table 1.4. Where there is overlap with another environmental issue BMP, it can be noted in the other BMP (e.g., soil sustainability and quality) that salinity issues are dealt with in the specific salinity BMP section.

A second approach is to incorporate specific salinity BMP components into one of the 17 environmental issues where it is relevant, such as incorporating the saline water quality issues under item 5, "Irrigation Water Alternative Sources and Quality Management"; the soil salinity aspects under item 9, "Soil Sustainability and Quality"; the nutrient issues under item 7, "Nutrients"; and so on. Regardless of which approach is selected in developing the overall EMS or sustainable management plan, Section 19.2, "Components of a Sustainable or EMS Plan on Salt-Affected Sites," will be useful in understanding how to formulate the salinity BMP plan. As previously noted, salinity has such a broad impact on sites that it affects many other environmental issues.

19.2 COMPONENTS OF A SUSTAINABLE OR EMS PLAN ON SALT-AFFECTED SITES

An overall salinity EMS or sustainability program involves a four-way interaction among irrigation water quality, the soil profile, the salinity tolerance level of the specific turfgrass cultivar, and the climate. A turfgrass manager has partial to full control over all of these components except for the climate. The manager must have a proper infrastructure in place on the recreational turfgrass site, and, coupled with experience and knowledge concerning salinity challenges and having flexibility in implementing changes in the management program, he or she should be able to adjust to climatic extremes in temperature and moisture to sustain acceptable turfgrass performance requirements. Implementing management changes will be governed by individual salinity BMP action plans that were designed for the particular recreational turfgrass site. The BMP approach is appropriate for every environmental issue, but the individual management strategies that are combined to develop a holistic, science-based BMP are specific to each environmental issue—and these are adjusted on a site-specific basis to achieve the best environmental plan for the site. Components of the critical salinity BMPs can be integrated into the overall EMS or sustainability program for implementation as outlined below. In Table 1.4, the individual components of salinity BMPs are summarized, including both site assessment and management aspects.

One "water quality" aspect that requires clarification is that water quality concerns in the Clean Water Act for protection of surface and subsurface waters are focused on pesticide, nutrient, and sediment pollutants and not what additional comprehensive components can be found in the irrigation water (i.e., except nutrients, but not total soluble salts or Na) (USEPA, 2003). But, with saline irrigation water, total water salinity, Na, nutrients, and nonnutrient salts are all additional water quality issues that must be addressed in the overall sustainable EMS plan. In Table 19.2, the traditional pollutants of pesticides, nutrients, and sediment are listed separately since the BMP strategies differ for each component.

The *initial site assessment* to determine the complex salinity impact on the recreational turfgrass environment is critical to formulating an implementation plan for the overall salinity EMS on a particular site. A qualified professional consultant will usually be required to analyze water supply sources encompassing supply adequacy, economic viability, engineering considerations, and both short-term and long-term environmental impacts including salinity when developing this individual BMP (Carrow and Fletcher, 2007c). Each irrigation source must be assessed for water quality, including any potential blending options that might reduce the total salt and specific salt ion (Na, Cl, bicarbonates, and sulfates) load, and that could minimize the long-term ecosystem salt accumulation and subsequent functional impact to the environment.

The *irrigation water quality* BMP test should be multidimensional and encompass the following components:

- Total salinity impact on turfgrass, landscape plants, soil profile, surface and subsurface waters, native flora in and around lakes and streams, drainage discharge, and soil sequestration of salts
- Sodic soil impacts on soil physical properties, and all plant communities
- Nutrient and element toxicities and imbalances on turfgrass and landscape plants, and native flora
- Nutrient and Element × Salinity interactions on fertilization, and water treatment
- Plant health issues, including Biotic Stress × Salinity interactions and Abiotic Stress × Salinity interactions
- Other environmental issues such as eutrophication, pesticide efficacy, erosion, and sediment control

Integrated pest management (IPM, where IPM is the same as BMPs, but evolved with a different terminology) guidelines governing pesticide chemical use reduction and safety that are encompassed in the water quality management BMP include volatilization, water solubility, sorption, plant uptake, degradation, runoff (buffer zones), leaching, biological control, pH × Chemical Efficacy interactions, cultural and mechanical practices, pest-resistant plants, invasive flora, proper disposal of containers, and wash-off areas (Carrow and Fletcher, 2007c).

Nutrient fate issues that are encompassed in the water quality management BMP include leaching control and prevention, runoff management, minimizing use (proactive soil and tissue testing, and prescription fertilization), volatilization, enhancing nutrient use efficiency (plant selection, genetic improvements in nutrient use efficiency, and the production of deep and viable roots and increased total functional and sustainable root volume), Salt Ion × Nutrient interactions, site contamination and runoff, fertilizer storage, and disposal of containers (Carrow and Fletcher, 2007c).

The *erosion and sediment control assessment* in the water quality management BMP includes construction; land renovation; sediment management (vegetative buffers, filters, and detainment or catchment facilities); quantification and analysis of pond, lake, and stream sediment; soil degradation by wind and water erosion; salt ion sequestration capability; and soil deposition onto landscape and adjacent native recreational areas (Carrow and Fletcher, 2007c).

Alternative irrigation water sources must be documented in the water quality management BMP (Carrow and Fletcher, 2007c; Duncan et al., 2009). These sources can include individual resources or combinations from the following: larger streams, rivers, or canals; natural or constructed surface lakes or ponds; harvested water from flood control diversion channels, surface runoff from surrounding terrain during normal rainfall events, or storm runoff from impervious surfaces that is captured in retention ponds or other catchment facilities; groundwater (both saline and nonsaline) from deep or shallow wells; groundwater from aquifers (from both saline and nonsaline strata); ultrapure water sources (snow melt, ice melt, and high-volume stormwater collection); reclaimed (water reuse, wastewater, and effluent) water from tertiary sewage treatment facilities; recycled water from drainage or stormwater lines, or wastewater collected from soil-filtered leaching fields; seawater or lagoon water blends; blending from desalinization (reverse-osmosis or similar) technologies; and proper permitted disposal of reject concentrates from desalinization treatment.

A listing of the considerations for inclusion into the water quality BMP under the overall salinity EMS includes the following (Carrow and Fletcher, 2007c):

- Comprehensive water quality testing (contaminants and nutrients: deficiencies, imbalances, and interactions with salt ions; total salinity; and specific salt ions)
- Water treatment costs (acidification; fertilizer injection; reverse-osmosis and desalinization, including permitted discharge of reject concentrate; chemigation such as gypsum injection; storage facilities; and blending options)
- Location of the alternative water sources, including possible storage facilities or features
- Water supply developmental needs, costs, and potential problems
- Design and installation costs for wells, ponds, pumping volume, distribution lines, treatment equipment, fertigation equipment, well-field layout, drainage recycle features, watershed harvesting including pollutants (sediment, nutrients, oil and grease, metals, organics, pesticides, bacteria and viruses, and filters for trash) and their treatments, onsite reclamation, onsite desalination (throughput, reject concentrate disposal, and salt load), and dry well and sump pump disposal of excess high-salinity leachate
- Pond and lake location, construction, and inflow and outflow features (multiple pond and lake piping connections); and aquatic management (acidity; nutrient loads; eutrophication potential and monitoring; nuisance algae and mycocystin toxin production; vascular plant weed control; oxygen depletion and anoxia potential; surface runoff additions, contamination constituents, and sedimentation; salt layer stratification by depth; depth of irrigation pump intake; and fish management and monitoring)

- Pond and lake seepage control measures (chemical, clay surface capping layer, and liners for both upward subsoil salt migration and downward salt intrusion into other water resources such as aquifers)
- Seasonal water-level fluctuations (well yield and drawdown, surface water collection and storage, stream flow, and wet and dry season changes)
- Recycled (reclaimed or effluent) water source: receipt on demand or continuous supply, storage facility requirements, blending options, additional onsite treatment, volume by seasons, and quality fluctuations
- Permitting, regulatory negotiations, water rights, and competition for (a) source(s)
- Regulatory habitat compliance and actual volume water use issues for the site
- Regulatory use, transportation to site, and storage onsite for effluent water
- Watershed runoff potential for harvesting, storage, and blending
- Buffer zone regulations governing water quality (e.g., sediment and erosion) protection

Water conservation (site-specific water budget) mandates for the location include a number of considerations (Carrow et al., 2009c):

- Water price regulations, structures, rebates, and use incentive plans—these are regulatory aspects that the facility may be able to incorporate that would be advantageous.
- Characterization of the underlying aquifer (contamination issues, recharge potential, drawdown, conservation regulations, and permitted pumping volume)
- Water removal interactions from a source that impacts wetlands (biofiltration including vegetative filter strips and infiltration trenches, bank stabilization, buffer zones, and detention areas), streams, sink-hole problems, catchment facilities, waste bunkers, and wildlife migration and survival—naturalized areas and connecting corridors, IPM concepts, protection, habitat restoration, and invasive flora or vertebrates
- Other strategies that impact water conservation and water-use efficiency as discussed by Carrow et al. (2009c), such as plant selection, landscape design, irrigation design for water application uniformity, irrigation-scheduling approaches and tools, and adjustments in management practices to favor water conservation such as cultivation, soil modification, fertilization, wetting agents, and so on
- Energy costs to move irrigation water to the turfgrass site, management and maintenance costs, backup water resource during climatic extremes (i.e., severe prolonged drought), floodwater diversions, off-peak pumping times, and sump pump disposal of leachate

The *soil quality* BMP component in the overall salinity EMS should be comprehensively assessed to determine initial salinization and potential for additional secondary salinization on the site when continually applying saline irrigation water to the perennial turfgrass ecosystem in the future. *Soil quality* is the capacity of a soil as a natural resource to function in a native, natural, or managed ecosystem in order to sustain plant (and animal) productivity, maintain or enhance water and air quality, and support human health and habitation (Carrow and Fletcher, 2007c). The environmentally functional attributes of soil quality include regulating water directional flow both on the surface and in the subsurface, sustaining plant and animal life, filtering potential pollutants (minerals and microbes filter, buffer, degrade, immobilize, and detoxify organic and inorganic materials as well as industrial, municipal, and atmospheric byproducts and deposits), cycling nutrients, particle size migrations, salt ion migrations, minimizing erosion, and supporting structures.

Soil quality degradation occurs from loss of organic matter; loss of microorganism and biotic diversity; wind and water erosion; compaction; reduced water infiltration, percolation, and drainage; loss of oxygen flux (air porosity) creating anaerobic conditions; saline, sodic, and saline-sodic conditions; excessive alkaline or acid soil pH; acid-sulfate formation; excessively coarse textured soils (sands, decomposed granite, and volcanic pumice) with low cation exchange capacity (CEC) and low water-holding capacity; excessive additions of nonbiodegradable inorganic amendments that are designed to sequester water and total soluble salts; and contamination from organic and inorganic constituents.

Key indicators of soil quality include soil organic matter content (increased organic matter equates to increased sequestration of CO_2 from the atmosphere); soil microbial biomass and functionality; aerobic versus anaerobic microorganisms; C and N cycling; CEC; nutrient availability status and balances; available water-holding capacity; hydraulic conductivity and infiltration for water collection; leaching potential; plant and microbial interaction, activity, and biodiversity with increasing salt accumulation; and physical properties related to plant rooting volume and root regeneration. Additional infrastructure components that govern soil quality functionality in saline turfgrass ecosystems include turfgrass genus, species, and cultivar selection and plant genetic salinity tolerance level (additional abiotic stress tolerances such as cold, heat, drought, and wear and traffic are important additions); irrigation design, control features, and distribution efficiency; irrigation scheduling and method (e.g., plant based, soil based, water budget approach, deficit, atmospheric, and pulse cycle); use of nonpotable water sources for blending; water and soil treatments; cultivation efficiency and effectiveness; and surface and subsurface drainage functionality to keep the salts moving and minimize their accumulation in the soil profile.

The *specific genus, species, and cultivar genetic salinity tolerance* BMP component in the overall salinity EMS is a critical consideration in developing and implementing the overall management program for the recreational turfgrass site. Salinity tolerance with most halophytic turfgrass species, such as seashore paspalum, means that the grass tolerates increasing levels of accumulated salts in the soil plus topical absorption of specific soluble salt ions from sprinkler irrigation of saline water up to the individual level of salt tolerance in the specific cultivar. As a result, many of the halophytic turfgrass species are not phytoaccumulators of excess salts and will leave those excess salts in the rhizosphere, which means that the overall salinity EMS program must be oriented continuously on managing accumulating salts via specific BMPs in both the soil and the turfgrass plant. Reducing the salt load in the irrigation water (such as by blending with other lower salinity water sources, using desalinized water, etc.) is an additional management consideration.

The higher the plant genetic salinity tolerance, the more flexibility and time that the turfgrass manager has in making specific BMP adjustments to account for increasing salinity impacts on the grass ecosystem. This whole-ecosystem management approach is essential for long-term environmental stewardship and sustainability. In developing this individual plant salinity tolerance BMP for the overall salinity EMS program, a number of components must be considered:

- Climatic adaptation (warm season vs. cool season, overseeding capability, and interseeding within and among species' salinity tolerance capability)
- Abiotic stress tolerances: salinity (use of nonpotable alternative and variable saline irrigation water), drought (water use efficiency, root volume and regeneration, and transitioning into dormancy or semidormancy), cold hardiness (actual genetically controlled cold-temperature tolerance and dormancy, and transitioning capability going into and emerging from cold temperatures), high temperatures (dry heat with low humidity, and wet heat with high humidity), wear and traffic, root volume regeneration and functional sustainability, and performance with soil compaction (increased soil strength, loss of pore–space continuity, and compression damage) issues
- Biotic stress tolerances: weeds (aggressiveness to outcompete weedy species), insects, diseases, and adaptability to IPM programs
- Site use and use-related stresses: mowing height tolerance (low-height-of-cut manicuring maintenance program, i.e., greens), wear and traffic recoverability, divot recovery, putting quality, playability traits, and overseeding program transitioning capabilities

- Soil and water quality impact stresses: nutrient use efficiency, specific macro- and micronutrient requirements, specific cultivar nutrient sufficiency levels, and total root volume
- Planting, establishment, grow-in requirements (most halophytic turfgrass species do not immediately activate salinity tolerance mechanisms at planting and require low-salinity water during establishment and early grow-in), growth rate from planting to full canopy density, and root volume sustainability and regeneration capability
- Cosmetic appearance: genetic color, efficiency of nutrient-facilitated color enhancers internally in the plant, leaf texture, canopy density, and seasonal transitioning adjustments
- Proactive tissue, soil, and water monitoring to sustain expected performance levels
- Capability for improving infrastructure components to sustain grass cultivar performance with available saline water source: upgrading irrigation system distribution efficiency to enhance leaching effectiveness, drainage installation, cultivation frequency and methods conducive to maintaining adequate pore space that is sustainably effective in moving salts, careful additions of nonbiodegradable inorganic amendments to improve CEC and reduce sequestration of soluble salts, careful use of organic amendments that affect microbial activity and soil stability, and "dialing in" prescription fertility adjustments to the specific cultivar's salt tolerance requirements onsite
- Due diligence based on best available science and knowledge of the ecosystem to know when to upgrade to a more-salt-tolerant cultivar, flexibility in adjusting the overall salinity management program for the new cultivar, and adoption of new technology (such as new sensors, etc.)

Other environmental issues as listed in Table 19.2 where irrigation water quality and salinity issues may impact the specific BMP are as follows:

1. *Onsite waste management*: recycling of wastes and associated byproducts, composting, deactivation treatments, filters, reuse, and reduction practices
2. *Energy management*: irrigation pump system efficiency (pumping at night when low energy costs are in effect); lightning and surge control; heating, ventilation, and air conditioning; hot-water supply; solar energy (such as pond or lake diffusers and aerators); and equipment energy requirements
3. *Climate and energy management*: carbon sequestration; community stormwater management; community water remediation and conservation; green landscapes (temperature, air filtration, glare reduction, and screening); conservation and revitalization of surrounding land areas; fire-wise landscaping; wildlife and plant biodiversity maintenance; low-impact design factors; energy-efficient maintenance facilities, clubhouse, and irrigation system; and environmentally safe spray equipment (such as shielded sprayers)

Even though the climatic BMP involved in the overall salinity EMS is unpredictable and generally uncontrollable, certain infrastructure components need to be in place to provide the turfgrass manager time and flexibility to adjust daily to climate extremes. Key points for consideration include the following:

- Irrigation system distribution uniformity and efficiency to minimize wet and dry spots while effectively leaching excess salts down below the turfgrass root system
- Drainage, drainage, drainage—you cannot manage salts in a perennial turfgrass ecosystem without functionally effective and efficient drainage systems installed over the entire site
- Soil profiles with adequate pore space to facilitate effective salt leaching; and careful organic and nonbiodegradable inorganic amendment additions that are conducive to salinity management

- Cultivation to facilitate salt management: deep aeration (Soil Reliever, Vertidrain, drill and fill, Hydroject, and Dryject), shallow aeration (spiking, slicing, and vertical mowing), site-specific frequency, and microsite management
- Chemical injection equipment attachment(s) on the irrigation system: acidification to reduce bicarbonates or adjust pH, liquid fertilizer nutrient injection, and dihydrate gypsum or hydrated lime injection
- Effective chemical amendment equipment for soil applications: granular gypsum, lime + sulfur, granular fertilizers, and dolomite

Environmental issues are complex and require many diverse and dynamic factors to be considered in conducting a comprehensive salinity impact BMP for each site and then in formulating the daily detailed BMP-facilitated plan to be implemented for the overall salinity EMS. The questions that need to be asked, then, include "Are the BMP details comprehensive enough for the specific site and salinity challenges?" "Can they be implemented effectively on a daily basis and within most economic and water budget requirements?" And "When implemented, are they effective in facilitating long-term site-specific sustainable turfgrass ecosystems when salt challenged in a dynamic ecosystem?"

19.3 A COMMONSENSE APPROACH TO LONG-TERM SUSTAINABILITY

An EMS on a salt-challenged site is an ongoing, cyclic, and dynamic process that requires constant monitoring of each individual BMP (water quality, soil quality, specific turfgrass cultivar salinity tolerance level, and climatic challenges) and encourages constant vigilance to facilitate specific component improvements over time. The overall goal of the EMS concept is long-term environmental stewardship and turfgrass sustainability for a specific turfgrass site. The success or failure of the program is predicated on (a) comprehensive site assessments involving each salinity BMP, (b) development of and flexibility in adjusting to new management paradigms, (c) absolute focus on the entire ecosystem and all components that affect turfgrass sustainability, (d) inclusion of all environmental issues, and (e) development of detailed comprehensive action plans and proactive daily implementation of all components of those plans for the entire facility on a site.

Comprehensive salinity BMP templates must be refined and redefined for each location and within each component of the location based on site-specific knowledge and science. Implementation of the salinity BMP action plans must be integrated into the daily operations of the facility, including agronomic, personnel-training, economic, upper management, and public relations operations. For complex issues, such as water-use efficiency and conservation, irrigation water quality, and salt-affected turfgrass sites, consultants with in-depth understanding of those complexities can be used to refine and readjust the specific BMP action plans.

Salinity BMPs must encompass all possible strategies to address the complex and multiple challenges that affect turfgrass ecosystems. Those strategies should be based on the best and most current scientific knowledge and resources that are available, but with continuous due diligence to integrate new emerging technologies as they develop. The salinity BMP strategies cannot be one-dimensional, involving a single practice, piece of equipment, or "magic bullet product that will solve all the problems."

The dynamic interactive nature of salinity in the environment dictates that only the holistic or whole-system approach will result in successful turfgrass sustainability. Each site is different when assessing complex salinity challenges; and adjustments in the BMP action plans must account for individual ecosystem changes over time. Regional differences in climate, soils, and water quality will modify the site-specific salinity BMPs. Each individual salinity BMP must be frequently monitored in order to adjust the BMPs for growing the turfgrass in a saline environment.

Site assessment is the critical first step in determining the complex salinity problems for the site; if assessment is not done properly and comprehensively, the necessary details needed to formulate the action plan for an individual salinity BMP will be inadequate to sustain the turfgrass ecosystem. If the salinity BMP is not fully implemented, the turfgrass site will be less environmentally and functionally sustainable, and the cascading effects of multiple salinity challenges may overwhelm the ability to manage the turfgrass ecosystem over the long term. Success or failure will be determined by the details in those individual salinity BMPs and action plans.

Section V

Nontraditional Use of Turfgrasses on Salt-Affected Sites

20 Reclamation, Drainage Water Reuse Schemes, and Halophytic Forage Sites

CONTENTS

20.1 NONTRADITIONAL USES OF HALOPHYTIC TURFGRASSES

With increased interest by breeders and geneticists in the development of more salt-tolerant grasses at the interspecific and intraspecific levels for turfgrass situations, these same grasses may have value for use on other salt-affected locations, as noted by Duncan and Carrow (2000). Often the superior plant salinity tolerance is in conjunction with other desirable traits such as excellent drought tolerance, high-temperature tolerance, acid soil tolerance, or other abiotic or biotic stress tolerances. Nontraditional uses of grasses initially developed for turfgrass use may include the following:

- *Reclamation of salt-affected areas* (salinity issues are primary stresses) that may or may not be used for a future turfgrass site, but where a halophytic grass would be an essential component of reclaiming the soil (Loch et al., 2003, 2006; Corwin and Bradford, 2008).
- *Reclamation of sites where other stresses are primary* (e.g., low pH, ion toxicities such as Al and Mn or other heavy metals, and erosion), but salinity is a secondary abiotic stress. Grasses with salinity tolerance and tolerance to the other (often multiple) stresses may be used in the reclamation processes.
- *Drainage water reuse schemes* (Corwin and Bradford, 2008; Dudley et al., 2008; O'Connor et al., 2008). Duncan et al. (2009) reported how halophytic grasses may be included in schemes to reuse drainage water from a site where the drainage water would become increasingly saline after each cropping event.

- *Halophytic forage sites.* Locations where forage production or grazing would be desirable, but the soil is already saline and would be a candidate for planting salt-tolerant grasses (Rogers et al., 2005). In these cases, reclamation of the land is not the focus, but rather the ability to use the abiotically stressed land in a productive manner.

These grass-use situations are not necessarily separate from each other. For example, a salt-tolerant forage grass would be beneficial for many drainage water reuse schemes and on many salt-affected soils requiring reclamation. However, each situation presents management considerations that may be distinct, so they are discussed individually. Also, in many of these cases, there is a high degree of spatial and temporal variation in soil chemical and physical properties related to the abiotic and biotic stress factors on the site that complicates site assessment and remediation (Semple et al., 2006; Corwin et al., 2008; Thomas et al., 2009). In the previous chapters of this book, the focus has been on salinity best management practices (BMPs) for traditional salt-affected recreational turfgrass sites such as golf courses, athletic fields, sod production farms, lawns, general grounds, and parks. Recreational turfgrasses and turfgrass areas in highly visible locations, such as an oceanside park with saline or sodic soils, are the most likely to receive intensive management to address the salinity challenges. The various salinity management BMPs in Chapter 9 through 18, however, would be appropriate to consider in many of the nontraditional uses of halophytic grasses as well as many of the site assessment concepts noted in Chapters 4 through 8.

Diverse land reclamation situations and drainage water-reuse schemes almost always involve the presence of multiple and severe environmental stresses (i.e., that is why reclamation is required). The various salinity stresses addressed in this book may be present as the primary stresses where they will create secondary stresses, especially with sodic soil conditions. Also, other primary stresses are normally present that may be more dominant than salinity stresses. Thus, site assessment is essential to clearly understand the challenges in order to develop appropriate BMPs and to select appropriate grasses. It is beyond the scope of this book to address the nonsalt stresses in detail, but common situations are discussed in Section 20.2, "Reclamation Situations and Site Assessment," to provide an overview of the challenges. Grasses used in the diverse land reclamation situations and drainage water-reuse schemes must be able to tolerate the multiple stresses that are present on a site-specific basis, and this topic is the subject of Section 20.3, "Grass Selection Issues." Knowledge of site stresses and of the specific characteristics of each grass to be used allows preplanting, planting, establishment, grow-in (Section 20.4), and sustainable BMP protocols (Section 20.5) to be developed.

20.2 RECLAMATION SITUATIONS AND SITE ASSESSMENT

Land reclamation situations are very diverse. The situations noted in this section will provide a reasonable overview of common land reclamation cases that may have a salinity component in conjunction with other environmental stresses.

20.2.1 Phytoremediation

One reclamation component that is common in many land reclamation situations is phytoremediation. A brief overview will aid in summarizing the uses and terminology of this technology. In land reclamation literature, various terms are used to indicate reclamation approaches that can be confusing, especially for plant-based remediation. The most common terms are presented here for clarity. In natural ecosystems, plant-based remediation (a form of bioremediation; refer to Shukla et al., 2010, for a comprehensive treatise of bioremediation) can be used to remove contaminants from soil and water, or a process that is collectively termed *phytoremediation*. Phytoremediation techniques depend on the particular contaminants present, including high total salts and Na. Five phytomanagement techniques are associated with phytoremediation (McCutcheon and Wolfe, 1998; Raskin and Ensley, 2000; Vidali, 2001):

- *Phytoextraction*, *phytoaccumulation*, or *hyperaccumulation*: plants sequester and accumulate low levels of contaminants (heavy metals and salt ions) in the roots and shoots from the soil. Plant material can be harvested to remove the salts.
- *Phytotransformation*: uptake of organic contaminants from soil, sediments, or water, and transformation to more stable, less toxic, or less mobile forms via reductive or oxidative enzymes that metabolize or biochemically detoxify certain compounds (enzymes include dehalogenase, nitroreductase, peroxidase, laccase, nitrilase, phosphatase, etc.).
- *Phytodegradation*, *phytochelation*, or *phytovolatilization*: volatile organic compounds are adsorbed, and metal species and other chemical compounds are chemically changed and are then either transpired into the air or toxically deactivated in the rhizosphere.
- *Phytostabilization*: plants control soil gases, pH, and redox conditions in the soil or water by changing their mobility or migration patterns, usually through root exudates (such as organic acids). Leachable constituents are adsorbed and bound into the plant architecture, forming a stable plant mass from which the contaminants will not reenter and contaminate the ecosystem.
- *Rhizodegradation*, *rhizofiltration*, or *phytostimulation*: symbiotic relationship between plant roots and microbes, where the plants supply nutrients, proteins, and enzymes for microbial (bacteria, yeasts, and mycorrhizal fungi) activity and the microbes enhance the soil environment by retarding altered chemical movement.
- *Phytocapping*: an alternative technique for landfill remediation involving either a thick cap of 1400 mm (56 inches) or a thin cap of 700 mm (28 inches) of soil. A compacted clay capping is also an alternative strategy to minimize percolation of water into buried solid waste sites or drinking water aquifers. The plant species can intercept about 30% of the rainfall and can lower methane emissions 4–5 times with subsequent deposition into the atmosphere, compared to nonvegetated sites (Ashwath and Venkatraman, 2010).

20.2.2 Dredged Salt-Affected Soils

A common land reclamation situation is the process of creating new land from sea or riverbeds with saline-sodic landfill (usually fine-textured soils, from dredging of ocean bottoms, ponds, rivers, or lakes) to create soil profiles for nonturfgrass uses or for turfgrass fairways and roughs or other sports facilities. Actions may include use of a dredged soil for capping over the top of rock, caliche, decomposed granite, or pulverized pumice native soils. Salinity management options that may be considered are presented by Loch et al. (2006) in their study of using dredged soils from the ocean for future use as recreational park sites in Australia, which would include grassing with halophytic grasses. Some form of amelioration must accompany these newly reclaimed lands in order to alter the site-specific salinity dynamics, namely:

- *Hydrological improvements* (leaching of dredged saline soils, drainage, irrigation distribution uniformity, flood control, water table control, and blending of water resources to reduce the salt load)
- *Soil topography stabilization and erosion control* (contouring, sand capping, aquifer and surface water protection, catchment or containment facilities, and buffer zones)
- *Soil structural and physical improvements* (chemical adjustments, fertilization, additional organic and/or inorganic amendment applications, and appropriate cultivation processes)
- *Decontamination* (chemical detoxification of excess salts, leaching of those excess salts, and removal of the salts from the site)
- *Rehabilitation* (grassing with salt- and/or alkaline-tolerant species, and sustainable long-term site-specific BMPs)

Reclaiming saline or saline-sodic soils must include characterization of the physicochemical soil properties that impact pH, hydraulic conductivity, aggregate stability, texture composition, and clay dispersion (Lebron et al., 1994). The impact of the reclamation process on saline-sodic soils includes changes in electrolyte concentrations; in the composition of cations in the exchange complex, especially in carbonate-ladened soils; and in pHs that are generally >8.5 unless the site also is an acid-sulfate one (see the next section), in which case excessive acidity is normal. A reduction in the hydraulic conductivity of a soil with poor soil structure is partially caused by migration of fines (silt, clay, and colloidal material), and the cause of this dispersion (which is affected by salt concentration, sodium adsorption ratio [SAR], and pH) will impact the success or failure of the remediation program. Chemical amendments are generally useful when the soil particles have colloidal properties (Lebron et al., 1994). Reclaimed sea sand that was dredged from the Yellow Sea was amended with 5 cm of "mountain soil" plus nitrogen–phosphorus–potassium (NPK) chemical fertilizer and a calcium amendment, and *Zoysia japonica* had the highest cover rate (Joo et al., 2008). This particular treatment of the dredged sea sand reduced the pH, electrical conductivity (EC), alkalinity, and soluble sodium when calcium displaced the excess sodium.

The elimination of soluble salts in a calcareous saline-sodic soil without the addition of amendments may cause an irreversible loss of soil structure since available water content is decreased and the soil becomes easily erodible (Lebron et al., 1994). This enhanced erosion is caused by the substitution of exchangeable Na by calcite Ca and the subsequent accumulation of alkalinity in the soil solution, which affects the bonding between particles to the point of loss in cohesion. When amendments are properly incorporated into the soil profile, the increase in *exchangeable Ca content* also increases the amount of water potentially available for plant adsorption, resulting in more stable soil structure and progress toward salinity reclamation.

20.2.3 Acid-Sulfate Soils

In the study noted in the previous section by Loch et al. (2006), they often encountered acid-sulfate conditions in reclamation programs using dredged soils from the ocean bays along with high-salinity and sodic conditions. We have also encountered this combination of stresses in coastal areas of the eastern United States, where seashore paspalum or other grasses were to be used in golf courses. Acid-sulfate soils bring an extra level of toxicity problems requiring reclamation using halophytic grasses in the remediation process. Waterlogged pre–acid-sulfate soils contain iron sulfides mixed with organic matter under anaerobic conditions (often areas previously covered by ocean water for thousands of years) that when drained and exposed to air, such as when native flora are removed and soils are disturbed for coastal golf course or recreational park construction, sulfuric acid (pH < 1.0) is produced and aluminum, iron, and sulfates plus other heavy metals are released into the soil profile at toxic concentrations. These iron sulfides (such as pyrites and FeS_2) are formed in coastal lowland sediments where sulfur has accumulated and are exposed to oxygen-reducing or anaerobic conditions. The sulfides are stable in anaerobic conditions, but will rapidly start producing sulfuric acid when oxygenated. Development of acidity occurs in a sequence (Dent, 1992; Bush and Sullivan, 1999):

- Pyrite spatial distribution and concentrations in the soil are quite variable; anaerobic soil profiles are disturbed, and oxygenation is initiated.
- Pyrite is rapidly oxidized.
- Fe^{3+} is reduced to Fe^{2+}, but the bacterium *Thiobacillus ferrooxidans* regenerates Fe^{3+} at $pH < 4.0$.
- Chemical reaction: $FeS_2 + 15/4\ O_2 + 7/2\ H_2O \rightarrow Fe(OH)_3 + 2SO_4^{-2} + 4H^+$.
- Iron monosulfides (mackinawite $FeS_{0.94}$, and greigite $FeS_{1.34}$) are more reactive than pyrite in estuarine sediments.

- Severe acidity weathers aluminosilicate materials, releasing Al^{3+} ions at pH < 4.0, and Al^{3+} activity increases tenfold with each unit decrease in pH.
- Agronomically, the following problems result: highly soluble and toxic concentrations of Al; unavailability of phosphates caused by iron- and Al-phosphate complexation interactions; low base status and nutrient deficiencies, especially with some micronutrients such as Mo, and available Ca as a macronutrient; increased salinity; and altered microbial activity. Additionally, these soils when high in clay content often exhibit sodic conditions for long periods of time unless corrective measures are implemented.

Disturbed and oxygenated acid-sulfate soils (ASS) have pHs < 3.0, are often yellow or off-yellow in color (jarosite [$KFe_3(SO_4)(OH)_6$] is the straw-yellow mineral byproduct of the oxidation process and a conclusive ASS indicator; pH requirements must be <3.7 for formation), and have a distinct "rotten egg" (hydrogen sulfide gas) smell. Hydrogen sulfide gas can originate in the subsoil at the same time that more aerobic conditions allow hydrogen sulfate to form in the surface zone. When ASS acid water runs off into catchment facilities from acid-sulfate soils, the water will turn a clear blue-green color because high levels of aluminum bind with particulates in the water and settle the compounds at the bottom of the lake, leaving clear bluish-colored water with pH < 4.0 (http://www.dnr.qld.gov.au).

Reclamation of acid-sulfate soils can be quite complex and generally does not occur rapidly. Strategies (Dent, 1992; Dear et al., 2002; http://www.nrm.qld.gov.au) where the reclaimed site may be used as a golf course or recreational turfgrass site construction should include the following:

1. Comprehensive geohydrology site assessment prior to construction to determine the severity and distribution of potential acid-sulfate soil conditions.
2. Determine water management issues (wet with water surplus, wet with no water surplus, seasonally dry with seasonal water surplus, or dry with no water surplus) and water table control measures.
3. Comprehensive deep and shallow drainage control program, including runoff, raised elevations (topography), and mulching issues (for soil transpiration stabilization and reductions in upward capillary salt movement to the soil surface).
4. Neutralization by liming to pH > 5.0 in order to immobilize Al and Fe, increase P and other nutrient availability, increase mineralization of N from organic matter, and enhance microbial activity. Re-liming (Ca addition) may be needed, depending on the rate of acid generation in the subsoil and upconing, or upward flux, of acids into the rhizosphere. One part by mass of pyrite sulfur is countered by three parts calcium carbonate. Only 0.5% mass sulfur can be neutralized by other exchangeable bases when buffering against acidity.
5. Determination of gypsum requirements to remediate sodic soil conditions at the surface and subsurface when sodic conditions exist. Normally, if the soil is high in Na, then it is also high in total soluble salts. Depending on the irrigation water and rainfall conditions, the total soluble salts can be leached prior to grassing, especially if gypsum is used to correct sodic soil structural deterioration problems.
6. Implementation of site-specific leaching programs may be required after lime and gypsum applications, prior to grassing, and periodically during the growing season in order to remove excess Na, soluble salts, Al, Mn, and possibly other metal ions or toxic soluble compounds.
7. In extreme ASS situations, total excavation and strategic reburial or reinternment of the ASS soils may be required. This process is extremely expensive, and the permitting process may prohibit this strategy since the reburial site must be anoxic and preferably anaerobic to permanently preclude any acid regeneration and secondary site contamination.
8. Another expensive strategy is hydraulic separation (sluicing or hydrocycloning) of sulfides individually or in conjunction with dredging. Pyrite has a density of 5 gm/cm^3 in solid crystals, while common sand and clays have densities < or = to 2.7 gm/cm^3, allowing physical separation of the two components.

Neutralizing agents for treating ASS should be slightly alkaline (pH 7–9) with low solubility to minimize contamination of surface water and groundwater sources with high-rainfall events. The preferred neutralizing agent is agricultural lime ($CaCO_3$) with incorporation into the soil profile. Other agents with low solubility include dolomite, $CaMg(CO_{3+})_2$; magnesite, $MgCO_3$; and burnt magnesite, MgO (http://www.nrm.qld.gov.au). Magnesium sulfate is produced during the neutralization process with these Mg products, and this compound is quite soluble; however, calcium sulfate or gypsum has lower solubility and does not have the potential environmental contamination problems that Mg sulfate has. Hydrated builders' lime or calcium hydroxide, $Ca(OH)_2$, and sodium bicarbonate, $NaHCO_3$, can be used to neutralize acidity, but these compounds are also soluble materials with potential environmental contamination implications (especially in aquifers or other water resources).

Estimated lime requirements for permanent pH control of ASS, assuming complete oxidation, are as follows (Costigan et al., 1981): (a) 1% pyrite requires 40 t/ha (17.86 ton/acre) for pH control to a 15-cm (5.9-in.) upper soil profile depth; (b) 1% acid-neutralizing capacity is equivalent to 23 t/ha (10.27 ton/acre) of limestone for 15-cm (5.9-in) soil depth; (c) limestone should be incorporated to 45–50 cm (17.7–19.7 in.) depth, and application rates increased accordingly, based on appropriate laboratory analyses; and (d) high-pyrite soils may have lime requirements in the 100–400 t/ha (44.65–178.60 ton/acre) range for pH control down to a 45-cm (17.7-in.) soil depth.

Physiochemical water quality parameters for proactive monitoring of ASS include pH, ECw, dissolved oxygen, Al, Fe, SO_4, and dissolved Fe. One additional consideration when applying acidifying fertilizers is that 100 kg (220 lb.) of N in MAP can result in a fivefold increase of the soil profile acidification rate in the first year of application (http://www.adl.clw.csiro.au); therefore, use of and proper volume applications of acidifying fertilizers on ASS must be carefully considered in the salinity BMP.

Average Acidity Produced by Fertilizers

Product	kg Calcium Carbonate/kg of N or S Supplied
Urea	1.8
Ammonium nitrate	1.8
Elemental sulfur	3.0
DAP	3.6
Ammonium sulfate	5.4
MAP	5.4

20.2.4 Scald Sites and Areas with Extreme Spatial Diversity of Salt Problems

There is not one salinity problem on a site, but multiple salinity stresses where scald sites are the ultimate in diversity of salinity issues and spatial variability (Semple et al., 2006; Thomas et al., 2009). Scald areas occur in nature as well as from changes due to site use. Seepage locations often become scalds whether in natural setting or on managed salt-affected sites. *Scalds* are areas of localized soluble salt and Na accumulations that have become depressed, with sparse or no vegetation, and with extreme diversity of chemical properties when comparing different scald areas and within the spatial confines of a single scald (Seelig, 2000; Wentz, 2000; Semple et al., 2006; Thomas et al., 2009). With both surface and subsurface salt migration in the soil coupled with diverse spatial as well as temporal accumulation of total soluble salts and specific salt ions, the correct site assessment is essential for not only formulating BMPs but also actually implementing the whole-system management program successfully (Grunstra and Van Auken, 2007). When there is considerable diversity of chemical and physical problems between scald sites or within the boundaries of a scald area, the authors suggest that BMPs be formulated that will address the widest number of issues common across the diverse areas. As these BMPs start to improve the areas, then more microareas not showing sufficient improvement can be addressed on a more individual basis.

It is not unusual for localized scald areas to be present on coastal golf courses or parks, and these present challenges in terms of correction (i.e., they are essentially microsites that must be assessed individually and treated on an individual basis, especially for final remediation). In larger scale land reclamation that may be a planned forage grass area, the same microsite challenges occur. Understanding the primary salinity limitations in concert with seasonal climatic fluctuations usually starts with the incoming saline irrigation water if the site is irrigated, including inefficient irrigation water distribution uniformity and improper irrigation scheduling. Other infrastructure contributors to increasing salt accumulation include topography issues, inadequate drainage, intolerance in the current plants being grown, lack of proper cultivation equipment, and/or water table fluctuations. Formulating an acceptable series of salt BMPs involves addressing the key causes as well as implementing remediation practices in order to be comprehensive and holistic in concept for the site. These challenges are often coupled with budget limitations (restricted purchase of amendments or equipment needed to effectively minimize salt accumulation) and administrative curtailment of salt BMP implementation (e.g., delayed cultivation or delayed leaching) programs. With salt scald areas, remediation can involve several strategies:

- Aggressive deep cultivation and a regimented schedule for recultivation
- Complete excavation and introduction of more effective salt management soil profiles
- Additional drainage installation plus verification that the previous drainage system is actually functioning
- Chemical amendments (with sodic and alkaline conditions) and continued applications based on proactively monitored science-based data: gypsum, phosphogypsum, calcium chloride, or a sulfur source + lime. Sometimes on acid-sulfate soils, the surface may exhibit Na and soluble salt scald areas, and in these cases, deep injection of lime may be needed to deal with and remediate acidity and excess sulfur limitations.
- Organic amendments as needed: well-decomposed composts, manures (green and farmyard), peat moss, biosolids, crop residues, and biochar
- Carefully programmed and effective leaching events
- Topography changes such as through contouring or elevation changes (sand capping)
- Regrassing and reestablishment with grasses that can tolerate initial soil salinity and expected long-term salinity. If the site is to be used for forage production, this is considered in the selection of reclamation-oriented plants, and may or may not be a traditional turfgrass species or cultivar.
- Reimplementation of microsite-specific salt BMPs
- Applying sand mulch is one remediation strategy that can improve the productivity of a bare salt scald area (Bakker et al., 2010).

Successful salt scald remediation can take from 2–3 months up to several years, depending on severity, intensity of BMPs for remediation, major salt ion limitations, soil profile physical constraints, irrigation water quality, and actual scald size, to completely reclaim the area and reestablish an acceptable turfgrass playing surface or alternative use such as forage or simply grass-based stabilization. An incomplete or wrong initial site assessment and subsequent failure to completely implement the whole system plan can prolong the reclamation program. Proper soil sampling and analyses are critical components in determining the sustainable reoccurring strategies required for the site-specific remediation program.

The remediation program may require different sequential phases. *Phase 1* may require additional drainage installation followed by deep cultivation and gypsum or lime and sand injection to enhance salt movement down to the new drain lines. As the salt scald area is reduced in size over months and sometimes years, seepage sites can sometimes reoccur on low-topography areas or high salt accumulation areas, which require a *Phase 2* remediation program involving complete excavation of the smaller sized problem area and reintroduction of a more favorable soil profile for future salt management in the problem area. Topography improvements and organic additions may be required at some sites to improve salt scald remediation and enhance salt management (Gill et al., 2008; Bakker et al., 2010; Treonis et al., 2010). Maintaining the physical properties of the upper soil profile is essential for sustaining a salt BMP program (Morris et al., 2009; Bakker et al., 2010).

On older arid environment golf courses with long-term saline irrigation water use and hard low-permeable soils, the authors have observed fairway areas where the original main and lateral line trenches that were dug for the primary irrigation system have become secondary salt accumulation sites, in some cases resulting in scald areas, from both surface and subsurface salt migration over several years. The remediation strategies included installation of interception drains or mini–dry well drainage holes, or implementation of recontouring practices (with sand capping where needed) in order to redirect the salt migrations away from these older and deeper irrigation mainline trenches. An additional strategy would be to install a shallow drainage line in the trench above the original irrigation line and tie this new drainage system into existing outlets or new interception drains.

20.2.5 Mine Spoils and Severely Eroded Sites

Turfgrasses developed with superior salinity tolerance, especially when coupled with good heavy metal, acidic, and drought tolerances, are candidates for reclamation of mine spoil areas as a ground cover, forage grass, or turfgrass depending on the projected use of the reclaimed area. In many mine reclamation cases, salinity stress tolerance is desirable but of less importance than the drought or acidity issues. For example, sulphidic ore mine tailings are extremely acid and similarly pollute surface water and groundwater sources with increasing acidity as well as deposit toxic levels of Al, Fe, S, and heavy metals (Dent, 1992). Coal mine spoil also contains pyrites, and pH levels <3.0 are common occurrences in the surrounding environment.

Reclamation of these mine spoil sites can be multidimensional, involving covering pyritic spoil tailings with weathered organic materials such as pulverized fuel ash from coal-burning power stations (the fuel ash combines with iron as Fe^{3+} silicate), wood waste such as pine bark, chicken manure, pulverized coconut shells, or other organic products that are designed to inhibit acid generation by preventing the conversion of Fe^{2+} to Fe^{3+} or preventing pyrite oxidation by Fe^{3+} ions (Dent, 1992). The initial reclamation strategy is designed to complex iron and render it unavailable for pyrite oxidation.

Liming is a conventional treatment option both for the mine spoil soils and for the treatment of acid runoff into lakes or other catchment facilities. The objectives of liming are to raise the pH > 5.0 and to stimulate bacterial reduction processes; however, because of the inherent lack of organic material in these spoils, additional organic products normally need to be added (Gill et al., 2008; Treonis et al., 2010). Sewage sludge has been used for this purpose in lakes and reservoirs (Dent, 1992).

Revegetation of these mine spoil areas is usually difficult due to a lack of available nutrients because of the toxic ions and nutrient deficiencies that are common limitations. P fixation, K immobilization by jarosite, and lack of micronutrients such as Mo and others are nutritional problems for establishing and maintaining sustainable grass vegetation (Dent, 1992). Shallow rooting of the planted grasses enhances cyclic drought stress responses. The physical limitations of the mine spoils often create either coarse-textured or very-fine-textured soils that are difficult to cultivate and result in difficult ecosystems for plants to root properly into the soil profile.

Severely eroded sites expose soil structural instability problems and are prone to rapid soil gravity-induced movement with high-rainfall events. Zeolite additions can be incorporated into different types of clay mineralogies (kaolinitic, smectic, and allophonic) and saline-sodic soils to improve saturated hydraulic conductivity, decrease aggregate dispersion after soil sodification, decrease exchangeable sodium percentage (ESP), and control soil erosion (refer to Chapter 17, Section 17.4.3, "Zeolite") (Moritani et al., 2010).

Proper contouring plus integration of organic amendments and planting of rooted plants are normally required to revegetate these areas. Irrigation and fertilization are required for successful plant establishment and sustained growth over seasons. Frequent flushing of the soil surface zones with good-quality irrigation water or rainfall in combination with shallow drainage water may be required for successful establishment, grow-in, and sustainable growth of grasses or other plants.

Erosion problems can sometimes be controlled with diversion or collection ditches or by channeling to other containment facilities. Liners and impervious pads can be used to minimize pollutant runoff as well as reduce the potential for erosion.

Alkaline, dry flue gas desulfurization (FGD) byproducts have been used to reclaim acidic mine spoils (Stehouwer et al., 1995). Kentucky 31 tall fescue (*Festuca arundinacea* Schreb.) was planted on spoils with FGD byproduct amendment applications equivalent to spoil neutralization requirements.

20.2.6 Additional Comments on Multiple Stresses and Reclamation

The spatial and temporal diversity of soil salinity conditions contributes to significant and complex challenges when managing salt-tolerant grasses in perennial ecosystems, especially when coupled with other stresses that also exhibit spatial and temporal variability. In some cases, salinity is a contributor to some of the other site stresses; while in other cases, the additional stresses are due to conditions not related to salinity. Examples of diversity of salinity situations and multiple stresses are those associated with soil pH, groundwater processes, irrigation water, and abiotic and biotic stresses on the site.

The range of salt-affected soil categories extends from extreme acidity to extreme alkalinity, as noted by Rengasamy (2010a):

Saline, acidic-saline, alkaline-saline
Saline-sodic, acidic saline-sodic, alkaline saline-sodic
Sodic, acidic, sodic, alkaline, sodic

The *types of salinity based on soil and groundwater processes* include (Kelly and Rengasamy, 2006; Rengasamy, 2010a) (a) *dryland salinity*: groundwater-associated salinity, which occurs in discharge areas of the landscape where soluble salts in water exit from groundwater to the soil surface; (b) *transient salinity*: dry saline land, where the concentration of salts fluctuates with seasons, rainfall patterns, and the accumulation of salts in variable soil profile layers; and (c) *irrigation salinity*: caused by salts introduced from repeated irrigation cycles, and the salts accumulate in the rhizosphere due to insufficient leaching and high evapotranspiration (ET).

Types of salinity responses associated with surface and irrigation water application processes include (a) waterlogging and hypoxia or anoxia problems, (b) grass foliar adsorption of salt ions, and (c) grass root desiccation and need for energy-depletion root regeneration.

Salinity interactions involving other abiotic environmental stresses include (a) cyclic drought stress, (b) cyclic waterlogging conditions and high capillary movement of water into the rootzone with elevated soluble salt concentrations, (c) variable and especially decreased oxygen (air porosity) fluctuations in the soil profile rhizosphere, (d) reduced tolerance to high heat stress, (e) reduced tolerance to cold temperatures and difficult transitioning into and out of cold winter temperatures, and (f) reduced wear and traffic tolerance.

Salinity interactions involving biotic stresses in perennial turfgrass field situations include (a) increased soil-borne pathogen problems, (b) predisposition to surface drought and desiccation, (c) consistently high soil moisture conditions (including physiological drought stress plant responses) and subsequent ideal microenvironmental conditions for foliar pathogen attack, (d) salt-tolerant nematodes, (e) weakened turfgrass predisposition to cyclic insect attacks, and (f) increased weed competition.

20.2.7 Subsoil Constraints in Perennial Grass Ecosystems

Subsoil constraints can result from a series of major limiting factors to long-term grass sustainability and performance, especially since many of these limitations directly affect root development and viability, and, thereby, drought resistance and nutrient uptake efficiency (Duncan and Carrow, 1999). The impact of these constraints can vary depending on grass genetic tolerance, soil type, agronomic practices, construction issues (such as topsoil excavations), cyclic growing seasons, and site-specific knowledge base. The main subsoil constraints can include the following (Duncan and Carrow, 1999; Price, 2010):

- Acidity (pH < 4.8), which results in toxic concentrations of aluminum or sulfur and heavy metals and deficiencies of other critical nutrients that impede or prevent grass root and plant growth.
- Sodicity (generally ESP > 15%), where excess sodium ions cause soil structural deterioration, loss of pore continuity, and water infiltration and percolation problems. Grass root penetration is often restricted, and boron toxicity can sometimes be an associated stress.
- Transient or rhizosphere salinity, where increasing accumulated total dissolved salt concentrations and rising osmotic potentials cause the grass plants to expend more energy to adsorb water from the saline soil solution; in extreme cases, physiological drought stress or wet wilt can result even though the soil profile contains adequate moisture; this condition is usually not associated with rising water tables.
- Excessive soil Na that directly becomes a root toxin due to displacement of Ca from plant root cell walls, which is a limiting factor for plants sensitive to Na root toxicity.
- High soil strength and physical impermeability where plants cannot penetrate the soil structural profile; this problem is often not related to sodicity but is inherent in the soil properties. However, sodic conditions enhance inherent high soil strength.
- Low nutrient uptake levels, which reduce the capability for root system exploration of deeper soil layers for water.
- Toxic concentrations of elements that are not directly related to pH problems.
- Alkalinity, where the subsoil pH is >9.0, resulting in toxic concentrations of carbonate and hydroxyl ionic species.
- Compaction of subsoil layers caused by past land use, site construction issues, repeated zonal tillage, wheel traffic, or removal of upper soil layers.

Chloride concentrations in the subsoil was a more effective indicator of reduced water extraction and reduced grass productivity than either salinity (ECe) or sodicity (ESP) (Dang et al., 2010); available subsoil P and Zn partially alleviated the negative impact of high subsoil Cl. Significant correlations were found for ECe and Cl ($r = 0.90$), ESP and B ($r = 0.82$), ESP and ECe ($r = 0.79$), and ESP and Cl ($r = 0.73$) (Nuttal and Armstrong, 2010).

20.2.8 Serpentine Soil Challenges

The association of serpentine soils with high salinity is not very common, but has been encountered by the authors and provides another example of how multiple stresses must be addressed in different reclamation scenarios (this section summarized in Brady et al., 2005). These unique soils are formed by weathering of peridotite bedrock (i.e., igneous or metamorphic rocks, sometimes referred to as *ultramafic rocks*). Characteristics of these soils include the following:

- At least 70% ferromagnesian minerals (high concentrations of iron and magnesium, but low concentrations of calcium, silica, and aluminum)
- Often a toxic presence of heavy metals such as nickel, cobalt, and chromium

- Generally shallow soils with minimal water retention and restricted plant-rooting depth
- Inherently low retention concentrations of potassium, phosphorus, nitrogen, and molybdenum
- Significant imbalance between calcium and magnesium
- Highly erodible due to minimal silt and clay composition
- Ubiquitous but patch-like distribution across sites that exhibit three collective traits: poor plant productivity, highly endemic to a site-specific location, and distinct vegetation ecotypes producing species-level responses (autecology) and a plant community-level effect (synecology) that must adapt to the serpentine syndrome, or the cumulative effects needed for adaptation to chemical, physical, and biotic soil components

Amelioration of these problem soils when applying saline irrigation water on perennial turfgrass ecosystems must be multifaceted to deal with site-specific and dynamic chemical, physical, and biotic limitations to sustainable plant growth. BMP strategies include the following:

- Calcium amendment applied to the soil to adjust for Mg^{2+} toxicity concentrations and maintenance of available calcium for plant root uptake over time; supplemental foliar-absorbed calcium products (calcium nitrate, calcium chloride, calcium acetate, and calcium chelated with amino acids or alcohols) may need to be periodically applied to shoots for nutritional stabilization internally.
- Application of organic amendments (Section 17.4.5) to provide some soil structural and biotic (microbial) stabilization.
- Application of an inorganic amendment such as zeolite (Section 17.4.3) to provide increased CEC for nutrient retention.
- Adjustment and readjustment of fertility programs that include micronutrients as well as macronutrients to sustain acceptable plant growth; regularly scheduled tissue sampling and wet chemistry analysis (see Chapter 7, "Plant Analyses for Turfgrass").
- Selection of heavy metal–tolerant grass species; some of these soils may range in pH from 4.87 to 6.89, and the plant species may require acid soil tolerance (LaForce et al., 2002).

Successful turfgrass or other plant species that have been grown on serpentine soils include *Lolium perenne*, *Poa pratensis*, *Elymus junceus*, *Bromus inermis*, *Trifolium hybridum*, *Melilotus alba*, and *Medicago sativa* (Moore and Zimmermann, 1977); *Agrostis stolonifera* (Shewry and Peterson, 1975); and *Festuca rubra* (Johnston and Proctor, 1981). A warm-season grass such as seashore paspalum (*Paspalum vaginatum*) that has inherently high overall salinity tolerance, has a capability for heavy metal phytoaccumulation, and is a luxury user of magnesium is an additional turfgrass that can be planted on these serpentine soils (Duncan and Carrow, 2000). Additional species and serpentine plant-tolerant adaptation research is summarized in Brady et al. (2005).

20.3 GRASS SELECTION ISSUES

20.3.1 Grass Salinity Tolerance Assessment

In this chapter, attention is focused on salt-tolerant grasses being used for nontraditional purposes where other severe environmental stresses are almost always present. But, if a salt-tolerant grass is to have such multiple-use potential, it must start with superior multiple-salinity tolerance. Development of plants that are tolerant to multiple salinity stresses must be functionally adapted to soil water dynamics, soil structural stability and dynamics, solubility of compounds in relation to pH and pE (electron concentration related to redox potential), nutrient availability, nutrient imbalances, and water movement dynamics, both spatially and temporally in the soil profile (Rengasamy, 2010a). Due to the complex and interactive stress components of multiple salinities in any ecosystem and the multiple salt tolerance mechanisms inherent in plant species, selection of salt-tolerant species and cultivars must be linked to BMPs of site-specific soils and appropriate local cyclic environmental constraints (Barrett-Lennard and Setter, 2010). For turfgrasses, quality traits must parallel the salinity tolerance levels for the grass to be grown and to perform at the level of performance expectations for the golf course or sports facility.

Grass crop growth responses to salinity occur sequentially in two phases. First, the response is a continuous *osmotic phase* that inhibits water uptake due to osmotic pressure of saline soil solutions lowering its potential energy (water always moves from a higher to a lower energy potential in the soil–water–plant continuum). The osmotic effect continuously operates as salinity is increased from 0.7 to 41.0 dS/m in soil solution, with a parallel ionic effect individually from Na, Ca, SO_4, and Cl salt ions (Rengasamy, 2010b). The osmotic effect potentially becomes dominant in soil solution and can severely restrict plant growth when ECe increases above 25 dS/m. The percentage of available soil *water that was not adsorbed* by plants as the ECe (dS/m) increased ranged from 0.7 dS/m = 0%, to 22.6 dS/m = 50.2%, 41 dS/m = 84.8%, and 63.9 dS/m = 95.5% (Rengasamy, 2010a). The effect of osmotic pressure in the soil solution (ECe) has been shown to greatly affect plant yield (turfgrass performance) when the pressure increased above 700 kPa or 19.4 dS/m (ECe=1 dS/m = 36 kPa of osmotic pressure) (Kelly and Rengasamy, 2006).

Second, the response is a slower *ionic phase* when the accumulation of specific ions in the soil and in the plant leads to ion cytotoxicity or ion imbalances over a period of time (Munns and Tester, 2008). The ionic effects apparently are more significant at ECe levels <30 dS/m, while the osmotic effects tend to be more dominant above that threshold value of soil solution ECe (Munns et al., 2006; Tavakkoli et al., 2010; Rengasamy, 2010b).

Interactions between the soil matrix and the soil solution can directly affect plant responses to salinity stresses (Tavakkoli et al., 2010). Therefore, the different salt tolerance mechanisms and degree of salinity damage to plants vary according to severity and duration of exposure to multiple soil and topical salt stresses. Initial screening for plant genetic diversity in salt tolerance responses has generally been conducted in hydroponic solutions with or without sand, with the assumption that the salt tolerance plant responses will essentially mimic the same stress conditions in the soil when similar electrical conductivity values and Na^+ plus Cl^- concentrations are used. However, research has discounted that assumption and these salinity screening protocols do not necessarily reflect field conditions (Tavakkoli et al., 2010).

In the soil, ion exclusion was more important at low to moderate ECe levels <10 dS/m, but osmotic stress became increasingly more important as salt stress levels increased (Tavakkoli et al., 2010). Plant growth reductions as well as actual internal Na and Cl concentrations were greater under hydroponics than in the soil under similar EC. Tissue tolerance, in which Na and Cl compartmentalization occurs at the cellular and intracellular levels, is a critical salt avoidance mechanism within the cytoplasm (especially mesophyll cells in leaves) that must be genetically functional for long-term field perennial plant survival. Other salt tolerance mechanisms must also be activated in the plant with increasing levels of salt stress (see Chapter 9).

The solid soil matrix affects salinity responses in plants by (a) CEC influences on cation and anion adsorption and variability in soil-buffering capacity activities in soil solution (which does not occur in hydroponic solutions), and (b) soil physical properties that, combined with soil solution characteristics, will determine soil water potential and water uptake (where plants in hydroponics are only affected by the osmotic potential in the nutrient solution and its effect on photosynthesis and growth rates). Rhizosphere soil solution salt concentrations change when mass flow exceeds uptake, with high transpirational demands, and with low unsaturated hydraulic conductivity (i.e., coarse sands); these changes do not occur in solution culture because the matrix potential is zero and there is no resistance to water movement. Therefore, the plant multigene complex that is either upregulated or downregulated in response to changing salinity stresses will not be fully activated in solution culture, and adaptation responses to variable total field salinity challenges cannot be completely identified and exploited (Tavakkoli et al., 2010). Refer to Chapter 9 for additional discussions on grass responses to salinity.

Additionally, there is greater discrimination of K and Na uptake in soil solution where adsorption and fixation interactions occur, compared to hydroponics (Tavakkoli et al., 2010). In screening for salt tolerance and eventual categorization of actual cultivar-specific salinity tolerance levels, multiple cascading mechanisms (genetic upregulation or downregulation of multiple genes with associated responses) will need to be activated with increasing severity of salt stress conditions and preferably using soil-based media (Flowers et al., 1986, 2010; Witcombe et al., 2008; Qadir et al., 2008) for long-term sustainable turfgrass productivity. Refer to Chapter 9 for additional plant salinity tolerance mechanisms.

20.3.2 Forage Grasses for Reclamation

Biosaline agriculture for perennial forage production is utilizing a range of plants that has the capability for growing under saline soil and water conditions to produce a feed resource for livestock (Masters et al., 2007). With ECe salinity levels < 15 dS/m, legumes and grasses with moderate salt tolerance can provide 5–10 tons edible dry matter per year when water is not limiting. The limitation to these forages has been the accumulation of elevated levels of sulfur and selenium that negatively affect ruminant productivity.

At salt concentrations ECe > 25 dS/m, production levels of halophytic grasses and shrubs have produced 0.5–5.0 tons edible dry matter per year with variable crude protein and digestible fiber concentrations (Masters et al., 2007). Some halophytic shrubs (such as the chenopods) have a tendency to accumulate elevated levels of sodium, potassium, chlorides, calcium, and magnesium, thereby depressing feed intake.

Beneficial forage components in these biosaline environments include vitamin E and betaine; but toxic compounds such as oxalate, coumarin, and nitrates (this latter one especially a problem with severe drought conditions) can also concentrate in the biomass (Masters et al., 2007). The forages can provide soil cover and shading to reduce ET, thereby reducing the capillary rise of salt ions to the soil surface, which would contribute to scald formation (http://www1.agric.gov.ab.ca/$department/deptdocs.nsf/all/agdex11501#controls). Perennial salt-tolerant forages are effective in helping to control saline seepage areas and their expansion; however, they do not reclaim the area unless the groundwater source is controlled first (http://www1.agric.gov.ab.ca/$department/deptdocs.nsf/all/agdex11501). Grass species recommended for seeding into saline seepage areas include creeping foxtail, meadow foxtail, smooth bromegrass, meadow bromegrass, slender wheatgrass, intermediate wheatgrass, pubescent wheatgrass, tall wheatgrass, western wheatgrass, NewHy wheatgrass, green wheatgrass, Russian wildrye, altai wildrye, beardless wildrye, and Nuttall's alkaligrass (http://www.agric.gov.ab.ca/agdex120/22-3).

Common Bermuda grass (*Cynodon dactylon* [L.] Pers. Var. *dactylon*) and weeping lovegrass (*Eragrostis tef* [Zuccagni] Trotter) have been planted on coal fine ash reclaimed soils in South Africa (Van Rensburg et al., 1998). Either compost or kraal manure was an effective amendment to remediate the reclaimed coal ash soil. Selenium accumulated to an average level of 4.4 mg/kg in the biomass, which can be toxic to some livestock.

Data in Table 9.6 list some plants to consider for revegetation on saline-alkaline soils. Several additional plants are listed in Table 9.7 for rehabilitation of salt-affected sites after site-specific salt reclamation. Seashore paspalum (*Paspalum vaginatum* Swartz) has been used for saline forage production (summarized in Chapter 4 of Duncan and Carrow, 2000). Dry matter production has ranged from 1.3 to 7.0 tons/ha/year for the coarse-textured ecotypes on saline fields. With fertilizer applications, dry matter yields have reached almost 25 tons/ha/year. In addition, seashore paspalum (*Paspalum vaginatum* Swartz) has been used for erosion control on physically degraded soils; for coastal environmental stabilization; for sand dune revegetation; for bioremediation of contaminated or unproductive mine spoil soils; for rehabilitation of flood-prone areas plagued by wet, saline seepage problems; for hyperaccumulation of heavy metals (Fe, Mn, and Cu); and for reclamation of highly alkaline soils (summarized in Duncan and Carrow, 2000).

Kallar grass (*Leptochloa fusca*) is tolerant to salinity, sodicity, as well as alkalinity and has been planted to improve the physical characteristics of, rehabilitate, and restore soil fertility of a saline-sodic soil on a sustainable basis (Akhter et al., 2004). Kallar grass significantly improved plant-available water with time in association with increased soil organic matter content, soil porosity, and the Soil Structural Stability index, particularly in the surface soil profile. Soil hydraulic conductivity increased in the top 20 cm of the soil over 5 years and was correlated with porosity, water retention, structural stability, and organic matter content.

20.3.3 Forage Grasses and Drainage Water Reuse

Drainage water can be reused if the characteristics of the water, soil, and salinity tolerance of plants are known and are properly managed (Shannon and Cervinka, 1997; Dudley et al., 2008; O'Connor et al., 2008; Duncan et al., 2009). Poor-quality water requires site-specific, targeted water management, and maintenance of soil structure and permeability (tilth and hydraulic conductivity). As salinity in the reuse water increases, the major degradation factor is high concentrations of salt ions and their interference with uptake of critical nutrients. Blending, drainage, and leaching are all management components when utilizing reuse drainage water. Scheduled cyclic applications with lower salinity water are generally recommended to minimize soil salinity accumulation problems, especially during plant germination, rooting, establishment, and grow-in periods.

Many of the most salt-tolerant forages are C4 grasses (Kaffka, 2001). In general, the most salt-tolerant species tend to have poorer forage quality (higher ash content, tendency toward steminess, slower digestibility, more lignified crude fiber, and reduced intake) than less-salt-tolerant species. Rotational grazing systems are used to improve forage quality, production, and livestock productivity. Tall wheatgrass (*Agropyron elongatum*), common Bermuda grass (*Cynodon dactylon*), sudangrass (*Sorghum sudanese*), tall fescue (*Festuca arundinaceae*), and perennial ryegrass (*Lolium perenne*) generally produced some forage at ECe < 22 dS/m (Kaffka, 2001). Germination and establishment of these grasses need to be managed with lower salinity water before cycling with more saline drainage water (Harivandi et al., 1982a, 1982b; Harivandi, 1984).

When reusing irrigation water (initially, ECw averaged 3.6 dS/m at 3.5 inches volume per event, and the drainage water at 33.9 dS/m at 0.10 inch per event), common Bermuda grass pastures remained productive after 5 years in California (Kaffka et al., 2004). Salinity-related properties declined in the top 2 feet of the soil profile, indicating that soil reclamation was working. Cattle performance should be monitored for adverse physiological effects when using saline drainage water due to the potential for trace element imbalances on a location-by-location basis.

20.4 ESTABLISHMENT CHALLENGES AND METHODS

The inhibitory effects of soil salinity on germination of seed and on plantlet and root regeneration in vegetatively propagated turfgrass cultivars involve multiple problems: delay in germination or propagule plantlet and root generation, loss of seed or propagule viability, reduced germination percentage or increased propagule senescence, accumulation of toxic salt ion (such as Na and Cl) concentrations and their direct interactions with uptake of critical nutrients required for initial growth and development, and the increase in osmotic pressure of saline soil solutions that lowers its potential energy (which restricts water uptake and can lead to desiccation or physiological drought stress). Seed germination (including plantlet initiation from vegetative propagules) and early seedling and plantlet growth are the most sensitive growth stages when exposed to salinity stress (Ashraf and Foolad, 2005). Even if seed germination is initiated under saline conditions, the juvenile plantlets do not genetically and physiologically initiate salinity tolerance mechanisms immediately on germination or plantlet activation. Consequently, salinity tolerance levels differ between seed germination or plantlet initiation and later vegetative growth (Dai, 2006).

Several seed-priming treatments (refer to Chapter 9.2 for an extensive discussion) and metabolic enhancers or biostimulants have been utilized in order to promote improved stand establishment and seedling and plantlet survival under saline conditions (Ashraf and Foolad, 2005; Shahba et al., 2008; Serena et al., 2010; Ertani et al., 2010). Other studies have compared cool- and warm-season species for salinity germination and subsequent vegetative growth (Dai, 2006; Johnson et al., 2007; Richardson and McCalla, 2008; Dai et al., 2009).

Regardless of overall plant salinity tolerance, unless the site-specific BMP salt management program is fully implemented on soils with previous salt accumulation problems, success with germination and plantlet stand establishment will be limited. For example, any salt scald areas must be completely reclaimed and renovated prior to planting in order to have any chance of permanent turfgrass establishment and long-term growth.

One additional establishment consideration is the impact of salinity on mycorrhizal fungi populations and potential for successful reclamation of mine spoils (Johnson, 1998) or salt scalds. Mycotrophy was favored by increasing soil organic matter (composted papermill sludge) and avoiding heavy P fertilization in taconite mine tailings when planting the grasses *Panicum virgatum* and *Salsola kali*.

20.5 MANAGEMENT CHALLENGES AND CONSIDERATIONS

Because of the dynamic nature of multiple soil and water salinities and their impact on perennial turfgrass ecosystems, successful management and achieving performance expectations are challenges that increase when coupled with other inherent site abiotic stresses. If salinity additions are continuous, such as due to seepage or saline irrigation water, then management must also be ongoing. The principles of managing the salinity issues within the context of other abiotic stresses remain the same as if salinity is the sole dominant stress present. Proactive monitoring of water, soils, and plants is a key strategy to implementing salt management BMPs. *Do not expect a highly salt-tolerant plant species to remediate a salt-affected site.* You must reclaim the site-specific soil infrastructure structural and chemical problems first before planting any turfgrass or reclamation grass under salinity stress conditions. In summary, the steps in reclamation of salinity-impacted areas are as follows:

1. Collect water, soil, and plant tissue samples and submit for comprehensive laboratory analyses, including a saturated paste extract salinity test.
2. Conduct a proper site assessment for primary and secondary salinity limitations, and determine the impact of soil moisture, salinity, and oxygen flux interactions in the soil profiles both spatially and temporally (climatic and seasonal dynamics) (Bush, 2006).
3. Determine the possible needs for water treatment or irrigation water blending options to reduce the salt load.
4. Conduct an irrigation system audit to assess water distribution system uniformity and efficiency (successful salt management will only be as effective as your capability to effectively apply irrigation water and leach salts to drainage lines) (Chapters 10 and 11).
5. Implement drainage improvements and verify functionality (Chapter 15).
6. Apply appropriate cultivation operations to alleviate soil structural problems.
7. Add chemical amendments to remediate soil salt accumulation problems.
8. Implement leaching programs as needed (Chapter 11).
9. Repeat any of the above strategies as needed.

References

Abrol, I. P., J. S. P. Yadav, and F. I. Massoud. 1988. *Salt-affected soils and their management* (FAO Soils Bull. 39). Rome: United Nations Food and Agriculture Organization.

Achard, P., H. Cheng, L. De Grauwe, J. Decat, H. Schoutteten, T. Moritz, D. Van Der Straeten, J. Peng, and N. P. Harberd. 2006. Integration of plant responses to environmentally activated phytohormonal signals. *Science* 311: 91–94.

Adams, W. A. 2008. An overview of organic and inorganic amendments for sand rootzones: With reference to their properties and potential to enhance performance. *Acta Hort. (ISHS)* 783: 105–114. http://www.actahort.org/books/783/783_10.htm

Adams, W. A., and R. J. Gibbs. 1994. *Natural turf for sport and amenity: Science and practice*. Wallingford, UK: CAB Inter.

Adcock, D., A. M. McNeill, G. K. McDonald, and R. D. Armstrong. 2007. Subsoil constraints to crop production on neutral and alkaline soils in south-eastern Australia: A review of current knowledge of management strategies. *Aust. J. of Experimental Agric.* 47: 1245–1261.

Ahrens, C.W. and C. Auer. 2009. Drought and salinity tolerance of common *Agrostis* species. *2008 Turfgrass Research Report* (Connecticut), p. 119. TGIF Record #148625.

Akhter, J., R. Murray, K. Mahmood, K. A. Malik, and S. Ahmed. 2004. Improvement of degraded physical properties of a saline-sodic soil by reclamation with kallar grass (*Leptochloa fusca*). *Plant and Soil* 258 (1/2): 207–216.

Alam, S. M. 1999. Nutrient uptake by plants under stress conditions. In *Handbook of plant and crop stress*, ed. M. Pessarakli. New York: Marcel Dekker.

Al-Busaidi, A., T. Yamamoto, M. Inoue, A. Egrinya Eneji, Y. Mori, and M. Irshad. 2008. Effects of zeolite on soil nutrients and growth of barley following irrigation with saline water. *Journal of Plant Nutrition* 31: 1159–1173.

Al-Humaid, A. I. 2002. Effects of osmotic priming on seed germination and seedling growth of bermudagrass *Cynodon dactylon* L. under saline conditions. *Bulletin Faculty Agriculture Cairo University* 53: 265–274.

Ali, B., S. Hayat, and A. Ahmad. 2007. 28-Homobrassinolide ameliorates the saline stress in chickpea (*Cicer arietinum* L.). *Environmental and Experimental Botany* 59 (2): 217–223.

Allen, G. J., A. Amtmann, and D. Sanders. 1998. Calcium-dependent and calcium-independent K^+ mobilization channels in *Vicia faba* guard cell vacuoles. *Journal Experimental Botany* 49: 305–318.

Allen, G. J. and D. Sanders. 1996. Control of ionic currents in guard cell vacuoles by cytosolic and luminal calcium. *Plant Journal* 10: 1055–1069.

Allen, R. G. 1998. *Crop evapotranspiration: Guidelines for computing crop water requirements* (FAO Irrigation and Drainage Paper No. 56). Rome: United Nations Food and Agriculture Organization.

Allred, B. J., J. J. Daniels, and M. R. Ehsani. 2008. *Handbook of agricultural geophysics*. Boca Raton, FL: CRC Press.

Amezketa, E., R. Aragues, and R. Gazol. 2005. Efficiency of sulfuric acid, gypsum, and two gypsum by-products in soil crusting prevention and sodic soil reclamation. *Agron. J.* 97: 983–989.

Andrews, R. D., A. J. Koski, J. A. Murphy, and A. M. Petrovic. 1999. Zeoponic materials allow rapid green grow-in. *Golf Course Management* 67 (2): 68–72.

Anonymous. 2008. Prevent turf disease with manganese. *SportsTurf* 24 (10): 24. TGIF Record #141092.

Anonymous. 2010. *Guide to no-till TifEagle management*. Athens, GA: TifEagle Growers Association. http://www.tifeagle.com/NO-TILL%20TE%20BOOKLET.pdf

Apel, K. and H. Hirt. 2004. Reactive oxygen species: metabolism, oxidative stress and signal transduction. *Annual Review Plant Biology* 55: 373–399.

Apse, M. P., G. S. Aharon, W. A. Snedden, and E. Blumwald. 1999. Salt tolerance conferred by overexpression of a vacuolar Na^+/H^+ antiport in *Arabidopsis*. *Science* 265: 1256–1258.

Apse, M. P. and E. Blumwald. 2007. Na^+ transport in plants. *FEBS Letters* 581 (12): 2247–2254.

Aqeel Ahmad, S.M., Q. Ali, M. Ashraf, M. Z. Haider and Q. Abbas. 2009. Involvement of polyamines, abscisic acid and anti-oxidative enzymes in adaptation of Blue Panicgrass (*Panicum antidotale* Retz.) to saline environments. *Environmental and Experimental Botany* 66 (3): 409–417.

Arauzo, M. and M. Valladoilid. 2003. Short-term harmful effects of unionized ammonia on natural populations of *Moina micrura* and *Brachionus rebens* in a deep waste treatment pond. *Water Research* 37: 2547–2554.

Arizona Meteorological Network (AZMET). 2010. Web site includes reference crop ETo data from a statewide system of weather stations. http://ag.arizona.edu/azmet/index.html

Arkley, T. H., D. N. Munns, and C. M. Johnson. 1960. Preparation of plant tissues for micronutrient analysis: Removal of dust and spray contamination. *Agric. and Food Chemistry* 8 (4): 318–321.

Artiola, J. F., H. Gebrekidan, and D. J. Carty. 2000. Use of Langbeinite to reclaim sodic and saline sodic soils. *Communication of Soil Science and Plant Analysis* 31 (17–18): 2829–2842.

Aschenbach, T. A. 2006. Variation in growth rates under saline conditions of *Pascopyrum smithii* (Western wheatgrass) and *Distichlis spicata* (Inland saltgrass) from different source populations in Kansas and Nebraska: Implications for the restoration of salt-affected plant communities. *Restoration Ecology* 14 (1): 21–27.

Ashraf, M. 2004. Some important physiological selection criteria for salt tolerance in plants. *Flora-Morphology, Distribution, Functional Ecology of Plants* 199 (5): 361–376.

Ashraf, M., N. A. Akram, R. N. Arteca, and M. R. Foolad. 2010. The physiological, biochemical and molecular roles of brassinosteriods and salicylic acid in plant processes and salt tolerance. *Critical Reviews in Plant Sci.* 29 (3): 162–190.

Ashraf, M., H. R. Athar, P. J. C. Harris, and T. R. Kwon. 2008. Some prospective strategies for improving crop salt tolerance. *Advances in Agronomy* 97: 45–110.

Ashraf, M. and M. R. Foolad. 2005. Pre-sowing seed treatment—a shotgun approach to improve germination, plant growth, and crop yield under saline and non-saline conditions. *Advances in Agronomy* 88: 223–271.

Ashraf, M. and M. R. Foolad. 2007. Roles of glycine betaine and proline in improving plant abiotic-stress resistance. *Environmental and Experimental Botany* 59 (2): 206–216.

Ashraf, M. and P. J. C. Harris. 2004. Potential biochemical indicators of salinity tolerance in plants. *Plant Science* 166: 3–16.

Ashwath, N. and K. Venkatraman. 2010. Phytocapping: an alternative technique for landfill remediation. *Internat. J. of Environment and Waste Management* 6 (1–2): 51–70.

Audubon International. 2011. Audubon International environmental programs. Selkirk, NY: Audubon International. http://auduboninternational.org

Audubon Lifestyles. 2011. Audubon Lifestyles sustainable programs. Palm Harbor, FL: Audubon Lifestyles. http://www.audubonlifestyles.org/index.php?option=com_content&task=view&id=22

Australian Water Authority (AWA). 2000. Primary industries. In *Australian and New Zealand guidelines for fresh and marine water quality* (Paper No. 4). Artarmon, NSW: Australian Water Authority. http://www.mincos.gov.au/publications/australian_and_new_zealand_guidelines_for_fresh_and_marine_water_quality/volume_3

Aydemir, S. and N. F. Najjar. 2005. Application of two amendments (gypsum and langbinite) to reclaim sodic soil using sodic irrigation water. *Aust. J. of Soil Res.* 43: 547–553.

Ayers, R. S., and D. W. Westcot. 1976. *Water quality for agriculture* (FAO Irrigation and Drainage Paper 29). Rome: United Nations Food and Agricultural Organization.

Ayers, R. S., and D. W. Westcot. 1994. *Water quality for agriculture* (FAO Irrigation and Drainage Paper, 29, Rev. 1). Rome: United Nations Food and Agricultural Organization. http://www.fao.org/DOCREP/003/T0234E/T0234E00.htm#TOC

Aylward, L. 2010, Big on bio. *Golfdom* 66 (12): 36–39.

Azam, F. and M. Ifzal. 2006. Microbial populations immobilizing NH_4^+-N and NO_3^--N differ in their sensitivity to sodium chloride salinity in the soil. *Soil Biology and Biochemistry* 38 (8): 2491–2494.

Azevedo Neto, A. D., J. T. Prisco, J. Eneas-Filho, J-V. R. Medeiros, and E. Gomes-Filho. 2005. Hydrogen peroxide pre-treatment induces salt-stress acclimation in maize plants. *Journal of Plant Physiology* 162 (10): 1114–1122.

Babaeva, E. Y., V. F. Volobueva, B. A. Yagodin, and G. I. Klimakhin. 1999. Sowing quality and productivity of *Echinacea purpurea* in relation to soaking the seed in manganese and zinc solutions. *Izvestiya Timiryazevskoi Sel'skokhozyaistvennoi Akademii* 4: 73–80.

Backman, P. A., E. D. Miltner, G. K. Stahnke, and T. W. Cook. 2002. Worming your way out of a turf situation. *USGA Green Section Record* 40 (4): 7–8. http://turf.lib.msu.edu/2000s/2002/020707.pdf

Badawi, G. H., Y. Yamauchi, E. Shimada, R. Sasaki, K. Naoyoshi, and K. Tanaka. 2004. Enhanced tolerance to salt stress and water deficit by overexpressing superoxide dismutase in tobacco (*Nicotiana tabacum*) chloroplasts. *Plant Science* 166: 919–928.

Bailey, D. and T. Bilderback. 1998. *Alkalinity control for irrigation water used in nurseries and greenhouses* (Hort. Information Leaflet no. 558). Raleigh: North Carolina Cooperative Extension Service, NC State.

Baird, J. H. 2005. Putting green drainage, drainage, drainage. *USGA Green Section Record* 43 (6): 16–21.

Baird, J. H. 2007. Soil fertility and turfgrass nutrition 101. *USGA Green Section Record* 45 (5): 1–8.

Baker, S. W. 2004. Construction methods for public sector and professional sports pitches: A review. *Acta Hort. (ISHS) 661*: 27–37. http://www.actahort.org/books/661/661_1.htm

Baker, S. W. 2006. *Rootzones, sands and top dressing materials for sports turf.* Bingley, West Yorkshire, England: The Sports Turf Research Institute.

Baker, S. W. and P. M. Canaway. 1990. The effect of sand top dressing on the performance of winter games pitches of different construction types: I. soil physical properties and ground cover. *J. Sports Turf Res. Inst.* 66: 21–27.

Bakker, D. M., G. J. Hamilton, R. Hetherington, and C. Spann. 2010. Salinity dynamics and the potential for improvement of waterlogged and saline land in a Mediterranean climate using permanent raised beds. *Soil and Tillage Research* 110 (1): 8–24.

Baldwin, N. A. and B. A. Whitton. 1992. Cyanobacteria and eukaryotic algae in sports turf and amenity grasslands: a review. *Journal of Applied Phycology* 4: 39–47.

Barker, A. V. and D. J. Pilbeam, eds. 2007. *Handbook of plant nutrition.* Boca Raton, FL: Taylor & Francis.

Barret, J., B. Vinchesi, R. Dobson, P. Roche, and D. Zoldoske. 2003. *Golf course irrigation: Environmental design and management practices.* Hoboken, NJ: John Wiley.

Barrett-Lennard, E. G. 2003. The interaction between waterlogging and salinity in higher plants: causes, consequences, and implications. *Plant and Soil* 253: 35–54.

Barrett-Lennard, E. G. and T. L. Setter. 2010. Developing saline agriculture: Moving from traits and genes to systems. *Functional Plant Biology* 37 (7): iii–iv.

Bartels, J. M., ed. 1996. *Methods of soil analysis: Part 3 Chemical methods* (3rd ed., ASA and SSSA Book Series 5). Madison, WI: American Society of Agronomy and Soil Science Society of America.

Bauder, T. 1999. Atmometers: A flexible tool for irrigation scheduling. *Agron. News.* 19 (6). Fort Collins: Coop. Extension, Colorado State University. http://www.etgage.com/articles/CSUagnewsJun99.htm

Bell, G. E., and Xiong, X. (2008). The history, role, and potential of optical sensing for practical turf management. In *Handbook of turfgrass management and physiology,* ed. P. Pessaraki, 641–660. Boca Raton, FL: CRC Press.

Beltran, J. M. 1999. Irrigation with saline water: benefits and environmental impact. *Agric. Water. Manage.* 40: 183–194.

Berndt, W. L. and J. M. Vargas, Jr. 1992. Elemental sulfur lowers redox potential and produces sulfide in putting green sand. *HortScience* 27 (11): 1188–1190.

Berndt, W. L. and J. M. Vargas, Jr. 1996. Preventing black layer with nitrate. *Journal of Turfgrass Management* 1: 11–22.

Berndt, W. L. and J. M. Vargas, Jr. 2006. Dissimilatory reduction of sulfate in black layer. *HortScience* 41: 815–817.

Berndt, W. L. and J. M. Vargas, Jr. 2007. A review of the nature and control of black layer. *Dynamic Soil, Dynamic Plant* 1 (1): 17–23.

Berndt, W. L. and J. M. Vargas, Jr. 2008. Elemental sulfur reduces to sulfide in black layer soil. *HortScience* 43: 1615–1618.

Berndt, W. L. and J. M. Vargas, Jr. 2010. The nature and control of black layer. *Golf Course Management* 78 (4): 104–108.

Berndt, W. L., J. M. Vargas, Jr., A. R. Detweiller, P. E. Rieke, and B. E. Branham. 1987. Black layer formation in highly maintained turfgrass soils. *Golf Course Management* 55 (6): 106–108.

Bigelow, C. A., D. C. Bowman, and D. K. Cassel. 2004. Physical properties of three sand size classes amended with inorganic materials or sphagnum peat moss for putting green rootzones. *Crop Science* 44: 900–907.

Bigelow, C. A., A. G. Wollum II, and D. C. Bowman. 2000. Soil microbial populations in sand-based root zones. *Golf Course Management* 68 (11): 65–69.

Bond, W. J. 1998. Effluent irrigation: An environmental challenge for soil science. *Austral. J. Soil Res.* 36: 543–555.

Borst, S. M., J. S. McElroy, and G. K. Breeden. 2007. Cultural practices as important as chemicals for blue-green algae control. *TurfGrass TRENDS* Feb.: 51–52, 54–55.

Bower, C. A., G. Ogata, and J. M. Tucker. 1968. Sodium hazard of irrigation waters as influenced by leaching fraction and by precipitation or solution of calcium carbonate. *Soil Sci.* 106: 29–34.

Bradley, P. M. and J. T. Morris. 1992. Effect of salinity on the critical nitrogen concentration of *Spartina alterniflora* Loisel. *Aquatic Botany* 43 (2):149–161.

Brady, K. U., A. R. Kruckeberg, and H. D. Bradshaw, Jr. 2005. Evolutionary ecology of plant adaptation to serpentine soils. *Annual Review Ecology and Evolution Systematics* 36: 243–266.

Braman, K. 2009. Natural enemies of golf course pests. *Golf Course Management* 77 (9):102–108.

Braman, K., R. R. Duncan, W. W. Hanna, and M. C. Engelke. 2003. Arthropod predator occurrence and performance of *Geocoris uliginosis* (Say) on pest-resistant or susceptibility turfgrasses. *Environmental Entomology* 32 (4): 907–914.

Braman, K., R. R. Duncan, W. W. Hanna, and M. C. Engelke. 2004. Turfgrass species and cultivar influences on survival and parasitism of fall armyworm [*Spodoptera frugiperda* (J.E. Smith)]. *Journal of Economic Entomology* 97 (6): 1993–1998.

Bramley, R. G. V. (2009). Lessons from nearly 20 years of precision agriculture research, development, and adoption as a guide to its appropriate application. *Crop and Pasture Science*, *60*, 197–217.

Branson, R. L. and M. Fireman. 1980. *Gypsum and other chemical amendments for soil improvement* (Leaflet 2149). Berkeley: Div. of Agric. Sci., Univ. of California.

Brede, D. 2000. *Turfgrass maintenance reduction handbook.* Hoboken, NJ: John Wiley.

Brede, D. 2001. Two more unconventional grass families to know and love. *TurfGrass Trends* 10 (5): 8–14.

Brockhoff, S. R. 2010. Sand-based turfgrass root-zone modification with biochar. M.S. thesis, Iowa State University. TGIF Record #165654.

Brown, C. E., and S. Reza Pezeshki. 2007. Threshold for recovery in the marsh halophyte *Spartina alterniflora* grown under the combined effects of salinity and soil drying. *Journal of Plant Physiology* 164 (3): 274–282.

Brown, P. I., A. D. Halvorson, F. H. Siddoway, H. F. Mayland, and M.R. Miller. 1983. *Saline-seep diagnosis, control, and reclamation* (USDA Conserv. Res. Rep. No. 30). http://www.wsi.nrcs.usda.gov/products/w2q/downloads/Salinity/Saline_Seeps.pdf

Brown, P. W. 1999. Concerned about weather station ET? Perhaps you should be. *USGA Green Section Record* 37 (1), 5–7.

Burgess, R. M., M. C. Pelletier, K. T. Ho, J. R. Serbst, S. A. Ryba, A. Kuhn, M. M. Perron, P. Raczelowski, and M. G. Cantwell. 2003. Removal of ammonia toxicity in marine sediment TIEs: A comparison of *Ulva lactuca*, zeolite, and aeration methods. *Marine Pollution Bulletin* 46: 607–618.

Bush, J. K. 2006. The role of soil moisture, salinity, and oxygen on the growth of *Helianthus paradoxus* (Asteraceae) in an inland salt marsh in west Texas. *Journal of Arid Environments* 64 (1): 22–36.

Bush, R. T. and L. A. Sullivan. 1999. Pyrite micromorphology in three Australian Holocene sediments. *Australian Journal of Soil Science* 37: 637–653.

California Assembly Bill 174 (October) 1991. *Water resources—reclaimed water—nonpotable use.* In Statutes of 1991–1992 Regular Session, State of California Legislative Counsel's Digest, Chapter 553, pp. 2321–2322.

California Irrigation Management Information System (CIMIS). 2010. Provides estimated reference crop ET (ETo) date from state-wide weather station system. http://wwwcimis.water.ca.gov/cimis/welcome.jsp

Callery, A. G. 2003. Disinfect with Sodium Hypochorite. *Chemical Engineering Progress (CEP) Magazine* 99 (3): 42–46.

Camberato, J. J., P. D. Peterson, and S.B. Martin. 2005. Salinity alters rapid blight disease occurrence. *USGA TERO* (Turfgrass and Environmental Research Online) 4 (16): 1–7. TGIF Record #105569.

Camberato, J. J., P. D. Peterson, and S. B. Martin. 2006. Salinity and salinity tolerance alter rapid blight in Kentucky bluegrass, perennial ryegrass, and slender creeping red fescue. *Applied Turfgrass Science* February: 1–14. TGIF Record #109874.

Campbell, C. R. 2000. *Reference sufficiency ranges for plant analysis in the southern region of the United States* (Southern Cooperative Series Bulletin no. 394). http://www.ncagr.gov/agronomi/saaesd/s394.htm

Campbell Scientific. 2010. *ETo and weather monitoring.* Logan, UT: Campbell Scientific. http://www.campbellsci.com/eto

Carrow, R. N. 1989. Physical problems of fine-textured soils (Chapter 8) and Physical problems of coarse-textured soils (Chapter 9). In *Handbook of integrated pest management for turf and ornamentals,* ed. A. R. Leslie. Boca Raton, FL: CRC Press.

Carrow, R. N. 1991. Physical problems of fine-textured soils. *Golf Course Management* 59 (1): 118–124. http://archive.lib.msu.edu/tic/gcman/article/1991jan118.pdf

Carrow, R. N. 1992a. Understanding layered and compacted soils. *Golf Course Management* 60 (2) : 52–58. http://archive.lib.msu.edu/tic/gcman/article/1992feb52.pdf

Carrow, R. N. 1992b. Physical problems of coarse-textured soils. *Golf Course Management* 60 (2): 28–40. http://archive.lib.msu.edu/tic/gcman/article/1992feb28.pdf

Carrow, R. N. 1993. Evaluating soil and turf conditioners. *Golf Course Management* October: 56, 58, 60, 64, 70. http://archive.lib.msu.edu/tic/gcman/article/1993oct56.pdf

Carrow, R. N. 1995. Soil testing for fertilizer recommendations. *Golf Course Management* 63 (11): 61–68.

Carrow, R. N. 2004a. Surface organic matter in bermudagrass greens: A primary stress? *Golf Course Management* 75 (5): 102–106. http://archive.lib.msu.edu/tic/gcman/article/2004may102.pdf

Carrow, R. N. 2004b. Surface organic matter in bentgrass greens. *Golf Course Management* 75 (5): 96–101. http://archive.lib.msu.edu/tic/gcman/article/2004may96.pdf

Carrow, R. N. 2008. Managing sports fields during water restrictions. *Sports Turf* 24 (8): 20–26.

Carrow, R. N. Forthcoming. Turfgrass nutrition and irrigation water quality. *Comm. of Soil Sci. and Plant Analysis*.

Carrow, R. N., and R. R. Duncan. 1998. *Salt-affected turfgrass sites: Assessment and management*. Hoboken, NJ: John Wiley.

Carrow, R. N., and R. R. Duncan. 2003. Improving drought resistance and persistance in turf-type tall fescue. *Crop Sci.* 43 (3): 978–984.

Carrow, R. N., and R. R. Duncan. 2008. Best management practices for turfgrass water resources: Holistic-systems approach. In *Water quality and quantity issues for turfgrasses in urban landscapes* (Special publication), eds. M. Kenna and J. B. Beard. Ames, IA: Council for Agricultural Science and Technology (CAST).

Carrow, R. N., and R. R. Duncan. 2011. Salinity in soils. In *Turfgrass water conservation* (2nd ed.), eds. B. Leinauer and S. Cockerham. Riverside: ANR Communications Service, University of California.

Carrow, R. N., R. R. Duncan, and M. Huck. 1999. Treating the cause, not the symptoms: Irrigation water treatment for better infiltration. *USGA Green Section Record* 37 (6): 11–15.

Carrow, R. N., R. R. Duncan, and C. Waltz. 2005. *Golf course water conservation: Best management practices (BMPs) and strategies*. Athens: Univ. of Georgia, College of Agric. and Enviro. Sci. http://www.commodities.caes.uga.edu/turfgrass/georgiaturf/Water/Articles/BMP_GCSAA_05_Chapt_ALL_ref.pdf

Carrow, R. N., R. R. Duncan, and C. Waltz. 2009. *BMPs and water-use efficiency/conservation plan for golf courses: Template and guidelines*. Athens: Univ. of Georgia, College of Agric. and Enviro. Sci. http://www.commodities.caes.uga.edu/turfgrass/georgiaturf/Water/Articles/BMPs_Water_Cons_07.pdf

Carrow, R. N., and K. A. Fletcher. 2007a. Environmental management systems: A new standard for environmental management is coming. *USGA Green Section Record* 45 (4): 23–27.

Carrow, R. N., and K. A. Fletcher. 2007b. The devil is in the details. Environmental management systems (EMS) and golf courses. *USGA Green Section Record* 45 (5): 26–31.

Carrow, R. N., and K. A. Fletcher. 2007c. *Environmental management systems (EMS) for golf courses* (an educational guidebook developed by the University of Georgia and Audubon International). http://www.commodities.caes.uga.edu/turfgrass/georgiaturf/Water/Articles/EMS_2007_Guide.pdf

Carrow, R. N., M. Huck, and R. R. Duncan. 2000. Leaching for salinity management on turfgrass sites. *USGA Green Section Record* 38 (6): 15–24.

Carrow, R. N., J. Krum, I. Flitcroft, and V. Cline. 2009. Precision turfgrass management: Challenges and field applications for mapping turfgrass soil and stress. *Precision Agriculture* 11 (2): 115–134, doi:10.1007/s11119-009-9136-y. http://www.springerlink.com/content/7317k0048334q766/fulltext.pdf

Carrow, R. N., J. Krum, and C. Hartwiger. 2009. Precision turfgrass management: A new concept for efficient application of inputs. *USGA Turfgrass and Environmental Research Online (TERO)* 8 (13): 1–12. http://usgatero.msu.edu/v08/n13.pdf

Carrow, R. N., and A. M. Petrovic. 1992. Effects of traffic on turfgrasses. In *Turfgrass* (American Society of Agronomy Monograph No. 32), eds. D. V. Waddington, R. N. Carrow, and R. C. Shearman, 285–330. Madison, WI: American Society of Agronomy.

Carrow, R. N., L. J. Stowell, S. D. Davis, M. A. Fidanza, J. B. Unruh, and W. Wells. 2001. Developing regional soil and water baseline information for golf course turf. *Agronomy Abstracts*. Madison, WI: American Society of Agronomy.

Carrow, R. N., L. Stowell, W. Gelernter, S. Davis, R. R. Duncan, and J. Skorulski. 2003. Clarifying soil testing. I. Saturated paste and dilute extracts. *Golf Course Management* 71 (9): 81–85. http://www.gcsaa.org/gcm/2003/sept03/PDFs/09Clarify.pdf

Carrow, R. N., L. Stowell, W. Gelernter, S. Davis, R. R. Duncan, and J. Skorulski. 2004a. Clarifying soil testing: II. Choosing SLAN extracts for macronutrients. *Golf Course Management* 72 (1): 189–193. http://www.gcsaa.org/gcm/2004/jan04/PDFs/01Clarify2.pdf

Carrow, R. N., L. Stowell, W. Gelernter, S. Davis, R. R. Duncan, and J. Skorulski. 2004b. Clarifying soil testing. III. SLAN sufficiency ranges and recommendation. *Golf Course Management* 72 (1): 194–198. http://www.gcsaa.org/gcm/2004/jan04/PDFs/01Clarify3.pdf

Carrow, R. N., D. V. Waddington, and P. E. Rieke. 2001. *Turfgrass soil fertility and chemical problems: Assessment and management*. Hoboken, NJ: John Wiley.

Carrow, R. N., F. C. Waltz, and K. Fletcher. 2008. Environmental stewardship requires a successful plan: Can the turfgrass industry state one? *USGA Green Section Record* 46 (2): 25–32.

Carson, T., V. Cline, I. Flitcroft, J. Krum, and R. Carrow. 2010. Mobile device for spatial mapping of soil salinity on a golf course (Abstrast). ASA, CSSA, and SSSA 2010 Annual Meetings, October 31–November 4, Long Beach, CA. http://a-c-s.confex.com/crops/2010am/webprogram/Paper61359.html

Causin, H. F., I. N. Roberts, V. Criado, S. M. Gallego, L. B. Pena, M. del Carmen Rios, and A. J. Barneix. 2009. Changes in hydrogen peroxide homeostasis and cytokinin levels contribute to the regulation of shade-induced senescence in wheat leaves. *Plant Science* 177 (6): 698–704.

Center for Irrigation Technology. 2010. *Agricultural pumping efficiency program pump efficiency seminar workbook*. http://www.pumpefficiency.org/About/literature/seminarbrochure.asp

Charlesworth, P. 2005. *Soil water monitoring*. 2nd ed. Irrigation Insights no. 1. Canberra, ACT, Australia: Land and Water Australia.

Cheeseman, J. M. 1982. Pump-leak sodium fluxes in low salt corn roots. *Journal Membrane Biology* 70: 157–164.

Chen, T. H. H. and N. Murata. 2002. Enhancement of tolerance of abiotic stress by metabolic engineering of betaines and other compatible solutes. *Current Opinion Plant Biology* 5: 250–257.

Chen, W., P. Cui, H. Sun, W. Guo, C. Yang, H. Jin, B. Fang, and D. Shi. 2009. Comparative effects of salt and alkali stresses on organic acid accumulation and ionic balance of seabuckthorn (Hippophae rhamnoides L.). *Industrial Crops and Products* 30 (3): 351–358.

Chen, W., Z. Hou, L. Wu, Y. Liang, and C. Wei. 2010. Evaluating salinity distribution in soil irrigated with saline water in arid regions of northwest China. *Agricultural Water Management* 97 (12): 2001–2008.

Chhabra, R. 1996. *Soil salinity and water quality,* Brookfield, VT: A. A. Balkema.

Christen, E. W. and J. E. Ayars. 2001. *Subsurface drainage system design and management in irrigation agriculture: Best management practices for reducing drainage volume and salt load* (CSIRO Land and Water Tech. Report 38/1). Griffith, NSW, Australia: CSIRO Land and Water. http://www.clw.csiro.au/publications/technical2001/tr38-01.pdf

Christen, E. W. and D. Skehan. 2001. Design and management of subsurface horizontal drainage to reduce salt loads. *J. of Irrigation and Drainage Eng.* 127 (3): 148–155.

Chrominski, A., D. J. Weber, B. N. Smith, and D. F. Hegerhorst. 1989. Is dimethylsulfonium propionate an osmoprotectant of terrestrial glycophytes? *Die Naturwissenschaften* 76: 473–475.

Clark, D. H., H. F. Mayland, and R. C. Lamb. 1987. Mineral analysis of forages by near infrared reflectance spectroscopy. *Agron. J.* 79: 485–490.

Clark, G. A. and A. G. Smajstria. 1999. Treating irrigation systems with chlorine. *Cooperative Extension Service, Circular 1039.* Gainesville: Inst. of Food and Agricultural Sciences, Univ. of Florida.

Clark, G. J., N. Dodgshun, P. W. G. Sale, and C. Tang. 2007. Changes in chemical and biological properties of a sodic clay subsoil by addition of organic amendments. *Soil Biology and Biochemistry* 39 (11): 2806–2817.

Clark, G. J., P. W. G. Sale, and C. Tang. 2009. Organic amendments initiate the formation and stabilization of macroaggregates in a high clay sodic soil. *Aust. J. of Soil Res.* 47: 770–780.

Clarke, F. E. 1980. *Corrosion and encrustation in water wells* (FAO Irrigation and Drainage Paper 34). Rome: United Nations Food and Agriculture Organization.

Clean Water News. 2007. Regional water board adopts new salinity effluent limits for Delta dischargers. Central Valley Clean Water Association, Grass Valley, CA. *CVCWA Clean Water News* 1 (3). http://www.cvcwa.org/pdf%20files/Newsletter_V1_I3.pdf

Colmer, T. D. and L. A. C. J. Voesenek. 2009. Flooding tolerance: suites of plant traits in variable environments. *Functional Plant Biology* 36: 665–681.

Connellan, G. 2002. *Efficient irrigation: A reference manual for turf and landscape*. Melbourne, VIC, Australia: Brunley College, University of Melbourne.

Corwin, D. L. 2008. Past, present, and future trends of soil electrical conductivity measurement using geophysical methods. In B. J. Allred, J. J. Daniels, and M. R. Ehsani (eds.), *Handbook of agricultural geophysics*. Boca Raton, FL: CRC Press.

Corwin, D. L., and S. A. Bradford. 2008. Environmental impacts and sustainability of degraded water reuse. *J. Environ. Qual.* 37: S-1–S-7.

Corwin, D. L., and S. M. Lesch. 2003. Application of soil electrical conductivity to precision agriculture: theory, principles, and guidelines. *Agron. J.* 95: 455–471.

Corwin, D. L., and S. M. Lesch. 2005. Apparent soil electrical conductivity measurements in agriculture. *Computers and Electronics in Agriculture, 46,* 11–43.

Corwin, D. L., S. M. Lesch, J. D. Oster, and S. R. Kaffka. 2008. Short-term sustainability of drainage water reuse: spatio-temporal impacts on soil chemical properties. *J. Environ. Qual.* 37: S-8–S-24.

Corwin, D. L., J. D. Rhoades, and J. Simunek. 2007. Leaching requirement for soil salinity control: Steady-state versus transient models. *Agric. Water Manage.* 90: 165–180.

Costigan, P. A., A. D. Bradshaw, and R. P. Gemmell. 1981. The reclamation of acidic colliery spoils. I. Acid production potential. *Journal of Applied Ecology* 18 (3): 865–878.

Criado, M. V., C. Caputo, I. N. Roberts, M. A. Castro, and A. J. Barneix. 2009. Cytokinin-induced changes of nitrogen remobilization and chloroplast ultrastructure in wheat (*Triticum aestivum*). *Journal of Plant Physiology* 166 (16): 1775–1785.

Crum, J. R., T. F. Wolff, and J. N. Rogers. 2003. Agronomic and engineering properties of USGA putting greens. *USGA Turfgrass and Environmental Research Online* 2 (15): 1–9.

Csonka, L. N., and A. D. Hanson. 1991. Prokaryotic osmoregulation: genetics and physiology. *Annual Review of Microbiology* 45: 569–606.

Cullimore, D. R., S. Nilson, S. Taylor, and K. Nelson. 1990. Structure of black plug layer in a turfgrass putting sand green. *Journal Soil and Water Conservation* Nov.–Dec.: 657–659.

Czempinski, K., S. Zimmermann, T. Ehrhardt, and B. Muller-Rober. 1997. New structure and function in plant K^+ channels: KCO1, an outward rectifier with a steep Ca^{2+} dependency. *EMBO Journal* 16: 2565–2575.

Dai, J. 2006. Salinity tolerance of greens-type *Poa annua*. M.S. thesis, The Pennsylvania State University.

Dai, J., D. R. Huff, and M. J. Schlossberg. 2009. Salinity effects on seed germination and vegetative growth of greens-type *Poa annua* relative to other cool-season turfgrass species. *Crop Science* 49 (2): 696–703.

Dang, Y. P., R. C. Dalal, S. R. Buck, B. Harms, R. Kelly, Z. Hochman, G. D. Schwenke, et al. 2010. Diagnosis, extent, impacts, and management of subsoil constraints in the northern grains cropping region of Australia. *Soil Research* 48 (2): 105–119.

Da Silva, E. C., R. J. M. C. Nogueira, F. P. de Araujo, N. F. de Melo, and A. D. de Azevedo Neto. 2008. Physiological responses to salt stress in young umbu plants. *Environmental and Experimental Botany* 63 (1–3): 147–157.

Davis, W. B. 1981. Sand green construction. *Calif. Turfgrass Culture* 31 (1): 4–7.

Dean, D. E. 1996. Physiological response of two turfgrass species to increasing drought and salinity stress using a line source gradient. M.S. thesis, University of Nevada, Las Vegas. TGIF Record #101561.

Dear, S. E., N. G. Moore, S. K. Dobos, K. M. Watling, and C. R. Ahern. 2002. *Queensland acid sulfate soil technical manual: Soil management guidelines, version 3.8.* Indooroopilly, QLD, Australia: Dept. Nat. Resources and Mines. http://www.derm.qld.gov.au/land/ass/pdfs/soil_mgmt_guidelines_v3_8.pdf

DEC/NSW. 2003. *Environmental guidelines: Use of effluent by irrigation.* Sydney South, NSW, Australia: Department of Environment and Conservation. http://www.environment.nsw.gov.au/water/effluent.htm

Dello Ioio, R., F. S. Linhares, and S. Sabatini. 2008. Emerging role of cytokinin as a regulator of cellular differentiation. *Current Opinion in Plant Biology* 11 (1): 23–27.

Dello Ioio, R., F. S. Linhares, E. Scacchi, E. Casamitjana-Martinez, R. Heidstra, P. Constantino, and S. Sabatini. 2007. Cytokinins determine Arabidopsis root meristem size by controlling cell differentiation. *Current Biology* 17: 678–682.

Demidchik, V. and R. J. Davenport. 2002. Nonselective cation channels. *Annual Review of Plant Biology* 53: 67–107.

Dent, D. 1992. Reclamation of acid sulphate soils. In *Advances in soil science*, eds. R. Lal and B. A. Stewart, 79–122. Boca Raton, FL: CRC Press.

Dent, D. L., and L. J. Pons. 1995. A world perspective on acid sulphate soils. *Geoderma* 67: 263–276.

DePew, M. 1998. Humates in SportsTurf management. *SportsTurf* 14 (7): 36–37, 39. http://www.sportsturfonline.com

DeSutter, T. M. and L. J. Cihacek. 2009. Potential agricultural used of flue gas desulfurization gypsum in the Northern Great Plains. *Agron. J.* 101: 817–825.

Devitt, D. A., D. C. Bowman, and R. L. Morris. 1991. Effects of irrigation frequency, salinity of irrigation water, and soil type on growth and response of bermudagrass. *Arid Land Research and Management* 5 (1): 35–46.

Devitt, D. A., M. Lockett, R. L. Morris, and B. M. Bird. 2007. Spatial and temporal distribution of salts on fairways and greens irrigated with reuse water. *Agron. J.* 99: 692–700.

Diedhiou, C. J., O. V. Popova, and D. Golidack. 2009. Transcript profiling of the salt-tolerant *Festuca rubra* spp. *litoralis* reveals a regulatory network controlling salt acclimation. *Journal of Plant Physiology* 166 (7): 697–711.

Ding, F., M. Chen, N. Sui, and B. S. Wang. 2010. Ca^{2+} significantly enhanced development and salt-secretion rate of salt glands of *Limonium bicolor* under NaCl treatment. *South African Journal of Botany* 76 (1): 95–101.

Doerner, P. 2007. Plant meristems: Cytokinins—the alpha and omega of the meristem. *Current Biology* 17 (9): R321–R323.

Douhan, G. W., M. W. Olsen, A. Herrell, C. Winder, F. Wong, and K. Entwistle. 2009. Genetic diversity of *Labyrinthula terrestris*, a newly emergent plant pathogen, and the discovery of new Labyrinthulid organisms. *Mycological Research* 113 (10): 1192–1199.

Doyle, D. A., J. M. Cabral, R. A. Pfuetzner, A. Kuo, J. M. Gilbis, S. L. Cohen, B. T. Chait, and R. MacKinnon. 1998. The structure of the potassium channel: molecular basis of K^+ conduction and selectivity. *Science* 280: 69–77.

Dreyer, I. and M. R. Blatt. 2009. What makes a gate? The ins and outs of Kv-like K^+ channels in plants. *Trends in Plant Science* 14 (7): 383–390.

Duda, F. J. 2008. Suppression of turfgrass diseases through manipulation of soil pH. *Journal of Natural Resources and Life Sciences Education* 37: 38–42.

Dudley, L. M, A. Ben-Gal, and N. Lazarovitch. 2008. Drainage water reuse: biological, physical, and technological considerations for system management. *J. Environ. Qual.* 37: S-25–S-35.

Duncan, R. A., M. G. Bethune, T. Thayalakumaran, E. W. Christen, and T. A. McMahon. 2008. Management of salt mobilization in the irrigated landscape: A review of selected irrigation regions. *J. of Hydrology* 351: 238–252.

Duncan, R. R. and R. N. Carrow. 1999. Turfgrass molecular genetic improvement for abiotic/edaphic stress resistance. *Advances in Agronomy* 67: 233–305.

Duncan, R. R. and R. N. Carrow. 2000. *Seashore paspalum: The environmental turfgrass.* Hoboken, NJ: John Wiley.

Duncan, R. R. and R. N. Carrow. 2005. Just a grain of salt: As salinity increases, turf management will need to increase too. *Turfgrass Trends* (July): 70–75.

Duncan, R. R., R. N. Carrow, and M. Huck. 2000. Effective use of seawater irrigation on turfgrass. *USGA Green Section Record* 38 (1):11–17.

Duncan, R. R., R. N. Carrow, and M. Huck. 2009. *Turfgrass and landscape irrigation water quality: Assessment and management.* Boca Raton, FL: Taylor & Francis.

Eaton, F. M. 1950. Significance of carbonates in irrigation water. *Soil Sci.* 69: 123–133.

EBS. 2011. Environmental Business Solutions e-Par program for golf courses. http://www.epar.com.au/brochure/Default.aspx

Egamberdieva, D. 2009. Alleviation of salt stress by plant growth regulators and IAA producing bacteria in wheat. *Acta Physiol. Plant* 31: 861–864.

El-Haddad, E-S., and M. M. Noaman. 2001. Leaching requirement and salinity threshold for the yield and agronomic characteristics of halophytes under salt stress. *Journal of Arid Environments* 49: 865–874.

Elinder, F. and P. Arhem. 1999. Role of individual surface charges of voltage-gated K channels. *Biophysical Journal* 77: 1358–1362.

Entwistle, C. A., M. W. Olsen, and D. M. Bigelow. 2005. First report of *Labyrinthula* spp. causing rapid blight of *Agrostis capillaris* and *Poa annua* on amenity turfgrass in the UK. *New Disease Reports*, Vol. 11. http://www.bspp.org.uk/ndr

Entwistle, K., and M. Olsen. 2007. A new disease comes to Europe. *Buckeye Turf* August 10: 1–4.

Epstein, E. 1994. The anomaly of silicon in plant biology. *Proceedings National Academy Science USA* 91: 11–17.

Ertani, A., L. Peserico, C. Franceschi, A. Altissimo, and S. Nardi. 2010. The effects of biostimulant on *Lolium perenne* salt-stressed plants. *2nd European Turfgrass Society Conference Proceedings*, Vol. 2. TGIF Record #164043.

Ervin, E. H. and X. Zhang. 2008. Applied physiology of natural and synthetic plant growth regulators on turfgrasses. In *Handbook of turfgrass management and physiology*, ed. M. Pessarakli, 171–200. Boca Raton, FL: CRC Press.

ESRI. 2004a. *ArcGIS 9: Using ArcGIS geostatistical analyst.* Redlands, CA: ESRI.

ESRI. 2004b. *ArcGIS 9: Using ArcGIS spatial analyst.* Redlands, CA: ESRI.

Ezlit, Y. D., R. J. Smith, and S. R. Raine. 2010. *A review of salinity and sodicity in irrigation* (Irr. Matters Series N. 01/10). Darling Heights, QLD, Australia: Coop. Res. Centre for Irr. Futures.

Fanning, D. S. and S. N. Burch. 2000. Coastal acid sulfate soils. In *Reclamation of drastically disturbed lands* (Agronomy Monograph No. 41), eds. R. I. Barnhiesel, R. G. Darmody, and W. L. Daniels. Madison, WI: American Society of Agronomy.

FAO. 1990. *Management of gypsiferous soils* (FAO Soils Bulletin 62). Rome: United Nations Food and Agriculture Organization. http://www.fao.org/docrep/t0323e/t0323e00.htm

FAO. 2009a. Salt-affected soils. ProSoil: Problem soils database. Rome: United Nations Food and Agriculture Organization. http://www.fao.org/ag/aGL/agll/prosoil/salt.htm

FAO. 2009b. *Global network on integrated soil management for sustainable use of salt-affected sites*. Rome: United Nations Food and Agriculture Organization. http://www.fao.org/landandwater/agll/spush/degrad.htm

Fauteux, F., W. Remus-Borel, J. G. Menzies, and R. R. Belanger. 2005. Silicon and plant disease resistance against pathogenic fungi. *FEMS Microbiology Letters* 249 (1): 1–6.

Fayyad M. K., and A. M. al-Sheikh. 2001. Determination of N-chloramines in As-samra chlorinated wastewater and their effect on the disinfection process. *Water Research* 35 (5): 1304–1310.

Fech, J. C. and R. E. Gaussoin. 2009. Root afflictions of turf. *Superintendent* 8 (9): 22, 24–25.

Feigin, A. 1985. Fertilization management of crops irrigated with saline waters. *Plant and Soil* 89: 285–299.

Feil, K., K. Kubick, R. Waters, and R. Wong. 1997. *Guidelines for the on-site retrofit of facilities using disinfected tertiary recycled water*. Ontario, CA: California-Nevada Section, American Water Works Association.

Felle, H. 1994. The H^+/Cl^- symporter in root-hair cells of *Sinapis alba*. An electrophysiological study using ion-selective microelectrodes. *Plant Physiology* 106: 1131–1136.

Ferguson, G. A., and I. L. Pepper. 1987. Ammonium retention in sand amended with clinoptilolite. *Soil Science Society America Journal* 51: 231–234.

Ferguson, G. A., I. L. Pepper, and W. R. Kneebone. 1986. Growth of creeping bentgrass on a new medium for turfgrass growth: Clinoptilolite zeolite-amended sand. *Agronomy Journal* 78: 1095–1098.

Ferreira, F. J., and J. J. Kieber. 2005. Cytokinin signaling. *Current Opinion in Plant Biology* 8 (5): 518–525.

Fitzpatrick, R. 1999. *Rising saline watertables and development of acid sulfate soils* (CSIRO Land and Water Research Project, Sheet No. 13). http://www.clw.csiro.au/publications/projects/projects13.pdf

Fitzpatrick, R., and P. Shand, eds. 2008. *Inland acid sulfate soil systems across Australia* (CRC LEME Open File Report No. 249). Perth, Australia: CRC LEME. http://www.clw.csiro.au/acidsulfatesoils/ass-book.html

Flowers, T. J., H. K. Galal, and L. Bromham. 2010. Evolution of halophytes: multiple origins of salt tolerance in land plants. *Functional Plant Biology* 37 (7): 604–612.

Flowers, T. J., M. A. Hajibagheri, and N. J. W. Clipson. 1986. Halophytes. *The Quarterly Review of Biology* 61: 313–337.

Flowers, T. J., P. F. Troke, and A. R. Yeo. 1977. The mechanism of salt tolerance in halophytes. *Annual Review of Plant Physiology* 28: 89–121.

Foley, W. J., A. Mc Ilivee, I. Lawler, L. Aragones, A. P. Woolnough, and N. Berding. 1998. Ecological applications of near infrared spectroscopy: A tool for rapid, cost-effective prediction of the composition of plant and animal tissues and aspects of animal performance. *Oecologia* 116: 293–305.

Foyer, C. H. and G. Noctor. 2005. Oxidant and antioxidant signaling in plants: A re-evaluation of the concept of oxidative stress in a physiological concept. *Plant Cell Environment* 28: 1056–1071.

Frazier, B. E., C. S. Walters, and E.M. Perry. 1999. Role of remote sensing in site-specific management. In *The state of site specific management for agriculture*, eds. F. J. Pierce and E. J. Sadler. Madison, WI: American Society of Agronomy.

Frensch, J. and T. C. Hsaio. 1994. Transient responses of cell turgor and growth in maize roots as affected by changes in water potential. *Plant Physiology* 104: 247–254.

Fricke, W., G. Akhiyarova, D. Veselov, G. Kudoyarova. 2004. Rapid and tissue-specific changes in ABA and in growth rate response to salinity in barley leaves. *Journal Experimental Botany* 55: 1115–1123.

Fricke, W., G. Akhiyarova, W. Wei, F. Alexandersson, A. Miller, P. O. Kjellbom, A. Richardson, et al. 2006. The short-term growth response to salt of the developing barley leaf. *Journal Experimental Botany* 57: 1079–1095.

Fruby, S., P. Caccetta, and J. Wallace. 2010. Salinity monitoring in Western Australia using remotely sensed and other spatial data. *J. Environ. Qual.* 39: 16–25.

Full Coverage Irrigation. 2010. Advanced irrigation nozzle technology. http://www.fcinozzles.com

Gadallah, M. A. A. 1999. Effects of kinetin on growth, grain yield and some mineral elements in wheat plants growing under excess salinity and oxygen deficiency. *Plant Growth Regulation* 27: 63–74.

Gafni, A., and Y. Zohar. 2001. Sodicity, conventional drainage and bio-drainage in Israel. *Aust. J. of Soil Res.* 39 (6): 1269–1278.

Gale, G., R. Koenig, and J. Barnhill. 2001. *Managing soil pH in Utah*. Logan: Utah State Univ.

Gandhi, C. S., E. Clark, E. Lotts, A. Pralle, and E. Y. Isacoff. 2003. The orientation and molecular movement of a K^+ channel voltage-sensing domain. *Cell* 40 (3): 515–525.

Garbow, G. 2008. *Tools for turfgrass irrigation water management*. Raleigh: Center for Turfgrass Environmental Research and Education. North Carolina State University. http://www.bae.ncsu.edu/topic/go_irrigation/docs/tools-turfgrass.pdf

Gaussion, R. 1999. Algae control in ponds with barley straw bales: on-site results in Nebraska. *Center for Grassland Studies* 5 (2): 3.

Gavlak, R., D. Horneck, R. O. Miller, and J. Kotuby-Amacher. 2003. *Soil, plant and water reference methods for the western region* (WREP-125 2nd ed., WCC-103 Publication). Corvallis: Oregon State University. http://cropandsoil.oregonstate.edu/wera103/soil_methods

Gaxiola, R. A., J. S. Li, S. Undurraga, L. M. Dang, G .J. Allen, S. L. Alper, and G. R. Fink. 2001. Drought- and salt-tolerant plants result from overexpression of the AVP1 H^+-pump. *Proceedings National Academy Sciences USA* 98: 11444–11449.

Gaxiola, R. A., A. Sherman, P. Grisafi, S. L. Alper, and G. R. Fink. 1999. The *Arabidopsis thaliana* proton transporters, Nhx1 and Avp1 can function in cation detoxification in yeast. *Proceedings National Academy Sciences USA* 96: 1480–1485.

Gaymard, F., G. Pilot, B. Lacombe, D. Bouchez, D. Bruneau, J. Boucherez, N. Michaux-Ferriere, J-B. Thibaud, and H. Sentenac. 1998. Identification and disruption of a plant Shaker-like outward channel involved in K^+ release to the xylem sap. *Cell* 94 (5): 647–655.

Gelernter, W. and L. J. Stowell. 2000. Cyanobacteria (A.K.A. blue-green algae): WANTED for causing serious damage to turf. *PACE Insights* 6 (8): 1–4. www.pace-ptri.com

Gelernter, W. and L. J. Stowell. 2002. Turfgrass tissue testing: pros and cons. *PACE Insights* 8 (1): 1–4. http://www.paceturf.org/PTRI/Documents/Soil_tis/0201.pdf

Gelernter, W. and L. J. Stowell. 2004. Summer disease update: Making take-all patch a "takeless" disease. *PACE Insights* 10 (6): 1. TGIF Record #105056.

Gerhart, V. J., R. Kane, and E. P. Glenn. 2006. Recycling industrial saline wastewater for landscape irrigation in a desert urban area. *J. Arid. Environ.* 67 (3): 473–486.

Gharaibeh, M. A., N. I Eltaif, and S. H. Shaah. 2010. Reclamation of a calcareous saline-sodic soil using phosphoric acid and by-product gypsum. *Soil Use and Management* 26: 141–148.

Gibeault, V. A. and S. T. Cockerham, eds. 1985. *Turfgrass water conservation* (Publication No. 21405). Oakland, CA: Cooperative Extension Service, Univ. of Calif.

Gierth, M. and P. Maser. 2007. Potassium transporters in plants—involvement in K^+ acquisition, redistribution and homeostasis. *FEB Letters* 581 (12): 2348–2356.

Gilhuly, L. 1999. A decade of piling it on. *USGA Green Section Record* 37 (6): 1–5.

Gill, J. S., P. W. G. Sale, R. R. Peries, and C. Tang. 2009. Changes in soil physical properties and crop root growth in dense sodic subsoil following incorporation of organic amendments. *Field Crops Res.* 114: 137–146.

Gill, J. S., P. W. G. Sale, and C. Tang. 2008. Amelioration of dense sodic subsoil using organic amendments increases wheat yield more than using gypsum in a high rainfall zone of southern Australia. *Field Crops Res.* 107: 265–275.

Gilliham, M. and M. Tester. 2005. The regulation of anion loading to the maize root xylem. *Plant Physiology* 137: 819–828.

Gleick, P. H. 1993. *Water in crisis: A guide to the world's fresh water resources*. New York: Oxford Univ. Press.

Glenn, E.P., J. J. Brown, and E. Blumwald. 1999. Salt tolerance and crop potential of halophytes. *Critical Reviews in Plant Science* 18: 227–255.

Gong, H. J., D. P. Randall, and T. J. Flowers. 2006. Silicon deposition in the root reduces sodium uptake in rice (*Oryza sativa* L.) seedlings by reducing bypass flow. *Plant Cell Environment* 29: 1970–1979.

Goss, P. 1999. Flood your greens: Not your bunkers. *USGA Green Section Record* 26 (3): 26.

Goss, P. 2003. Making the right spending decisions when tackling soil and water quality problems. *USGA Green Section Record* 30 (3): 17–20.

Gosset, D. R., E. P. Millhollon, and M. C. Lucas. 1994. Antioxidant response to NaCl stress in salt-tolerant and salt-sensitive cultivars of cotton. *Crop Science* 34: 706–714.

Grandlic, C. J., M. W. Palmer, and R. M. Maier. 2009. Optimization of plant growth-promoting bacteria-assisted phytostabilization of mine tailings. *Soil Biology and Biochemistry* 41 (8): 1734–1740.

Grattan, S. R., and C. M. Grieve. 1999. Mineral nutrient acquisition and response by plants grown in saline environments. In *Handbook of plant and crop stress*, ed. M. Pessarakli, 203–229. New York: Marcel Dekker.

Grattan, S. R., and J. D. Oster. 2003. Use and reuse of saline-sodic waters for irrigation of crops. In *Crop production in saline environments: Global and integrative perspectives*, eds. S. S. Goyal, S. K. Sharma, and D. W. Rains. New York: Food Products Press/Haworth Press.

Gray, N. F. 1997. Environmental impact and remediation of acid mine drainage: A management problem. *Environ. Geol.* 30 (1/2): 62–71.

Green, R. L., L. Wu, and G. J. Grant. 2001. Summer cultivation increases field infiltration rates of water and reduces soil electrical conductivity on annual bluegrass golf greens. *HortSci.* 36: 776–779.

Greenway, H. and R. Munns. 1980. Mechanisms of salt tolerance in non-halophytes. *Annual Review of Plant Physiology* 31: 149–190.

Gross, P. J. 1999. Flood your greens—not your bunkers. *USGA Green Section Record,* 26 (3): 26.

Gross, P. J. 2008. A step-by-step guide for using recycled water. *USGA Green Section Record* 46 (2): 1–8.

Grunstra, M., and O. W. Van Auken. 2007. Using GIS to display complex soil salinity patterns in an inland salt marsh (Chapter 19). *Developments in Environmental Sciences* 5: 407–431. Concepts and Applications in Environmental Geochemistry.

Guertal, E. A., and J. N. Shaw. 2004. Multispectral radiometer signatures for stress evaluation in compacted bermudagrass turf. *Hortscience* 39: 403–407.

Haby, V. A., M. P. Russelle, and E. O. Skogley. 1990. Testing soils for potassium, calcium and magnesium. In *Soil testing and plant analysis* (3rd ed., Soil Sci. Soc. of Amer. Book Series, No. 3), ed. R. L. Westerman. Madison, WI: Soil Science Society of America.

Hagstrom, G. R. 1986. Fertilizer sources of sulfur and their use. In *Sulfur in agriculture,* ed. M. A. Tabatabai, 567–581. Madison, WI: American Society of Agronomy.

Halfter, U., M. Ishitani, and J. K. Zhu. 2000. The Arabidopsis SOS2 protein kinase physically interacts with and is activated by the calcium-binding protein SOS3. *Proceedings National Academy Science USA* 97: 3735–3740.

Halliwell, B., and J. M. C. Gutteridge. 1986. Oxygen free radicals and iron in relation to biology and medicine: some problems and concepts. *Arch. Biochemistry and Biophysics* 246: 501–514.

Handreck, K. A., and N. D. Black. 2010. *Growing media for ornamental plants and turf.* 4th ed.. Randwick, NSW, Australia: Univ. of NSW Press.

Hanson, B., S. R. Grattan, and A. Fulton. 1999. *Agricultural salinity and drainage* (Division of Agriculture and Natural Resources Publication No. 3375). Davis: University of California.

Happ, K. 1994. Tissue testing: questions and answers. *USGA Green Section Record* 32 (4): 9–11.

Hare, P. D., W. A. Cress, and J. Van Staden. 1997. The involvement of cytokinins in plant responses to environmental stress. *Plant Growth Regulation* 23: 79–103.

Hare, P. D., W. A. Cress, and J. Van Staden. 1998. Dissecting the roles of osmolyte accumulation during stress. *Plant Cell Environment* 21: 535–553.

Harivandi, M. A. 1984. Managing saline, sodic or saline-sodic soils for turfgrasses. *California Turfgrass Culture* 34 (2–3): 9–10.

Harivandi, M. A. 1991. *Effluent water for turfgrass irrigation* (Leaflet 21500). Oakland: Cooperative Extension, Division of Agriculture and Natural Resources, University of California.

Harivandi, M. A., Butler, J. D., and Lin, W. 1992. Salinity and turfgrass culture. In *Turfgrass Monograph no. 32*, eds. D. V. Waddington, R. N. Carrow, and R. C. Shearman, 207–229. Madison, WI: American Society of Agronomy.

Harivandi, M. A., J. D. Butler, and P. N. Soltanpour. 1982a. Effects of seawater concentrations on germination and ion accumulation in alkaligrass (*Puccinella* spp.). *Comm. in Soil Science and Plant Analysis* 13 (7): 507–517.

Harivandi, M. A., J. D. Butler, and P. N. Soltanpour. 1982b. Salt influence on germination and seedling survival of six cool season turfgrass species. *Comm. in Soil Science and Plant Analysis* 13 (7): 519–529.

Haro, R., M. A. Banuelos, M. A. F. Senn, J. Barrero-Gil, and A. Rodriguez-Navarro. 2005. HKT1 mediates sodium uniport in roots. Pitfalls in the expression of HKT1 in yeast. *Plant Physiology* 139: 1495–1506.

Hart, B. T., P. Bailey, R. Edwards, K. Hortle, K. James, A. McMahon, C. Meredith, and K. Swadling. 1990. Effects of salinity on river, stream, and wetland ecosystems in Victoria, Australia. *Water Research* 24 (9): 1103–1117.

Hartwiger, C. 2007. No till in no time. *USGA Green Section Record* 45 (6): 22–26.

Harvel Plastics. 2010. *Chemical resistance of Harvel PVC and CPVC piping products.* Easton, PA: Harvel Plastics. http://www.harvel.com/downloads/chemical-resistance.pdf

Hasegawa, M., R. Bressan, and J. M. Pardo. 2000. The dawn of plant salt tolerance genetics. *Trends in Plant Science* 5 (8): 317–319.

Hayat, R., S. Ali, U. Amara, R. Khalid, and I. Ahmed. 2010. Soil beneficial bacteria and their role in plant growth promotion: a review. *Ann. Microbiol* 60 (4): 579–598.

Hayat, S., S. A. Hasan, M. Yusul, Q. Hayat, and A. Ahmad. 2010. Effect of 28-homobrassinolide on photosynthesis, fluorescence and antioxidant system in the presence or absence of salinity and temperature in *Vigna radiata. Environmental and Experimental Botany* 69 (2): 105–112.

Hazelton, P., and B. Murphy. 2007. *Interpreting soil test results—What do the numbers mean?* Collingwood, VIC, Australia: CSIRO.

He, Y., Z. Zhu, J. Yang, X. Ni, and B. Zhu. 2009. Grafting increases the salt tolerance of tomato by improvement of photosynthesis and enhancement of antioxidant enzymes activity. *Environmental and Experimental Botany* 66 (2): 270–278.

Healy, M. J. 2008. Toxin trail (August): 60–64. http://www.golfcourseindustry.com

Heckman, J. R., B. B. Clarke, and J. A. Murphy. 2003. Optimizing manganese fertilization for the suppression of Take-All patch disease on creeping bentgrass. *Crop Science* 43: 1395–1398.

Hedley, C. B. and I. J. Yule. 2009. A method for spatial prediction of daily soil water status for precise irrigation scheduling. *Agric. Water Manage.* 96: 1737–1745.

Hedrich, R. and E. Neher. 1987. Cytoplasmic calcium regulates voltage-dependent ion channels in plant vacuoles. *Nature* 329: 833–836.

Helfrich, L. A., J. Parkhurst, and R. Neves. 2001. *Liming acidified lakes and ponds* (Pub. 420-254). Blacksburg, VA: Virginia Coop. Extension, Virginia Polytech. Institute and State University. http://www.ext.vt.edu/pubs/fisheries/420-254/420-254.html#L3

Hendershot, W. H., H. Lalande, and M. Duquette. 1993. Ion exchange and exchangeable cations. In *Soil sampling and methods of analysis*, ed. M. R. Carter, 167–176. Boca Raton, FL: Lewis.

Hendrick, J. M. H., J. M. Wraith, D. L. Corwin, and R. G. Kachanoski. 2002. Miscible solute transport. In *Methods of soil analysis. Part 4. Physical methods* (SSSA Book Series no. 5), eds. J. H. Dane and G. C. Topp, 1253–1321. Madison, WI: Soil Science Society of America.

Hernandez, J. A., A. Jimenez, P. Mullineaux, and F. Sevilla. 2000. Tolerance of pea (*Pisum sativum* L.) to long-term salt stress is associated with induction of antioxidant defenses. *Plant Cell Environment* 23: 853–862.

Hodges, C. F. 1987. Blue-green algae and black layer. Part I. *Landscape Management* 26 (10): 38–44.

Hodges, C. F. 1989a. Blue-green algae and black layer. Part II. *Landscape Management* 26 (11): 30–31.

Hodges, C. F. 1989b. Another look at black layer. *Golf Course Management* 57 (3): 54–58.

Hodges, C. F. 1992a. Pathogenicity of *Pythium torulosum* to roots of *Agrostis palustris* in black-layered sand produced by the interaction of the cyanobacteria species *Lyngbya*, *Phormidium*, and *Nostoc* with *Desulfovibrio desulfuricans*. *Canadian J. Bot.* 70: 2193–2197.

Hodges, C. F. 1992b. The biology of algae in turf. *Golf Course Management* 61 (8): 44–56.

Hodges, C. F., and D. A. Campbell. 1997. Nutrient salts and the toxicity of black-layer induced by cyanobacteria and *Desulfovibrio desulfuricans* to *Agrostis palustris*. *Plant and Soil* 195: 53–60.

Hoffman, G. J., and D. S. Durnford. 1999. Drainage design for salinity control. In *Agriculture drainage* (No. 38 Series in Agronomy), eds. R. W. Skaggs and J. van Schilfaarde, 539–614. Madison, WI: American Society of Agronomy.

Horie, T., A. Costa, T. H. Kim, M. J. Han, R. Horie, H. Y. Leung, A. Miyao A, H. Hirochika, G. An, and J. I. Schroeder. 2007. Rice OsHKT2:1 transporter mediates large Na^+ influx components into K^+-starved roots for growth. *EMBO Journal* 26: 300–314.

Huang, Z. T., and A. M. Petrovic. 1991. Clinoptilolite zeolite amendment of sand influences on water use efficiency of creeping bentgrass and nitrate leaching. *CSSA/ASA Annual Meetings Abstracts*, p. 177.

Huang, Z. T., and A. M. Petrovic. 1994. Cliniptilolite zeolite influence on nitrate leaching and nitrogen use efficiency in simulated sand based golf greens. *Journal of Environmental Quality* 23: 1190–1194.

Huang, Z. T., and A. M. Petrovic. 1995. Physical properties of sand as affected by clinoptilolite zeolite particle size and quantity. *Journal Turfgrass Management* 1: 1–15.

Huck, M. 1997. Irrigation design, rocket science, and the SPACE Program. *USGA Green Section Record* 35 (1): 1–7.

Huck, M., R. N. Carrow, and R. R. Duncan. 2000. Effluent water: nightmare or dream come true? *USGA Green Section Record* 38 (2): 15–29.

Hudson, R. 2010. Chloramine resistance. *Hudson Tech Files*. Broken Arrow, OK: R. L. Hudson. http://www.rlhudson.com/publications/techfiles/chloramine.htm

Hull, R. J. 2004. Less familiar nutrients also deserve spotlight. *TurfGrass Trends* 60 (6): 65–69.

Hur, S. N. 1991. Effect of osmoconditioning on the productivity of Italian ryegrass and sorghum under suboptimal conditions. *Korean Journal Animal Science* 33: 101–105.

Hurst, A. C., T. Meckel, S. Tayefeh, G. Thiel, and U. Homann. 2004. Trafficing of the plant potassium inward rectifier KAT1 in guard cell protoplasts of *Vicia faba*. *Plant Journal* 37: 391–397.

Hussain, T. M., T. Chandrasekhar, M. Hazara, Z. Sultan, B. K. Saleh, and G. R. Gopal. 2008. Recent advances in salt stress biology—a review. *Biotechnology and Molecular Biology Review* 3 (1): 8–13.

Ichida, A. M., Z. M. Pei, V. M. Baizabal-Aguirre, K. J. Turner, and J. I. Schroeder. 1997. Expression of a Cs^+-resistant guard cell K^+ channel confers Cs^+-resistant, light–induced stomatal opening in transgenic *Arabidopsis*. *Plant Cell* 9: 1843–1857.

Ikenaga, S., and T. Inamura. (2008). Evaluation of site-specific management zones on a farm with 124 contiguous small paddy fields in a multiple-cropping system. *Precision Agriculture* 9: 147–159.

Iqbal, M., M. Ashraf, and A. Jamil. 2006. Seed enhancement with cytokinins: changes in growth and grain yield in salt stressed wheat plants. *Plant Growth Regulation* 50 (1): 29–39. doi:10.1007/s10725-006-9123-5.

Irrigation Association. 2003. *Certified golf course auditor training seminar and manual*. Falls Church, VA: Irrigation Association.

Irrigation Association. 2005, April. *Turf and landscape irrigation best management practices*. http://www.irrigation.org/Resources/Turf—Landscape_BMPs.aspx

Irrigation Association. 2010. http://www.irrigation.org

Irrigation Assoc. of Australia. 2010. Hornsby, NSW, Australia. http://www.irrigation.org.au

Jalali, M., and F. Ranjbar. 2009. Effects of sodic water on soil sodicity and nutrient leaching in poultry and sheep manure amended soils. *Geoderma* 153 (1–2): 194–204.

Jaleel, C. A., G. M. A. Lakshmanan, M. Gomathinayhagam, and R. Panneerselvam. 2008. Triadimefon induced salt stress tolerance in *Withania somnifera* and its relations to antioxidant defense system. *South African Journal of Botany* 74 (1): 126–132.

James, R. A., R. Munns, and S. von Caemmerer. 2006. Photosynthetic capacity is related to cellular and subcellular partitioning of Na^+, K^+ and Cl^- in salt-affected barley and durum wheat: *Nax1* and *Nax2*. *Plant Physiology* 142: 1537–1547.

James, R. A., A. R. Rivelli, R. Munns, and S. von Caemmerer. 2002. Factors affecting CO_2 assimilation, leaf injury and growth in salt-stressed durum wheat. *Functional Plant Biology* 29: 1393–1403.

Jayawardane, N. S. and K. Y. Chan. 1994. The management of soil physical properties limiting crop production in Australian sodic soils: A review. *Aust. J. Soil Res.* 32: 13–44.

Jiang, Y., R. N. Carrow, and R. R. Duncan. 2003. Correlation analysis procedures for canopy spectral reflectance data of seashore paspalum under traffic stress. *J. American Society HortScience* 128 (3): 343–348.

Johnson, C. J., B. Leinauer, A. L. Ulery, D. E. Karcher, and R. M. Goss. 2007. Moderate salinity does not affect germination of several cool- and warm-season turfgrasses. *Applied Turfgrass Science* September: 1–7.

Johnson, C. K., Doran, J. W., Duke, H. R., Wienhold, B. J., Eskridge, K. M., and Shanahan, J. F. (2001). Field-scale electrical conductivity mapping for delineating soil conditions. *Soil Science Society of America Journal* 65: 1829–1837.

Johnson, N.C. 1998. Responses of *Salsola kali* and *Panicum virgatum* to mycorrhizal fungi, phosphorus and soil organic matter: implication for reclamation. *Journal of Applied Ecology* 35 (1): 86–94.

Johnston, W. R., and J. Proctor. 1981. Growth of serpentine and non-serpentine races of *Festuca rubra* in solutions simulating the chemical conditions in a toxic serpentine soil. *Journal of Ecology* 69: 855–869.

Jones, J. B., Jr. 1998. Soil test methods: Past, present, and future, use of soil extractants. *Commun. Soil Sci. Plant Anal.* 29 (11–14): 1543–1552.

Jones, J. B., and Y. P. Kalra. 1992. Soil testing and plant analysis activities: the United States and Canada. *Communication in Soil Science and Plant Analysis* 23 (17–20): 2015–2027.

Joo, Y. K., S. K. Lee, Y. S. Jung, and N. E. Christians. 2008. Turfgrass revegetation of amended sea sand dredged from the Yellow Sea. *Comm. in Soil Science and Plant Analysis* 39 (11–12): 1571–1582.

Jorenush, M. H., and A. R. Sepaskhah. 2003. Modeling capillary rise and soil salinity for shallow saline water table under irrigated and non-irrigated conditions. *Agric. Water Management* 61: 125–141.

Kadiri, M. and M. A. Hussaini. 1999. Effect of hardening pretreatments on vegetative growth, enzyme activities and yield of *Pennisetum americanum* and *Sorghum bicolor*. *Global Journal of Pure Applied Science* 5: 179–183.

Kaewmano, C., I. Kheoruenromme, A. Suddhiprakam, and R. J. Gilk. 2009. Aggregate stability of salt-affected kaolinitic soil Northeast Plateau, Thailand. *Soil Res.* 47 (7): 697–706.

Kaffka, S. 2001. Salt tolerant forages for the reuse of saline drainage water. In *Proceedings 31st California Alfalfa and Forage Symposium,* December 12–13, Modesto, CA. http://alfalfa.ucdavis.edu

Kaffka, S., J. Oster, and D. Corwin. 2004. Forage production and soil reclamation using saline drainage water. In *Proceedings National Alfalfa Symposium*, December 13–15, San Diego, CA. http://alfalfa.ucdavis.edu

Kah, G., and W. C. Willig. 1993. Irrigation management by the numbers: Put your system to the test. *Golf Course Irrigation Magazine* 1 (1): 8–13.

Kalra, Y. 1997. *Handbook of reference methods for plant analysis*. Boca Raton, FL: CRC Press.

Kammerer, S., and P. F. Harmon. 2009. A new *Rhizoctonia* sp. pathogenic to seashore paspalum turfgrass. *Phytopathology* 99 (6, June supplement): S61.

Karnok, K. J., K. Xia, and K. A. Tucker. 2004. Wetting agents: what are they, and how do they work. *Golf Course Management* 72 (6): 84–87.

Katembe, W. J., I. A. Ungar and J. P. Mitchell. 1998. Effect of salinity on germination and seedling growth of two *Atriplex* species (Chenopodiaceae). *Annals of Botany* 82 (2): 167–175.

Kaur, J., O. P. Choudhary, and B. Singh. 2008. Microbial biomass carbon and some soil properties as influenced by long-term sodic-water irrigation, gypsum, and organic amendments. *Aust. J. of Soil Res.* 46: 141–151.

Kaya, C., A. L. Tuna, M. Ashraf, and H. Altunlu. 2007. Improved salt tolerance of melon (*Cucumis melo* L.) by the addition of proline and potassium nitrate. *Environmental and Experimental Botany* 60 (3): 397–403.

Kelly, J., and P. Rengasamy. 2006. Diagnosis and management of soil constraints: Transient salinity, sodicity, and alkalinity. The University of Adelaide and Grain Research and Development Corporation, Australia. http://www.arris.com.au/index.php?id=39

Kelly, J., M. Unkovich, and D. Stevens. 2006. Crop nutrition considerations in reclaimed water irrigation systems. In *Growing crops with reclaimed wastewater,* ed. D. Stevens. Collingwood, VIC, Australia: CSIRO.

Kenna, M., and J. B. Beard. 2008. *Water quality and quantity issues for turfgrasses in urban landscapes*. Ames, IA: Council for Agricultural Science and Technology (CAST).

Keren, R. 2000. Salinity. In *Handbook of soil science*, ed. M. E. Sumner, G3–G25. Boca Raton, FL: CRC Press.

Ketterings, Q. M., B. C. Bellows, K. J. Czymmek, and W. S. Reid. 2001. Conversion equations Part 2: Do Mehlich III K, Ca, and Mg have Morgan equilvalents? *Cornell Cooperative Extension. What's Cropping Up*? 11 (4). http://css.cals.cornell.edu/cals/css/extension/cropping-up-archive/wcu_vol11no4_2001a2mehlichvsmorgan.pdf

Khan, M. A., R. Ansari, H. Ali, B. Gul, and B. L. Nielsen. 2009. *Panicum turgidum*, a potentially sustainable cattle feed alternative to maize for saline areas. *Agriculture, Ecosystems and Environment* 129 (4): 542–546.

Khan, M. A., and S. Gulzar. 2003. Germination responses of *Sporobolus ioclados*: A saline desert grass. *Journal of Arid Environments* 53 (3): 387–394.

Khan, W., U. P. Rayirath, S. Subramanian, M. N. Jithesh, P. Rayorath, D. M. Hodges, A. T. Critchley, J. S. Craigie, J. Norrie, and B. Prithiviraj. 2009. Seaweed extracts as biostimulants of plant growth and development. *J. Plant Growth Regul.* 28: 386–399.

Kidder, G. and E. A. Hanlon, Jr. 2009. Neutralizing excess bicarbonates from irrigation water. *Cooperative Extension Service Pub. SL-142*. Gainesville: IFAS, University of Florida.

Kiegle, E., C. Moore, J. Haseloff, M. Tester, and M. Knight. 2000. Cell-type specific calcium responses to drought, NaCl, and cold in *Arabidopsis* root: a role for endodermis and pericycle in stress signal transduction. *Plant Journal* 23: 267–278.

Kilic, C. C., Y. S. Kukul, and D. Anac. 2008. Performance of purslane (*Portulaca oleracea* L.) as a salt-removing crop. *Agricultural Water Management* 95 (7): 854–858.

King, K. W., J. C. Balogh, and R. D. Harmel. 2000. Feeding turf with wastewater. *Golf Course Mgt.* 68 (1): 59–62.

Knight, H. 2000. Calcium signaling during abiotic stress in plants. *International Review of Cytology* 192: 269–324.

Knight, H., A. J. Trewavas, and M. R. Knight. 1997. Calcium signaling in *Arabidopsis thaliana* responding to drought and salinity. *Plant Journal*. 12: 1067–1078.

Kopec, D. M., J. Gilbert, and M. Olsen. 2003. Evaluation of different cool season grasses for resistance or tolerance to rapid blight: When used for overseeding Tifgreen bermudagrass. *Cactus Clippings*, July: 11, 13.

Kopec, D., M. W. Olsen, J. J. Gilbert, D. M. Bigelow, and M. J. Kohout. 2004. Cool-season grass response to rapid blight disease: Field observations of rapid blight were confirmed in the lab. *Golf Course Management* 72 (12): 78–81.

Kopittke, P. M., and N. W. Menzies. 2007. A review of the use of the basic cation saturation ration and the "ideal" soil. *Soil Sci. Soc. Amer. J.* 71 (2): 259–265.

Kostka, S. J., L. W. Dekker, C. J. Ritsema, J. L. Cisar, and M. K. Franklin. 2007. Surfactants as management tools for ameliorating soil water repellency in turfgrass systems. In *Proc. 8th Inter. Symposium on Adjuvants for Agrochemicals,* ed. R. E. Gaskin, 1–7. Columbus, OH: Inter. Soc. for Agrochemical Adjuvants.

Krum, J. M., R. N. Carrow, and K. Karnok. 2010. Spatial mapping of complex turfgrass sites: Site-specific management units and protocols. *Crop Science* 50 (1): 301–315.

Krum, J. M., I. Flitcroft, P. Gerber, and R. N. Carrow. 2011. Performance of mobile salinity monitoring device for turfgrass situations. *Agron. J.* 103: 23–31.

Kumar Parida, A., and A. Bandhu Das. 2005. Salt tolerance and salinity effects on plants: A review. *Ecotoxicology and Environmental Safety* 60 (3): 324–349.

Laegdsmand, M., P. Moldrup, and P. Schjonning. 2010. Multitracer and filter-separated half-cell method for measuring solute diffusion in undisturbed soil. *Soil Science Society of America Journal* 74 (4): 1084–1091.

LaForce, M. J., J. Neiss, and C. Domrose. 2002. Geochemical properties of three serpentine soils: Presidio, CA. *2002 Annual Meeting Abstracts* [ASA/CSSA/SSSA]. TGIF Record #83466.

Lagarde, D., M. Basset, M. Lepetit, G. Conejero, F. Gaymard, S. Astruc, and C. Grignon. 1996. Tissue-specific expression of Arabidopsis AKT1 gene is consistent with a role in K^+ nutrition. *Plant Journal* 9: 195–203.

Lauchli, A., and U. Luttge. 2002. *Salinity: Environment-plants-molecules*. Boston: Kluwer Academic.

Laurie, S., K. A. Feeney, F .J. M. Maathuis, P .J. Heard, S. J. Brown, and R. A. Leigh. 2002. A role for HKT1 in sodium uptake by wheat roots. *Plant Journal* 32: 139–149.

Lebaudy, A., A-A. Very, and H. Sentenac. 2007. K^+ channel activity in plants: genes, regulations and functions. *FEBS Letters* 581 (12): 2357–2366.

Lebron, I., D. L. Suarez, and F. Alberto. 1994. Stability of a calcareous saline-sodic soil during reclamation. *Soil Science Society of America Journal* 58 (6): 1753–1762.

Lee, G-J., R. N. Carrow, and R. R. Duncan. 2005. Criteria for assessing salinity tolerance of the halophytic turfgrass seashore paspalum. *Crop Science* 45: 251–258.

Lee, G., R. N. Carrow, R. R. Duncan, M. A. Eiteman, and M. W. Rieger. 2008. Synthesis of organic osmolytes and salt tolerance mechanisms in *Paspalum vaginatum*. *Enviro. and Experi. Botany* 63: 19–27.

Lee, G. L., R. R. Duncan, and R. N. Carrow. 2007. Nutrient uptake responses and inorganic ion contribution to solute potential under salinity stress in halophytic seashore paspalums. *Crop Sci.* 47: 2504–2512.

Lee, S. S., J. H. Kim, S. B. Hong, S. H. Yuu, and E. H. Park. 1998. Priming effect of rice seeds on seedling establishment under adverse soil conditions. *Korean Journal of Crop Science* 43: 194–198.

Lehmann, J., and S. Joseph (eds.). 2009. *Biochar for environmental management: Science and technology*. London and Sterling, VA: Earthscan.

Leinauer, B. and S. Cockerham, eds. 2011. *Turfgrass water conservation*. 2nd ed. Riverside: ANR Communications Service, University of California.

Lembi, C. A. 2002. Aquatic plant management: Barley straw for algae control. *APM-1-W*. Lafayette, IN: Purdue University, Cooperative Extension Service.

Lesch, S., D. Strauss, and J. Rhoades. 1995a. Spatial prediction of soil salinity using electromagnetic induction techniques. 1. Statistical prediction models: A comparison of multiple linear regression and cokriging. *Water Resources Research* 31 (2): 373–386.

Lesch, S., D. Strauss, and J. Rhoades. 1995b. Spatial prediction of soil salinity using electromagnetic induction techniques. 2. An efficient spatial sampling algorithm suitable for multiple linear regression model identification and estimation. *Water Resources Research* 31 (2): 387–398.

Levelift L.L.C. 2010. Lifting and leveling system for sprinklers. www.levelift.net

Levent Tuna, A., C. Kaya, D. Higgs, B. Murillo-Amador, S. Aydemir, and A.R. Girgin. 2008. Silicon improves salinity tolerance in wheat plants. *Environmental and Experimental Botany* 62 (1): 10–16.

Levy, G. J. 2000. Sodicity. In *Handbook of soil science*, ed. M. E. Sumner, G27–G63. Boca Raton, FL: CRC Press.

Li, D. 2004. How to categorize organic materials in turfgrass root zones. *TurfGrass Trends*: April 1. www.turfgrasstrends.com

Li, R., F. Shi, and K. Fukuda. 2010a. Interactive effects of various salt and alkali stresses on growth, organic solutes, and cation accumulation in a halophyte Spartina alterniflora (Poaceae). *Environmental and Experimental Botany* 68 (1):66–74.

Li, R., F. Shi, and K. Fukuda. 2010b. Interactive effects of salt and alkali stresses on seed germination, germination recovery, and seedling growth of halophyte Spartina alterniflora (Poaceae). *South African Journal of Botany* 76: 380–387.

Liang, Y., Q. Chen, Q. Liu, W. Zhang, and R. Ding. 2003. Exogenous silicon (Si) increases antioxidant enzyme activity and reduces lipid peroxidation in roots of salt-stressed barley (*Hordeum vulgare* L.). *Journal of Plant Physiology* 160 (10): 1157–1164.

Liang, Y., W. Zhang, Q. Chen, Y. Liu, and R. Ding. 2006. Effect of exogenous silicon (Si) on H^+-ATPase activity, phospholipids, and fluidity of plasma membrane in leaves of salt-stressed barley (*Hordeum vulgare* L.). *Environmental and Experimental Botany* 57 (3): 212–219.

Liang, Y., J. Zhu, Z. Li, G. Chu, Y. Ding, J. Zhang, and W. Sun. 2008. Role of silicon in enhancing resistance to freezing stress in two contrasting winter wheat cultivars. *Environmental and Experimental Botany* 64 (3): 286–294.

Lindenbach, S. K., and D. R. Cullimore. 1989. Preliminary *in vitro* observations on the bacteriology of the black plug layer phenomenon associated with the biofowling of golf greens. *Journal of Applied Bacteriology* 67: 11–17.

Loch, D. S., E. Barrett-Lennard, and P. Truong. 2003. Role of salt tolerant plant for production, prevention of salinity and amenity values. *Proceedings of 9th National Conf. on Productive Use of Saline Lands (PURSL Conference)*, September 29–October 2, Rockhampton, QLD, Australia.

Loch, D. S., R. E. Poulter, M. B. Roche, C. J. Carson, T. W. Lees, L. O'Brien, and C. R. Durant. 2006. Amenity grasses for salt-affected parks in coastal Australia. *Final project report, horticulture Australia.* http://nla.gov.au/anbd.bib-an000041030712

Logan, B. A. 2005. Reactive oxygen species and photosynthesis. In *Antioxidants and reactive oxygen species in plants*, ed. N. Smirnoff, 250–267. Oxford: Blackwell.

Lone, M. I., J. S. H. Kueh, R. G. Wyn Jones, and S. W. J. Bright. 1987. Influence of proline and glycine betaine on salt tolerance of cultured barley embryos. *Journal of Experimental Botany* 38: 479–490.

Lubin, T. 1995. Controlling soil pH with irrigation water. *Golf Course Management* 63 (11): 56–60.

Maas, E. V. 1978. Crop salt tolerance. In *Crop tolerance to suboptimal land conditions* (ASA Spec. Publ. 32), ed. G. A. Jung. Madison, WI: American Society of Agronomy.

Maas, E. V. 1984. Salt tolerance of plants. In *The handbook of plant science in agriculture*, ed. B. R. Christie. Boca Raton, FL: CRC Press.

Maas, E. V. 1994. Testing crops for salinity tolerance. In *Adaptation of plants to soil stress* (INTSORMIL Pub. 94-2), chair J. W. Maranville, 234-247. Lincoln: Univ. of Nebraska, International Programs, Div., Institute of Agric. and Natural Resources.

Maas, E. V., and G. J. Hoffman. 1977. Crop salt tolerance: Current assessment. *J. Irrig. Drainage. Div. ASCE* 103 (IRZ): 115–132.

MacKinnon, R., and C. Miller. 1989. Mutant potassium channels with altered binding of charybdotoxin, a pore-blocking peptide inhibitor. *Science* 245: 1382–1385.

Madigan, M. T., and D. O. Jung. 2009. An overview of purple bacteria: Systematic, physiology, and habits. In *The purple phototropic bacteria*, eds. N. Hunter, F. Daldal, M. C. Thurnauer, and J. Thomas Beatty, 1–15. New York: Springer Science + Business Media.

Marcum, K. B. 1999. Salinity tolerance mechanisms of grasses in the subfamily Chloridoidae. *Crop Science* 39: 1153–1160.

Marcum, K. B. 2002. Growth and physiological adaptations of grasses to salinity stress. In *Handbook of plant and crop physiology* (2nd ed.), ed. M. Pessarakli, 623–636. New York: Marcel Dekker.

Marcum, K. B. 2006. Use of saline and non-potable water in the turfgrass industry: Constraints and developments. *Agric. Water Manage.* 80: 132–146.

Marcum, K. B., and C. L. Murdock. 1994. Salinity tolerance mechanisms of six C_4 turfgrasses. *Journal American Society Horticultural Science* 119 (4): 779–784.

Marschner, H. 1995. *Mineral nutrition of higher plants.* New York: Academic Press.

Martin, S. B., J. J. Camberato, and P. D. Peterson. 2009. Irrigation water composition affects rapid blight of perennial ryegrass. *Phytopathology* 99 (6, June Supplement): S80.

Masoni, A., L. Ercoli, and M. Mariotti. 1996. Spectral properties of leaves deficient in iron, sulfur, magnesium, and manganese. *Agron, J.* 88: 937–943.

Masters, D. G., S. E. Benes, and H. C. Norman. 2007. Biosaline agriculture for forage and livestock production. *Agriculture, Ecosystems and Environment* 119 (3–4): 234–248.

Mbagwu, J. S. C. and A. Piccolo. 1989. Changes in soil aggregate stability induced by amendment with humic substances. *Soil Technology* 2 (1): 49–57.

McAuliffe, K. 2009. A further look at sand capping to drain fairways in southeast Asia. *Asian Golf Business* 13: 60–64.

McCarty, L. B., J. W. Everest, D. W. Hall, T. R. Murphy, and F. Yelverton. 2001. *Color atlas of turfgrass weeds.* Chelsea, MI: Ann Arbor Press.

McCormick, P. L. 2006. Not your typical drainage: Passive capillary system drains putting surfaces slowly. *Golfweek's SuperNEWS* 8 (16): 40–41. TGIF Record #116363.

McCormick, S., C. Jordan, and J. S. Bailey. 2009. Within and between-field spatial variation in soil phosphorus in perennial grassland. *Precision Agriculture*, 10: 262–276.

McCoy, E. 2005. OSU innovation: passive capillary drainage of turf soil profiles. *Buckeye Turf* (Aug. 30): 1–3. TGIF Record #126528.

McCoy, E., and K. Danneberger. 2008. Thoughts on sand capping. *Buckeye Turf* (Feb. 8). http://buckeyeturf.osu.edu/index.php?option=com_intsportsnotes&Itemid=85¬eid=1299

McCrimmon, J. N. 1994. Comparison of washed and unwashed plant tissue samples utilized to monitor the nutrient status of creeping bentgrass putting greens. *Comm. in Soil Sci. and Plant Anal.* 25 (7–8): 967–988.

McCrimmon, J. N. 1998. Effect of nitrogen and potassium on the macronutrient content of fifteen bermudagrass cultivars. *Comm. in Soil Sci. and Plant Anal.* 29 (11–14): 1851–1861.

McCrimmon, J. N. 2000. Nitrogen and potassium effects on the macronutrient and micronutrient content of zoysiagrasses. *J. of Plant Nutr.* 23 (5): 683–696.

McCrimmon, J. N. 2002. Macronutrient and micronutrient concentrations of seeded bermudagrasses. *Comm. in Soil Sci. and Plant Anal.* 33 (15–18): 2739–2758.

McCrimmon, J. N. 2004. Effects of mowing height nitrogen rate, and potassium rate on Palmetto and Raleigh St. Augustinegrass. *J. of Plant Nutr.* 27 (1): 1–13.

McCutcheon, S. C., and N. L. Wolfe. 1998. Phytoremediation: The vital role of plant biochemistry in cleaning up organic compounds. *PBI Bulletin* September: 17–20.

McGrew, J. C., and C. B. Monroe. 2000. *An introduction to statistical problem solving in geography*. 2nd ed. Dubuque, IA: McGraw-Hill.

McIntyre, K. and B. Jakobsen. 2000. *Practical drainage for golf, Sportsturf, and horticulture*. Hoboken, NJ: John Wiley.

Mennen, H., B. Jacoby, and H. Marschner. 1990. Is sodium proton antiport ubiquitous in plant cells? *Journal Plant Physiology* 137: 180–183.

Metternicht, G., and J. A. Zinck. 1997. Spatial discrimination of salt- and sodium-affected soil surfaces. *International Journal of Remote Sensing* 18 (12): 2571–2586.

Metternicht, G., and J. A. Zinck, eds. 2009. *Remote sensing of soil salinization: Impact on land management.* Boca Raton, FL: CRC Press.

Mevarech, M., F. Frolow, and L. M. Gloss. 2000. Halophilic enzymes: Proteins with a grain of salt. *Biophysical Chemistry* 86 (2–3): 155–164.

Micro Surface Corporation. 2010. Corrosion solutions. Morris, IL: Micro Surface Corporation. http://www.microsurfacecorp.com/corrosion.php

Miller, G., N. Pressler, and M. Dukes. 2003. How uniform is coverage from your irrigation system. *Golf Course Management* 71 (8): 100–102.

Miller, G. L., and A. Thomas. 2003. Using near infrared reflectance spectroscopy to evaluate phosphorus, potassium, calcium, and magnesium concentrations in bermudagrass. *HortSci.* 38 (6): 1247–1250.

Ming, D. W., and E. R. Allen. 2001. Use of natural zeolites in agronomy, horticulture, and environmental soil remediation. *Reviews in Mineralogy and Geochemistry* 45(1): 619–654.

Miyamoto, S., and A. Chacon. 2006. Soil salinity of urban turf areas irrigated with saline water II. Soil factors. *Landscape and Urban Planning* 77: 28–36.

Miyamoto, S., A. Chacon, M. Hossain, and I. Martinez. 2005. Soil salinity of urban turf areas irrigated with saline water I. Spatial variability. *Landscape and Urban Planning* 71: 233–241.

Miyamoto, S., and J. L. Stroehlein. 1975a. Potentially beneficial use of sulfuric acid in southwestern agriculture. *J. Environ. Qual.* 4: 431–437.

Miyamoto, S., and J. L. Stroehlein. 1975b. Sulfuric acid for controlling calcite precipitation. *Soil Science* 120: 264–271.

Miyamoto, S., and J. L. Stroehlein. 1986. Sulfuric acid effects on water infiltration and chemical properties of alkaline soils and water. *Trans ASAE* 29 (5): 1288–1296.

Mok, D. W. S., and M. C. Mok. 1994. *Cytokinins: Chemistry, activity, and function*. Boca Raton, FL: CRC Press.

Moller, I. S., M. Gilliham, D. Jha, G. M. Mayo, S. J. Roy, J. C. Coates, J. Haseloff, and M. Tester. 2009. Shoot Na^+ exclusion and increased salinity tolerance engineered by cell type-specific alteration of Na^+ transport in *Arabidopsis*. *Plant Cell* 21: 2163–2178.

Moore, D., S. J. Kostka, T. J. Boerth, M. Franklin, C. J. Ritsema, L. W. Dekker, K. Oostindie, C. Stoof, and J. Wesseling. 2010. The effect of soil surfactants on soil hydrological behavior, the plant growth environment, irrigation efficiency, and water conservation. *J. Hydrol. Hydromech.* 58: 142–148.

Moore, T. R. and R. C. Ziommermann. 1977. Establishment of vegetation on serpentine asbestos mine wastes, southeastern Quebec, Canada. *Journal of Applied Ecology* 14: 589–599.

Moritani, S., T. Yamamoto, H. Andry, M. Inoue, A., Yuya, and T. Kaneuchi. 2010. Effectiveness of artificial zeolite amendment in improving the physicochemical properties of saline-sodic soils characterized by different clay mineralogies. *Australian Journal of Soil Research* 48: 470–479.

Morris, K., R. Zobel, and A. Hass. 2009. Engineering the best soils for turfgrass applications. *USGA TERO* 8 (19): 1–8.

Mortverdt, J. J., ed. 1991. *Micronutrients in agriculture*. 2nd ed. Madison, WI: Soil Science Society of America.

Muehlstein, L. K., D. Porter, and F. T. Short. 1991. *Labyrinthula zosterae* sp. nov., the causative agent of wasting disease of eelgrass, *Zostera marina*. *Mycologia* 83: 180–191.

Munns, R. 2002. Comparative physiology of salt and water stress. *Plant Cell Environment* 25: 239–250.

Munns, R. 2005. Genes and salt tolerance: bringing them together. *New Phytology* 167: 645–663.

Munns, R., R. A. James, and A. Lauchli. 2006. Approaches to increasing the salt tolerance of wheat and other cereals. *Journal of Experimental Botany* 57: 1025–1043.

Munns, R., and M. Tester. 2008. Mechanisms of salinity tolerance. *Annual Review Plant Biology* 59: 651–681.

Munoz-Carpena, R. 2009. Field devices for monitoring soil water content. *IFAS Extension Bulletin 343*. Gainesville: Dept. of Agric. and Biol. Engineering, University of Florida.

Murata, N., S. Takahashi, Y. Nishiyama, and S. I. Allakhverdiev. 2007. Photoinhibition of photosystem II under environmental stress. *Biochimica et Biophysica Acta (BBA)-Bioenergetics* 1767(6):414–421. Structure and Function of Photosystems.

Murphy, J. A. 2007. Rootzone amendments for putting green construction. *USGA Green Section Record* 45 (3): 8–13.

Murphy, J. A., H. Samaranayake, J. A. Honig, T. J. Lawson, and S. L. Murphy. 2005. Creeping bentgrass establishment on amended-sand root zones in two microenvironments. *Crop Science* 45: 1511–1520.

Nabati, D. A., R. E. Schmidt, and D. J. Parrish. 1994. Alleviation of salinity stress in Kentucky bluegrass by plant growth regulators and iron. *Crop Sci.* 34: 198–202.

Nadeem, S. M., Z. A. Zahir, M. Naveed, H. N. Asghar, and M. Arshad. 2010. Rhizobacteria capable of producing ACC-deaminase may mitigate salt stress in wheat. *Soil Sci. Soc. Amer. J.* 74 (2): 533–542.

Naidoo, G., R. Somaru, and P. Achar. 2008. Morphological and physiological responses of the halophyte, *Odyssea paucinervis* (Staph) (Poaceae), to salinity. *Flora—Morphology, Distribution, Functional Ecology of Plants* 203 (5): 437–447.

Naidu, R., and P. Rengasamy. 1993. Ion interactions and constraints to plant nutrition in Australian sodic soils. *Aust. J. Soil Res.* 31: 801–819.

Naidu, R., M. E. Sumner, and P. Regengasamy, eds. 1995. *Australian sodic soils: Distribution, properties and management.* East Melbourne, VIC, Australia: CSIRO.

Nakamura, R. L., W. L. McKendree, Jr., R. E. Hirsch, J. C. Sedbrook, R. F. Gaber, and M. R. Sussman. 1995. Expression of an *Arabidopsis* potassium channel gene in guard cells. *Plant Physiology* 109: 371–374.

NCR. 1998. Recommended chemical soil test procedures for the North Central Region. *NCR Research Publ. No. 221* (revised). Columbia: Missouri Agric. Experiment Station SB 1001. http://extension.missouri.edu/explorepdf/specialb/sb1001.pdf

Nedjimi, B. 2009. Salt tolerance strategies of *Lygeum spartum* L.: a new fodder crop for Algerian saline steppes. *Flora—Morphology, Distribution, Functional Ecology of Plants* 204 (10): 747–754.

Nektarios, P. A., A. M. Petrovic, and T. S. Streenhuis. 2002. Effect of surfactant on fingered flow in laboratory golf greens. *Soil Sci.* 167 (9): 572–579.

Nektarios, P. A., A. M. Petrovic, and T. S. Steenhuis. 2004. Aeration type affects preferential flow in golf putting greens. *Acta Hort. (ISHS)* 661: 421–425. http://www.actahort.org/books/661/661_58.htm

Neylan, J. 1994. Sand profiles and their long-term performance. *Golf and Sports Turf Australia* Aug.: 22–37.

Neylan, J. 1997. Irrigation management tools. *Golf and Sports Turf Australia* Feb.: 21–28.

Nielsen, D. L., M. A. Brock, G. N. Rees, and D. S. Baldwin. 2003. Effects of increasing salinity on freshwater ecosystems in Australia. *Aust. J. of Botany* 51: 655–665.

Nielsen, D. M., G. L. Nielsen, and L. M. Preslo. 2006. Environmental site characterization. In *Environmental site characterization and ground-water monitoring* (2nd ed.), ed. D. M. Nielsen, 35–205. Boca Raton, FL: CRC Press.

NIWQP. 1998. *Guidelines for interpretation of the biological effects of selected constituents in biota, water, and sediment.* Denver: National Irrigation Water Quality Program, U.S. Bureau of Reclamation. http://www.usbr.gov/niwqp/guidelines/index.htm

Norum, E. 1999. *Sand problems call for irrigation technology* (CATI Pub. 990801). Fresno: Center for Irr. Tech., California State University.

Novak, J. M., W. J. Busscher, D. L. Laird, M. Ahmedna, D. W. Watts, and M. A. S. Niandou. 2009. Impact of biochar amendment on fertility of a southeastern coastal plain soil. *Soil Science* 174 (2): 105–112.

NRP. 2009. *Recommended soil testing procedures for the Northeastern United States*. 3rd ed. Northeastern Regional Publication (NRP) No. 493. Newark: Cooperative Extension, University of Delaware. http://ag.udel.edu/extension/agnr/soiltesting.htm

Nus, J. 1994. Algae control. *Golf Course Management* 62 (5): 53–57.

Nus, J. L., and Brauen, S.E. 1991. Clinopltilolite zeolite as an amendment for establishment of creeping bentgrass on sandy media. *HortScience* 26: 117–119.

Nuttal, J. G., and R. D. Armstrong. 2010. Impact of subsoil physicochemical constraints on crops grown in Wimmera and Mallee is reduced during dry seasonal conditions. *Australian Journal of Soil Research* 48 (2): 125–139.

O'Brien, P., and C. Hartwiger. 2001. Aeration and topdressing for the 21st century: Two old concepts are linked together to offer up-to-date recommendations. *USGA Greens Section Record* 41: 1–7.

O'Brien, P. M. 2005. Planning a golf course drainage project. *USGA Green Section Record* 43 (5): 16–20.

O'Connor, G. A., H. A. Elliott, and R. K Bastian. 2008. Degraded water reuse: An overview. *J. Environ. Qual.* 37: S-157–S-168.

Ok, C-H., S. H. Anderson, and E. H. Ervin. 2003. Amendments and construction systems for improving the performance of sand-based putting greens. *Agronomy Journal* 95: 1583–1590.

Okello, W., V. Ostermaier, C. Portmann, K. Gademann, and R. Durmayer. 2010. Spatial isolation favours the divergence in mycrocystin net production by *Microcystis* in Ugandan freshwater lakes. *Water Research* 44 (9): 2803–2814.

Oliver, M. A. 2010. *Geostatistical applications for precision agriculture*. New York: Springer.

Olsen, M. W. 2007. *Labyrinthula terrestris*: a new pathogen of cool-season turfgrasses. *Molecular Plant Pathology* 8 (6): 817–820.

Olsen, M. W. 2008. Timely fungicide applications, salinity reduction help control rapid blight. *Turfgrass Trends* September: 47–49. www.golfdom/turfgrass-trends

Olsen, M. W., D. M. Bigelow, R. L. Gilbertson, L. J. Stowell, and W. D. Gelernter. 2003. First report of a *Labyrinthula* spp. causing rapid blight disease on rough bluegrass and perennial ryegrass. *Plant Disease* 87: 1267.

Olsen, M. W., D. M. Bigelow, M. J. Kohout, J. Gilbert, and D. Kopec. 2004. Rapid blight: A new disease of cool-season turf: Rapid blight is caused by an organism that usually affects turfgrass irrigated with poor-quality water. *Golf Course Management* 72 (8): 87–91.

Olsen, M. W., A. Herrell, and J. Gilbert. 2008b. Detection of the rapid blight pathogen *Labyrinthula terrestris* on non-symptomatic Poa trivalis. *2007–2008 Turfgrass, Landscape and Urban IPM Research Summary (Arizona),* p. 13–18. TGIF Record #139379.

Olsen, M. W., M. J. Kohout, and D. M. Bigelow. 2006. Effect of salt and PEG induced water stress on rapid blight disease of turfgrass caused by *Labyrinthula* sp. *Phytopathology* 96 (6, June Supplement): S170. TGIF Record #112297.

Olsen, M. W., G. Towers, and J. Gilbert. 2008a. Evaluation of fungicides for control of rapid blight of *Poa trivalis* in fall 2006. *2007–2008 Turfgrass, Landscape and Urban IPM Research Summary (Arizona),* p. 10–12. TGIF Record #139378.

Olsen, M. W., G. Towers, and J. Gilbert. 2008c. Control of rapid blight in rough bluegrass overseeded bermudagrass, 2007. *PDMR: Plant Disease Management Reports* 2: T030. TGIF Record #163422.

Olson, R. A., K. D. Frank, P. H. Grabouski, and G. W. Rehm. 1982. Economic and agronomic impacts of varied philosophies of soil testing. *Agron. J.* 74: 492–499.

Onwugbuta-Enyi, J. A., and B. A. Onuegbu. 2008. Remediation of dredge spoils with organic soil amendments using *Paspalum vaginatum* L. as a test crop. *Advances in Environmental Biology* 2 (3): 121–123.

Oster, J. D. 1994. Irrigation with poor quality water. *Agric. Water Manage*. 25: 271–297.

Oster, J. D., and H. Frenkel. 1980. The chemistry of the reclamation of sodic soils with gypsum and lime. *Soil Sci. Soc. Am. J.* 44: 41–45.

Oster, J. D., and S. R. Grattan. 2002. Drainage water reuse. *Irrig. Drain. Syst.* 16: 297–310.

Oster J. D., and Jayawardane, N. S. (1998) Agricultural management of sodic soils. In *Sodic soils: Distribution, properties, management and environmental consequences,* eds. M. E. Sumner and R. Naidu, 125–147. New York: Oxford University Press.

Oster, J. D., and F. W. Schroer. 1979. Infiltration as influenced by irrigation water quality. *Soil Sci. Soc. Amer. J.* 43: 444–447.

Oster, J. D., M. J. Singer, A. Fulton, W. Richardson, and T. Prichard. 1992. *Water penetration problems in California soils: Diagnosis and solutions*. Riverside: University of California, Kearney Foundation of Soil Science, Division of Agricultural and Natural Resources.

Pair, C. H. , W. H. Hinz, K. R. Frost, R. E. Sneed, and T. J. Schiltz, 1983. *Irrigation*. Arlington, VA: The Irrigation Association.

Pannell, D. J., and M. A. Ewing. 2006. Managing secondary dryland salinity: Options and challenges. *Agric. Water Manage*. 80: 41–56.

Pardo, J. M., B. Cubero, E. O. Leidi, and F. J. Quintero. 2006. Alkali cation exchangers: Roles in cellular homeostasis and stress tolerance. *Journal Experimental Botany* 57: 1181–1199.

Paul, M. J., and C. H. Foyer. 2001. Sink regulation of photosynthesis. *Journal Experimental Botany* 52: 1383–1400.

Pei, Z-M., J. M. Ward, and J. I. Schroeder. 1999. Magnesium sensitizes slow vacuolar channels to physiological cytosolic calcium and inhibits fast vacuolar channels in fava bean guard cell vacuoles. *Plant Physiology* 121: 977–986.

Peiter, E., F. J. M. Maathuis, L. N. Mills, H. Knight, J. Pelloux, A. M. Hetherington, and D. Sanders. 2005. The vacuolar Ca^{2+}-activated channel TPC1 regulates germination and stomatal movement. *Nature* 434: 404–408.

Pepper, I. L., and C. F. Mancino. 1994. Irrigation of turf with effluent water. In *Handbook of plant and crop stress*, ed. M. Pessarakli. New York: Marcel Dekker.

Pescod, M. B. 1992. *Wastewater treatment and use in agriculture* (FAO Irrigation and Drainage Paper 47). Rome: United Nations Food and Agriculture Organization. http://www.bvsde.ops-oms.org/bvsair/e/repindex/repi84/vleh/fulltext/acrobat/wastew.pdf

Pessarakli, M., and D. M. Kopec. 2008. Competitive growth responses of three cool season grasses to salinity and drought stresses. *Acta Horticulturae* 783: 169–174.

Pessarakli, M., and I. Szabolcs. 1999. Soil salinity and sodicity as particular plant/crop stress factors. In *Handbook of plant and crop stress* (2nd ed.), ed. M. Pessarakli, 1–15. New York: Marcel Dekker.

Peterson, P. D., S. B. Martin, J. J. Camberato, and D. E. Fraser. 2005. The effect of temperature and salinity on growth of the turfgrass pathogen, *Labyrinthula terrestris*. (Abstr.) *Phytopathology* 95.

Peterson, P., S. B. Martin, and J. Camberato. 2006. Biology and integrated management of rapid blight: A new disease of rough bluegrass, perennial ryegrass, annual bluegrass, and creeping bentgrass. Clemson, SC: Clemson University. http://turf.lib.msu.edu/ressum/2005/8.pdf

Petrovic, A. M. 1975. The effects of several chemical soil conditioners and the algal polymer on compacted soil and growth of cool-season turfgrasses. M.S. thesis, University of Massachusetts. Amherst.

Petrovic, A. M. 1993. Potential for natural zeolite uses on golf courses. *USGA Green Section Record* 31 (1): 11–14.

Pettygrove, G. S and T. Asano. 1985. *Irrigation with reclaimed municipal wastewater: A guidance manual*. Chelsea, MI: Lewis Publ.

Peverill, K. I., L. A. Sparrow, and D. J. Reuter, eds. 1999. *Soil analysis: An interpretation manual*. Collingwood, VIC, Australia: CSIRO.

Pill, W. G., and A. D. Necker. 2001. The effects of seed treatments on germination and establishment of Kentucky bluegrass (*Poa pratensis* L.). *Seed Science and Technology* 29: 65–72.

Pillsbury, A. F. 1981. The salinity of rivers. *Scientific American*. 245 (1): 54–65.

Pilon-Smits, E. 2005. Phytoremediation. *Annual Review of Plant Biology* 56: 15–39.

Pira, E. S. 1997. *A guide to golf course irrigation design and drainage*. Hoboken, NJ: John Wiley.

Pitman, M. G., and A. Lauchi. 2002. Global impact of salinity and agriculture ecosystems. In *Salinity: Environment-plants-molecules*, ed. A. Lauchi and U. Luttge. Boston: Kluwer Academic.

Plank, C. O., and R. N. Carrow. 2010. *Plant analysis: An important tool in turf production*. Athens: College of Agricultural and Environmental Science, University of Georgia. http://www.commodities.caes.uga.edu/turfgrass/georgiaturf/SoilTesting/analysis.html

Plank, C. O., and D. E. Kissel. 2010. *Plant analysis handbook for Georgia*. Athens: Agric. and Environ. Services Laboratories, Coll. of Agric. and Environ. Sci., University of Georgia. http://aesl.ces.uga.edu/publications/plant

Porter, D. 1990. Phylum Labyrinthulomycota. In *The handbook of protoctista*, eds. L. Margulis, J. O. Corliss, M. Melkonian, and D. J. Chapman, 388–398. Boston: Jones and Bartlett.

Potter, D. A. 1998. *Destructive turfgrass insects: Biology, diagnosis, and control*. Chelsea, MI: Ann Arbor Press.

Prathapar, S. A., M. Aslam, M. A. Kahlown, Z. Iqbal and A. S. Qureshi. 2005. Gypsum slotting to ameliorate sodic soils in Pakistan. *Irrigation and Drainage* 54: 509–517.

Price, P. 2010. Combating subsoil constraints: R&D for the Australian grains industry. *Soil Research* 48 (2): i–iii.

Prior, L. D., A. M. Grieve, K. B. Bevington, and P. G. Slavich. 2007. Long-term effects of saline irrigation water on 'Valencia' orange trees: Relationships between growth and yield, and salt levels in soil and leaves. *Australian Journal of Agricultural Research* 58: 349–358.

Qadir, M., A. Ghafoor, and G. Murtaza. 2000. Amelioration strategies for saline soils: A review. *Land Degradation and Develop*. 11: 501–521.

Qadir, M., A. D. Noble, S. Schubert, R. J., Thomas, and A. Arslan. 2006. Sodicity-induced land degradation and its sustainable management: problems and prospects. *Land Degradation and Development* 17 (6): 661–678.

Qadir, M. and J. D. Oster. 2004. Crop and irrigation management strategies for saline-sodic soils and waters aimed at environmentally sustainable agriculture. *Science of the Total Enviro.* 323: 1–19.

Qadir, M. J. D. Oster, S. Schubert, and G. Murtaza. 2006. Vegetative bioremediation of sodic and saline-sodic soils for productivity enhancement and environment conservation. In *Biosaline agriculture and salinity tolerance in plants*, eds. M. Ozturk, Y. Waisel, M. A. Khan and G. Gork, 137–146. Basel, Switzerland: Birkhauser.

Qadir, M., J. D. Oster, S. Schubert, A. D. Noble, and K. L. Sahrawat. 2007. Phytoremediation of sodic and saline-sodic soils. *Advances in Agronomy* 96: 197–247.

Qadir, M., R. H. Qureshi, N. Ahmad, and M. Ilyas. 1996. Salt tolerant forage cultivation on a saline-sodic field for biomass production and soil reclamation. *Land Degradation and Development* 7: 11–18.

Qadir, M., A. Tubeileh, J. Akhtar, A. Larbi, P. S. Minhas, and M. A. Khan. 2008. Productivity enhancement of salt-affected environments through crop diversification. *Land Degradation and Development* 19: 429–453.

QASSIT. 2009. Acid sulfate soils website by the Queensland Acid Sulfate Soils Investigation Team (QASSIT). Indooroopilly, QLD, Australia: QASSIT. http://www.derm.qld.gov.au/land/ass/index.html

Qian, Y. L., W. Bandaranayake, W. J. Parton, B. Mecham, M. A. Harivandi, and A. R. Mosier. 2003. Long-term effects of clipping and nitrogen management in turfgrass on soil organic carbon and nitrogen dynamics: The CENTURY Model Simulation. *J. Environ. Qual.* 32: 1694–1700.

Qian, Y. L., A. J. Koski, and R. Welton. 2001. Amending sand with Isolite and zeolite under saline conditions: leachate composition and salt deposition. *HortScience* 36 (4): 717–720.

Qiao, W., S. Xiao, L. Yu, and L-M. Fan. 2009. Expression of a rice gene OsNOA1 re-establishes nitric oxide synthesis and stress-related gene expression for salt tolerance in *Arabidopsis* nitric oxide-associated 1 mutant Atnoa1. *Environmental and Experimental Botany* 65 (1): 90–98.

Qui, Q. S., Y. Guo, M. A. Dietrich, K. S. Schumaker, and J. K. Zhu. 2002. Regulation of SOS1, a plasma membrane Na^+/H^+ exchanger in *Arabidopsis thaliana*, by SOS2 and SOS3. *Proceeding National Academy Sciences USA* 99: 8436–8441.

Rabhi, M., S. Ferchichi, J. Jouini, M. H. Hamrouni, H-W. Koyro, A. Ranieri, C. Abdelly, and A. Smaoui. 2010. Phytodesalination of a salt-affected soil with the halophyte *Sesuvium portulacastrum* L. to arrange in advance the requirements for the successful growth of a glycophytic crop. *BioResource Technology* 101: 6822–6828.

Rahman, M. S., H. Miyake, and Y. Takeoka. 2002. Effects of exogenous glycine betaine on growth and ultrastructure of slat-stressed rice seedlings (*Oryza sativa* L.). *Plant Production Science* 5: 33–44.

Raskin, I. and B. D. Ensley. 2000. *Phytoremediation of toxic metals: Using plants to clean up the environment.* New York: John Wiley.

Raven, J. A. 1985. Regulation of pH and generation of osmolarity in vascular plants: A cost-benefit analysis in relation to efficiency of use of energy, nitrogen and water. *New Phytology* 101: 25–77.

Reiber, S. 1993. *Chloramine effects on distribution system materials.* Denver: Water Works Research Foundation.

Rengasamy, P. 2002. Transient salinity and subsoil constraints to dryland farming in Australian sodic soils: an overview. *Aust. J. of Exper. Agric.* 42: 351–361.

Rengasamy, P. 2010a. Soil processes affecting crop production in salt-affected soils. *Functional Plant Biol.* 37: 613–620.

Rengasamy, P. 2010b. Osmotic and ionic effects of various electrolytes on the growth of wheat. *Soil Research* 48: 120–124.

Rengasamy, P., and K. A. Olsson. 1991. Sodicity and soil structure. *Aust. J. Soil Res.* 29: 935–952.

Reuter, D. J. and J. B. Robinson, eds. 1997. *Plant analysis and interpretation manual.* 2nd ed. Collingwood, VIC, Australia: CISRO.

Reuter, D. J., J. B. Robinson, K. I. Peverill, G. H. Price and M. J. Lambert. 1997. Guidelines for collecting, handling, and analyzing plant materials. In *Plant analysis and interpretation manual* (2nd ed.), eds. D. J. Reuter and J. B. Robinson. Collingwood, VIC, Australia: CISRO.

Rhoades, J. D. 1974. Drainage for salinity control. In. *Drainage for agriculture* (Agronomy Monograph No. 17), ed. J. van Schilfgaarde, 433–461. Madison, WI: Soil Science Society of America.

Rhoades, J. D., and F. Chanduvi, and S. Lesch. 1999. *Soil salinity assessment: Methods and interpretation of electrical conductivity measurements* (FAO Irr. and Drain. Paper 57). Rome: United Nations Food and Agriculture Organization.

Rhoades, J. D., and J. Loveday. 1990. Salinity in irrigated agriculture. In *Irrigation of agricultural crops* (Agronomy Monograph No. 30), ed. B. A. Stewart and D. R. Nielson, 1089–1142. Madison, WI: American Society of Agronomy.

Rhoades, J. D., A. Kandiah, and A. M. Mashali. 1992. *The use of saline waters for crop production* (FAO Irrigation and Drainage Paper #48). Rome: United Nations Food and Agriculture Organization.

Rhoades, J. D., and S. Miyamoto. 1990. Testing soils for salinity and sodicity. In *Soil testing and plant analysis* (3rd ed., Book series no. 3), ed. R. L. Westerman, 299–336. Madison, WI: Soil Science Society of America.

Rhodes, D., A. Nadolska-Orczyk, and P. J. Rich. 2002. Salinity, osmolytes and compatible solutes. In *Salinity: Environment-plants-molecules*, eds. A. Lauchli and U. Luttge, 181–204. Dordrecht, the Netherlands: Kluwer.

Richardson, M., and J. McCalla. 2008. Germination of three ryegrass species and meadow fescue under saline conditions. *Arkansas Turfgrass Report 2007*: 63–65. TGIF Record #135318.

Richmond, K. E., and M. Sussman. 2003. Got silicon? The non-essential beneficial plant nutrient. *Current Opinion in Plant Biology* 6 (3): 268–272.

Rider, D., R. J. Zasoski, and V. P. Claassen. 2005. Ammonium fixation in sub-grade decomposed granite substrates. *Plant and Soil* 277 (1–2): 73–84.

Ritchie, W. E., R. L. Green, and V. A. Gibeault. 1997. Using ET_O (reference evapotranspiration) for turfgrass irrigation efficiency. *California Turfgrass Culture Volume* 47 (3–4): 9–15.

Ritsema, C. J., M. E. F. van Mensvoort, D. L. Dent, Y. Tan, H. van den Bosch, and A. L. M. van Wijk. 2000. In *Handbook of soil science*, ed. M. Sumner, G-121–154. Boca Raton, FL. CRC Press.

Robinson, D. A., C. S. Campbell, J. W. Hopmans, B. K. Hornbuckle, S. B. Jones, R. Knight, F. Ogden, J. Selker, and O. Wendroth. 2008. Soil moisture measurement for ecological and hydrological watershed-scale observations: A review. *Vadose Zone J.* 7: 358–389.

Robinson, M. and J. Neylan. 2001. Sand amendments for turf construction. *Golf Course Management* 69 (1): 65–69.

Rodriguez, I. A., and G. L. Miller. 2000. Using near infrared reflectance spectroscopy to schedule nitrogen applications on dwarf-type bermudagrass. *Agron. J.* 92: 423–427.

Roelfsema, M. R., and R. Hedrich. 2005. In the light of stomatal opening: new insights into 'the Watergate'. *New Phytology* 167: 665–691.

Rogers, M. E., A. D. Graig, R. E. Mumns, T. D. Colmer, P. G. H. Nichols, C. V. Malcolm, E. G. Barrett-Lennard, A. J. Brown, W. S. Semple, P. M. Evans, K. Cowley, S. J. Hughes, R. Snowball, S. J. Bennett, G. C. Sweeney, B. S. Dear, and M. A. Ewing. 2005. The potential for developing fodder plants for salt-affected areas of southern and eastern Australia: An overview. *Aust. J. Exper. Agric.* 45: 301–329.

Rosicky, M.A., P. Slavich, L.A. Sullivan, and M. Hughes. 2006. Surface and sub-surface salinity in and around acid sulfate soil scalds in the coastal floodplains of New South Wales, Australia. *Australian Journal of Soil Science* 44: 17–25.

Sakakibara, H., K. Takei, and N. Hirose. 2006. Interactions between nitrogen and cytokinin in the regulation of metabolism and development. *Trends in Plant Science* 11 (9): 440–448.

Sammut, J. 2000. *An introduction to acid sulfate soils*. Canberra, ACT, Australia: National Heritage Trust.

Sanden, B., A. Fulton, and L. Ferguson. 2010. Managing salinity, soil and water amendments. Kern County: University of California. http://cekern.ucdavis.edu/Irrigation_Management/MANAGING_SALINITY,_SOIL_AND_WATER_AMENDMENTS.htm

Sanden, B., T. L. Prichard, and A. E. Fulton. 2010. Improving water penetration. Kern County: University of California. http://cekern.ucdavis.edu/Irrigation_Management/IMPROVING_WATER_PENETRATION.htm

Sanders, D. 1980. The mechanism of Cl^- transport at the plasma-membrane of chara-corallina. 1. Cotransport with H^+. *Journal Membrane Biology* 53: 129–141.

Sands, Z., A. Grottesi, and M. S. P. Sansom. 2005. Voltage-gated ion channels. *Current Biology* 15 (2): R44–R47.

Sannazzaro, A. I., M. Echeverria, E. O. Alberto, O. A. Ruiz, and A. B. Menendez. 2007. Modulation of polyamine balance in *Lotus glaber* by salinity and arbuscular mycorrhiza. *Plant Physiology and Biochemistry* 45 (1): 39–46.

Sartain, J. B. 2008. *Soil and tissue testing and interpretation for Florida turfgrasses* (IFAS Extension Publ. SL 181). Gainesville: Institute of Food and Agricultural Sciences, University of Florida. http://edis.ifas.ufl.edu/pdffiles/SS/SS31700.pdf

Schmitz, M., and H. Sourell. 2000. Variability in soil moisture measurements. *Irrig. Sci.* 19: 147–151.

Schroder, J. L., H. Zhang, J. R. Richards, and M. E. Payton. 2009. Interlaboratory validation of the Mehlich 3 method as a universal extractant for plant nutrients. *J. of AOAC International* 92 (4): 995–1007.

Scott, C. A., N. I. Faruqui, and L. Raschid-Sally. 2004. *Wastewater use in irrigated agriculture*. Wallingford, Oxfordshire, UK: CABI Publishing.

SCSB. 2009. Procedures used by state soil testing laboratories in the southern region of the United States. *Southern Cooperative Series Bulletin (SCSB) # 190-D*. Clemson, SC: Clemson Experiment Station. http://www.clemson.edu/agsrvlb/sera6

Seelig, B. D. 2000. *Salinity and sodicity in North Dakota soils* (ND State Univ. Extension Pub. EB 57). Fargo: North Dakota State University Agriculture and University Extension. http://www.ag.ndsu.edu/pubs/plantsci/soilfert/eb57-2.htm#location

Semple, W. S., T. B. Koen, D. J. Eldridge, K. M. Duttmer, and B. Parker. 2006. Variation in soil properties on two partially vegetated saline scalds in south-eastern Australia. *Aust. J. of Experimental Agric.* 46: 1279–1289.

Serena, M., B. Leinauer, S. Macolino, and M. Gill. 2010. Germination of coated and uncoated turfgrass seed under saline conditions. *2nd European Turfgrass Society Conference Proceedings*. Vol. 2. TGIF Record #164213.

Serrano, R. and A. Rodriguez-Navarro. 2001. Ion homeostasis during salt stress in plants. *Current Opinion in Cell Biology* 13 (4): 399–404.

Seth-Carley, D., S. Davis, D. Bowman, L. Tredway, T. Rufty, and C. H. Peacock. 2009. Effluent application to creeping bentgrass in the transition zone: Effluent analysis and potential negative effects from salinity and low oxygen. *International Turfgrass Society Research Journal* 11 (2): 1023–1031.

Shahba, M. A., Y. L. Qian, and K. D. Lair. 2008. Improving seed germination of saltgrass under saline conditions. *Crop Science* 48 (2): 756–762.

Shalata, A., V. Mittova, M. Volokita, M. Guy, and M. Tal. 2001. Response of the cultivated tomato and its wild salt-tolerant relative *Lycopersicon pennellii* to salt-dependent oxidative stress: the root antioxidative system. *Plant Physiology* 112: 487–494.

Shalhevet, 1994. Using water of marginal quality for crop production: major issues. *Agric. Water Manage.* 25: 233–269.

Shaner, D. L., Khosla, R., Brodahl, M. K., Buchleiter, G. W., and Farahani, H. J. 2008. How well does zone sampling based in soil electrical conductivity maps represent soil variability? *Agronomy Journal* 100: 1472–1480.

Shannon, M. C. and V. Cervinka. 1997. Drainage water re-use. *Management of agricultural drainage water quality* (FAO Water Report 13). http://www.fao.org/docrep/W7224E/w7224e08.htm

Shaw, R. J. 1999. Soil salinity—electrical conductivity and chloride. 1999. In *Soil analysis and interpretation manual,* eds. K. I. Peverill, L. A. Sparrow, and D. J. Reuter. Collingwood, VIC, Australia: CSIRO.

Shewry, P. R., and P. J. Peterson. 1975. Calcium and magnesium in plants and soils from a serpentine area on Unst, Shetland. *Journal of Applied Ecology* 12 (1): 381–391.

Shi, D., and Y. Sheng. 2005. Effect of various salt-alkaline mixed stress conditions on sunflower seedlings and analysis of their stress factors. *Environmental and Experimental Botany* 54 (1): 8–21.

Shi, H., F. J. Quintero, J. M. Pardo, and J. K. Zhu. 2002. The putative plasma membrane Na^+/H^+ antiporter SOS1 controls long-distance Na^+ transport in plants. *Plant Cell* 14: 465–477.

Shukla, K. P., N. K. Singh, and S. Sharma. 2010. Bioremediation: Developments, current practices and perspectives. *Genetic Engineering and Biotechnology Journal 2010*: GEBJ-3: 1–22. http://astonjournals/gebj

Sickler, C. M. , G. E. Edwards, O. Kiirats, Z. Gao, and W. Loescher. 2007. Response of mannitol-producing *Arabidopsis thaliana* to abiotic stress. *Functional Plant Biology* 34: 382–391.

Sims, J. T. 2000. Soil fertility evaluations. In *Handbook of soil science*, ed. M. E. Sumner. Boca Raton, FL: CRC Press.

Singh, B., B. P. Singh, and A. L. Cowie. 2010. Characterization and evaluation of biochars for their application as a soil amendment. *Soil Research* 48 (7): 516–525.

Singh, N. K. and D. W. Dhar. 2010. Cyanobacterial reclamation of salt-affected soil. In *Genetic engineering, biofertilisation, soil quality and organic farming*, ed. E. Lichtfouse. Sustainable Agric. Review 4. New York: Springer Science + Business Media.

Skaggs, R. W, and J. van Schilfgaarde, eds. 1999. *Agricultural drainage* (Agronomy No. 30). Madison, WI: American Society of Agronomy.

Skipton, S., and B. Dvorak. 2002. *Drinking water: Chloramines water disinfection in Omaha Metropolitan Utilities District* (Neb Facts University of Nebraska Cooperative Extension Bulletin #NF02-505). Lincoln: University of Nebraska.

Skoog, F., and C. O. Miller. 1957. Chemical regulation of growth and organ formation in plant tissue cultures *in vitro*. *Symposium Society of Experimental Biology* 11: 118–131.

Skorulski, J. 2003. Micro-managing. *USGA Green Section Record* 41 (5): 13–17.

Skorulski, J., J. Henderson, and N. A. Miller. 2010. Topdressing fairways: more is better. *USGA Green Section Record* 48 (2): 15–17.

Slavens, M., A. M. Petrovic, J. Lehmann, and K. B. Heymann. 2009. Biochar as an organic amendment alternative in sand-based turfgrass sites or golf course putting greens. *Abstract: 2009 International Annual Meetings ASA-CSSA-SSSA*, p. 55649. TGIF Record #158198.

Smiley, R. W., P. H. Dernoeden, and B. B. Clarke. 2005. *Compendium of turfgrass diseases*. 3rd ed. St. Paul, MN: APS Press.

Smith, J. N. G. 2001. Managing black layer: although heavy rains and soils rich in organic matter can promote black layer, the condition can be controlled. *Golf Course Management* 69 (12): 59–62.

Snow, J. T., ed. 1994. *Wastewater reuse for golf course irrigation*. Boca Raton, FL: CRC Press.

Snow, J., ed. 2004. USGA recommendations for putting green construction. *USGA Green Section Record*, March/April. http://www.usga.org/course_care/articles/construction/greens/USGA-Recommendations-For-A-Method-Of-Putting-Green-Construction%282%29/

So, H. B., N. W. Menzies, R. Bigwood, and P. M. Kopittke. 2006. Examination into the accuracy of exchangeable cation measurements in saline soils. *Comm. in Soil Science and Plant Analysis* 37: 1819–1832.

Sohi, S. P., E. Krull, E. Lopez-Capel, and R. Bol. 2010. A review of biochar and its use and function in soil. *Advances in Agronomy* 105: 47–82.

Soldat, D. J., B. Lowery, and W. R. Kussow. 2010. Surfactants increase uniformity of soil water content and reduce water repellency on sand-based golf greens. *Soil Sci.* 175 (3): 111–117.

Solomon, K. H. 1988. A new way to view sprinkler patterns. *Irrigation Notes* (CATI Publication No. 880802). Fresno: Center for Irrigation Technology at California State University.

Soltanpour, P. N., and R. H. Follett. 1996. *Soil test interpretation* (no. 502). Fort Collins: Colorado State Univ. Coop. Extension. http://www.extsoilcrop.colostate.edu/SoilLab/documents/0502SoilTestExplaination.PDF

SPACE. 2007. Sprinkler profile and coverage evaluation (SPACE) program. Fresno: Center for Irrigation Technology, California State University. http://cati.csufresno.edu/cit/

Sparks, D. L., ed. 1996. *Methods of soil analysis. Part 3 – Chemical methods* (SSSA Book Series No. 5). Madison, WI: Soil Science Society of America.

Spellman, F. R. 2008. *The science of water: Concepts and applications*. Boca Raton, FL: CRC Press.

Spring, C. A., J. A. Wheater, and S. W. Baker. 2007. Fertiliser, sand dressing and aeration programmes for football pitches: I. performance characteristics under simulated wear. *J. Turf. and Sports Surface Science* 83: 40–55.

Starr, J. L., D. J. Timlin, P. M. Downey, and L. R. McCann. 2009. Laboratory evaluation of dual-frequency multisensory capacitance probes to monitor soil water and salinity. *Irrig. Sci.* 27: 393–400.

Stehouwer, R. C., P. Sutton, and W. A. Dick. 1995. Minespoil amendment with dry flue gas desulfurization by-products: Plant growth. *Journal of Environmental Quality* 24 (5): 861–869.

Stewart, B. 2009. Study result: Use of natural siliceous mineral on turf. *SportsTurf* December: 18–20. http://www.stma.org and http://www.sportsturfonline.com

Stewart, B. 2010. Amending fairways. *Golf Course Industry* 22 (3): 68–71.

Stevens, D. 2006. *Growing crops with reclaimed wastewater*. Collingwood, VIC, Australia: CSIRO.

St. John, R., and N. Christians. 2006. Soil testing methods for sand-based putting greens. *USGA Turf. and Enviro. Res. Online* 5 (13): 1–5.

St. John, R., and N. Christians. 2007. Soil testing methods for sand-based putting greens. *Golf Course Management* 75 (1): 154–156.

Stowell, L. J., and S. Davis. 1993. Direct measurement of electrical conductivity (EC) in golf course, high-sand-content soils. *Phytopathology* 83: 693.

Stowell, L. J., and W. Gelernter. 1998a. Tissue analyses; Guidelines and NIRS revisited. *Pace Insights* 4 (11): 1–4. San Diego, CA: PACE Consulting.

Stowell, L. J., and W. Gelernter. 1998b. Water amendments: Gypsum, acid, and sulfur burners. *PACE Insights* 4 (3): 1–4. San Diego, CA: PACE Consulting.

Stowell, L. J., and W. Gelernter. 2001. Negotiating reclaimed water contracts: Agronomic considerations. *Pace Insights* 7 (3): 1–4. San Diego, CA: PACE Consulting.

Stowell, L. J., and W. Gelernter. 2009. Evaluation of gypsum and mined calcium sulfate anhydrite as pre-leaching soil amendments for sodium management. *Pace Turf Super Journal*, August 31. San Diego, CA: PACE Consulting. http://www.paceturf.org/index.php/journal/gypsum_vs_mined_calcium_anhydrite_for_sodium_ management

Stowell, L. J., and W. Gelernter. 2010a. Tools for estimating sodium hazard based on irrigation water quality reports. *Pace Turf Super Journal*, June 28. San Diego, CA: PACE Consulting. http://www.paceturf.org/PTRI/Documents/1006sj.pdf

Stowell, L. J., and W. Gelernter. 2010b. Water quality guidelines. San Diego, CA: PACE Consulting. http://www.paceturf.org/PTRI/Documents/water/0502water.pdf

Stowell, L., W. Gelernter, B. Williams, and P. Reedy. 2008. Evaluation of Eximo plus Dispatch for reduction of sodium in USGA specification golf course greens (Docket #08060603). http://www.paceturforg/PTRI/Documents/0806sjp.pdf

Suarez, D. L. 1981. Relation between pHc and sodium adsorption ratio (SAR) and an alternative method of estimating SAR of soil or drainage waters. *Soil Sci. Soc. of Am. J.* 45: 469–475.

Suarez, D. L. 2001. Sodic soil reclamation: modeling and field study. *Aust. J. Soil Res.* 39: 1225–1246.

Subbarao, G. V., and C. Johansen. 1994. Strategies and scope for improving salinity tolerance in crop plants. In *Handbook of plant and crop stress*, ed. M. Pessarakli, 559–579. New York: Marcel Dekker.

Sumner, M. E. 1993. Sodic soils: New perspectives. *Aust. J. Soil Res.* 31: 683–750.

Sumner, M. E. and W. P. Miller. 1996. Cation exchange capacity and exchange coefficients. In *Methods of soil analysis. Part 3. Chemical methods* (Soil Sci. Soc. of Amer. Book Series No. 5), ed. D. L. Sparks, 1201–1229. Madison, WI: Soil Science Society of America.

Sumner, M. E. and Naidu, R., eds. 1998. *Sodic soils: Distribution, properties, management and environmental consequences*. New York: Oxford University Press.

Sunarpi, H. T., J. Motoda, M. Kubo, H. Yang, K. Yoda, R. Horie, W. Y. Chan, H. Y. Leung, K. Hattori, M. Konomi, M. Osumi, M. Yamagami, J. I. Schroeder, and N. Uozumi. 2005. Enhanced salt tolerance mediated by AtHKT1 transporter-induced Na^+ unloading from xylem vessels to xylem parenchyma cells. *Plant Journal* 44: 928–938.

Sundstrom, R., M. Astrom, and P. Osterholm. 2002. Comparison of the metal content of acid sulfate soil runoff and industrial effluents in Finland. *Environ. Sci. Technol.* 36 (20): 4269–4272.

Sunkar, R., V. Chinnusamy, J. Zhu, and J-K. Zhu. 2007. Small RNAs as big players in plant abiotic stress responses and nutrient deprivation. *Trends in Plant Science* 12 (7): 301–309.

Szczerba, M. W., D. T. Britto, and H. J. Kronzucker. 2009. K^+ transport in plants: physiology and molecular biology. *Journal of Plant Physiology* 166 (5): 447–466.

Takahashi, S., and N. Murata. 2008. How do environmental stresses accelerate photoinhibition? *Trends in Plant Science* 13 (4): 178–182.

Talke, I. N., D. Blaudez, F. J. M. Maathuis, and D. Sanders. 2003. CNGCs: prime targets of plant cyclic nucleotide signaling? *Trends in Plant Science* 8: 286–293.

Tanaka, Y., T. Hibino, Y. Hayashi, A. Tanaka, S. Kishitani, T. Takabe, S. Yokota, and T. Takabe. 1999. Salt tolerance of transgenic rice overexpressing yeast mitochondria Mn-SOD in chloroplasts. *Plant Science* 148: 131–138.

Tani, T., and J. B. Beard. 1997. *Color atlas of turfgrass diseases: Disease characteristics and control.* New York: John Wiley.

Tang, M., M. Sheng, H. Chen, and F. F. Zhang. 2009. In vitro salinity resistance of three ectomycorrhizal fungi. *Soil Biology and Biochemistry* 41 (5): 948–953.

Tanji, K. K. 1996. *Agricultural salinity assessment and management.* New York: American Society of Civil Engineers.

Tanji, K. K., and N. C. Kielen. 2002. *Agricultural drainage water management in arid and semi-arid regions* (FAO Irrigation and Drainage Paper 61). Rome: United Nations Food and Agriculture Organization.

Tanou, G., A. Molassiotis, and G. Diamantidis. 2009. Induction of reactive oxygen species and necrotic death-like destruction in strawberry leaves by salinity. *Environmental and Experimental Botany* 65 (2–3): 270–281.

Tavakkoli, E., P. Rengasamy, and G. K. McDonald. 2010. The response of barley to salinity stress differs between hydroponic and soil systems. *Functional Plant Biology* 37: 621–633.

Tejada, M., Garcia, C., Gonzales, J. L., and Hernandez, M. T. 2006. Use of organic amendment as a strategy for saline soil remediation: influence on the physical, chemical and biological properties of soil. *Soil Biol. Biochem* 38: 1413–1421.

Tester, M. and R. J. Davenport. 2003. Na+ transport and Na+ tolerance in higher plants. *Annals of Botany* 91: 503–527.

Thigpen, M. (2007). Precision turfing … coming of age, part one. *Turfnet: The Newsletter* 14 (9): 1–4. http://www.nutecsoil.com/precision_turfing.pdf

Thomas, M., R. W. Fitzpatrick, and G. S. Heinson. 2009. Distribution and causes of intricate saline-sodic patterns in an upland South Australian hillslope. *Aust. J. of Soil Research* 47: 328–339.

Throssell, C. S., R. N. Carrow, and G. A. Milliken. 1987. Canopy temperature based irrigation scheduling indices for Kentucky bluegrass turf. *Crop Sci.* 27: 126–131.

Tikhonova, L. I., I. I. Pottosin, K-J. Dietz, and G. Schonknecht. 1997. Fast-acting cation channel in barley mesophyll vacuoles: inhibition by calcium. *Plant Journal* 11: 1059–1070.

Tisdale, S. L., W. L. Nelson, J. D. Beaton, and J. L. Havlin. 1993. *Soil fertility and fertilizers.* New York: Macmillan.

To, J. P. C. and J. J. Kieber. 2008. Cytokinin signaling: Two-components and more. *Trends in Plant Science* 13(2): 85–92.

Trenholm, L. E., Carrow, R. N., and Duncan, R. R. 1999. Relationship of multispectral radiometry data to qualitative data in turfgrass research. *Crop Science* 39: 763–769.

Trenholm, L. E., R. N. Carrow, and R. R. Duncan. 2001. Wear tolerance, growth, and quality of seashore paspalum in response to nitrogen and potassium. *HortScience* 36 (4): 780–783.

Trenholm, L. E., R. R. Duncan, and R. N. Carrow. 1999. Wear tolerance, shoot performance, and spectral reflectance of seashore paspalum and bermudagrass. *Crop Science* 39: 1147–1153.

Trenholm, L. E., R. R. Duncan, R. N. Carrow, and G. H. Snyder. 2001. Influence of silica on growth, quality, and wear tolerance of seashore paspalum. *J. Plant Nutrition* 24 (2): 245–259.

Treonis, A. M., E. E. Austin, J. S. Buyer, J. E. Maul, L. Spicer, and I. A. Zasada. 2010. Effects of organic amendment and tillage on soil microorganisms and microfauna. *Applied Soil Ecology* 46 (1): 103–110.

Tucker, B. M. 1985. Active and exchangeable cations in soils. *Australian Journal of Soil Research* 23: 195–209.

Turkan, I., and T. Demiral. 2009. Recent developments in understanding salinity tolerance. *Environmental and Experimental Botany* 67 (1): 2–9.

Turner, T. R., and N. W. Hummel, Jr. 1992. Nutritional requirements and fertilization. In *Turfgrass Monograph, No. 32*, eds. D. V. Waddington, R. N. Carrow, and R. C. Shearman. Madison, WI: American Society of Agronomy.

Tuteja, N. 2007. Mechanisms of high salinity tolerance in plants. *Methods in Enzymology: Osmosensing and Osmosignaling* 428: 419–438.

Udomchalothorn, T., S. Maneeprasobsuk, E. Bangyeekhun, P. Boon-Long, and S. Chadchawan. 2009. The role of the bifunctional enzyme, fructose-6-phosphate-2-kinase/fructose-2,6-bisphosphatase, in carbon partitioning during salt stress and salt tolerance in rice (*Oryza sativa* L.). *Plant Science* 176 (3): 334–341.

USDA-ARS. 2008. Research databases. Bibliography on salt tolerance. Riverside, CA: US Dep. Agric., Agric. Res. Serv., U.S. Salinity Laboratory. http://www.ars.usda.gov/Services/docs.htm?docid=8908

USDA-NRCS. 1996. *Plant materials for saline-alkaline soils.* Technical Notes Plant Materials No. 26.

USDA-NRCS. 2001. *Plant materials and techniques for brine site reclamation.* Technical Notes Plant Materials No. 26.

USEPA. 2003. *National management measures to control nonpoint source pollution from agriculture* (EPA 841-B-03-004). Washington, DC: USEPA, Office of Water.

USEPA. 2004. *Guidelines for water reuse* (EPA/625/R-04/108). Washington, DC: USEPA, Office of Water.

USEPA. 2007a. *Environmental management systems.* Washington, DC: USEPA, Office of Water. http://www.epa.gov/ems/index.html

USEPA. 2007b. Key elements of an EMS. Washington, DC: USEPA, Office of Water. http://www.epa.gov/ema./info/elements.html

USEPA. 2011. Washington, DC: USEPA. http://www.epa.gov/sustainability/basicinfo.htm.

USGA. 2010. Green section recommendations for a method of putting green construction. Far Hills, NJ: United States Golf Association. http://www.usga.org/Content.aspx?id=25872

U.S. Salinity Laboratory (USSL). 1954. *Diagnosis and improvement of saline and alkali soils: Handbook 60.* Washington, DC: U.S. Government Printing Office. http://www.ussl.ars.usda.gov

Van Dyke, A., and P. G. Johnson. 2007. Do humic substances bolster water and nutrient availability? *TurfGrass Trends*, April 1. http://www.turfgrasstrends.com

Van Rensburg, L., R. I. De Sousa Correia, J. Booysen, and M. Ginster. 1998. Land reclamation: Revegetation of a coal fine ash disposal site in South Africa. *Journal of Environmental Quality* 27 (6): 1479–1486.

Vanton Pump and Equipment Co. 2010. Polypropylene pumps critical to pumping seawater for marine research. Hillside, NJ: Vanton Pump and Equipment. http://www.vanton.com/lib_poly.php

Vargas, J. M., Jr. 1994. *Management of turfgrass diseases.* 2nd ed. Boca Raton, FL: CRC Press.

Vazquez de Aldana, B. R., B. G. Criado, A. G. Civdadi, and M. E. P. Corona. 1995. Estimation of mineral content in natural grasslands by near infrared reflectance spectroscopy. *Comm. Soil Sci. and Plant Anal.* 26 (9,10): 1383–1396.

Vermeulen, P. H. 1997. Know when to over-irrigate. *USGA Green Section Record* 35 (5): 16.

Vickers, A. 2001. *Handbook of water use and conservation.* Amherst, MA: Waterplow Press.

Vidali, M. 2001. Bioremediation. An overview. *Pure Applied Chemistry* 73 (7): 1163–1172.

Vinidex Systems and Solutions. 2010. *Chemical resistance guide.* North Rocks, NSW, Australia: Vinidex Systems and Solutions. http://www.vinidex.com.au/page/chemical_resistance_guide.html

Voleti, S. R., A. P. Padmakumari, V. S. Raju, S. Mallikarjuna Babu, and S. Ranganathan. 2008. Effect of silicon solubilizers on silica transportation, induced pest and disease resistance in rice (*Oryza sativa* L.). *Crop Protection* 27 (10): 1398–1402.

Waddington, D. V., A. E. Gover, and D. B Beegle. 1994. Nutrient concentrations of turfgrass and soil test levels as affected by soil media and fertilizer rate and placement. *Comm. in Soil Sci. and Plant Anal.* 25 (11–12): 1957–1990.

Walker, J. P., G. R. Willgoose, and J. D. Kalma. 2004. In situ measurement of soil moisture: A comparison of techniques. *J. Hydrol.* 293: 85–99.

Walker, R., M. Lehmkuhl, and G. Kah. 1995. *Irrigation: Landscape water management principles version 1.01.* San Luis Obispo: Training and Research Center at California State Polytechnic University.

Walker, R., M. Lehmkuhl, G. Kah, and P. Corr. 1995. *Landscape water management auditing.* San Luis Obispo: Cal Poly, Irrigation and Training Research Center.

Walla-Walla Sprinkler Company. 2010. The MP rotator: A water conservation device. http://www.mprotator.com

Wang, Y., and F. Chen. 2008. Decomposition and phosphorus release from four different size fractions of *Microcystis* spp. taken from Lake Taihu, China. *Journal of Environmental Sciences* 20 (7): 891–896.

Ward, J. M., and J. I. Schroeder. 1994. Calcium-activated K^+ channels and calcium-induced calcium release by slow vacuolar ion channels in guard cell vacuoles implicated in the control of stomatal closure. *Plant Cell* 6: 669–683.

Warnock, D. D., J. Lehmann, T. W. Kuyper, and M. C. Rillig. 2007. Mycorrhizal responses to biochar in soil—concepts and mechanisms. *Plant and Soil* 300: 9–20.

Warnock, D. D., D. L. Mummey, B. McBride, J. Major, J. Lehmann, and M. C. Rillig. 2010. Influences of non-herbaceous biochar on arbuscular mycorrhizal fungal abundances in roots and soils: results from growth-chamber and field experiments. *Applied Soil Ecology* 46 (3): 450–456.

Warrington, D. N., D. Goldstein, and G. J. Levy. 2007. Clay translocation within the soil profile as affected by intensive irrigation with treated wasterwater. *Soil Sci.* 172 (9): 692–700.

Water on the Web. 2008. Understanding: Lark ecology primer. Minneapolis, MN: Water on the Web. http://waterontheweb.org/under/lakeecology/lakeecology.pdf

Weeaks, J. 2009. Creeping bentgrass (*Agrostis stolonifera*) establishment in a greenhouse using subsurface drip irrigation. Dissertation, Texas Tech University.

Weeaks, J. D., M. A. Maurer, R. E. Zartman, and J .G. Surles. 2008. Creeping bentgrass establishment using subsurface drip irrigation. *Joint Annual Meeting (Abstract) GSA/SSSA/ASA/CSSA/HGS*, p. 41860.

Weeaks, J., R. Zartman, and M. Maurer. 2007. Establishing bentgrass can be difficult with subsurface irrigation. *Turfgrass Trends* June: 57–58. http://www.turfgrasstrends.com

Wehtje, G. R., J. N. Shaw, R. H. Walker, and W. Williams. 2003. Using inorganic soil amendments to improve a native soil. *Golf Course Management* 71 (11): 95–99.

Wentz, D. 2000. *Dryland saline seeps: types and causes.* Agri-Facts, Agdex 518-12. Lethbridge, Alberta, Canada: Alberta Agric. Food and Rural Dev. http://www1.agric.gov.ab.ca/$department/deptdocs.nsf/all/agdex167/$file/518_12.pdf?OpenElement

Werner, T., V. Motyka, V. Laucou, R. Smets, H. Van Onckelen, and T. Schmulling. 2003. Cytokinin-deficient transgenic Arabidopsis plants show multiple developmental alterations indicating opposite functions of cytokinins in the regulation of shoot and root meristem activity. *Plant Cell* 15: 2532–2550.

Werner, T., V. Motyka, M. Strand, and T. Schmulling. 2001. Regulation of plant growth by cytokinin. *Proceedings National Academy of Sciences USA* 98: 10487–10492.

Werner, T., and T. Schmulling. 2009. Cytokinin action in plant development. *Current Opinion in Plant Biology* 12 (5): 527–538.

Westcot, D. W., and R. S. Ayers. (1985). Irrigation water quality criteria. In *Irrigation with reclaimed municipal wastewater: A guidance manual*, eds. G. S. Pettygrove and T. Asano. Chelsea, MI: Lewis Publishers.

Westerman, R. I., ed. 1990. *Soil testing and plant analysis* (3rd ed., Soil Sci. Soc. of Amer., Book Series No. 3). Madison, WI: Soil Science Society of America.

Whitlark, B. 2010. Acid substitutes and pH reduction. *USGA Green Section Record* 48 (2):18–22.

Wiecko, G. 2003. Ocean water as a substitute for post emergence herbicides in Tropical turf. *Weed Tech* 17: 788–791.

Wilcox, L. V., G. Y. Blair, and C. A. Bower. 1954. Effect of bicarbonate on suitability of water for irrigation. *Soil Sci.* 77: 259–266.

Wilson, S. M. 2003. Understanding and preventing impacts of salinity on infrastructure in rural and urban landscapes. 9th PUR$L National Conference, September 29–October 2, Yappon, QLD, Australia.

Wink Fasteners Inc. 2010. Fighting the endless battle of corrosion. *Wink's Words* 2 (1). Richmond, VA: Wink Fasteners. http://www.winkfast.com/documents/WinkWordsVol2No1.pdf

Winsley, P. 2007. Biochar and bioenergy production for climate change modification. *New Zealand Science Review* 5: 5–10.

Witcombe, J. R., P. A. Hollington, C. J. Howarth, S. Reader, and K. A. Steele. 2008. Breeding for abiotic stresses for sustainable agriculture. Philosophical Transactions of the Royal Society of London, Series B. *Biological Sciences* 363: 703–716.

Wong, V. N. L., R. S. B. Greene, R. C. Dalal and B. W. Murphy. 2010. Soil carbon dynamics in saline and sodic soils: a review. *Soil Use and Management* 26: 2–11.

Woolf, D. 2008, January. Biochar as a soil amendment: A review of the environmental implications. http://orgprints.org/13268/1/Biochar_as_a_soil_amendment_-_a_review.pdf

Wu, L., J. Chen, H. Lin, P. Van Mantgem, M. A. Harivandi, and J. A. Harding. 1995. Effects of regenerant wastewater irrigation on growth and ion uptake of landscape plants. *J. Environ. Hort.* 13 (2): 92–96.

Wu, L., and D. R. Huff. 1983. Characteristics of creeping bentgrass clones (*Agrostis stolonifera* L.) from a salinity-tolerant population after surviving drought stress. *HortScience* 18 (6): 883–885.

Wyn Jones, R. G., R. Storey, R. A. Leigh, N. Ahmad, and A. Pollard. 1977. A hypothesis on cytoplasmic osmoregulation. In *Regulation of cell membrane activities in plants*, eds. E. Marre and O. Cifferi, 121–136. Amsterdam: Elsevier.

Xu, J., H-D. Li, L-Q. Chen, Y. Wang, L-L. Liu, L. He, and W-H. Wu. 2006. A protein kinase, interacting with two calcineurin B-like proteins, regulates K^+ transporter AKT1 in *Arabidopsis*. *Cell* 125 (7): 1347–1360.

Yadagiri, K. and J. L. Kerrigan. 2010. Hispopathology of 'rapid blight', a disease caused by *Labyrinthula terrestris* on cool-season turfgrasses. *Phytopathology* 100 (6, Suppl. 1): S204.

Yamada, M., M. Uehira, L. S. Hun, K. Asahara, T. Endo, A. E. Eneji, S. Yamamoto, T. Honna, T. Yamamoto, and H. Fujiyama. 2002. Ameliorative effect of K-type and Ca-type artificial zeolites on the growth of beets in saline and sodic soils. *Soil Science and Plant Nutrition* 48 (5): 651–658.

Yan, H., L. Z. Gang, C. Y. Zhao, and W. Y. Guo. 2000. Effects of exogenous proline on the physiology of soyabean plantlets regenerated from embryos *in vitro* and on ultrastructure of their mitochondria under NaCl stress. *Soybean Science* 19: 314–319.

Yan, L., S. Zhou, L. Feng, and L. Hong-Yi. 2007. Delineation of site-specific management zones using fuzzy clustering analysis in a coastal saline land. *Computers and Electronics in Agriculture* 56: 174–186.

Yancey, P. H. 1994. Compatible and counteracting solutes. In *Cellular and molecular physiology of cell volume regulation*, ed. K. Strange, 81–109. Boca Raton, FL: CRC Press.

Yang, M-H., R. Gibbs, and M. Wrigley. 1998. Laboratory investigation of New Zealand produced zeolite as an inorganic amendment for sand-based root zones. *Proceedings 6th NZ Sports Turf Convention,* pp. 27–31.

Yao, H., and W. Shi. 2010. Soil organic matter stabilization in turfgrass ecosystems: importance of microbial processing. *Soil Biology and Biochemistry* 42 (4): 642–648.

Yao, L., Z. Wu, Y. Zheng, I. Kaleem, and C. Li. 2010. Growth promotion and protection against salt stress by *Pseudomonas putida* Rs-198 on cotton. *European Journal Soil Biology* 46 (1): 49–54.

Yao, R., and J. Yang. 2010. Quantitative evaluation of soil salinity and its spatial distribution using electromagnetic induction method. *Agricultural Water Management* 97 (12): 1961–1970

Yenny, R. 1994. Salinity management. *USGA Green Section Record* 32 (6):7–10.

Yensen, N. P. 2008. Halophyte uses for the twenty-first century. In *Ecophysiology of high salinity tolerant plants*, ed. M. A. Khan and D. J. Webster, 367–396. Dordrecht, the Netherlands: Springer.

Yeo, A. R., M. Yeo, and T. Flowers. 1987. The contribution of an apoplastic *pathway to sodium uptake by rice roots in saline conditions. Journal Experimental Botany* 38: 1141–1153.

Yi, B. A., and L. Y. Jan. 2000. Taking apart the gating of voltage-gated K^+ channels. *Neuron* 27 (3): 423–425.

Yiasoumi, B., L. Evans, and L. Rogers. 2005. Farm water quality and treatment. Agfact AC.2, 9th ed. http://www.dpi.nsw.gov.au/__data/assets/pdf_file/0013/164101/farm-water-quality.pdf

Yuan, B-C., Z-Z. Li, H. Liu, M. Gao, and Y-Y. Zhang. 2007. Microbial biomass and activity in salt affected soils under arid conditions. *Applied Soil Ecology* 35 (2): 319–328.

Zahran, H. H. 1997. Diversity, adaptation and activity of the bacterial flora in saline environments. *Biol. Fertil. Soils* 25: 211–223.

Zaveleta-Mancera, H. A., H. Lopez-Delgado, H. Loza-Tavera, M. Mora-Herrera, C. Trevilla-Garcia, M. Vargas-Suarez, and H. Ougham. 2007. Cytokinin promotes catalase and ascorbate peroxidase activities and preserves the chloroplast integrity during dark-senescence. *Journal of Plant Physiology* 164 (12): 1572–1582.

Zhang, J., W. Jia, J. Yang, and A. M. Ismail. 2006. Role of ABA in integrating plant responses to drought and salt stresses. *Field Crops Research* 97 (1): 111–119.

Zhao, X., Y-L. Wang, Y-J. Wang, X-L. Wang, and X. Zhang. 2008. Extracellular Ca^{2+} regulating stomatal movement and plasma membrane K^+ channels in guard cells of *Vicia faba* under salt stress. *Acta Agronomica Sinica* 34 (11): 1970–1976.

Zhu, J-K. 2002. Salt and drought signal transduction in plants. *Annual Review of Plant Biology* 53: 247–273.

Zhu, J-K. 2003. Regulation of ion homeostasis under salt stress. *Current Opinions in Plant Biology* 6: 441–445.

Zhu, Z., G. Wei, J. Li, Q. Qian, and J. Yu. 2004. Silicon alleviates salt stress and increases antioxidant enzymes activity in leaves of salt-stressed cucumber (*Cucumis sativus* L.). *Plant Science* 167 (3): 527–533.

Zia, M. H., A. Ghafoor, Saifullah, and T. M. Boers. 2006. Comparison of sulfurous acid generator and alternative amendments to improve the quality of saline-sodic water for sustainable rice. *Paddy and Water Environment* 4: 153–162.

Zinck, J. A., and Metternicht, G. 2009. Soil salinity and salinization hazard. In *Remote sensing of soil salinization*, ed. G. Metternicht and J. A. Zinck, 3–20. Boca Raton, FL: CRC Press.

Zoldoske, D. F., E. L. Bundy, and M. Y. Miyasaki. 1987. *Large turf irrigation systems: Design and management.* Fresno: Center for Irrigation Technology, California State University.

Zoldoske, D. F. 2003. *Improving golf course irrigation uniformity: A California case study* (CATI Publication No. 030901). Fresno: California Agricultural Technology Institute for the California Department of Water Resources. http://cati.csufresno.edu/cit/Golf%20Course%20Irrigation%20Nozzle%20Study.pdf

Zoldoske, D. F., K. H. Solomon, and E. M. Norum. 1994. Uniformity measurements for turfgrass: What's best? *CATI Notes* (Publication No. 941102). Fresno: Center for Irrigation Technology, California State University.

Index

J

K

L

R

T

U

V

W

Z